ECOLOGY AND BIOGEOGRAPHY
IN SRI LANKA

MONOGRAPHIAE BIOLOGICAE

VOLUME 57

Series Editor

H.J. Dumont

1984 **DR W. JUNK PUBLISHERS**
a member of the KLUWER ACADEMIC PUBLISHERS GROUP
THE HAGUE / BOSTON / LANCASTER

ECOLOGY AND BIOGEOGRAPHY
IN SRI LANKA

Edited by

C. H. FERNANDO

1984 **DR W. JUNK PUBLISHERS**
a member of the KLUWER ACADEMIC PUBLISHERS GROUP
THE HAGUE / BOSTON / LANCASTER

Distributors

for the United States and Canada: Kluwer Academic Publishers, 190 Old Derby Street, Hingham, MA 02043, USA
for the UK and Ireland: Kluwer Academic Publishers, MTP Press Limited, Falcon House, Queen Square, Lancaster LA1 1RN, England
for all other countries: Kluwer Academic Publishers Group, Distribution Center, P.O. Box 322, 3300 AH Dordrecht, The Netherlands

Library of Congress Cataloging in Publication Data

Main entry under title:

Ecology and biogeography of Sri Lanka.

   (Monographie biologicae ; v. 57)
   Includes bibliographies.
   1. Ecology--Sri Lanka. 2. Biogeography--Sri Lanka.
I. Fernando, C. H. II. Series.
QP1.P37 vol. 57 [QH183.5] 574s [574.5'09549'3] 83-26784
ISBN-13:978-94-009-6547-8   e-ISBN-13:978-94-009-6545-4
DOI: 10.1007/978-94-009-6545-4

ISBN-13:978-94-009-6547-8

Copyright

Preface

When the late Professor Joachim Illies suggested in 1980 that I edit a volume of the Monographiae Biologicae on Sri Lanka, I was glad to accept the challenge. Although I had spent only six years of my research and teaching career in Sri Lanka, I had made personal contact or corresponded with many scientists who had worked in, still work in, or who have studied material from Sri Lanka. The present domicile of the authors of the chapters in this volume shows the wide geographic spread of interest in Sri Lanka, and indicates also the dispersion of Sri Lankan scientists like myself.

Sri Lanka has had a relatively long history of indigenous scientific research in the natural sciences. From the early work of Kelaart (1852, Prodromous Fauna Zeylanicae, Ceylon Govt. Press, 250 pp.) to the present time, there has been a more or less sustained research effort in the natural sciences. The Colombo Museum, which celebrated its centenary only a few years ago, and the world famous Peradeniya Botanical Gardens, served as repositories and bases for continued research on the fauna and flora. There are a number of landmarks in these studies. Linnaeus' (1747) Flora Zeylanica, which was based on the collection of a German Physician and Botanist, Paul Hermann, was the first notable contribution. The monumental work of Trimen (1839-1900, A Handbook to the Flora of Ceylon, Vol. 1-5) has not yet been fully superseded. Tennant (1891) gave a fairly detailed account of many animals, and lists of fishes and insects. The vertebrates have received detailed attention from researchers working mainly in the Colombo Museum. There are reliable and up to date monographs on the fishes by I.S.R. Munro (1955), amphibians by P. Kirtisinghe (1957), birds by G.M. Henry (1955) and mammals by W.W.A. Philips (1935).

P.E.P. Deraniyagala, in his long career at the Colombo Museum, made many contributions to the systematics of fishes and reptiles, and palaeontological studies on mammals. P.H.D.H. de Silva, who succeeded Deraniyagala, carried on this work, and recently published a monograph on the snakes. The publication of 'A guide to the freshwater fauna of Ceylon' by Mendis & Fernando (1962) was the first for any tropical country encompassing the whole spectrum of freshwater fauna with a listing of species. This work has been updated with four supplements.

Commercial interests have contributed to the study of the living organisms from a relatively early date. Herdman (1903, Report to the Government by Ceylon on the Pearl Oyster fishery off the gulf of Manaar. Roy. Soc. Lond. 1:1-146) was one of the earliest such contributors. The work of the Tea, Rubber and Coconut Research Institutes, the Agricultural and Medical Research Institutes and the Department of Fisheries contributed much to the study of living organisms of economic importance which could be exploited or were pests of economically important plants and animals.

Sri Lanka, called Serendib, Taprobane and later Ceylon, has from very early times attracted the attention of many travellers, among them naturalists and scientists. Because of its small size, explorers have not been involved in 'discovering' the country, which was settled fairly densely around 600 B.C.

Sri Lanka has a land area of only 65,000 km². It's latitudinal position (6-10° N) is clearly in the equatorial zone, but the climate in the mountains (to 2,300 m) is almost 'temperate'. The dry and wet zones mark the major climate division but within these zones are many gradations. Although geologically not very varied, soils, altitude and rainfall have combined to give Sri Lanka a wide variety of topography and vegetation. The range of climates in Sri Lanka is unrivalled for a country of its size.

A major feature of the natural landscape is the abundance of rivers and streams. Human enterprise has added another feature which is equally striking, namely the man-made lakes, of which there are >10,000. These reservoirs comprised a sophisticated irrigation system for rice cultivation in ancient times. It is mainly restricted to the dry zone. This irrigation system supported a resilient civilization (600 B.C.-1200 A.D.) and a peak population of about 6 million, a very high carrying capacity for that time. Since then Sri Lanka has undergone profound environmental changes due to dense human settlement; first in the foothills, followed by plantation agriculture in the highlands, urbanization of the lowlands and a revitalization of the agriculture of the dry zone. The impact of high human densities on grasslands, forests, water resources and the flora and fauna in general has been very pervasive. This volume refers to some of the effects of these high human densities.

In preparing this volume I have endeavoured to cover as wide a range of subjects as possible. It was also my intention to invite both senior and junior scientists to contribute. I was sorry that some potential contributors who had done extensive work in Sri Lanka, notably Professors D. Mueller-Dombois, Honolulu; Per Brinck, Lund; B.A. Abeywickrema, Colombo and Drs. R. Fosberg and K.V. Krombein, Washington, were already heavily committed to other work. However, the contributors I have been able to assemble come from a wide spectrum of disciplines and backgrounds.

This volume consists of 22 chapters. The three chapters dealing with geology, land forms and geochemistry set the stage for later chapters which include vegetation, aquatic fauna, terrestrial fauna and human impact. Besides the systematics and ecology of a wide range of faunal groups, certain specific environments have been dealt with in detail, e.g. grasslands and ricefields, both heavily influenced by human settlements, and rocky shores and lagoons. A recurring theme is the high endemicity in some groups of organisms. Only one chapter, however, deals with this subject specifically.

Sri Lanka also is a well researched area for the role of ecological factors on parasitic diseases. A chapter is devoted to this subject. Another theme in several chapters is the effect of high human densities; the final chapter, on settlements, is devoted to this subject. The emphasis on freshwater may be a reflection of my own speciality. However, this biotope has received a relatively high level of scientific attention in recent years because of the perceived danger of irreversible pollution and the prospects of harvesting fish from inland waters.

I received much help and co-operation from the authors. Mrs Dawn Sephton assisted me in editing, as did Dr Russell Shiel, who also put the chapters into final form on a word processor.

This volume should interest a very wide audience, both professional and non-professional. There is a great deal of information and syntheses useful to biologists, agricultural scientists, environmental biologists and medical scientists. There is a substantial body of information on the flora and fauna, on the ecology of a wide range of habitats, and on the impact of humans on the environment.

VIII

To A. C. J. Weerekoon, a bold and innovative biologist
and a courageous person

Addresses of Authors

K.D. Arudpragasam, Department of Zoology, University of Colombo, P.O. Box 1490, Colombo 3, Sri Lanka.

P.G. Cooray, Institute of Applied Geology, King Abdulaziz University, P.O. Box 1744, Jeddah, Saudi Arabia.

H.H. Costa, Department of Zoology, University of Kelaniya, Kelaniya, Sri Lanka.

H. Crusz, Department of Zoology, University of Peradeniya, Peradeniya, Sri Lanka.

D.M. Davies, Department of Biology, McMaster University, 1280 Main Street West, Hamilton, Ontario, Canada, L8S 4K1.

C.D. de Silva, Department of Zoology, Ruhuna University College, Medawatte, Matara, Sri Lanka.

S.S. de Silva, Department of Zoology, Ruhuna University College, Medawatte, Matara, Sri Lanka.

A.S. Dissanaike, Parasitic Diseases Programme, World Health Organization, 1211 Geneva 27, Switzerland.

C.B. Dissanayake, Department of Geology, University of Peradeniya, Peradeniya, Sri Lanka.

D.K. Erb, Department of Geography, University of Waterloo, Waterloo, Ontario, Canada, N2L 3G1.

T.L. Erwin, Curator, Coleoptera, National Museum of Natural History, Smithsonian Institute, Washington, D.C. 20560, U.S.A.

C.H. Fernando, Department of Biology, University of Waterloo, Waterloo, Ontario, Canada, N2L 3G1.

A.V. Gussev, Zoological Institute, Academy of Sciences, Leningrad, B-164, U.S.S.R.

M.D. Hubbard, Department of Entomology, Florida A & M University, College of Science and Technology, Tallahassee, Florida 32307, U.S.A.

G.M. McKay, School of Biological Sciences, Macquarie University, North Ryde, New South Wales, 2113, Australia.

J.W. Neale, Department of Geology, University of Hull, Hull, HU6 7R, England

M.A. Pemadasa, Department of Botany, Ruhuna University College, Medawatte, Matara, Sri Lanka.

N.P. Perera, FAO Land Use Planning Unit, UNDP, P.O. Box 301, Maseru, Lesotho.

W.L. Peters, Department of Entomology, Florida A & M University, College of Science and Technology, Tallahassee, Florida 32307, U.S.A.

R. Rajapaksa, Department of Biology, University of Waterloo, Waterloo, Ontario, Canada, N2L 3G1.

R. Ratnapala, Anti-Malaria Campaign, Colombo 5, Sri Lanka.

F. Starmühlner, 11 Zoologisches Institut der Universität Wien, A-1010 Karl Lueger Ring 1, Vienna, Austria.

S. Weeraratna, Department of Agronomy, Faculty of Agriculture, Ruhuna University College, Mapalana, Matara, Sri Lanka.

Table of contents

XII

XVI

1. Geology, with special reference to the Precambrian

P.G. Cooray

Introduction

Geologically, the island of Sri Lanka is an extension of Peninsular India and forms part of the Indian Shield, one of the oldest and most stable parts of the earth's crust. Early ideas on the island's geology by pioneers like F.D. Adams (1929), J.S. Coates (1935) and N.D. Wadia (1945) were therefore largely influenced by concepts of the evolution of the crust of south India current at the time. It was only after systematic geological mapping of the island began in 1952 that more recent ideas on the island's geology and geological history began to emerge. Within the last 15 years or so, the completion of the preliminary geological mapping and of the gravity survey of the island, on-shore and off-shore exploration for oil, and individual research have contributed to a greater understanding of the geology of Sri Lanka, though many major problems remain unsolved.

Four-fifths of the island is made up of Precambrian crystalline rocks, which, on the basis of their lithology, structure and age, have been subdivided into three major units (Fig. 1). These are, as shown in Table 1:
a) the Highland Series of metasediments, hypersthene-bearing (charnockitic) gneisses, and hornblende-biotite gneisses, metamorphosed, for the most part, under granulite facies conditions;
b) the Southwestern group, lithologically somewhat dissimilar to the Highland Series, but still bearing the imprint of granulite facies metamorphism, and
c) the Vijayan complex of migmatites, granitic gneisses and granitoid rocks on the east and west of the centrally situated Highland Series belt.

Two small basins of Mesozoic (Jurassic) deposits are faulted into the Precambrian basement in the northwestern part of the island. One is the Tabbowa basin which is exposed at the surface, the other the Andigama-Pallegama basin to the south of it, covered by later deposits. The rest of the island, chiefly in the extreme north and northwest is underlain by sedimentary Miocene limestone (Fig. 1). Belts of uncon-solidated post-Miocene deposits, mainly sands and clays, occupy the coastal areas, especially of the northwest and the east.

2

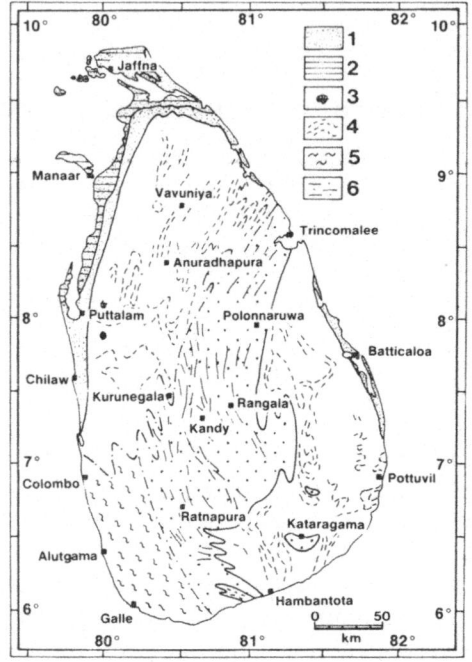

Fig. 1. Simplified geological map of Sri Lanka. 1. post-Miocene; 2. Miocene; 3. Jurassic; 4. Vijayan Complex with trend lines; 5. Southwestern group; 6. Highland Series, with major fold axes.

Precambrian

Highland series

Extent

Rocks of the Highland Series occupy the entire central Hill Country of Sri Lanka, including the Rakwana-Ratnapura hills in the southwest and the Knuckles Massif in the northeast. These extend north-eastwards as a gradually narrowing belt to the coast at Trincomalee (Fig. 1). Enclaves of Highland Series rocks are found within the eastern Vijayan, at Kataragama and Maligawila. The eastern boundary of the Highland Series lies along the foothils of the hill country and takes the form of a 'transitional zone' containing rocks typical of both major units (Cooray 1962). This boundary is sheared and tectonised in places (Vitanage 1959). The northwestern boundary is more difficult to determine. It is tentatively located where typical Highland Series rocks are increasingly replaced by Vijayan-type granitic and migmatitic gneisses, as seen in the Horowupotana and Kaudulla areas.

Table 1. Formations present in Sri Lanka.

Era	System	Epoch	Formation
MESOZOIC		Recent	'Younger' group: coral reefs, alluvium lagoonal and estuarine clays, beach and dune sands, beachrock
	QUATERNARY	? Pliocene-Pleistocene	'Older' group: Red Earth, terrace gravels, ferruginous gravels; Ratnapura beds
	TERTIARY	Miocene	Jaffna Limestone; Minihagalkanda beds
CENOZOIC	CRETACEOUS		? Dolerite dykes
	JURASSIC		Tabbowa beds; Andigama-Pallama beds
	PRE-CAMBRIAN		Tonigala and other granites, with associated migmatites. VIJAYAN COMPLEX – gneisses, migmatities and granitoid rocks; scattered metasediments and charnockitic gneisses. SOUTHWESTERN GROUP – metasediments, charnockitic gneisses, granitic gneisses. HIGHLAND SERIES, Kataragama and Maligawila Complexes – metasediments and charnockitic gneisses. ? Basement rocks – not seen

Lithology

The Highland Series is made up of a well stratified and interbanded succession of metasedimentary schists, granulites and gneisses such as quartzites, quarzt schists, garnet-sillimanite gneisses and schists, marbles, calciphyres (impure calcareous rocks and garnet-biotite gneisses, together with other rock types whose origin is uncertain. The latter include charnockitic, hypersthene-bearing rocks, which are predominant in many parts of the Highland Series, as well as hornblende-biotite gneisses and amphibolites. The metasedimentary rocks are clearly the high-grade metamorphic equivalents of sandstones, arkoses, shales, limestones, and various intermediate types typical of a miogeosynclinal facies of sedimentation. It has recently been suggested that the more basic charnockitic rocks and amphibolites may be the metamorphosed equivalents of volcanic rocks of basaltic composition erupted during the course of sedimentation (Munasinghe & Dissanayake 1980a). Typical mineral assemblages present in the Highland Series rocks are shown in Table 2, and the chemical composition of the more common rock types given in Table 3.

Published descriptions of Highland Series rocks are available for only two areas, viz. Polonnaruwa (Vitanage 1959) and Rangala (Cooray 1961). These rocks can, however, be recognized over most of the Central Highlands by the close associa-

4

Table 2. Some common mineral assemblages in Highland Series-Southwestern group rocks.

PELITIC:	Quartz-microperthite ± garnet ± sillimanite ± plagioclase ± biotite Quartz-microperthite-plagioclase-cordierite-garnet ± sillimanite ± biotite ± hypersthene
QUARTZO-FELDSPATHIC:	Quartz-microperthite-garnet ± biotite ± sillimanite Quartz-microperthite-plagioclase-garnet-hornblende Quartz-microperthite-plagioclase-hypersthene ± diopside ± garnet ± biotite
PSAMMITIC:	Quartz ± sillimanite
CALCAREOUS:	Calcite-dolomite ± phlogopite ± diospide ± forsterite Diopside-scapolite Wollastonite-scapolite-diopside
BASIC:	Plagioclase-hypersthene-diopside ± hornblende ± garnet ± biotite Plagioclase-hornblende ± garnet ± biotite

tion of metasediments and charnockitic rocks on all scales, by their regular interbanding (Fig. 2), by the continuity of some rock types, e.g. quartzites over long distances, and by their reddish brown lithomargic weathering products, particularly in the wetter parts of the island.

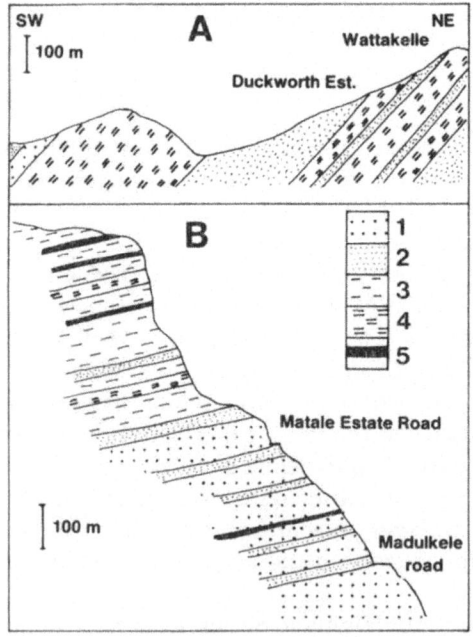

Fig.2. Cross-sections showing typical associations of Highland Series rocks in the Rangala area. 1. garnet-sillimanite schist; 2. quartzite; 3. quartzo-feldspathic gneiss; 4. charnockitic gneiss; 5. basic rock.

The charnockitic rocks probably are the most common and distinctive rock types, recognizable by their bluish grey or greenish grey colour when fresh, and their greasy appearance on exposed surfaces. They often form large masses, as for example along the Haputale escarpment, the Bulutota Pass and in the Hatton area. The quartzites also are distinctive as they generally form long continuous ridges and escarpments with whitish weathered slopes, e.g. the Sudukande ridge, which extends for over 80 km north and south of Polonnaruwa. The marbles largely are dolomitic, and although they do not form conspicuous natural features, their presence often is indicated by the occurrence of lime-burning kilns along the outcrops. The largest and most persistent of the marble bands is that which runs for many tens of kilometres northwards from Matale (Fig. 3). The garnet-sillimanite schists (the 'khondalites' of India) are recognized by the presence of large garnets; these rocks weather easily and outcrops of fresh rock are encountered rarely. The

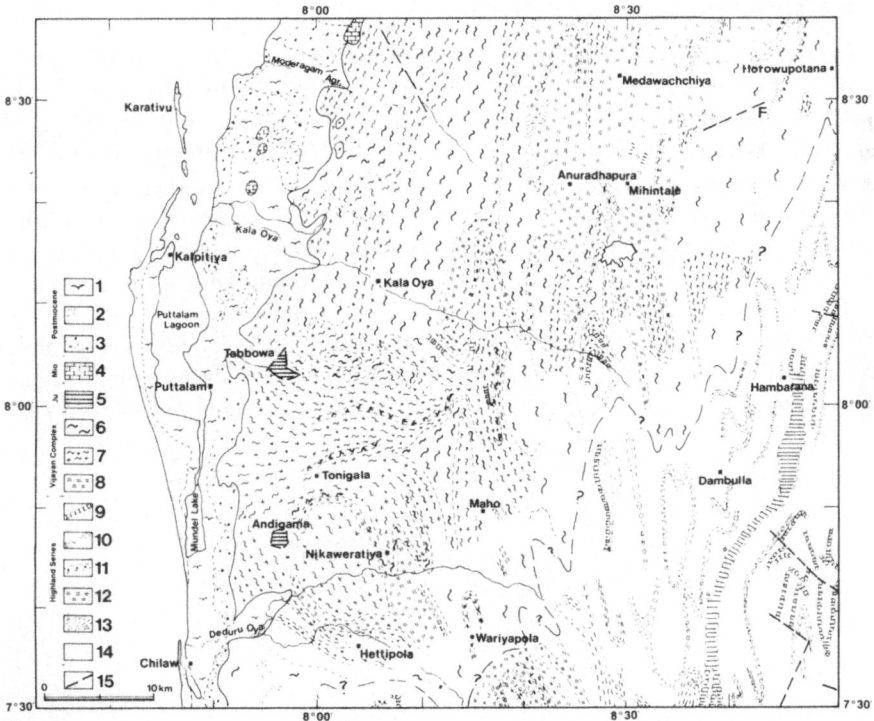

Fig.3. Geology of part of the north-western sector, Sri Lanka. *Post-Miocene:* 1. alluvial and lagoonal clays; 2. unconsolidated sands; 3. Red Earth. *Miocene:* 4. Jaffna Limestone. *Jurassic:* 5. Tabbowa Beds, Andigama Beds. *Vijayan Complex:* 6. granitic gneiss and migmatite, with pink microline; 7. leucocratic gneiss, migmatite; 8. charnockitic gneiss. *Highland Series:* 9. marble, calciphyre; 10. quartzite; 11. cordierite gneiss; 12. charnockitic gneiss; 13. hornblende-biotite gneiss; 14. undifferentiated Highland Series; 15. fault.

6

calciphyres are essentially greenish white rocks and they occur mostly as relatively narrow bands. Quartz-feldspar granulites and gneisses, generally with garnet and sometimes with biotite, also are common in the Highland Series.

Around Kandy and occupying a succession of elongate, basin-like synformal structures (Fig. 4) known as 'arenas' (Vitanage 1970, 1972) are a group of hornblende-biotite gneisses which have been formed under lower grade metamorphic conditions than the rest of the Highland Series. Their origin is uncertain, one suggestion being that the synforms were affected by migmatisation and partial melting, and that they reacted to the regional stresses in a different manner to their external counterparts (Berger & Jayasinghe 1976). Another view is that they are Vijayan rocks intruded into the Highland Series (Munasinghe & Dissanayake 1980b).

Metamorphism

The majority of mineral assemblages present in Highland Series rocks (Table 2) indicate that they were formed in the granulite (or high grade) facies of metamorphism in a relatively dry environment (Cooray 1962). This accounts for the general absence of OH-bearing minerals such as biotite and hornblende. Where these minerals are present it suggests that the containing rocks were either formed under different metamorphic conditions, in the presence of water, after their initial formation. Structural and mineralogical evidence does in fact suggest that the Highland Series did undergo more than one metamorphic event (Hapuarachchi 1968, Berger & Jayasinghe 1976).

The mineral assemblages suggest that the Highland Series rocks were formed at temperatures of 700° to 800° C at about 6 kb pressure (corresponding to crustal depths of roughly 17 to 25 km) (see Katz 1972a, Jayawardena & Carswell 1976).

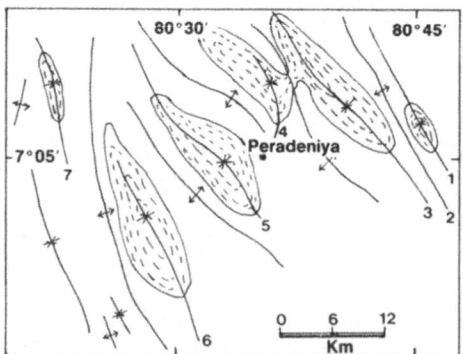

Fig. 4. Sketch map of synformal basins or 'areas' around Peradeniya (after Munasinghe & Dissanayake 1980). 1. Hula Ganga synform; 2. Digana antiform; 3. Dumbara synform; 4. Getambe synform; 5 Gadaladeniya synform; 6. Aranayake synform; 7. Maha Oya synform.

More recent calculations by workers in many parts of the world indicate, however, that the granulite facies rocks could have been formed at higher temperatures and pressures, e.g. 700° to 1,000 ° C and 7 to 12 kb (with implied depths of 25 to 42 km) (see Tarney & Windley 1977).

Table 3.   Chemical analyses of some typical rocks from the Precambrian of Sri Lanka.

	Highland Series															
	1	2	3	4	5	6	7	8	9	10	11	12	13	14	15	16
	Weight Percentage															
SiO_2	11.2	45.7	58.8	61.6	43.6	45.8	66.5	48.2	74.7	68.4	65.3	52.1	45.0	70.1	75.8	72.4
TiO_2	0.2	0.1	1.0	1.7	1.8	0.8	0.7	0.9	0.0	0.4	0.7	0.8	1.3	0.1	0.0	tr.
Al_2O_3	1.1	22.7	18.2	13.3	16.8	16.0	16.0	9.8	14.7	17.4	13.4	14.4	13.0	15.1	15.2	15.1
Fe_2O_3	0.0	1.1	2.5	0.6	2.0	5.0	1.4	0.1	0.0	0.0	1.0	1.7	0.7	0.4	1.2	1.1
FeO	0.7	0.9	7.1	9.3	11.9	9.3	4.8	5.3	1.5	4.4	6.0	10.1	11.0	2.4	1.0	0.7
MnO	0.1	0.2	0.3	0.6	1.1	1.0	0.1	0.0	tr.	0.1	0.1	0.2	0.4	1.0	0.0	0.2
MgO	18.3	4.8	4.0	1.7	6.6	5.6	0.1	3.4	0.5	1.5	3.0	5.7	11.8	0.0	0.0	0.1
CaO	35.2	20.2	4.1	6.9	12.3	10.9	2.6	28.2	2.0	1.7	4.3	9.2	12.8	3.4	0.8	1.3
Na_2O	0.3	1.7	2.5	1.8	2.2	2.9	2.4	1.0	3.4	3.6	2.7	3.1	1.5	4.8	1.3	4.3
K_2O		0.9	2.0	1.4	0.7	1.5	4.7	0.3	3.3	2.5	2.6	1.1	0.8	1.9	5.4	4.9
H_2O^+	n.d.	0.5	0.1	0.5	0.3	0.5	0.4	0.1	0.1	0.1	0.5	1.1	1.1	0.4	0.3	0.1
H_2O^-	n.d.	0.1		0.1	0.1	0.1	0.1	0.0	0.0	0.0	0.1	0.1	0.0	0.1	0.0	0.1
CO_2	32.5	1.0	n.d.	n.d.	n.d.	0.6	n.d.	0.9	n.d.	n.d.	n.d.	n.d.	n.d.	n.d.	n.d.	n.d.
P_2O_5	0.2	0.5	tr.	o.4	0.2	0.0	0.3	0.1	0.0	0.1	0.2	0.2	0.3	0.3	0.0	tr.
Total	99.6	100.3	100.7	99.8	99.5	100.1	100.3	98.3	100.2	100.2	99.9	99.9	99.8	99.9	101.3	100.3
S.G.			2.8	2.9	3.2	3.1	2.8	3.1	2.7	2.8		3.1	3.3	2.7	2.8	2.6

1. Dolomite marble near mst. 41, Kandy Weragantota road, Rangala. (Anal. J.P.R. Fonseka, 1958)
2. Calciphyre, 1 mile south of Wilgomuwa on Udawelwela footpath, Rangala. (Anal. J.P.R. Fonseka, 1958)
3. Quartzo-feldspathic granulite, Hattota Amuna, Pallegama, Rangala. (Anal. J.P.R. Fonseka, 1958)
4. Charnockitic gneiss, Galbida Estate, Rangala. (Anal. J.P.R. Fonseka, 1959)
5. Pyroxite granulite ('basic charnockite'), Dikpatana, Laggala, Rangala. (Anal. J.P.R. Fonseka, 1959)
6. Amphibolite, culvert 36/8, Kandy-Weragantota road, Rangala. (Anal. J.P.R. Fonseka, 1958)
7. Hornblende-garnet granulite, Lethenty Estate, Castlereagh, Hatton. (Anal. J.P.R. Fonseka, 1966)
8. Calciphyre, culvert 12/4, Migahatenne-Pelawatte road, Alutgama. (Anal. R. Thomas, 1960)
9. Quartzo-feldspathic granulite, mst. 2, Pelawatte-Nelluwa road, Alutgama. (Anal. O.C. Wickremasinghe, 1963)
0. Garnet-sillimanite-cordierite gneiss, Pelawatte-Nelluwa road, Alutgama. (Anal. O.C. Wickremasinghe, 1963)
1. Charnockitic gneiss, Pallegoda road junction, Alutgama. (Anal. R. Thomas, 1963)
2. Pyroxite granulite ('basic charnockite'), St George's Estate, Matugama. (Anal. J.P.R. Fonseka, 1964)
3. Pyroxene amphibolite, Bodinagoda, 1 mile east of Induruwa, Alutgama. (Anal. N.R. de Silva, 1964)
4. Microcline-biotite gneiss, Horiwila, Polonnaruwa. (Anal. O.C. Wickremasinghe)
5. Microcline-biotite granitic gneiss, Kalkudah quarry. (Anal. O.C. Wickremasinghe)
6. Tonigala Granite (average of 4), Tonigala, Galgamuwa. (Anal. J.P.R. Fonseka, 1958)

Sources: 1-6, Cooray 1961; 7, Cooray 1972; 8-13, Cooray 1965; 14, 15, Vitenage 1959; 16, Cooray 1971.

8

Structure

Highland Series rocks appear, at first sight, to be folded into a simple system of parallel folds trending from SE-NW in the southern part of the belt, through N-S in the central part, to NE-SW in the northern part (Fig. 1). A detailed study of the Kandy area (Berger & Jayasinghe 1976) has shown, however, that the Highland Series rocks have undergone a much more complex deformational history in which at least two, but possibly three periods of deformation can be recognized. It is the latest of these deformations that has given rise to the existing major structures which dominate the Highland Series. They include open folds, overturned and recumbent folds, some of considerable magnitude (e.g. in the Knuckles, Ramboda and Sri Pada areas) (Fig. 5), refolded folds, and large-scale interference patterns as seen in the Ambalangoda and Buttala area (Seneviratne & Herath 1969, Jayawardena & Seneviratne 1971).

Jointing is well developed in the Highland Series rocks and it is these joints as well as the configuration of major folds that often control drainage patterns. In the Hatton area, for example, the trellis drainage pattern is the result of the main streams flowing along parallel, lithologically controlled strike valleys and the tributaries along joint directions. In the Kandy area, the courses of the major streams often follow the outline of the basin-like synforms in large loops.

Many faults appear to be present in the Highland Series belt but most of those shown on the geological maps have been photogeologically located. The few that can be located on the ground, e.g. north of Polonnaruwa and south of Kaudulla Wewa, occur where quartzite ridges can be seen to be displaced. Most of the apparent faults trend either NE-SW or NW-SE.

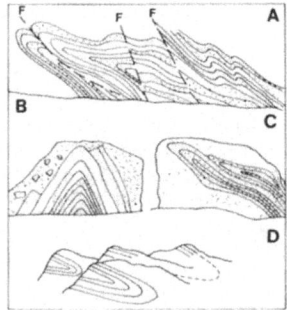

Fig. 5. Typical structures in Highland Series rocks. A: Overfolds and high-angled thrusts, Bindunuwewa Farm road, Bandarawela; B: Symmetrical fold in quartzite, Nyanza Estate, Maskeliya; C: Small-scale overfold, near Bandarawela; D: Large-scale recumbent fold, Knuckles Massif, looking south from Mimure.

Southwestern group

The name 'Southwestern group' has been given to a succession of crystalline rocks in the south-western sector of the island which, though similar to the Highland Series in some respects, are significantly enough different in others to justify their separate identity. Although they have been mapped in some detail, the rocks of this group have been studied only in the Alutgama area (Cooray 1965) and much of what follows is based on what is known about the rocks of that area.

Extent

Rocks of the Southwestern group occur in an area extending from the coast eastwards to the Sinharaja Forest, the eastern boundary more or less coinciding with a zone of basic rocks lying within the Forest (Fig. 6). Its southern boundary is about the latitude of Galle and it runs northwards approximately to a line joining Gampaha and Narammala. The exact limits of the group have not yet been determined.

Lithology

The commonest rock types are:
a) thin, persistent quartzites and quartz schists;
b) narrow, continuous bands of calciphyre, sometimes containing wollastonite;
c) cordierite-bearing granulites and gneisses, sometimes with andalusite, sillimanite, and hypersthene, and with rare sapphire and corundum as accessories (Katz 1972a);
d) charnockitic hypersthene gneisses and charnockitoid rocks with fayalite and quartz;
e) 'charnockitic-looking' gneisses with hornblende, biotite and garnet, but lacking hypersthene;
f) coarse-grained, intrusive hypersthene granite (charnockite) bodies and cross-cutting pegmatites;
g) garnet-biotite gneisses with relic charnockite patches;
h) abundant granitic gneisses and augen gneisses
i) small granite plutons and granitoid rocks;
j) a variety of basic rocks ranging in composition from two-pyroxene granulites to amphibolites.

The Sinharaja Basic zone near the eastern boundary coincides with a marked aeromagnetic anomaly extending for over 30 km in length. The dominant rock types in this zone are basic rocks (as above) with minor quartzites, garnet-biotite, gneisses and charnockitic gneisses (Hapuarchchi & Herath 1969). The relatively high magnetic intensity of the zone is due to the high percentage of magnetite in the basic rocks (Pattiaratchi 1959, Hapuarachchi *et al.* 1964).

The different rock types of the Southwestern group are interbanded with each other (Fig. 6) and, except for the granites and pegmatites, appear to represent a sedimentary-volcanic sequence similar to that represented by the Highland Series, but probably of different composition and metamorphic history. Chemical analyses of some typical rocks from the Alutgama area are given in Table 3.

Metamorphism

Probably the most significant rocks in the group are the cordierite- and wollastonite-bearing rocks. Although the former have been found in the Highland Series, these are scattered, minor occurrences, whereas the cordierite-bearing gneisses of the Southwestern group are a major and critical member of the group (Fig. 6). In a study of the metamorphic history of the south-western region (Hapuarachchi 1968), it was suggested that the region underwent a second period of metamorphism under pressure conditions that affected it only and not the rest of the Highland Series.

One of the main differences between the Southwestern group and the Highland Series is the presence of several bands of granitic gneisses in the former and their relative absence in the latter (Fig. 6). This is another indication that the two units have had somewhat different metamorphic histories.

Structure

Geological mapping has revealed the presence of several major folds running NW-SE within the region. Recumbent folds also are present. The streakiness and small-folding in the gneisses (Fig. 7) is a measure of the ductility of the rocks at the time of deformation and is in contrast to the styles of folding in the Highland Series. A major fault is postulated running NW-SE from near Kalutara to Akuressa, mainly on geophysical grounds.

Vijayan complex

Practically the whole of the northern, eastern and south-eastern parts of the island are made up of 'a heterogeneous association of gneisses, migmatites, granites and granitic rocks' (Fernando 1949), to which the name Vijayan Complex has now been given. Because these rocks lie on either side of the Highland Series belt they usually are referred to as the 'western' Vijayan and the 'eastern' Vijayan, the boundaries of which generally follow the foothills of the Central Highlands.

No detailed description of any part of the Vijayan Complex has yet been published, but the rocks are known to be mainly microcline-bearing biotite and hornblende-biotite gneisses, migmatites and granitoid rocks in which are found scattered layers and bands of amphibolite, quartzite, marble and calciphyre as well

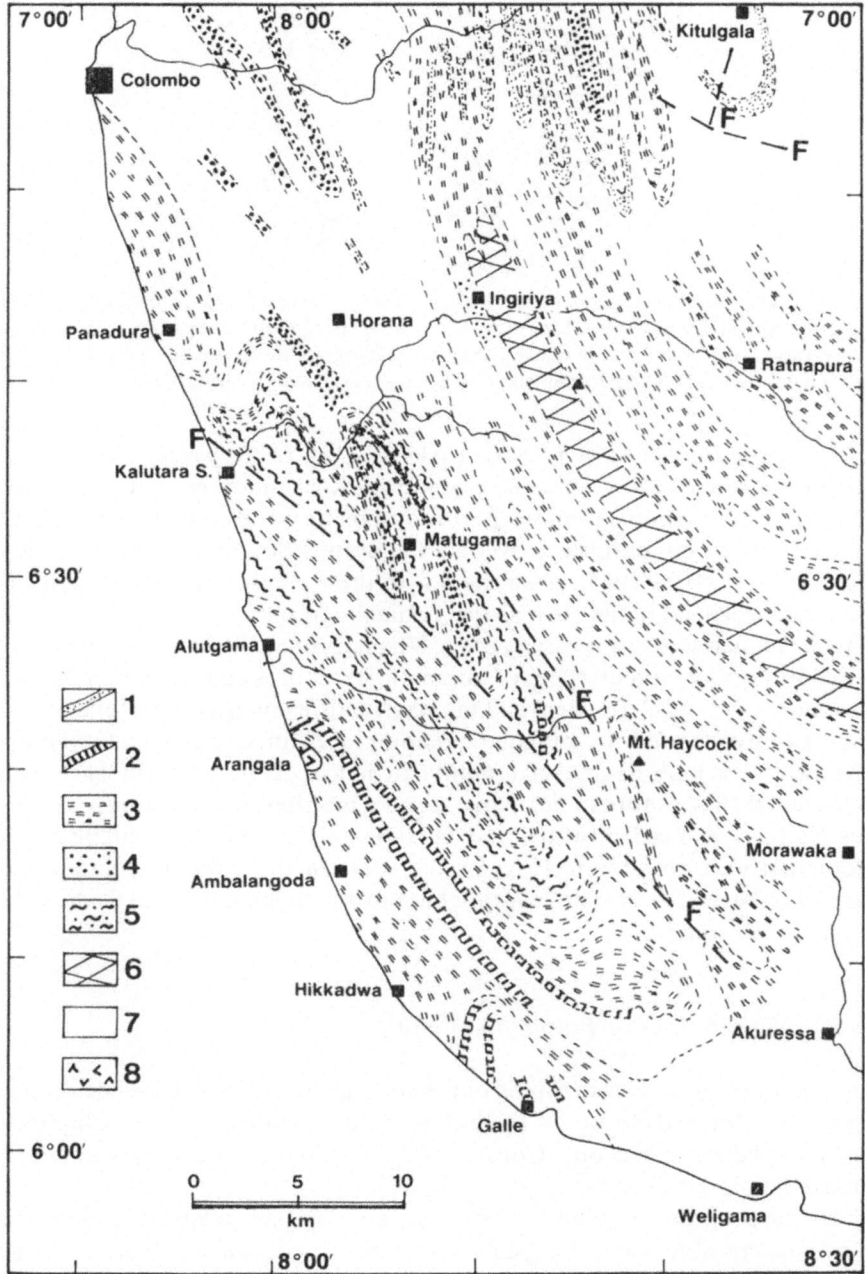

Fig.6. Geology of part of the south-western sector, Sri Lanka. *Highland Series and Southwestern Group:* 1. quartzite, quartz schist; 2. calciphyre; 3. charnockite gneiss; 4. cordierite gneiss; 5. garnet-biotite gneiss; 6. Sinharaja basic zone; 7. undifferentiated H.S. and Swg. *Intrusives:* 8. hornblende granite. F. fault.

12

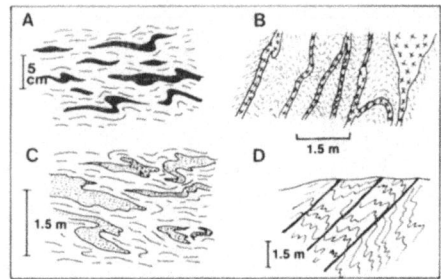

Fig. 7. Structural styles in gneisses of the Alutgama area. A: Small-folded garnet-biotite gneiss with quartzo-feldspathic segregations (solid), Vogan Estate; B: Irregularly contorted garnet-biotite gneiss with pegmatitic veins (crosses), Owitigala; C: Syntectonic pegmatitic segregations (stippled) in granitic gneiss, Vogan Estate; D: Quartz veins (solid) in slip planes of small-folded gneiss.

as larger occurrences of charnockitic rocks (Fig. 8). In the western Vijayan of the Galgamuwa area, for example, the granitic gneisses and migmatites consist of an early, leucocratic group, granodioritic in composition and complexly folded, and a later, pinkish group, granitic to syenitic in composition and partly replacing the earlier group (Cooray 1964). Associated with these gneissic rocks are small bodies of pink, microcline granite such as the Tonigala granite.

There is much structural variety in the Vijayan rocks. Macroscopically they are banded to streaky, and all types of migmatic structures can be seen in them. The rocks generally are well foliated, and minor structures such as small folds, corrugations, crumpling, rodding, ptygmatic veining and lineations are common. The rocks exhibit a high degree of mobility in places. Boudinage of basic bands, pinch-and-swell structure in pegmatites, and agmatites, are often seen. In certain areas, as near Komari, Pottuvil and Okanda in the southeast, circular, domal tectonic patterns are present (Fig. 9), the cores of these domes being occupied either by hypersthene granite (charnockite) or by augen gneiss (D.E. de S. Jayawardena pers. comm.).

Stratigraphic relations of Precambrian units

Although the preliminary mapping of the geology of Sri Lanka has been completed, very few detailed studies of the different units and their relationship to each other have been carried out. Consequently, stratigraphic concepts about these units are largely conjectural.

Early ideas on the Highland Series/Vijayan Complex relationship were based on the concept of an early Vijayan gneissic 'basement' on which later sediments were deposited and subsequently metamorphosed and folded into a synclinorium represented by the present Highland Series (see Adams 1929, Coates, 1935, Wadia 1945, Fernando, 1948). Detailed mapping of the Polonnaruwa and Rangala areas

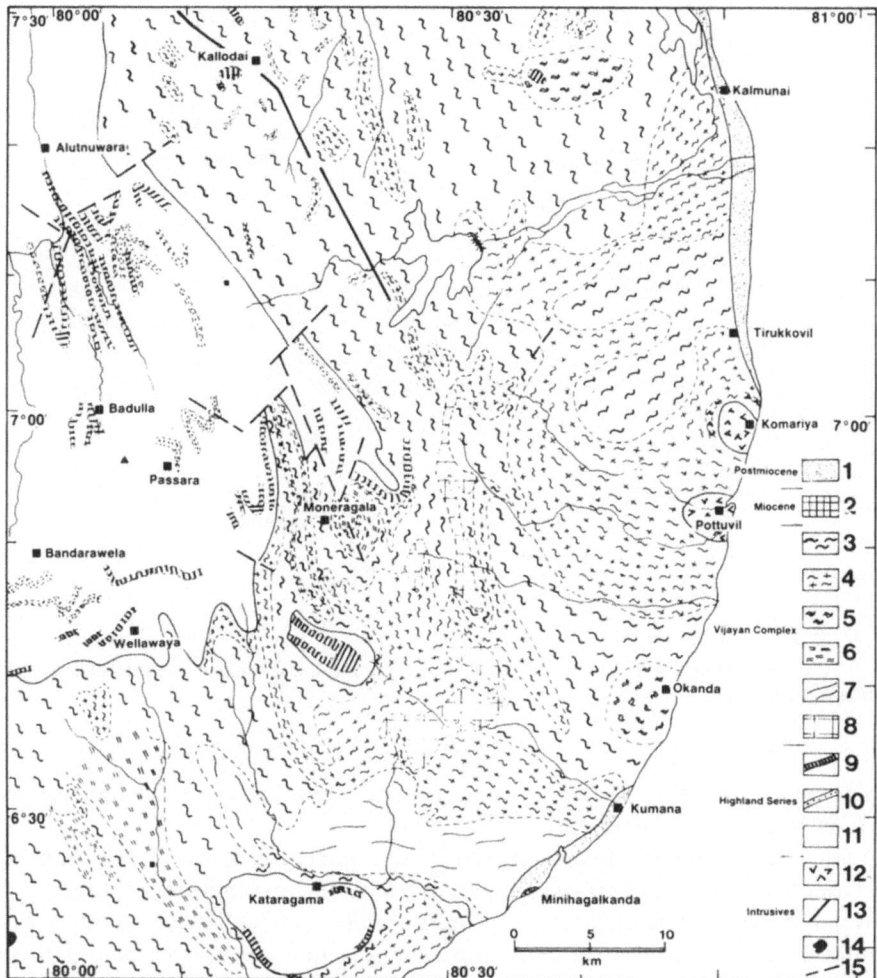

Fig.8. Geology of part of the south-eastern sector, Sri Lanka. *Post-Miocene:* 1. unconsolidated sand. *Miocene:* 2. Minihagalkanda beds. *Vijayan Complex:* 3. granitic gneiss with pink microcline; 4. leucocratic gneiss and migmatite; 5. augen gneiss; 6. charnockitic gneiss; 7. undifferentiated Vijayan gneiss with trend lines; 8. undifferentiated metasediments. *Highland Series:* 9. marble and calciphyre; 10. quartzite; 11. undifferentiated Highland Series. *Intrusives:* 12. hypersthene granite; 13. dolerite; 14. serpentine; 15. fault.

(Vitanage 1959, Cooray 1961) resulted, however, in the opposite view, i.e. the Vijayan rocks probably were younger than the Highland Series, being in fact originally Highland Series rocks subsequently migmatised and granitised to their present state. The boundary between the two, especially on the eastern side, was seen as a Transitional Zone, about 10 km wide and extending for about 100 km, in

14

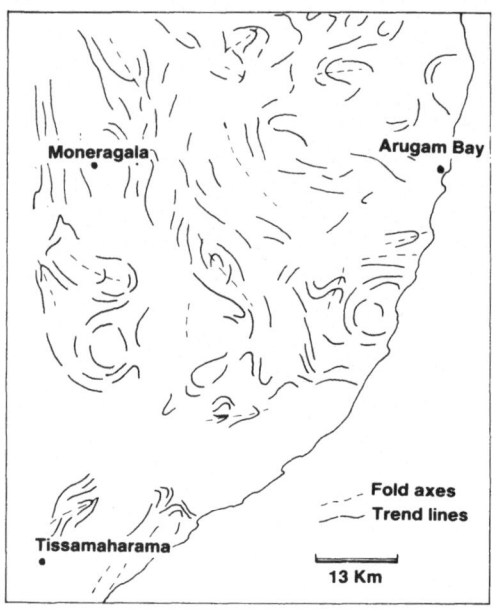

Fig. 9. Tectonic pattern in the eastern Vijayan showing trend lines, fold axes and domal structures, south-eastern Sri Lanka.

which rock types of both units were present. Geochronological data seem to support this view (see below).

A later view (Katz 1971) was a return to the earlier concept of a Vijayan basement with an overlying sedimentary succession, both of which underwent three periods of deep-seated metamorphism (at 2,000, 1,200 and 650-450 m.y. ago) under granulite facies conditions, yet retaining their present mineral assemblages through all the high-grade metamorphic episodes. The present lithological differences between the two units were attributed by Katz to the original differences in chemical composition.

The most recent view (Munasinghe & Dissanayake 1982) is that the Highland Series, originally a volcanic-sedimentary succession, was formed by the collision of two continental blocks, now represented by the eastern and western Vijayan. This idea is based on the presence of shear zones, hot springs, and serpentines along the eastern contact. The presence of a gravity 'low' below the eastern Vijayan also suggests thickening of the crust by overthrusting (Munasinghe & Dissanayake 1982).

The Highland Series/Southwestern group relationship also is unclear. Either the latter is part of the Highland Series geosynclinal succession, but showing a marked change in sedimentary facies westwards, or it represents a separate succession. The latter view has been expanded by Katz (1972a, b) into the concept of a paired metamorphic belt. The Highland Series is said to represent a miogeosynclinal type of sequence of limestones and sandstones which forms an intermediate pressure-temperature series of the Barrovian type. The southwestern group represents a eugeosynclinal greywacke-shale sequence forming an Abukuma-type low pressure-intermediate temperature facies series. Correlation with similar paired belts in adjacent parts of Gondwanaland has been suggested by Katz.

Age of Precambrian units

Although the crystalline rocks of Sri Lanka always were assumed to be of Precambrian age (i.e. more than 550 m.y.), the first reliable geochronological data were provided by Crawford (1969) and Crawford & Oliver (1969). Crawford's work was based on 59 Rb/Sr analyses of whole rock and mineral samples and 9 K/Ar analyses of hornblendes from localities thought to be representative of the Precambrian.

According to these authors (Table 4), the HighlandSeries and the Kataragama Complex were metamorphosed more than 2,000 m.y. ago, the isochron age for the latter being 2,055 ± 55 m.y. Confirmation of this age is given by an isochron age of 2,170 m.y. for charnockitic gneisses of the Highland Series (Wickremasinghe 1969). It must be noted, however, that several samples within the Highland Series have given older ages, of the order of 2,300 to 2,600 m.y., which suggests that a widespread metamorphic event affected the Highland Series rocks much earlier. A similar suggestion was made by Berger & Jayasinghe (1976) on the tectonic evidence.

The Vijayan retrogressive metamorphism ended about 1,150 ± 60 m.y., and the Tonigala Granite, with an isochron age of 985 ± 30 m.y. post-dates the Vijayan metamorphism (Crawford & Oliver 1969). Two other events, regarded as overprints, affected the Precambrian rocks. The first of these, at 650 m.y., corresponds to the Pan-African thermal or tecto-thermal event that has left its mark over most

Table 4. Geochronology of the Precambrian of Sri Lanka (in millions of years).

	A	B	C
? Earlier relics	2,600-3,000		
Highland Series	>2,100	2,170	
Kataragama Complex	2,055 ± 55		
'Charnockite' formation	1,250		
Vijayan Complex	1,150 ± 60		
Tonigala Granite	985 ± 30		
Tecto-thermal event	650		
Metamorphic event	450		540-350

A: Crawford & Oliver 1969, B: O.C. Wickremasinghe 1969, C: Cooray 1965.

of former Gondwanaland. The second, at 450 m.y. (Cooray 1969), cannot be correlated with any event in the field or petrographically, but it is recorded in a variety of rocks over a large area.

Comparisons with the Indian Shield show that the Proterozoic 2,100 m.y. event in Sri Lanka is comparable to the 2,000 m.y. regional ages in the mobile belt of granulites and gneisses surrounding the central craton of greenstones, granites and gneisses of south India. Older ages also may be present here. Some biotite ages of 1,150 m.y. may be a reflection of the Vijayan event, and evidence of later 'overprints' can be seen as far north as the Mysore boundary (Crawford & Oliver 1969).

Influence of the Precambrian on ecology and biogeography

It is probably true to say that over the island as a whole, and hence in the Precambrian generally, climatic factors are the most important influence on ecology. Although lithology does play a part, this latter factor is subordinate, and for ecological purposes the subdivision of the island into Wet, Intermediate and Dry Zones (in terms of rainfall) is more meaningful than into respective Precambrian units.

At the same time it must be noted that most of the Highland Series and the Southwestern group fall within the Wet and Intermediate Zones, and the Vijayan entirely within the Dry Zone. Consequently, the main Precambrian units do affect the ecology in the manner in which they have reacted to the climatic conditions of the different zones, as seen in the soils, vegetation, groundwater conditions, landforms, drainage patterns and settlements in the respective areas.

Soils and vegetation

In a recently published map of the Agro-Ecological regions of Sri Lanka (Dept. of Agriculture 1979), the predominant soils developed on Vijayan rocks are shown to be Reddish Brown Earths and Non-Calcic Brown Soils (Fig. 10). The Reddish-Brown Earths are derived from rocks that have a sufficiency of iron- and manganese-bearing minerals, and hence are developed on the biotite gneisses and hornblende-biotite gneisses that form the major part of the Vijayan in the northern and

Fig. 10. Map showing geology, climate (rainfall) regions, and major soil groups, Sri Lanka (Dept of Agriculture 1979). 1. Red Yellow Podsolic Soils; 2. hard and soft laterite; 3. Reddish Brown Latosolic Soils and Immature Brown Loams; 4. Reddish Brown Earths and Low Humic Gley Soils; 5. Non-calcic Brown Soils and Reddish Brown Earths; 6. Red Yellow Latosols; 7. Regosols; 8. Solodized Solonetz, Solonchaks; 9. Reddish Brown Earths and Immature Brown Loams. *Geology:* horizontal lines – Miocene and post-Miocene sedimentary formations; vertical lines – Vijayan Complex; blank – Highland Series and Southwestern group; thick dashed lines – main geological boundaries.

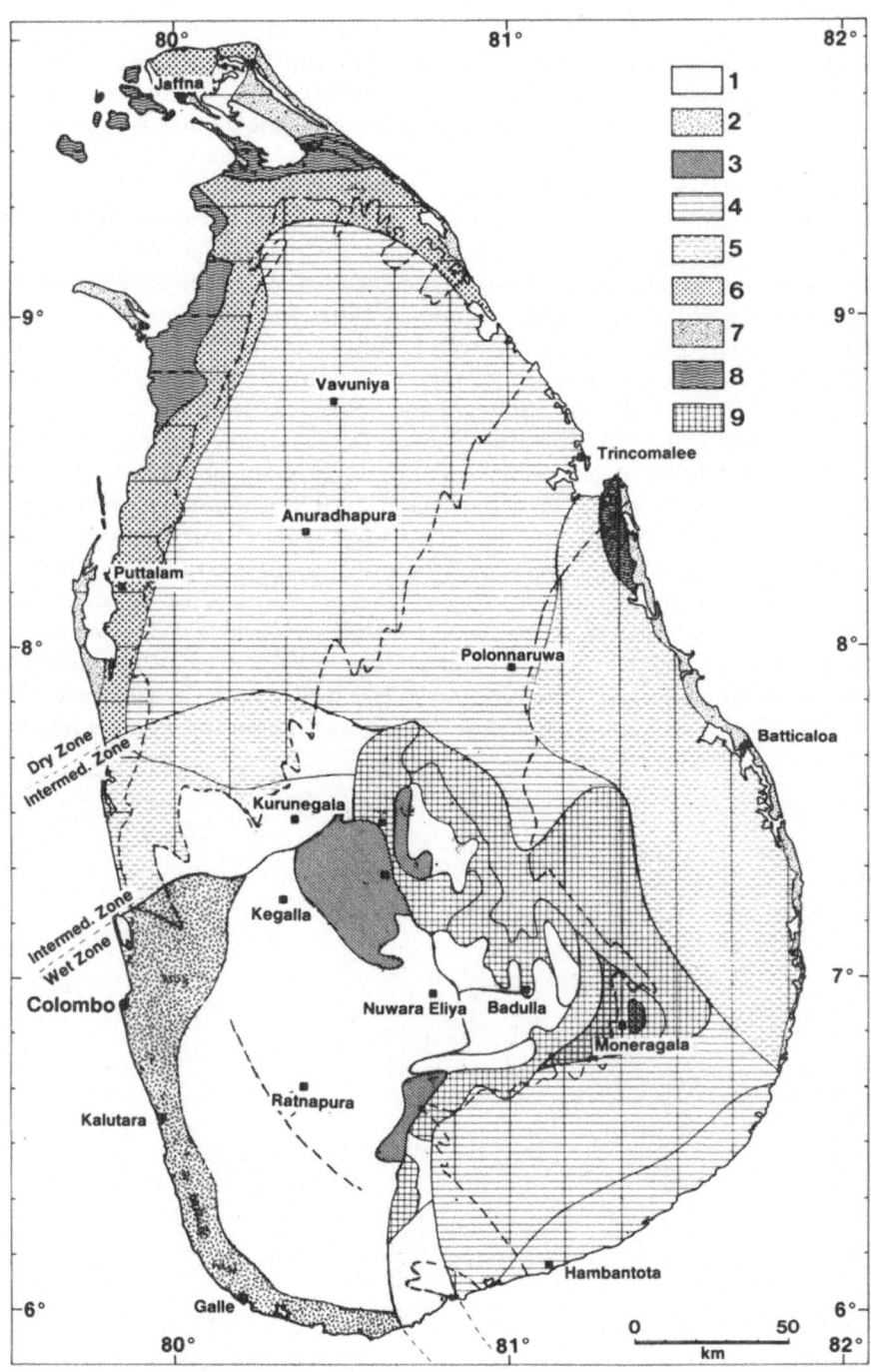

18

southern parts of the island. The Non-Calcic Brown Soils are found over much of the same area, but only where there is a deficiency of Fe-Mg minerals in the parent rock. They are especially present in the Eastern province, where highly acidic, quartzo-feldspathic gneisses and granitoid rocks with little biotite and hornblende are common.

These soils sustain a natural vegetation known as Tropical Dry Mixed Evergreen Forest consisting of a mixture of evergreen and deciduous trees, with different tree species growing on different soils (Abeywickrema 1961). Two main associations have been recognized, namely the *Manilkara-Hemicyclia-Chloroxylon* association, and the *Alsaeodaphne-Berrya-Diospyros* association, best developed near rivers and streams (de Rosayro 1950).

In the Wet Zone, the major soil groups are the Red-Yellow Podzolic Soils, derived from a variety of metasediments and charnockitic rocks, with Reddish-Brown Latosolic soils and Immature Broan Loams found mainly within the Kandy Plateau. The latter soils display signs of immaturity, possibly due to recent uplift and consequent rapid erosion and dissection. The dominant forest type here is the Tropical Wet Evergreen Forest in which several communities, largely dependent on the physical properties of the soils, are found (de Rosayro 1942). These are: *Dipterocarpus* consociation, *Campnosperma* lociation, *Mesua-Doona* association, and the *Vitex-Wormia-Chaetocarpus-Anisophyllea* association.

The manner in which human and plant ecology is influenced by a combination of relief, landforms, climate, lithology and soils is well seen in the area covered by the Rangala One-inch Topographical Sheet (Fig. 11). The eastern part of the sheet

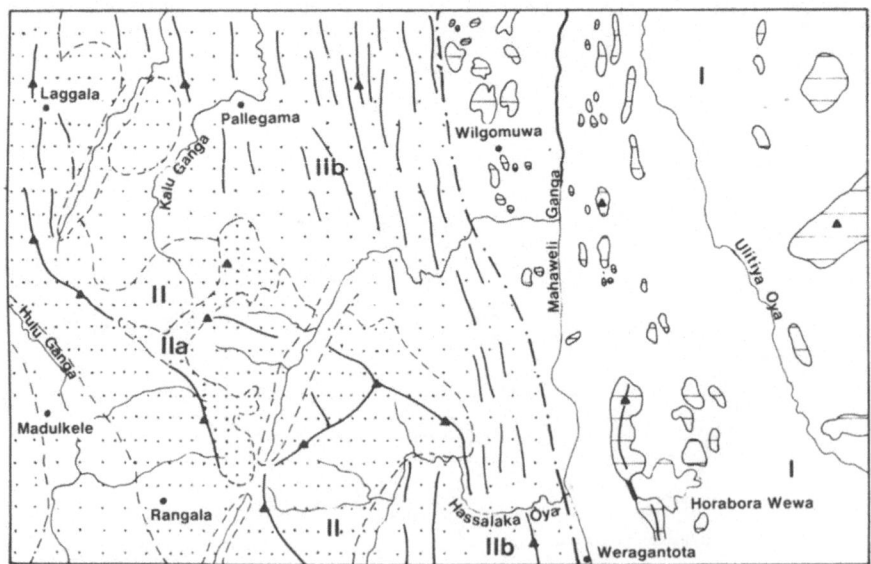

Fig. 11. Physiographic regions of the Rangala area (for explanation see text).

(Zone 1), which is part of the lowest peneplain and lies within the Intermediate and Dry Zones, is underlain by Vijayan gneisses on which have developed Reddish Brown Earths and Non-Calcic Brown Soils. The area is covered by relatively open forest of the moist facies of the dry mixed evergreen type (Holmes 1951), within which are patches of open grassy plains or damanas, which occur abruptly in the midst of the forest. Further east are patches of savannah forest type (Holmes 1951) in which there is a relatively sparsely distributed tree element of fire-resistant species in permanent grassland. The area is sparsely populated, except along both banks of the Mahaweli Ganga (Ganga = River in Sinhalese), where resettlement has brought extensive areas under rice cultivation in the last few decades.

The western half of the sheet (Zone II) is occupied by the Knuckles Massif and its outlying foothills, made up entirely of Highland Series rocks. The area lies within the Intermediate Zone, and has Red Yellow Podzolic Soils on the higher, wetter parts and Reddish Brown Earths on the lower slopes. The vegetation on the Massif is characterized by two important features, viz., the complete sequence of vegetational types from dry evergreen to wet evergreen montane forests within a relatively small compass, and an intermingling of species typical of the several forest types (de Rosayro 1958).

In the valleys, the lowland tropical wet semi-evergreen forest is dominant, with a mixture of species from the dry forest, the lowland wet forest, and the montane forest (de Rosayro, in Cooray 1956). The lower slopes of the Massif (450 to 1,050 m) are characterized by the sub-montane tropical wet semi-evergreen forest containing the earlier mentioned species but with a higher proportion of some montane species. The upper slopes (1,050 to 1,860 m) have a montane tropical wet evergreen forest with a unique 'dry' character on the eastern and northern (drier) sides. Here, semi-evergreen species are mixed with the montane species.

Occupying the centre of the Knuckles Massif is the Selvakanda Plateau (Zone IIa) where a striking tropical pygmy forest is found. It contains montane tree species stunted to a metre or less in height and forming a closely entangled, waist high, virtually impenetrable mass, with patches of grass in which are found flowering shrubs like *Osbeckia walkeri* and *Heydotis* spp. This pygmy forest is the result of exposure to severe, continuous winds, and similar pygmy forests are found in other parts of the central mountains of Sri Lanka, e.g. Land's End on Horton Plains.

The parallel ridges of the eastern part of the Massif (Zone IIb) are covered mostly by montane grassland. According to de Rosayro (1958) some of this grassland is abandoned coffee land, but a fair proportion is due to the coincidence of surface slope with dip of bedrock. In the northern foothills, some control of vegetation by lithology is seen in the development of dry forest type on quartzites; it stands out in strong contrast to the surrounding wet forest type on aerial photographs.

The Alutgama area in the southwest of the island is typical of the Wet Zone. It is underlain by rocks of the Southwestern group, upon which Red Yellow Podzolic

20

Soils have developed, with laterite along the coastal belt (Fig. 10). Four major physiographic regions (Fig. 12) have been recognized in the area (Cooray 1965), namely, the coastal plain (Zone I, less than 30 m in height), the central plateau (Zone II, average elevation 120-150 m), the Pelawatte Ganga valley (Zone III), and the eastern ranges (Zone IV, more than 300 m in height).

The metasediments, gneisses and charnockitic rocks in Zone IV form parallel strike ridges running NW-SE. Together with the prominant dip-joint pattern, this has resulted in a marked trellis pattern of drainage which has had a strong influence on settlement and communication routes. In contrast, the western part of the area (Zone I) is dominated by the Bentota Ganga, with a marked dendritic drainage pattern and a vast floodplain on which rice cultivation is predominant. The sandy coastal strip (Zone Ia) is characterized by a high density of fishing population and, because of the numerous bays and white, sandy beaches, by the growth of the tourist industry in recent years. The central plateau (Zone II) is one of the major rubber-growing areas of the island, and the heavily forested eastern region (Zone IV) is the locus of a major timber-extraction industry. Population in the eastern part of the sheet is sparse, owing to poor accessibility and lack of cultivable land.

The northeastern corner of the area, east of the Maguru Ganga (Fig. 12) forms

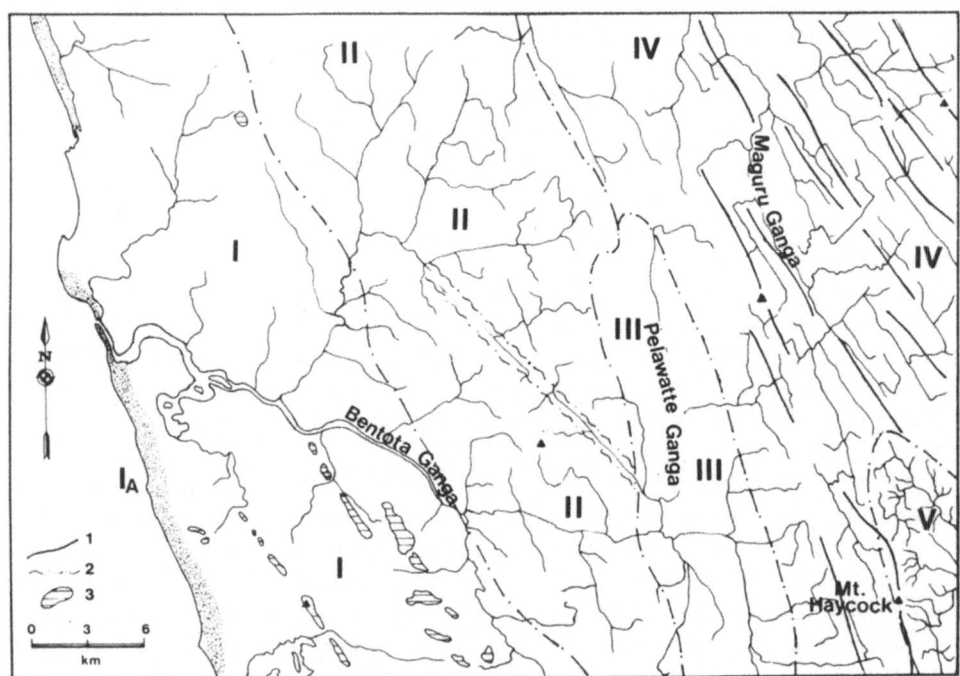

Fig. 12. Physiographic regions of the Alutgama area (for explanation see text). 1. strike ridge; 2. fault valley; 3. erosion remnant.

part of the Sinharajah Rain Forest, which is typical of the Wet Evergreen Forest found in many parts of the island (de Rosayro 1954). Here too, the trees assume almost stunted proportions near the tops of ridges. The valleys and lower slopes are covered mostly in scrub, fernland and grassland. The fernlands (kekilla) take their names from the fern *Dicranopteris linearis*, occur side by side with the forest and the grasslands, and have sharp boundaries with both. *D. linearis* is a semi-scandent fern which forms a dense, thick growth, varying from a few centimetres to two metres or more in height, depending on the length of time it has escaped burning.

In the southeastern corner of the area (Zone IV), around the Hiniduma range, the vegetation is markedly different, and the smooth, rolling topography is covered by large extents of open grassland, tree growth being confined to the valleys and tops of ridges. The striking dendritic pattern of the drainage is in marked contrast to the trellis pattern of the surrounding area. Lithological and structural control is evident here, the area being underlain by metasediments, comparable to those of the Highland Series, outcropping in the core of the Hiniduma antiform.

Several places in the hilly regions of the area are occupied by marshes, these being the result of river capture and drainage changes in recent times. Secondary scrub is dominant in these marshes. In the lowlands, however, reeds and sedges are characteristic of the marshy tracts, where black, peaty soils have developed.

Groundwater

The presence or absence of water, particularly groundwater, is probably the most critical factor influencing human settlement. Groundwater must always be considered a 'mineral' resource, but in the Precambrian of Sri Lanka, climate rather than geology determines its availability.

There are no real groundwater problems in the Wet Zone, only in the Dry. In the latter is was long believed that groundwater was relatively scarce, being restricted to isolated pockets of deeply weathered rocks. Since 1978, however, systematic exploration for groundwater in the Vijayan rocks of the Kurunegala, Moneragala and Hambantota Districts by the Water Resources Board has considerably altered the groundwater picture in those area (Dr. L. Herath, pers. comm.). It has become evident in the course of this work that there are, scattered throughout the Vijayan Complex, zones of highly jointed, fissured and fractured rocks, both gneisses and metasediments, with moderate to low permeability, giving average yields of 1,000 g.p.h. Furthermore, certain lithological types have higher yields. Marbles, for example, with their solution cavities and moderate permeability, can have average yields of 2,000 g.p.h., and quartzites, with a high degree of jointing and hence high permeability, have an average yield of 3,000 g.p.h., even in the Dry Zone. Another important property of quartzites is that lines of springs often mark the contacts between them and relatively impermeable rocks below, as seen in the Polonnaruwa area (Vitanage 1959). Large perennial springs (bubula or ulpotha) also occur in marbles, the water from such springs being characteristically 'hard'.

Villages close to such springs often bear the prefix 'kiwul' (kiwul (Sinhalese) or kivul = hard), as in the name Kiwulwadiya.

Irrigation tanks of all sizes, besides providing water for irrigation, also are a source of groundwater recharge by seepage, and this explains the location of the village wells in the Dry Zone generally below the tank bunds and by the sides of irrigation channels.

Landforms

The Precambrian rocks, which generally are banded and well foliated, by their very nature give rise to landforms that are characteristic of the conditions under which they weather. These landforms in turn have a considerable influence on human ecology. The parallel strike ridges and valleys that have developed on the Highland Series rocks in the Wet Zone; the escarpments, waterfalls and devlopment of hydro-electric power with their fresh-water reservoirs in the Central Highlands; the long, persistent quartzite ridges of the northern Dry Zone - these are only a few of the characteristic landforms resulting from a combination of lithology and climate in the Precambrian of Sri Lanka. However, as this subject is dealt with elsewhere in this volume (see Ch. 2), no further mention will be made of it here.

Mesozoic and Tertiary

Jurassic

Two small occurrences of Jurassic sediments have been preserved in faulted basins within the Precambrian basement in the northwestern part of the island at Tabbowa, 15 km northeast of Puttalam, and at Andigama, 30 km south of Tabbowa (Fig. 3).

The Tabbowa Beds (Wayland 1925) are exposed on the surface within and around Tabbowa Wewa, and especially below the tank bund and in channel-cuts. Several faults occur within the basin itself, and considerable tilting of the segments has taken place. The main rock types are arkose, feldspathic sandstone, siltstone and mudstone, the arkose in particular displaying such sedimentary features as cross-bedding, channel-fill and cyclothems (Money & Cooray 1966). These and other features suggest that the Tabbowa Beds represent a deltaic facies of sedimentation.

Fossil plant impressions of *Cladophebis zeylanica, C. reversa, Sphenopteris wadiai, Taenopteris spatulata, Nilssonia fissa* and *Ptilophyllum* sp. in the siltstones and mudstones establish the Upper Gondwana (Upper Jurassic) age of the Tabbowa Beds (Sitholey 1944). Several of these forms are indentical with those found in the Upper Gondwana deposits of Ramjahal in the Damodar Valley of India, and in coastal patches near Madras. Simlar plant remains have more recently been found in the Terani Beds, Tiruchirapally District, and in the Sivaganga Beds of the Ramnad District (Ayyasami & Gururaja 1977).

The Andigama beds are covered by more recent deposits but are seen in well sections over an area of about 30 km². There is evidence that the basin extends to Pallama, 11 km south of Andigama. There also is evidence that the basin is fragmented into a number of minor basins by faulting towards the south. Drill-hole data show that the succession at Andigama and Pallama consists predominantly of an alternating sequence of dense, hard, black carbonaceous shale and calcareous or arkosic sandstone. Brown shales occur at the top of the succession, and thin mudstone and coal seams are present sometimes. Analyses of samples from the carbonaceous shales show that they contain montmorillonite, that their carbon contents range from 4.6 to 6.3%, and that they are high in extractable hydrocarbon content (Fernando 1968, 1969, 1970). The microfloral assemblages of the shales (e.g. pteridophytic and other trilete spores and gymnospermous forms) suggest an age ranging from Middle to Upper Jurassic, and therefore partly contemporaneous with the Tabbowa Beds. Geophysical data (Hatherton *et al.* 1975) indicate that the Tabbowa Beds are at least 500 m thick, and those at Andigama 900 to 1,200 m thick. The Andigama basin is about 20 km long.

Tertiary

The extreme north and northwestern parts of Sri Lanka, comprising the Jaffna Peninsula, the surrounding islands, and the northwestern coastal strip narrowing gradually to the vicinity of Puttalam are underlain by nearly flat-lying limestone of lower Miocene age.

The typical Jaffna Limestone is hard, partly crystalline, compact, indistinctly bedded, creamy coloured rock. It is massive in parts, but some layers are richly fossiliferous and weather out into a honeycombed mass. Sandy layers containing grains of magnetite, garnet, zircon, monazite and mica also are present. The limestone becomes more variable towards the south, as near Aruakalu, north of Puttalam, where the lithology varies from pure limestone and siliceous limestone to calcareous sandstone and impure calcareous muds (Herath *et al.* 1961). Though generally horizontal, the beds dip slightly to the west in places. The limestone is well jointed with a NW-SE and NE-SW trending conjugate system of joints.

A varied assemblage of fossils is found in the Jaffna Limestone, and it includes Foraminifera, Lamellibranchiata, Gastropoda, Echinoidea, corals, calcareous algae, Bryozoa and Anthozoa (see Cooray 1967, Appendix 3, for full list of fossils). The presence of index microfossils such as *Austrotrillina howchini* and *Taberina malabarica* (Tewari & Tandon 1960) place the Jaffna Limestone in the upper part of the Lower Miocene, and specifically in the Burdigalian stage (zones f_1-f_2) (Rao *et al.* 1957).

The results of the gravity survey (Hatherton *et al.* 1975) suggest that an elongat-

24

ed NW-SE trending basin underlies the Jaffna Peninsula. It varies in depth from about 300 m in the east, where the floor is relatively flat, to about 4,000 m near Misalai in the northwest. The sudden steepening of the floor may be due to tilting, down-warping or to faulting.

Exploration drilling for oil on Manaar island and offshore in the southern parts of the Cauvery Basin shows that there is a succession of Mesozoic and Tertiary sediments of over 2,400 m thickness (Cantwell *et al.* 1978). Lying unconformably on Precambrian gneisses are 360 m of Lower Cretaceous shales with intercalations of limestones and sandstones (Fig. 13). These are separated by a regional unconformity from the overlying 750 m of Upper Cretaceous calcareous claystones and siltstones. The overlying Tertiary succession varies from limestones and calcareous siltstones to argillaceous sandstones. The presence of several unconformities, both regional and local, and the absence of parts of the succession in different wells, and the variations in lithology from shallow continental to deep marine deposits in different parts of the basin indicate periodic subsidence and uplift as well as tilting during the basin's history.

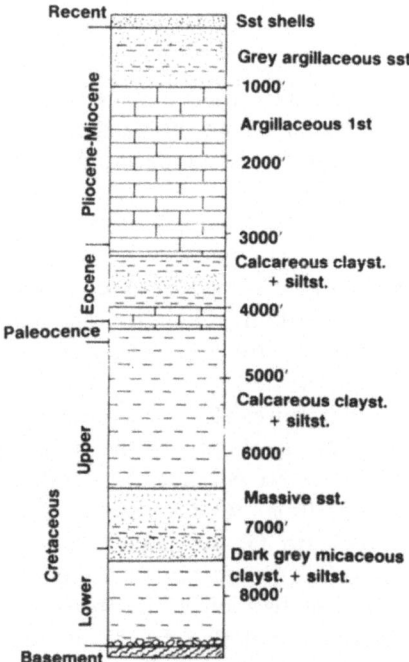

Fig. 13. Section showing succession from Cretaceous to Recent, Pesalai No. 1 exploration well (from Cantwell *et al.* 1978).

Post-Miocene

Extensive stretches of the northwestern, northern and eastern coastal belts of the island, as well as the lower reaches of the major rivers, are occupied by a variety of clays, sands and gravels. These deposits, in the absence of more specific data, can best be classified as being of 'post-Miocene' age as they rest on Miocene limestone in the northwestern coastal belt (Fig. 3). Furthermore, it is only in that area that these deposits have been mapped in any detail and an attempt made to subdivide them into two groups, viz., 'Older' and 'Younger' as shown in Table 1 (Seneviratne *et al.* 1964).

'Older' group

The basal ferruginous gravel appears to be the oldest post-Miocene deposit; it is fairly widespread but generally is concealed by later deposits (Fig. 14). It lies unconformably on Miocene limestone or on Precambrian gneisses, and is thought to be partly marine in origin, deposited on the floor of a shallow, post-Miocene sea.

Overlying the gravel in places is the Red Earth formation, a distinctive, unstratified, brick-red loamy sand, generally whitish or grey on the lower slopes of the elongate ridges which it usually forms. The Red Earth varies from 3 to 30 m in thickness. The quartz grains are well rounded, frosted and pitted, and their similarity to dune sand suggests that the ridges once formed barrier beaches backed by sand dunes (Cooray 1968a). Heavy minerals such as ilmenite, magnetite, zircon, rutile, monazite, garnet and spinel constitute about 9 % of the sands.

Terrace gravels, lying from 4 to 9 m above abandoned river courses in the Battulu Oya area, rest directly on Precambrian gneisses. They consist of well rounded pebbles and cobbles in a matrix of sandy clay (Cooray 1963). Most of these gravels are ferruginized to nodular ironstone in their upper layers. The gravels are thought to be fluviatile in origin and to have been laid down during a more pluvial climatic phase in the island's history. The high-level gravels of the Kelani Ganga at Malwana and Ranale (Coates 1913) probably are of the same age.

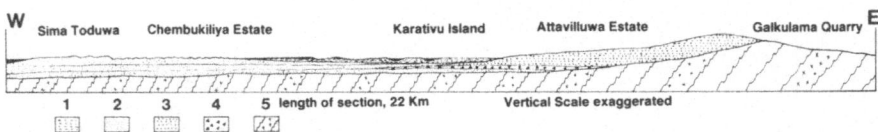

Fig. 14. Geological cross-section across the southern part of the Puttalam area. 1. Lagoonal clays; 2. unconsolidated beach and dune sands; 3. Red Earth formation; 4. basal ferruginized gravel; 5. Vijayan basement of granitic gneisses.

26

'Younger' group

Lagoonal, lacustrine and estuarine beds occupy most of the zone between the present shoreline and the Red Earth ridges in the Negombo-Puttalam area, and the area between the barrier bars and beaches and Precambrian gneissic basement in other parts of the island where the Red Earth is absent, e.g. the east coast north and south of Batticaloa. These beds consist of alternating layers of variously coloured clays and sandy clays with intercalations of sand (Fig. 15). Gastropod and pelecypod shells are common in the lagoonal clays, and oyster beds occur at several localities inland from the coast. Typical estuarine deposits can be seen in the bed of the Beira Lake in Colombo, where non-fossiliferous river sands and lacustrine clays alternate with marine silts and clays (Wadia 1941).

Sand dunes occupy a well defined zone (or zones) behind the berm along most of the coastline of the island, except in the southwest, between Colombo and Galle, and in the northeast, near Trincomalee; a zone of low dunes in places precedes the zone of high dunes inland from the shoreline (Fig. 18). The dune sands are mostly of sub-rounded quartz grains with varying amounts of ilmenite, zircon, rutile and monazite which occur in laminations and streaks. Zones of curved, old beach ridges, generally less than one metre high, occur in some places, e.g. south of Puttalam lagoon, and stand out prominently on aerial photographs. Thick alluvial deposits are found in the flood plains of all the major rivers.

Beachrock, or cemented beach sand, formed by a fluctuating water-table, occurs at several places along the west coast, e.g. Chilaw, Negombo, Pamunugama, Beruwela) as well as on the east coast. Beachrock is composed mainly of quartz grains and shell fragments cemented together by calcium carbonate (Cooray 1968b).

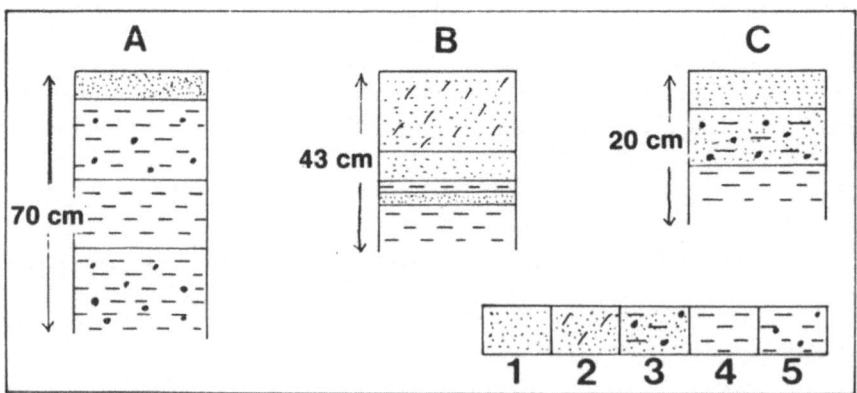

Fig. 15. Typical sections of lagoonal beds, Puttalam area. 1. sand; 2. sand and silt; 3.mottled sandy clay; 4. clay; 5. mottled clay.

Fringing coral reefs are present along the Jaffna Peninsula and at many points along the coast. Some raised reefs, e.g. at Hikkaduwa, also are present.

Secondary formations

Among the most important secondary formations are laterite, confined to the southwestern sector of the island (see Fig. 10), and ferricrete, which is present throughout most of the Dry Zone. The former is derived from the underlying crystalline rocks by weathering, the latter by cementation of the decomposed bedrock by iron oxides carried by a fluctuating water-table. Other secondary deposits such as chert, flint, travertine and kankar are of local and restricted occurrence.

Influence of the sedimentary formations on ecology and biogeography

Soils and vegetation

The soils that have developed on the Miocene and post-Miocene deposits of the northwestern coastal belt, the Jaffna Peninsula, and the eastern coastal strip on either side of Batticaloa are of three main types, viz., Regosols, Solodized Solonetz and Solonchaks, and Red-Yellow Latosols (Fig. 16).

Regosols are soils that are so young that a recognizable soil profile has hardly had time to develop on them. These soils are found on the extensive stretches of unconsolidated coastal sands and sand dunes, and include almost the entire Kalpitiya Peninsula,the island of Manaar, and the southern and eastern parts of the Jaffna Peninsula. Rapid percolation of rain water is a feature of these soils, and they support a characteristic vegetation. Ground creepers like *Ipomoea pes-caprae* and *Spinifex littoreus* occupy the berms and barrier bars, low shrubs like *Vernonia cinerea, Heydotis corymbosa* and *Crotolatia retusa* are found growing on the low dunes, and 'littoral jungle', with such shrubs and low tree species as *Pandanus tectoris, Calophyllum inophyllum* and *Barringtonia asiatica,* occupies the high dunes (Tansley & Fritsch 1905). The principal plant of economic value that has grown on these soils is the sturdy *Borassus flabellifer* (or palmyrah) with its deep-feeding roots (Perera 1978), but the coconut palm has been increasingly cultivated on these sandy soils along the northwestern coast.

Soils developed on the extensive lagoonal and tidal flats, estuaries and marshes of these same coastal areas are known as Solodized Solonetz and Solonchaks, characterized by their high salt content, and very often their high clay content as well. These soils can support only a very specialized xerophytic plant community, with *Sueda monoica* and *Zoysia matrella* as the commonest plant (Abeywickrema 1961); mangrove species such as *Avicennia* and *Rhizophora* also are found on these soils.

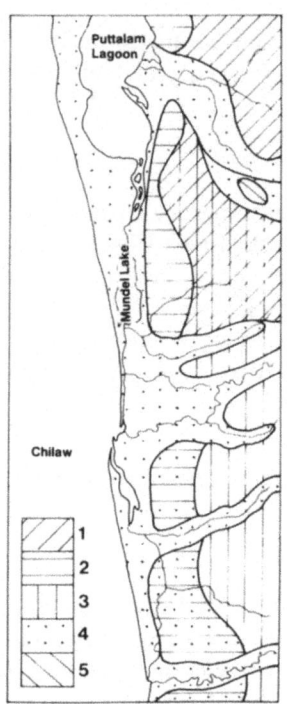

Fig. 16. Soil map of part of the north-western coastal plain south of Puttalam. 1- Reddish Brown Earths; 2. Red Yellow Latosols; 3. Red-Yellow Podsolic soils; 4. Regosols and alluvial soils; 5. lacustrine clays on tank bed.

The Red-Yellow Latosols occupy a well defined belt running from south of Puttalam across the north of the island to the east coast near Mullaitivu; it more or less coincides with the outcrop of the Red Earth formation. The soils are bright red in colour on the crests and middle slopes of the ridges, changing to yellowish on the lower slopes. They are deep, relatively well drained soils, thought to be old or relic soils developed under an earlier climatic regime and since preserved by the absence of erosion and the relative stability of the existing landscape. The parent material of these soils are sediments that have undergone previous cycles of weathering, and they are thus poor in weatherable minerals but rich in quartz. The typical vegetation growing on these soils is the Dry Mixed Evergreen Forest, which forms a dense, thorny jungle of *Acacia, Zizyphus* and other species (Abeywickrema 1961).

Groundwater

The unconsolidated sands of the coastal areas are important carriers of groundwater. The sands have high porosity and permeability, and hence store fresh rainwater generally in the form of lens like bodies resting on salt water, with a zone of brackish water between. Fresh water is thus usually available a few feet below the surface, as is evident from the large number of shallow wells that provide water for domestic use and for drinking to the large fishing populations living on the coasts of the island. However, over-extraction of this fresh water is a very real danger, as this can lead to the intrusion of salt water from below, thus contaminating the well supples and rendering them unfit for human consumption. Control of groundwater extraction in the coastal sands is thus a necessity, especially in the Dry Zone, if this valuable resource is not to be exhausted.

Special conditions govern the occurrence of groundwater in Miocene limestones. Owing to its cavernous nature and to the presence of innumerable joints, fissures, solution channels and chambers, the limestone is a prominent source of groundwater. Most of these openings are constantly being enlarged by solution, the slightly acidic waters which circulate in them as an underground drainage system using the fissures and joints as river courses and the chambers and caverns as fresh-water reservoirs. Where the limestone is at depth and is covered by later clay deposits, artesian conditions exist.

Artesian and unconfined aquifers at depths ranging from 30 to 200 m or more below ground level are encountered along this coastal belt, within which artesian tube wells have recorded yields of 200 to 600 g.p.m., and non-artesian wells yields of about 200 g.p.m. Investigations by the Irrigation Department (Wijesinghe 1973, 1977) and by the Water Resources Board have proved the existence of several deep artesian aquifers in the northwestern coastal region, chief of which, from south to north, are: Madurankuli, Vanthavillu, Manaar-Murunkan, Kondachchi, Mulankavil and Paranthan groundwater basins (Fig. 17). In the Manaar, Pooneryn and Jaffna areas, however, the limestone beds are close to the surface and fresh-water lenses, limited in extent and recharged entirely by direct infiltration of rainfall, are scattered over the area. Study of the hydrological conditions of the groundwater basins continues, and it is clear that they are an important factor that will assist considerablythe development of this part of Sri Lanka.

Settlement

The coastal areas of the island are the natural habitat of the fishing population and, as would be expected, the post-Miocene deposits have a strong influence on settlement and cultivation. The clearing of the 'littoral jungle' on the dunes and the planting of coconut and of small market gardens, especially along the northwestern coast, has resulted in the more settled existence of the fishing population, hitherto largely migrant, in that area. Middle-class colonization also has developed there, particularly within the sand-dune and beach-ridge zones. Owing to its good drainage properties, the Red-Earth ridges are occupied largely by small- to medium-sized coconut plantations. The older (and slightly higher) lagoonal flats often are planted with rice, but the salinity of the present flats prevents cultivation of any kind (Fig. 18). Fishing for crabs, prawns and oysters is the main occupation of the

Fig. 17. Map showing artesian groundwater basins in Miocene Limestone, N.W. Sri Lanka (after Wijesinghe 1980). 1. isohyets; 2. deep groundwater basins; 3. artesian wells; 4. limestone outcrops.

population along the lagoonal and 'lake' shores. The alluvium of the river flood plains is extensively cultivated with rice.

The post-Miocene deposits are the source of much mineral wealth, providing the main bulk of the 'industrial' minerals with which the island is generously endowed. They contain, for example, the heavy mineral concentrates of ilmenite, monazite, rutile and zircon found on many of the beaches; quartz sands for the glass industry; alluvial clays for the brick, tile, ceramic and cement industries; gem-bearing gravels of the river valleys; ferricrete for road metal in the Dry Zone; sand from river beds for the building industry; and laterite as building stone in the southwest.

Intrusive rocks

No large igneous intrusions have so far been recorded in Sri Lanka, but small bodies of dolerite, granite, carbonatite and serpentinite are known to occur.

Dolerite dykes are common on the eastern side of the island, the best known being the Kallodai Dyke, which can be traced for almost 100 km, running in a SW-NE direction (Fig. 8). Several large dykes running parallel to the Kallodai Dyke occur near Vakaneri, and smaller dolerite dyles are found in the Trincomalee, Kantalai, Kalkudah, Buttala and Uda Potana areas. Most of these have been classed as quartz-dolerites (Pattiaratchi 1961).

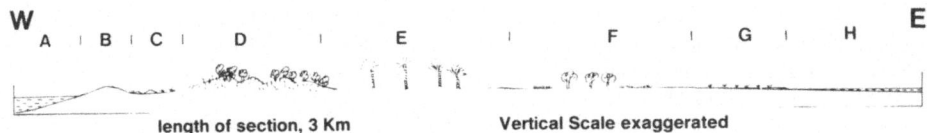

W A | B | C | D | E | F | G | H E

length of section, 3 Km   Vertical Scale exaggerated

Fig. 18. Generalised section across Udappu sand spit, through Ottopani, Battulu Oya area, showing zonation. A: sea; B: beach-ridge and barrier-bar; C: berm, with low dunes, spinifex and marshy patches; D: high dunes with beach jungle; E: blown sand with coconut cultivation; F: old lagoonal flat, dominantly clayey with blown sand, rice and garden crops; G: lagoonal flat with grass; H: lagoon.

The best known granite is the pink, microcline granite at Tonigala, southeast of Puttalam (Fig. 3). It occurs as two sheet-like bodies running in a WSW-ENE direction for 15 to 20 km (Cooray 1971), and is noted for the archaeologically interesting Tonigala inscription, located near where the Puttalam-Kurunegala road crosses the granite. Other granites occur at Gampaha, Arangala (Fig. 6), Balangoda, Beruwela and Loluwa. Numerous small granitic bodies also are present in the eastern Vijayan, among them the hypersthene-granite intrusions on the southeastern coast (Fig. 8).

Pegmatites are fairly common in the Precambrian, some of which are economically important as sources of mica and feldspar. Pegmatites in the southwestern part of the island carry radioactive and rare-earth minerals such as thorionite, thorite, monazite, zircon and allanite. The source rocks of some of the island's gemstones also are believed to be pegmatites, though few have been located.

An intrusion of carbonatite was discovered in 1971 at Eppawala, 25 km from the Kekirawa junction in the Anuradhapura area. It is classed as an 'apatite-magnetite carbonatite' (Jayawardena 1976) containing over 30% apatite in its 'leached' zone. Large apatite crystals, some over 30 cm long, give the carbonatite a striking appearance. Over 40 million tonnes of ore have been proved, and the deposit is now being mined as a fertilizer mineral.

References

Abeywickrema, B.A. 1961. The vegetation of the lowlands of Ceylon in relation to the soil. pp. 87-92 In: Proc. UNESCO Symposium on Tropical Soils and Vegetation, Abdijan.
Adams, F.D. 1929. The geology of Ceylon. Can. J. Res. 1:425-511.
Ayyasami, K. & M.N. Gururaja. 1977. Plant fossils from the east coast Gondwana Beds of Tamil Nadu, with a note on their age. J. Geol. Soc. India 18:398-400.
Berger, A.R. & N.R. Jayasinghe. 1976. Precambrian structure and chronology in the Highland Series of Sri Lanka. Precamb. Res. 3:559-576.
Cantwell, T., T.E. Brown and D.G. Matthews. 1978. Petroleum geology of the northwest offshore area of Sri Lanka. Seapex Conference, Singapore.
Coates, J.S. 1913. The high-level gravels of the lower Kelani valley. Ceylon Mineral Surv. Prog. Rept. Imp. Inst. London (unpubl.).
Coates, J.S. 1935. The geology of Ceylon. Ceylon J. Sci. 19B:101-187.

Cooray, P.G. 1956 The Knuckles Expedition. Ceylon Geog. Sci. Bull. 10:47-59

Cooray, P.G. 1961. The geology of the country around Rangala. Ceylon Dept. Minerol. Mem. 2:1-138

Cooray, P.G. 1962. Charnockites and their associated gneisses in the Precambrian of Ceylon. Quart. J. Geol. Soc. Lond. 118:239-273.

Cooray, P.G. 1963. The Erunwala Gravel and the significance of its ferricrete cap. Ceylon Geographer 17:39-48.

Cooray, P.G. 1964. The geology of Ceylon, some recent advances in knowledge. Proc. 20th Ann. Sess. Ceylon Assoc. Adv. Sci. part 2:87-120.

Cooray, P.G. 1965. The geology of the country around Alutgama. Ceylon Geol. Surv. Mem. 3:1-111.

Cooray, P.G. 1967. An introduction to the geology of Ceylon. Ceylon Nat. Mus. Dept. 324 pp.

Cooray, P.G. 1968a. The geomorphology of part of the north-western coastal plain of Ceylon. Zeits. Geomorph. 7:95-113.

Cooray, P.G. 1968b. A note on the occurrence of beachrock along the western coast of Ceylon. J. Sed. Pet. 38:650-654.

Cooray, P.G. 1969. The significance of mica ages from the crystalline rocks of Ceylon. Geol. Soc. Canada Sp. Pap.5:47-57.

Cooray, P.G. 1971. The Tonigala Granite, N.W. Ceylon. Bull. Geol. Soc. Finland 43:19-37.

Cooray, P.G. 1972a. Hornblende-garnet granulites from Hatton, Ceylon. Geol. Mag. 109:37-44.

Cooray, P.G. 1972b. One-inch geological map series no. 28, Puttalam. Ceylon Geol. Surv. Dept. Colombo.

Cooray, P.G. 1978. Geology of Sri Lanka. pp. 701-710 In: Proc. 3rd Conf. Geol. Min. Res. S.E. Asia, Bangkok.

Crawford, A.R. 1969. India, Ceylon and Pakistan: new age data and comparisons with Australia. Nature 223:380-384.

Crawford, A.R. & R.L. Oliver. 1969. The Precambrian geochronology of Ceylon. Geol. Soc. Aust. Spec. Publ. 2:283-316.

Dept. of Agriculture. 1979. Map of agro-ecological regions of Sri Lanka.

de Rosayro, R.A. 1942. The soils and the ecology of the wet evergreen forests of Ceylon. Trop. Agric. 98:4-14, 13-35.

de Rosayro, R.A. 1950. Ecological conceptions and vegetational types with special reference to Ceylon. Trop. Agric. 151:108-121.

de Rosayro, R.A. 1954. A reconnaissance of the Sinharaja rain forest. Ceylon Forester 1, (N.S.).

de Rosayro, R.A. 1958. The climate and vegetation of the Knuckles region of Ceylon. Ceylon Forester 3:201-260.

Fernando, L.J.D. 1948. The geology and mineral resources of Ceylon. Bull. Imp. Inst. Lond. 46:303-325.

Fernando, L.J.D. 1949. Admin. Report, Ceylon Govt. Minerologist for 1948, Ceylon Govt. Press, Colombo.

Fernando, L.J.D. 1968. Admin. Report, Dir. Geol. Surv. Dept. Ceylon for 1966-1967. Ceylon Govt. Press, Colombo.

Fernando L.J.D. 1969. Admin. Report, Dir. Geol. Surv. Dept. Ceylon for 1967-1968. Ceylon Govt. Press, Colombo.

Fernando, L.J.D. 1970. Admin. Report, Dir. Geol. Surv. Dept. Ceylon for 1968-1969. Ceylon Govt. Press, Colombo.

Hapuarachchi, D.J.A.C. 1968. Cordierite- and wollastonite-bearing rocks of south-western Ceylon. Geol. Mag. 105:317-324.

Hapuarachchi, D.J.A.C. & J.W. Herath. 1969. Brief explanations of the Rakwana Sheet. Ceylon Geol. Surv. Rept. No. RG/009/69 (unpubl.).

Hapuarachchi, D.J.A.C., J.W. Herath & V.V.C. Ranasinghe. 1964. The geological and geophysical investigations of the Sinharaja Forest area (abstr.). Proc. 20th Ann. Sess. Ceylon Assoc. Adv. Sci. part 1:23.

Hatherton, T., D.B. Pattiaratchi & V.V.C. Ranasinghe. 1975. Gravity map of Sri Lanka, 1:1,000,000. Sri Lanka Geol. Surv. Dept. Prof. Pap. 3:39 pp.

Herath, J.W., D.B. Pattiaratchi & L.J.D. Fernando. 1961. Report on the cement raw material investigations in the Puttalam North District. Ceylon Geol. Surv. Dept. Rept. (unpubl.).

Holmes, C.H. 1951. The grass, fern and savannah lands of Ceylon, their ecological significance. Imp. Forestry Inst. Oxford, Pap. no. 28.

Jayawardena, D.E. de S. 1976. The Eppawala carbonatite complex in north-west Sri Lanka. Sri Lanka Geol. Surv. Dept. Econ. Bull. 3:41 pp.

Jayawardena, D.E. de S. & D.A. Carswell. 1976. The geochemistry of 'charnockites' and their constituent ferromagnesian minerals from the Precambrian of south-east Sri Lanka. Minerol. Mag. 40:541-554.

Jayawardena, D.E. de S. & L.K. Seneviratne. 1971. Brief explanation of the geology of the Buttala Sheet. Ceylon Geol. Surv. Dept. Rept. No. RG/011/71 (unpubl.).

Katz, M.B. 1971. The Precambrian metamorphic rocks of Ceylon. Ceylon Geol. Rund. 60:1523-1549.

Katz, M.B. 1972a. Paired metamorphic belts of the Gondwanaland Precambrian and plate tectonics. Nature 239:272-273.

Katz, M.B. 1972b. Facies series of the high-grade metamorphic rocks of the Ceylon Precambrian. 24th Int. Geol. Congr. 2:43-51.

Money, N.J. & P.G. Cooray. 1966. Sedimentation in the Tabbowa rocks of Ceylon. J. Geol. Soc. India 7:134-141.

Munasinghe, T. & C.B. Dissanayake. 1980a. Are the charnockites metamorphosed Archaean volcanic rocks? A case study from Sri Lanka. Precamb. Res. 12:459-470.

Munasinghe, T. & C.B. Dissanayake. 1980b. Pink granites in the Highland Series of Sri Lanka, a case study. J. Geol. Soc. India 21:4460452.

Munasinge, T. & C.B. Dissanayake. 1982. A plate-tectonic model on the geologic evolution of Sri Lanka. J. Geol. Soc. India 23:369-380.

Pattiaratchi, D.B. 1959. In Admin. Report of Ceylon Govt. Minerologist for 1958. Ceylon Govt. Press, Colombo.

Pattiaratchi, D.B. 1961. A preliminary account of the dolerite dykes of Ceylon (abstr.). Proc. 17th Sess. Ceylon Assoc. Adv. Sci. part 1:26.

Perera, N.P. 1978. Early agricultural settlements in Sri Lanka in relation to natural resources. Ceylon Hist. J. 25:58-73.

Rao, S.R.N., B.T. Tewari, K. Mohan & A.K. Chatterji. 1957. The Miocene of western India. Geol. Mag. 94:81-82.

Seneviratne, L.K. & J.W. Herath. 1969. Brief explanation of the geology of the Ambalangoda Sheet. Ceylon Geol. Surv. Dept. Rept. No. RG/101/69 (unpubl.).

Seneviratne, L.K., P.S. Kumarapeli & P.G. Cooray. 1964. The Quaternary deposits of Ceylon (abstr.). Proc. 20th Ann. Sess. Ceylon. Assoc. Adv. Sci. part 1:22.

Sitholey, R.V. 1944. Jurassic plants from the Tabbowa Series in Ceylon. Spol. Zeylan. 24:3-17.

Tansley, A.G. & F.E. Fritsch. 1905. The flora of the Ceylon littoral. New Phytol. 4:1-17, 27-55.

Tarney, J. & B.F. Windley. 1977. Chemistry, thermal gradients and the evolution of the lower continental crust. J. Geol. Soc. Lond. 134:153-172.

Tewari, B.S. & K.K. Tandon. 1960. On the microfauna and age of the Jaffna limestone, Ceylon. Proc. 47th Sess.Indian Sci. Congr. 5:281.

Vitanage, P.W. 1959. The geology of the country around Polonnaruwa. Ceylon Dept. Minerol. Mem. 1:1-75.

Vitanage, P.W. 1970. A study of the geomorphology and morphotectonics of Ceylon. pp. 391-406 In: Proc. Seminar on geochemical prospecting methods and techniques, Peradeniya. U.N., N.Y.

Vitanage, P.W. 1972. Post-Precambrian uplift and regional neotectonic movements in Ceylon. 24th Int. Geol. Congr. Montreal 3:642-654.

Wadia, D.N. 1941. The geology of Colombo and its environs. Spol. Zeylan. 23.

34

Wadia, D.N. 1945. The three superposed peneplains of Ceylon. Ceylon Dept. Minerol. Prof. Pap. 1:25-32.

Wayland, E.J. 1925. The Jurassic rocks of Tabbowa. Ceylon. J. Sci. 13:195-208.

Wickremasinghe, O.C. 1969. Recent studies on the geochemistry of Ceylon charnockites (abstr.). Proc. 24th Ann. Sess. Ceylon. Assoc. Adv. Sci. part 1:68.

2. Land forms and drainage

D.K. Erb

Introduction

Historically the description and discussion of the landforms of Sri Lanka was restricted almost entirely to minor segments of geological reports and papers, or to specific comments on unusual geomorphologic features with local significance (Erb 1963).

Subsequent, more systematic studies of the geology and landforms of the Island discussed: general geology and plains of denudation (Adams 1929, Coates 1935); superposed peneplains (Wadia 1945); the geology and mineral deposits (Fernando 1948); geological foundation of Ceylon's scenery (Cooray 1956); geological structure of Ceylon (Oliver 1957); the Pleistocene of Ceylon (Deraniyagala 1958); geomorphology and geomorphic history (Erb 1963); the geology of Ceylon (Cooray 1967); and landforms and drainage (Erb 1966-1970).

The island is characterized by a wide variety of landforms, from an essentially flat erosion surface or peneplain, to a very complex assemblage of mountains, ridges, plateaux and valleys, all of which have had a significant effect on the drainage systems, soils, and related components of the environment which have developed over time.

The following discussion will present an overview of the landforms and associated drainage systems of Sri Lanka, with the specific objective of providing a foundation for subsequent physical, ecological and biological presentations. Due to space limitations no attempt will be made to include a discussion of past or current water resource or environmental management programmes involving the above landforms and drainage systems.

Land forms and drainage

General

Fig. 1 shows the seven major landform units of Sri Lanka: Coastal Plain, Continental Shelf, Circum-Island Peneplain, Central Massif, Sabaragamuwa Hills, Gal Oya Hills and Elahera Ridges.

The Coastal Plain has formed along the seaward margin of the Circum-Island Peneplain. The Continental Shelf is an extension of this peneplain beneath the sea, approximately to the 100 fathom line. It has a width of from 8 to 32 km except in the north and northwest, where it forms part of India's continental shelf.

The Circum-Island Peneplain consists of a broad, flat to gently undulating erosion surface rising from sea level at the coast to a maximum of 305 m ASL at the base of the Central Massif. The Central Massif dominates the south-central part of the Island. It consists of a complex assemblage of mountains, mountain ranges, ridges; high, medium and low level plateaux and associated valleys and alluvial plains, and contains the highest point of the Island, Pidurutalagala at 2,524 m ASL.

Folded around the southwest 'corner' of the Central Massif are the structurally controlled Sabaragamuwa Hills, with peaks rising to 1,385 m ASL, and scattered plateaux surfaces from 305 to 1,219 m ASL.

Paralleling the eastern edge of the Central Massif are the somewhat similar, structurally controlled ridges and erosion remnants of the Gal Oya Hills. The Elahera Ridges are formed by a series of resistant strike ridges extending north-northeastwards from the Central Massif structure.

Fig. 1. General landform units of Sri Lanka.

Circum-Island Peneplain

The Circum-Island Peneplain is here considered to extend from the edge of the Continental Shelf at 100 fathoms through the Coastal Plain, sea level to 30 m ASL, to the base of the Central Massif at approximately 305 m ASL. The boundary elevation of the base of the Sabaragamuwa Hills in the south-west, the Gal Oya Hills in the east, and the Elahera Ridges in the north, is at approximately 152 m ASL.

The classification of peneplain is based on the concept that this surface is essentially an erosional surface cut into the well weathered, partially weathered, and more or less unweathered bedrock of the island. The surface truncates varying rock types and structures, is underlain by variable thicknesses of residual and alluvial soil, and is marked by a large number of bedrock erosion remnants varying from low, rounded 'whale backs' (Fig. 2), through isolated ridge fragments, to massive inselberg or bornhardt type features (Fig. 3). The rivers and streams which cross this plain often have their headwaters in the higher level features, Central Massif, Sabaragamuwa Hills and Gal Oya Hills, mentioned above (Fig. 4).

As an example, one of the main branches of the Mahaweli Ganga, which drainsa large proportion of the area within the Central Massif boundaries, is selected to illustrate the variation in terrain and related stream profile characterstics of the Central Massif. The profile contains the following segments:

Segment 1 (1.3 km):
   Dambagastalawa Oya - headwaters, high gradient, north and west flowing, beginning on the north slope of Totupola at 2,360 m ASL;
Segment 2 (10.3 km):
   Dambagastalawa Oya - mid-portion, low gradient, flowing across the Pattipola Plateau erosion surface at 1,890-1,920 m ASL;
Segment 3 (18.7 km):
   Dambagastalawa Oya - lower reaches, high gradient, descending from Pattipola Plateau margin to its junction with the Agra Oya at 1,317 m ASL, via Elgin Falls at 1,768 m ASL;
Segment 4 (27.9 km):
   Agra Oya - medium to low gradient to junction with Kotmale Oya at 1,201 m ASL, flow is west and northwest;
Segment 5 (32.3 km):
   Kotmale Oya - high gradient to 975 m ASL, via St Clair Minor Falls at 1,1158 m ASL, and St Clair Major Falls at 1,079 m ASL, flow is northwest;
Segment 6 (46.9 km):
   Kotmale Oya - medium gradient to 671 m ASL, flow is northwest;
Segment 7 (60.9 km):
   Kotmale Oya - low gradient to junction with main Mahaweli Ganga channel at 561 m ASL;

38

Fig. 2. Circum-Island Peneplain, associated erosion remnants and tors, from Magulmaha Vihare Pokuna (SE).

Fig. 3. Sigiriya, a bornhardt or inselberg type erosion remnant with the Elahera Ridges unit. A tank or water storage reservoir is visible in the foreground.

Segment 8 (150 km):

> Mahaweli Ganga - essentially low gradient with occasional short, medium gradient segments from this point to the margin of the Central Massif at Minipe just above the Badula Oya at 91 m ASL, flow is north, southeast and east;

Segment 9:

> Mahaweli Ganga - very low gradient from this point to the sea at Koddiyar Bay (Trincomalee), 200 km downstream, flow is north and northeastward, gradient is approximately 0.9 m km^{-1} (Fig. 5).

In general, the rivers draining the marginal slopes of the Central Massif, Sabaragamuwa Hills, and Gal Oya Hills, have less complex profiles. They have medium to high gradients in their headwaters and low gradients as they traverse the Circum-Island Peneplain to the sea.

Weathering and erosion, localized by faults, joints, and the strike of metamorphic bedrock ridges, may control the drainage where the rock is at or near the surface, but elsewhere thick residual or alluvial deposits, which may exceed 30 m in depth, result in a general dendritic pattern. The peneplain surface is one of low relief which is broken occasionally by erosion remnants, inselbergs or bornhardts, which rarely exceed 762 m in elevation (Fig. 6).

With the exception of the southwest segment, the Circum-Island Peneplain is covered by a relatively dense growth of dry jungle (Fig. 7 and Fig. 2). This growth varies from scrub in the drier areas along the coast, to relatively well developed forest on the lower slopes of the Central Massif and Gal Oya Hills. This type of forest is classified as Monsoon Forest (Fernando & Fernando 1968).

In the southwest, much higher rainfall results in a more luxuriant and well developed vegetative cover, the Lowland Rain Forest of Fernando & Fernando (*op. cit.*)

Coastal plain and continental shelf

The Coastal Plain and Continental Shelf are essentially marginal elements of the Circum-Island Peneplain. The Continental Shelf varies in width from 8 to 32 km. Its outer margin occurs approximately at the 100 fathom line. Beyond this depth the bottom drops off rapidly to over 1,000 fathoms. The shelf is narrowest at Trincomalee, Batticaloa and Panadura, where submarine canyons have been incised into its surface, and is continuous with the Indian continental shelf in the Palk Strait area, where Adam's Bridge and Pedro Banks are located (Fig. 1). Along the southwest coast from Negombo to Dondra, strike ridges of the local rock extend onto the shelf at a low angle. Similar bedrock ridges in the Hambantota-Yala area come to the surface as the Great and Little Basses Rocks.

The Coastal Plain has been delineated as the zone between the sea and the 30 m contour line. This classification is designed to include all the terrain characteristic

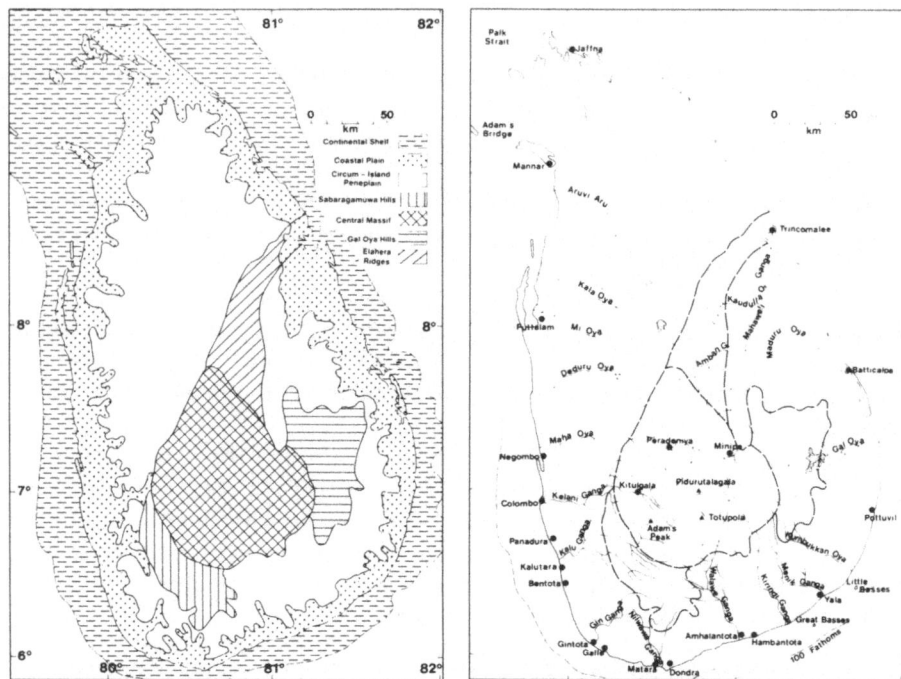

Fig. 4. Main river systems of Sri Lanka and their relationship to the general landform units.

of a coastal area without including unnecessary portions of the Circum-Island Peneplain inland.

Although raised beaches, wave-cut terraces, and elevated coral reefs are present in the coastal zone, the general geomorphologic characteristics indicate that the area is one of submergence. The presence of broad, flat alluvial plains, virtually at sea level, behind partly silted-up lagoons shut off from the sea by low sand bars and spits, is indicative of this sequence of events (Fig. 8). Promontories undergoing truncation by wave action also are characteristic of a retrograding shoreline.

The Coastal Plain's seaward margin is marked by very few irregularities. Long-shore drifting of sand, as a result of wave induced littoral currents, has tended to fill in depressions or indentations. At the same time, the wave action has tended to plane off projecting features, thus reinforcing the overall process (Dassanaike 1928). Coastal features related to these processes are sand spits and bars which in some cases, virtually close off the mouths of even major streams from the sea. As might be expected, bedrock is at or near the surface along much of the coastal plain. On the west coast, a number of lagoon-enclosing spits, such as those at Negombo, Puttalam and Manaar, may have bedrock cores. Further north, in the Jaffna area, limestone bedrock forms sea cliffs up to 15 m in height (Cooray 1967). On the southeast coast, rock ridges project into the sea at varying angles and result

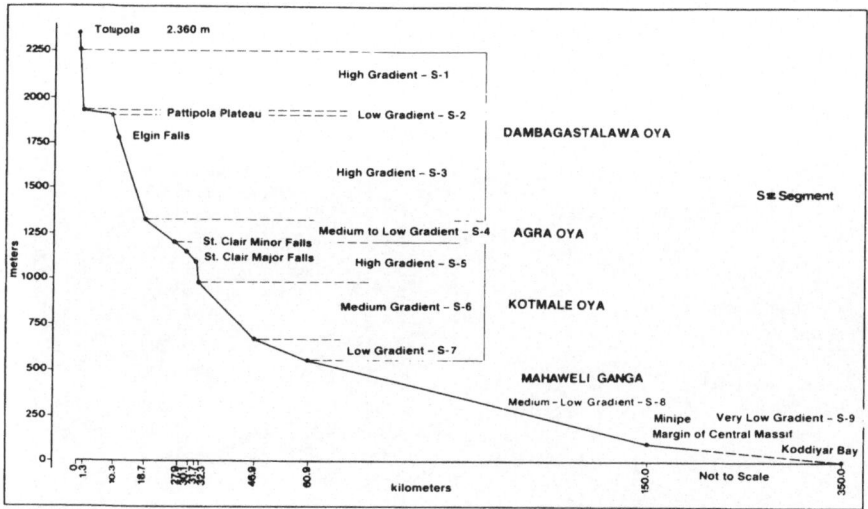

Fig. 5. Diagrammic profile of one element of the Mahaweli Ganga system from its initiation on the north slope of Totupola, to the sea at Koddiyar Bay, Trincomalee.

in a more scalloped shoreline. To the north, from Yala to Batticaloa, sandbars and spits are common, especially across the mouths of all but the major rivers. Some of these bars consist of a series of parallel beach ridges up to 5.6 m in width, apparently in a prograding condition (Fig. 9). Sand dunes are relatively common landforms along the coast, where strong monsoon winds combined with extensive deposits of beach sand provide ideal conditions for their formation (Fig. 10). However, very little inland migration of dunes occurs either in the southeast or northwest coastal areas, due to the growth of drought resistant, stabiliziing vegetation on their lee slopes.

In the southwest, the Coastal Plain consists of a network of drowned, strike, fault or joint controlled river valleys, distributed among bedrock ridges (Fig. 11). The presence of 30 m or more of alluvium in valleys 64 to 80 km inland places the level of valley bottom erosion well below present sea level. Coral reef ridges and beach sand deposits at an elevation of from 12 to 18 m ASL, well inland, when combined with the valley bottoms mentioned above, indicate relative changes in sea level of approximately 50 m (Cooray 1967). These changes in level may be attributable in part to tectonic and in part to eustatic activity.

Central Massif

The Central Massif is composed of four main ranges of mountains: Adam's Peak Range on the west, Namunamuli Range on the east, Pidurutalagala Range in the

42

Fig. 6. Circum-Island Peneplain with inselbergs in the distance and a tank in the foreground. View to the south from a small inselberg at Maha Iluppallama, 27.4 km south of Anuradhapura.

Fig. 7. Dry jungle, tank, and patches of jungle cleared for Chena type agriculture; southeast Circum-Island Peneplain, from a helicopter.

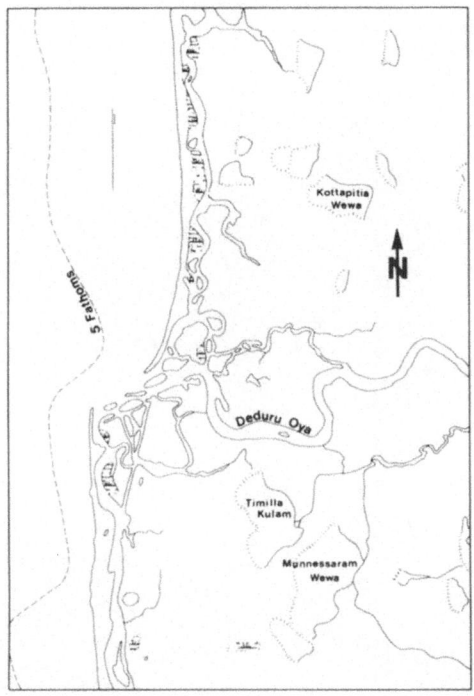

Fig. 8. Drowned mouth of the Deduru Oya with associated spits and bars; northwest Coastal Plain.

centre, and Knuckles Range on the north (Fig. 12). World's End Ridge and Haputale Ridge extend from Adam's Peak Range on the west, almost to Namuna-kuli Range in the east, and are joined to Pidurutalagala Range at its most southern point to form World's End Plateau. These structurally controlled mountain ranges and ridges are in part surrounded by, and in part surround or include, an assemblage of high level (1,200 to 2,200 m), medium level (500 to 1,060 m) and low level (270 to 480 m) plateaux.

All of the above have been cut by a complex pattern of faults, joints, and unclassified fractures; have been subjected to a high temperature, high precipitation climatic regime, with its associated intense weathering and erosion activity; have been dissected by several major and many intermediate and minor rivers and streams; and have, as a result, developed a terrain often characterized by a great range in local as well as regional relief.

Adam's Peak Range includes folded mountain ridges in the northwest, block-faulted mountains in the Adam's Peak area, and a broad, synclinal plateau in the south. The Range is bordered on the east by a vertical fault scarp varying in height from 450 to 760 m, on the south by a similar major fault scarp, and on the west by steep slopes being dissected by numerous actively eroding streams (Fig. 13).

44

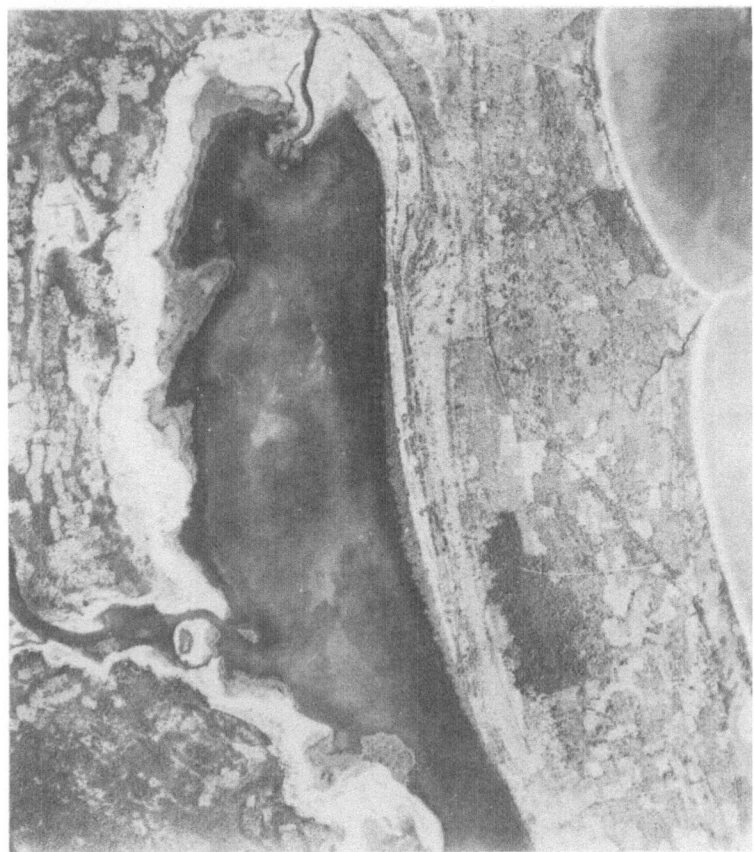

Fig. 9. Beach ridges and spits controlling the mouth of the Bodigoda Aru, and scalloped pattern shoreline due to rock cored points; northeast Coastal Plain.

Namunakuli Range on the east is marked by very rough, angular topography based on the weathering and erosion of steeply folded beds with a north-northwest axial strike, Knife-edge ridges, smooth dip slopes, deeply eroded scarp faces and sharp peaks are characteristic. Both west and east sides of the Range are cut by deep, fault controlled, V-shaped valleys and vertical-sided gorges. On the west, active weathering has resulted in a thick mantle of residual soil which tends to obscure the bedrock to some extent. On the east, more intense orographic rainfall has resulted in more intense erosion with a consequent reduction in residual soil.

Pidurutalagala Range consists of the weathered and eroded ends of a series of northwestward trending anticlines and synclines. The main peaks of the range are the eroded remnants of numerous strike ridges which make up these structures.

In general, marginal erosion in this complex structural region is marked by high

Fig. 10. Coastal sand dunes, wind oriented vegetation, and low relief peneplain surface inland; south coast, between Hambantota and Kirinda (air photo stereogram).

gradient streams flowing in steep sided V-shaped valleys, or vertical-sided gorges, which are controlled by the strike of the bedrock lineation, or by faults and joints cutting the bedrock (Fig. 14). In areas where structure, combined with weathering and mass wasting, have produced thick residual and colluvial depostis, lower gradient, open V-valleys with rounded interfluves have developed.

World' End Ridge joins the southeastern end of Adam's Peak Range to the western margin of World's End Plateau. It is formed by the uplifted ends of a series of northwest trending anticlines and synclines, which also form the foundation of the Hatton segment of the Hatton-Diyatalawa Plateau. Faulting has truncated these anticlines and synclines to form the southern face or margin of the Ridge (Fig. 15). This face is much steeper than the northern face and is marked by actively eroding, high gradient streams which hinder the development of soils in some valleys. Active mass-wasting, with associated deep talus and colluvial deposits, is

46

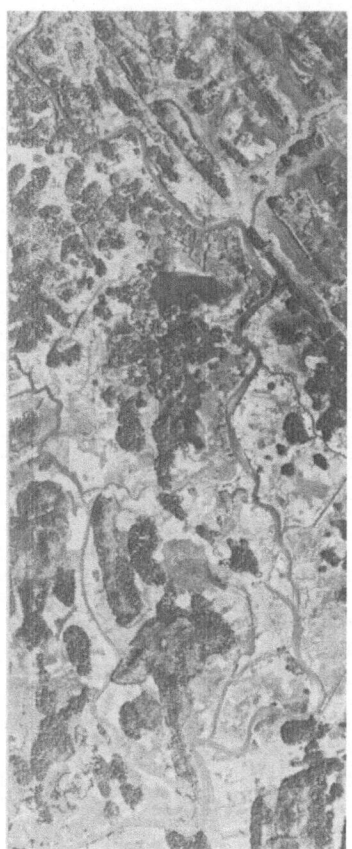

Fig. 11. Drowned, strike controlled valleys; Gin Ganga, 3.2 to 9.7 km upstream from its mouth at Gintota; southwest Coastal Plain (air photo stereogram).

present along the entire face of the scarp. These deposits have buried the lower slopes to the extent that only prominent ridges project. For this reason, the boundary fault scarp is visible only at the western and eastern ends of the Ridge.

Haputale Ridge is a continuation of World's End Ridge to the east of World's End Plateau and terminates in the group of peaks and ridges surrounding Bandera Eliya Plateau. As a result of deep weathering and regional structure, the northern face of this strike ridge is deeply mantled with residual soils. The southern, fault controlled face (910 to 1,220 m) however, has been subjected to more active erosion and mass-wasting which has resulted in a combination of bare rock faces, patches of residual soil, and thick talus deposits, especially on the lower slopes.

The Belihul Oya and Kiriketi Oya, with their associated tributaries, flow in excellent examples of deep, V-shaped fault controlled valleys which are indicative

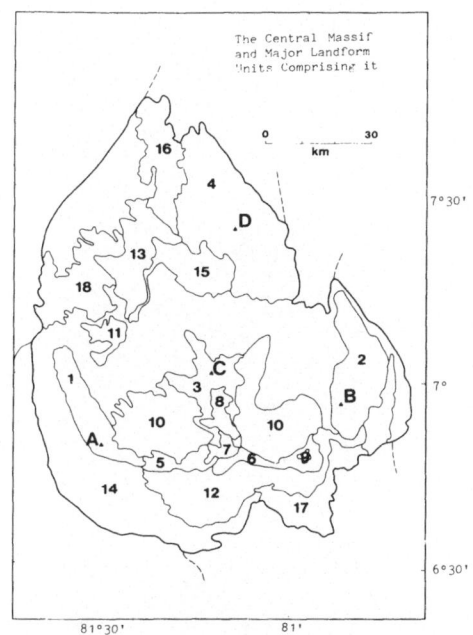

Fig. 12. The Central Massif and the major landform units comprising it. *Legend:* 1. Adam's Peak Range, A: Adam's Peak – 2,243 m; 2. Namunajuli Range, B: Namunakuli Peak – 2,033 m; 3. Pidurutalagala Range, C: Pidurutalagala Peak – 2,524 m; 4. Knuckles Range, D: Knuckles Peak – 1,863 m; 5. World's End Ridge; 6. Haputale; 7. World's End Plateau – 2,195 m; 8. Pattipola Plateau – 1,92 m; 9. Bandera Eliya Plateau – 1,829 m; 10. Hatton-Diyatalawa Plateau – 1,372-1,463 m; 11. Ulapana Shelf – 793-853 m; 12. Balangoda Shelf – 610-762 m; 13. Peradeniya Basin – 549 m; 14. Molamure Shelf – 457 m; 15. Kondesalle Basin – 457 m; 16. Nalanda Shelf – 427-457 m; 17. Wekada Shelf – 366-396 m; 18. Mawanella Shelf – 274-305 m.

of the active erosion characteristic of the south face of the World's End Ridge-World's End Plateau-Haputale Ridge-Bandera Eliya Plateau complex (Fig. 16). Heavier rainfall along this face is due in part to the effect of the southwest monsoon, and in part to orographic factors.

Knuckles Range consists of a complex of block faulted strike ridges, deeply eroded on all sides. Part of the range is marked by a rough plateau with steep marginal scarps. Extensive faulting and jointing, together with variations in bedrock lithology, have localized weathering, erosion, and mass-wasting, to produce thick residual soils, exposed bedrock faces, and deep talus deposits (Fig. 17). Thick forest covers many areas in the Knuckles Range as a result of the heavy rainfall brought by the northeast monsoon.

In addition to the main mountain ranges discussed above, the Central Massif contains numerous high level, medium level, and low level plateaux, which are considered to represent the remnants of ancient surfaces of erosion (Fig. 12). In

48

Fig. 13. Eastern fault scarp, synclinal plateau, and southern fault scarp; Adam's Peak Range (air photo stereogram).

general, each surface is characterized by a rough accordance of summit levels, deep weathering, rounded hills separated by broad rounded valleys, and low gradient streams. In addition, these erosion surfaces may have a variety of characteristics depending on their geomorphologic history. They may be bounded on one or more sides by higher level mountains; they may have higher level mountains projecting from their surfaces; they may be bounded on one or more sides by scarps or steep slopes dropping to a lower elevation surface; or they may have a combination of these characteristics (Fig. 18).

Streams draining such areas may thus have their headwaters at a higher elevation, follow a medium to high gradient course until they reach the plateau surface, change to a low gradient course during their traverse of the plateau, change to a very high gradient course as they descend the scarp face to a lower level, and there regain low gradient status as they traverse the Circum-Island Peneplain to the sea

Fig. 14. Strike, fault and joint controlled erosion; Pidurutalagala Range (air photo stereogram).

(Fig. 5). The associated stream valleys may thus have a steep sided, V-shape in the headwater zone, a broad rounded form on the plateau surface, a steep-sided V-shape or gorge form in the scarp zone, and a broad, flat-bottomed form at the lower level (Fig. 18). Rapids and falls often are associated with the steep gradient segments of these streams.

The erosion surfaces or plateaux within the Central Massif vary in elevation from the highest, World's End Plateau at 2,200 m, through Pattipola Plateau at 1,920 m, Hatton-Diyatalawa Plateau at 1,370 to 1,460 m, Tangamale Plateau at 1,070 m, Balngoda Shelf at 610 to 760 m to the lowest, Mawanella Shelf at 270 to 300 m. The local relief of these surfaces generally is low, the weathering is active, and the residual soils relatively thick. In some cases, such as the Mawanella Shelf (Fig. 19), the valley bottoms are alluvium filled and their pattern controlled by faults, joints and strike ridges associated with the bedrock structure. Resistant ridges and inselberg-type domes occasionally project above the general level.

Fig. 15. Southern scarp face and northern, deeply weathered slope of World's End Ridge (air photo stereogram).

Sabaragamuwa Hills

The Sabaragamuwa Hills are formed by a series of sub-parallel strike ridges which have been etched out of a more massive metamorphic bedrock structure by the differential erosion of lithologically less resistant beds (Fig. 20). Heavy rainfall (greater than 500 cm yr^{-1}), brought by the Southwest Monsoon, has resulted in deep weathering and active erosion, virtually to base level in west-flowing rivers, whereas less rainfall to the east has resulted in less weathering, less erosion, and the east-flowing rivers being marked by falls and rapids.

This area contains a number of remnants of ancient erosion surfaces, including the Handapan Ella Plateau at 1,250 m ASL (Fig. 21) and the Tangamale Plateau at 1,060 m ASL.

Drainage of this area is mainly to the west and northwest along strike, fault and joint controlled valleys. The Madalagama Ganga, Delwala Ganga, Karawita Ganga, Kiriella Ganga, Hangomuwa Ganga, Koswatta Ganga, Delgoda Ganga, Wewa Ganga, Kukulu Ganga and Kuda Ganga all drain northwestward until they

Fig. 16.   Deep, V-shaped, fault controlled valleys of the Belihul Oya, characteristic of the south face of the World's End Ridge – World's End Plateau – Haputale Ridge – Bandera Eliya Plateau complex (air photo stereogram).

reach the Kalu Ganga, which then flows westward to the sea at Kalutara. The Gin Ganga and its tributaries drains a small portion of the area southewestward to the sea at Gintota, and the Nilwala Ganga a similar area southward to the sea at Matara. A somewhat larger area of the Sabaragamuwa Hills is drained eastward to the Walawe Ganga and thence southeastward to the sea at Amablantota (Fig. 4).

It is interesting to note, as mentioned previously, that the valley of the Kalu Ganga (and probably other rivers in the region) has been eroded well below present sea level, an indicator of eustatic changes in sea level, or isostatic changes in the land surface during the geologic history of the island.

Weathering in the Sabaragamuwa Hills area is intense, and together with extensive deposits of talus and colluvial material resulting from active mass-wasting, has produced deep residual soils. Almost the entire lower slopes of the major ridges, and often all of some of the minor ridges, are buried beneath these deposits. In

Fig. 17. Complex, structurally controlled weathering and erosion; Knuckles Range (air photo stereogram).

addition, except where erosion is active, the valley bottoms are filled with alluvium, graded to local base levels of resistant bedrock, or, in the case of the major valleys, to present sea level.

Gal Oya Hills

The Gal Oya Hills comprise a group of well defined erosion remnants rising from the Circum-Island Peneplain to the east of the Central Massif (Fig. 1). As with the Sabaragamuwa Hills to the southwest of the Central Massif, these ridges and residual hills are cut by numerous faults and joints. Some of the major northeast trending faults are traceable through the southeast section of the Central Massif to the Balangoda Shelf and southern portion of the Circum-Island Peneplain beyond. In the northeast, the hills have a more massive structure, and jointing rather than faulting seems dominant (Fig. 22).

Weathering in the Gal Oya Hills appears to be extensive, resulting in deep

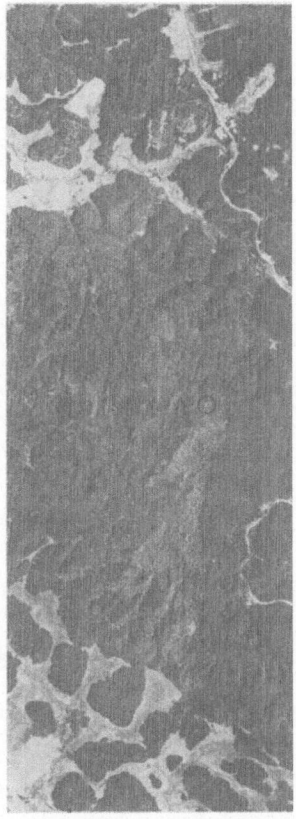

Fig. 18. Low, rounded interfluves, broad valleys, low gradient meandering streams, projecting mountain peaks, and marginal scarps or slopes characteristics of the deeply weathered erosion surface of World's End Plateau. Pattipola Plateau is the lower level erosion surface to the north (air photo stereogram).

residual soils on the peneplain surface and in the fault and joint controlled depressions in the hills where moisture is concentrated. Drainage in this area is controlled partly by linear metamorphic bedrock structure and partly by joints and faults. The former is most noticeable in the area where strike ridges are well developed, and the latter in the areas of more massive rocks.

In this region, vegetation reflects the generally arid climatic regime and takes the form of scrub jungle on the peneplain. Heavier forests occur on the slopes of the hills where the northeast monsoon rains are concentrated. In places, scattered patches of savanna type forest occur.

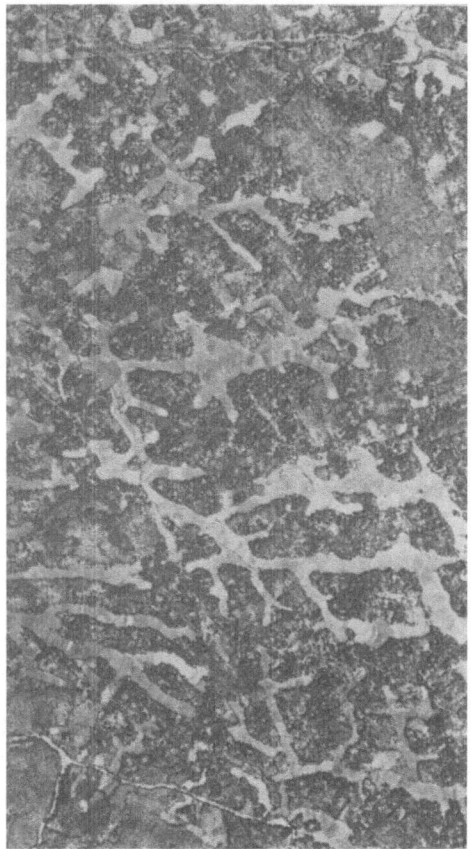

Fig. 19. Structure controlled, alluvium filled valley bottoms, and erosion remnants; Mawanelle Shelf (air photo stereogram).

Elahera Ridges

The Elahera Ridges comprise a series of well marked strike ridges extending from the northern margin of the Knuckles Massif, in a broad curve to the sea at Trincomalee (Fig. 1). These ridges and ridge fragments are the erosion remnants of steeply dipping, resistant metamorphic beds which are the northward extensions of the synclinal and anticlinal structures of the Central Massif (Fig. 23).

Weathering of these beds has been intense and all but the most resistant have been reduced to a low, flat to gently undulating peneplain continuous with the Circum-Island Peneplain to the west and east. Except for the crests, the ridges are mantled by residual and colluvial deposits of varying thickness. Vegetation is thick and little erosion is visible. Local variations in structure have resulted in some areas

Fig. 20. Differential weathering and erosion controlled by strike ridges, faults and joints characteristic of the Sabaragamuwa Hills (air photo stereogram).

of more massive erosion remnants which, as was the case with the Gal Oya Hills, can be classified as inselbergs or bornhardts (Fig. 24).

At their northeast end, the Elahera Ridges project into the sea and form the dominant relief of Trincomalee Harbour (Fig. 25). The development of Koddiyar Bay and its associated submarine canyon may be related to similar resistant ridges now almost entirely buried by alluvium brought down by the Mahaweli Ganga.

General discussion

The landforms which make up the complex terrain of Sri Lanka owe their characteristics to a variety of geomorphologic processes (Erb 1963), acting upon a diverse

56

Fig. 21. Handapan Ella Plateau, a structurally controlled erosion surface remnant within the Sabara-gamuwa Hills complex (air photo stereogram).

geological foundation (Cooray 1967), under a relatively pure equatorial climatic regime (Tricart 1972). This foundation, because of variations in the resistance of the rock types comprising it to these geomorphological processes, combined with its structural characteristics - folding, faulting and jointing - and often major relative displacement due to tectonic forces, controls the form of the terrain, the course of the drainageways, the nature of surficial soils, the depth of these soils, and ultimately, the potential of the area to support vegetation, and man[1].

The general drainage pattern of the island (Fig. 4), and a portion of the drainage of the Mahaweli Ganga in the Central Massif (Fig. 26), illustrate the control exercised by the landforms and geologic foundation mentioned above. The significance of these landforms in terms of precipitation and runoff may be illustrated using the Central Massif as an example.

Fig. 22. Massive fault, joint and strike controlled erosion remnants; northern margin of the Gal Oya Hills (air photo stereogram).

As indicated by the drainage map, a major portion of the interior of the Central Massif is drained to the east and north by the Mahaweli Ganga and its tributaries above Yatiyantota. The northwest margin of the Central Massif is drained by the Deduru Oya and Maha Oya, the west by tributaries of the Kelani Ganga below Yatiyantota, the southwest by the Kalu Ganga, the south by the Walawe Ganga and Kirindi Oya, the southeast and east by the Menik Ganga, Kumbukkan Oya and tributaries, and the northeast by the Mahaweli and its tributaries below Minipe.

All these rivers have flow characteristics which are dependent on their geographic position, the physical characteristics of the terrain, and the regional climatic parameters in effect (Table 1). The rivers of Sri Lanka fall into four categories: those with headwaters in the mountains, plateaux and plains of the inner region of

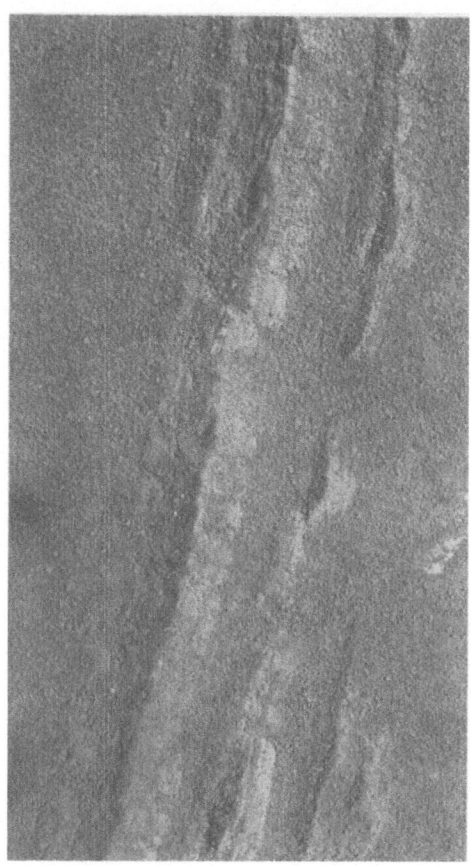

Fig. 23. Sudakanda ridge, a resistant erosion remnant which forms part of the eastern margin of the Elahera Ridges unit (air photo stereogram).

the Central Massif and then flow outward across the Circum-Island Peneplain and Coastal Plain to the sea; Geology, climate soils and their agricultural significance, and vegetation are treated elsewhere in this volume. Those with headwaters in the marginal hills, plateaux and mountains of the Central Massif and then flow outward across the Circum-Island Peneplain and Coastal Plain to the sea; those with headwaters in the mountains, hills and plateaux of the Sabaragamuwa Hills, Gal Oya Hills, or Elahera Ridges, and then traverse the Circum-Island Peneplain and coastal Plain to the sea; and with headwaters on the Circum-Island Peneplain and traverse it and the Coastal Plain to the sea.

Rivers of the first category have the most complex characteristics since, as is the case with Mahaweli Ganga (Fig. 5), they may have high gradient headwaters on the slopes of the highest central peaks; may traverse low gradient, relict erosion surfaces at various elevations; may have their profiles interrupted by numerous

Fig. 24. Sigiriya, a bornhardt or inselberg type erosion remnant located in the west-central part of the Elahera Ridges unit; viewed from the west.

falls or rapids; may flow in low gradient, alluvium filled, bedrock controlled channels at lower levels; and then eventually reach grade as they flow out on to the Circum-Island Peneplain and thence across the Coastal Plain to the sea.

Rivers of the second category, such as the Walawe Ganga, tend to have high gradient, actively eroding headwaters (Fig. 27), greater variability in flow resulting from orographic precipitation in the headwaters, and low gradient on the Circum-Island Peneplain and Coastal Plain.

Rivers of the third category have similar characteristics to those of the second category and, as is the case with the Kalu Ganga, may combine tributaries draining the margin of the Central Massif, with the main system draining the Gal Oya Hills. The complexity of such streams varies considerably.

Rivers of the fourth category e.g. the Mi Oya, originate on the Circum-Island Peneplain and flow in low gradient channels across it and the Coastal Plain.

Many other rivers, although they may have their headwaters draining the lower slopes of the major landform units discussed in this chapter, or the lower slopes of the innumerable isolated or semi-isolated erosion remnants or inselbergs which rise from the Circum-Island Peneplain, in effect have the majority of their drainage basin on the plain and should therefore be placed in category four.

The physical characteristics of Sri Lanka's rivers, when combined with the associated landforms and regional climatic patterns, determines to a considerable

60

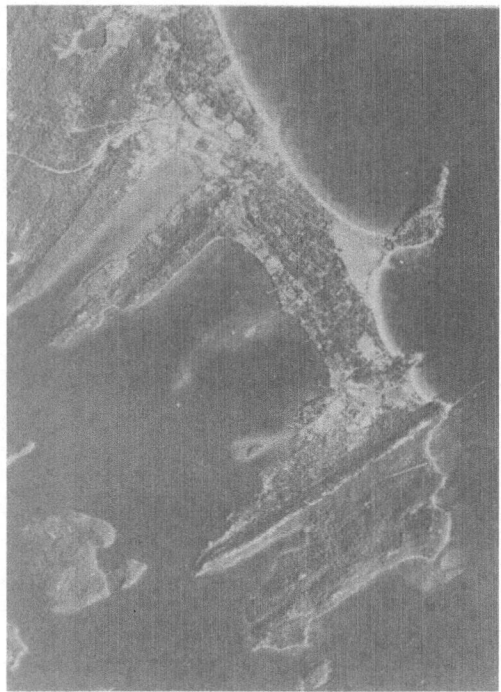

Fig. 25. Elahera Ridges controlling the form of Tincomalee harbour; part of Koddiyar Bay.

extent their significance to man. The development of hydro-electric power and the provision of water for irrigation are two of the most important river-related problems being dealt with by the Government of Sri Lanka[1]. The former problem involves, almost entirely, the high gradient headwater segments of these drainage systems; the latter, the lower gradient, lower reaches of the systems, especially those that occupy the dry zones of the Circum-Island Peneplain and Coastal Plain.

With the exception of the fouth category mentioned above, Sri Lanka's rivers may thus have a dual potential in terms of development. Establishing, evaluating and utilizing this potential has been the subject of an on-going series of programmes over the last 124 years[2]. As well, many similar programmes to utilize these resources are planned for the future.

[1] Numerous other river/drainage-related problems, such as the maintenance of flow, fresh water fisheries, pollution, etc., exist, but are not treated in this brief discussion.

[2] After the collapse of the ancient Kingdoms of Sri Lanka in the 13th Century, little was done with respect to the great system of irrigation works (tanks and canals) which were distributed along the rivers of the dry zone, until the British began a programme of restoration in 1857. In 1900, the Department of Irrigation was created and in 1932 it became part of the Ministry of Agriculture and Lands. From 1900 to 1931, work was concentrated on repairing, maintaining and operating the ancient irrigation works. From 1947 to date, major and minor works have expanded, surveys have been carried out, projects have been planned, and new works have been developed.

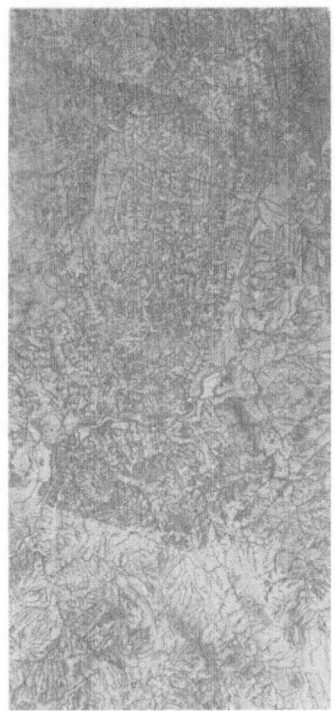 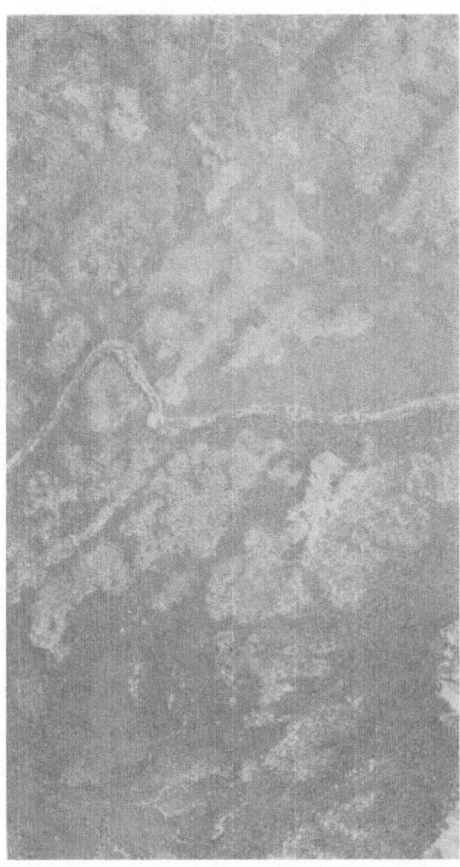

Fig. 26. Structurally controlled drainage pattern of a portion of the Mahaweli Ganga; Peradeniya Basin area of the Central Massif.

Fig. 27. Incised channel and falls of the Walawe Ganga at the margin of the Balangoda Shelf erosion (air photo stereogram).

Table 1. Basin area and flow characteristics of the main rivers of Sri Lanka based on data prepa
the Hydrologic Section of the Irrigation Department.

River	Source (region drained)	Drainage area km²	Mean annual runoff m³ × 10⁶	m³/km
Mahaweli Ganga	Central Massif and NE Peneplain (Dry Zone)	10,488	3,984	0.38
	Central Massif above Peradeniya	1,193	2,159	1.81
	Central Massif below Peradeniya & NE Peneplain (Dry Zone)	9,925	1,826	0.2
Aruvi Aru	NW Peneplain (Dry Zone)	3,297	506	0.15
Kalu Ganga	Sabaragamuwa Hills	2,730	7,260	2.7
Kala Oya	NW Peneplain (Dry Zone)	2,714	444	0.16
Deduru Oya	NW margin Central Massif & NW Peneplain	2,657	1,135	0.43
Walawe Ganga	S margin Central Massif & E margin Sabaragamuwa Hills & SE Peneplain (Dry Zone)	2,486	1,998	0.8
Kelani Ganga	Central Massif above Yatiyantota & W margin Central Massif & SW Peneplain	2,301	6,120	2.66
Gal Oya	Gal Oya Hills & SE Peneplain (Dry Zone)	1,820	1,209	0.66
Mi Oya	NW Peneplain (Dry Zone)	1,755	296	0.17
Kumbukkan Oya	Gal Oya Hills & SE Peneplain (Dry Zone)	1,706	605	0.36
Maduru Oya	NW margin Gal Oya Hills & NE Peneplain (Dry Zone)	1,565	518	0.33
Maha Oya	NW margin Central Massif & W Peneplain	1,534	1,159	0.76
Mink Ganga	Gal Oya Hills & SE Peneplain (Dry Zone)	1,292	494	0.38
Kirindi Oya	S margin Central Massif & S Peneplain (Dry Zone)	1,183	567	0.48
Nilwala Ganga	W margin Sabaragamuwa Hills & S Peneplain	975	1,480	1.52
Gin Ganga	Sabaragamuwa Hills & SW Peneplain	962	2,516	2.62

Hydrological Section, Irrigation Dept., Ceylon.

References

Adams, F.D. 1929. The geology of Ceylon. Can. J. Res 1:425-511.
Cooray, P.J. 1956. Geological foundations of Ceylon's scenery. Bull. Cey. Geog. Soc. 10:20-30.
Cooray, P.J. 1967. The Geology of Ceylon. Nat. Mus. Cey. Pub. Colombo. 324 pp.
Coates, J.S. 1935. Geology of Ceylon. Cey. J. Sci. 19:101-187.
Dassenaike, S.W. 1928. Coast erosion in Ceylon. Eng. Assoc. Cey. Colombo.
Deraniyagala, P.E.P. 1958. The Pleistocene of Ceylon. Cey. Nat. Mus. Pub. Colombo.
Erb, D.K. 1963. The geomorphology of Ceylon: a study of tropical terrain based on aerial photographs. Unpublished Ph.D. thesis, McGill University, Montreal. 431 pp.
Erb, D.K. 1966-70. Landforms and drainage of Ceylon. Bull. Cey. Geog. Soc. 20:1-38.
Fernando, L.J.D. 1948. The geology and mineral deposits of Ceylon. Bull. Imp. Inst. 46:303-325.
Fernando, S.N.V. & A.D.N. Fernando 1968. Natural vegetation map of Ceylon. In: Resource maps of Sri Lanka Pt 1. Integrated Devt Assoc. Sri Lanka, 1st edit. Aug. 1975. Colombo.
International Bank for Reconstruction and Development. 1953. Water resources and development. Ch. 13. In: The economic development of Ceylon. Report of a mission of the I.B.R.D. Johns Hopkins Press, Baltimore. 829 pp.
Oliver, R.L. 1957. The geological structure of Ceylon. Bull. Cey. Geog. Soc. 11:9-16.
Wadia, D.N. 1945. The three superposed peneplains of Ceylon: their physiographic and geological structure. Rec. Cey. Dept. Minerol. Prof. Pap. 1:25-32.

3. Environmental geochemistry and its impact on humans

C.B. Dissanayake

Introduction

Environmental geochemistry essentially deals with the geographical distribution of elements and forms the basis for a variety of interdisciplinary studies involving human and animal health, agriculture and nutrition, soil fertility, pollution and mineral exploration. The study of the abundance and distribution of some trace elements and the resulting biological manifestations involves geochemists, public health workers, soil scientists, ecologists and nutritionists.

The relationship between the distribution of trace elements in the surface environment and health in plants, animals and man is now receiving world-wide attention. Webb (1975) emphasized the significance of the variation in geographical distribution of the elements as a central factor in a wide spectrum of environmental problems. Hamilton (1976) stressed that problems facing the environmentalist are essentially interdisciplinary, and any attempt at erecting national or international acceptable levels for the chemical elements must necessarily consider those which are natural to defined environments. Problems of trace element deficiency or excess affecting health and production of agricultural crops and livestock have been reported, and many of the occurrences can be relatedto the geology of the area or region affected. For example, areas underlain by acid igneous rocks such as granite or arenaceous sedimentary rocks generally contain lower levels of essential trace elements - particularly the first row transition elements - than areas underlain by ultrabasic and igneous rocks or shales. These however, may sometimes contain sufficient concentration of a potentially toxic element to be hazardous (Thornton & Plant 1980).

Regional geochemical (ideally, trace element) maps are best suited for application to agriculture, ecology and human health investigations. Such an approach has been taken by the Applied Geochemistry Research Group (AGRG) of Imperial College, London, and the Institute of Geological Sciences (IGS), and has proved to be of immense value in a large number of disciplines. Even though there is still no proper geochemical map of Sri Lanka, the available geochemical data can be used to advantage to study the geographic distribution of essential elements, and applied in many disciplines (Dissanayake 1980).

The importance of trace elements in human health and nutrition is well known and there appear to be geographical patterns in the incidence of certain diseases. Well established relationships with the geochemical environment include iodine deficiency with goitre, fluoride deficiency with dental caries and fluoride excess with fluorosis. Webb (1975) points out that there are numerous and controversial correlations with no proven causal relationship such as water hardness and cardio-vascular diseases, lead and multiple sclerosis, cadmium and hypertension and artherosclerosis, and a range of trace elements with cancer. Cancer in particular has been the subject of intensive study, and the overwhelming number of causative factors which have been isolated are in one way or another environmental. Epstein (1974) has shown that in industrialized societies environmental agents (air, water, working place, lifestyle) have been proved to be responsible for 30-40% of human cancers and much research into the causes of cancer is now based on the hypothesis that all cancers are environmentally caused until the contrary is proven.

The delineation of areas of different trace element concentrations aids in the initial demarcation of geographical areas liable to be affected by the diseases concerned. The production of national geochemical maps and the establishment of geochemical data banks are thus of prime importance to a country. This is more apparent in a developing country such as Sri Lanka, where a large percentage of the population lives close to the soil and depends on the natural environment for necessary amenities. The geochemical status of trace elements in such environments governs to a great extent the general health of the community and much recent research has been devoted to the study of the effect of trace elements in environmental health (Hemphill 1967-71, Underwood 1971, Cannon & Hopps 1971, Shacklette et al. 1971, Kubota & Allaway 1972, Fortescue 1972, 1973, 1974).

The role of geochemistry in agriculture and nutrition is no less important than in heath and disease. Table 1 shows the concentration of essential plant nutrients in natural materials. It is apparent that the presence or absence of nutrients depends to a very great extent on the geochemistry of the soil and underlying rocks. As pointed out by Fortescue (1979), geochemists often are particularly interested in the circulation of the micronutrient elements in nature for a number of reasons. First, because with the exception of Fe, Mn and F, these elements are found in the earth's crust at levels of less than 100 ppm (Levinson 1974), and are thus relatively rare in nature. Second, the variation in abundance of particular trace elements is related to the genesis of particular rocks. The element cobalt, for example, has an abundance in the earth's crust of 25 ppm, in ultrabasic rocks, of 150 ppm, in granites of 1 ppm and in soils of 40 ppm (Levinson 1974). Fortescue (1979) stresses the need for geochemists to study the trace elements and the micronutrients together pertaining to geographically significant regions. Biogeochemical cycles which are defined as the pathways which chemical elements or compounds (e.g. D.D.T.) follow in ecosystems may involve abiotic as well as biotic components. In this context the geochemist plays an important role by:

Table 1. The concentration of essential plant nutrients in natural materials (ppm).

| Element | Rocks | | | | Soil (ppm) | Fresh Water (ppm) | Nutrient Solution (ppm) | Plant matter (Oven dry) | | Relative no. of atoms with respect to molybdenum (plant) |
	Igneous	Shale	Sandstone	Limestone				All plants (ppm)	Conifers (ppm)	
Micronutrients										
Molybdenum	1.5	2.6	0.2	0.4	2	0.00035	0.05	0.1	1	1
Copper	55	45	5	4	20	0.01	0.064	6	4	100
Zinc	70	95	16	20	50	0.01	0.065	20	30	300
Manganese	950	850	50	1,100	850	0.012	0.55	50	200	1,000
Iron	56,300	47,200	9,800	3,800	38,000	0.67	5.6	100	200	2,000
Boron	10	100	35	20	10	0.13	0.50	20	20	2,000
Macronutrients										
Chlorine	130	180	10	150	100	7.8	3.5	100	1,000	3,000
Sulphur	260	2,400	240	1,200	700	3.7	48	1,000	800	30,000
Phosphorus	1,050	700	170	400	650	0.00541	41	2,000	1,300	60,000
Magnesium	23,300	15,000	10,700	2,700	5,000	4.1	36	2,000	1,000	80,000
Calcium	41,500	22,100	39,100	302,000	13,700	15.0	134-300	5,000	2,500	125,000
Potassium	21,000	26,700	11,000	2,700	14,000	2.3	130-295	10,000	8,000	250,000
Nitrogen	20	—	—	—	1,000	0.23	140-284	15,000	15,000	1,000,000
Oxygen	464,000	483,000	492,000	497,000	490,000	889,000	—	450,000	—	30,000,000
Carbon	200	15,000	14,000	11,400	20,000	11,000	—	450,000	—	35,000,000
Phosphorus	1,400	5,600	1,800	860	15,000	111,000	—	60,000	—	60,000,000

(After Fortescue 1979)

a) studying the nutrient availability for plants in a particular area at a given time.
b) describing the reserves available for each element for plant growth within the landscape for a given length of time and using this data to predict which element (or elements) will be in short supplywithin given lengths of time, providing that the landscape is not disturbed by man's activities or by some natural catastrophe.
c) describing the aspects of the landscape which result from its evolution and which may be expected to interfere with the growth of plants, now or in the future.
d) predicting problems involving the geochemistry of the environment which may result in long-term or short-term variations in the growth of plants in particular landscapes.

The accumulation of geochemical data banks on the above mentioned lines would be of great value to agriculturalists and nutritionists in developing countries faced with crop failures related to the geochemistry of the environment.

General geology and geochemistry

The island of Sri Lanka, with an area of 69,450 km^2, is primarily a part of the shield which comprises Peninsular India. Geologically and physically Sri Lanka is a southern continuation of India, geologically recently separated from the mainland by the shallow sea covering the Palk Strait and the Gulf of Manaar. On the basis of height and slope characteristics, the island can be divided into 3 main morphological regions (Fig. 1) (Vitanage 1970, see also Cooray, this volume).
  I. The coastal lowlands with elevations from sea level to 270 m with a few inselbergs. Slopes generally are flat lying in the narrow marshy belt along the coastal fringe while further inland low 'turtle backs' appear.
 II. Uplands with elevations from 270 m to 1,060 m consisting of ridge and valley topography and highly dissected plateaux with narrow 'arenas' and domes occupying nearly 30% of the island. The average degree of slope varies from 10°-35° along the upland ridges depending on the lithology and structure. Well developed steep scarps are common. The arena floors are flat and undulating with gentle slopes ranging from 0°-10°.
III. Highlands with a series of well defined high plains and plateaus rimmed with mountain peaks and ridges with elevations from 1,060 to 2,420 m characterize the central part of Sri Lanka. High level topographic discontinuities are common and these form the boundary of a series of high plains, plateaus and structural terraces. Laterites and laterization are common in these places.

Sri Lanka, which has a typical humid tropical climate, lies in the monsoon region of south-east Asia. The island is characterized by clearly demarcated dry and wet

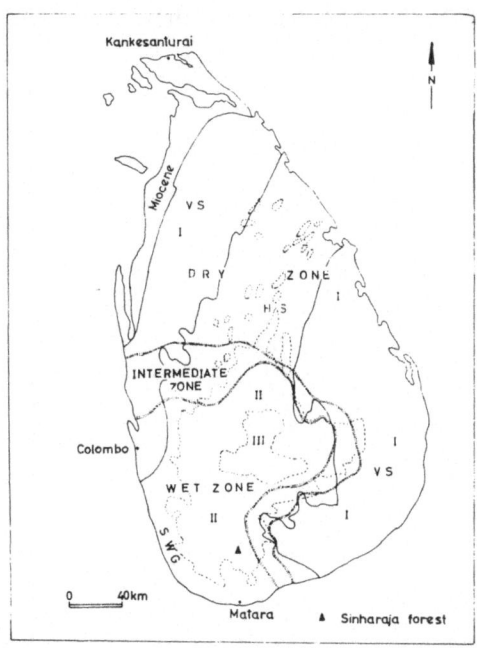

Fig. 1. The main morphological regions and the climatic zones of VS-Vijayan Series; HS-Highland Series; SWG-Southwest Group.

zones as shown in Fig. 1. The average mean temperature of the wet zone lies between 70°-85° F and in the dry zone it is approximately 90° F. Depending on the altitude the mean temperatures of the Highlands vary between 58° F and 78° F.

Geochemistry in agriculture

Fig. 2 illustrates the agro-ecological regions of Sri Lanka with their rainfall patterns, major soil groups and the geographical regions. The wet zone is characterized by red-yellow podzolic soils and latosolic soils whereas the dry zone has reddish brown earths and non-calcic brown soils. The intermediate zone consists of some features seen in both dry and wet zones.

Geologically, the greater part of the country (92%) is made up of rocks of Precambrian age. The various geological formations are summarized by Cooray (this volume).The Precambrian has been divided in a three-fold manner into the Highland Series occupying the central part of the island extending towards the northeast, the Vijayan Series on either side termed the eastern and western Vijayan, and the Southwest Group. For a more detailed account of the geology of Sri Lanka, see Cooray (1967 and this volume).

70

AGRO-ECOLOGICAL REGIONS OF SRI LANKA

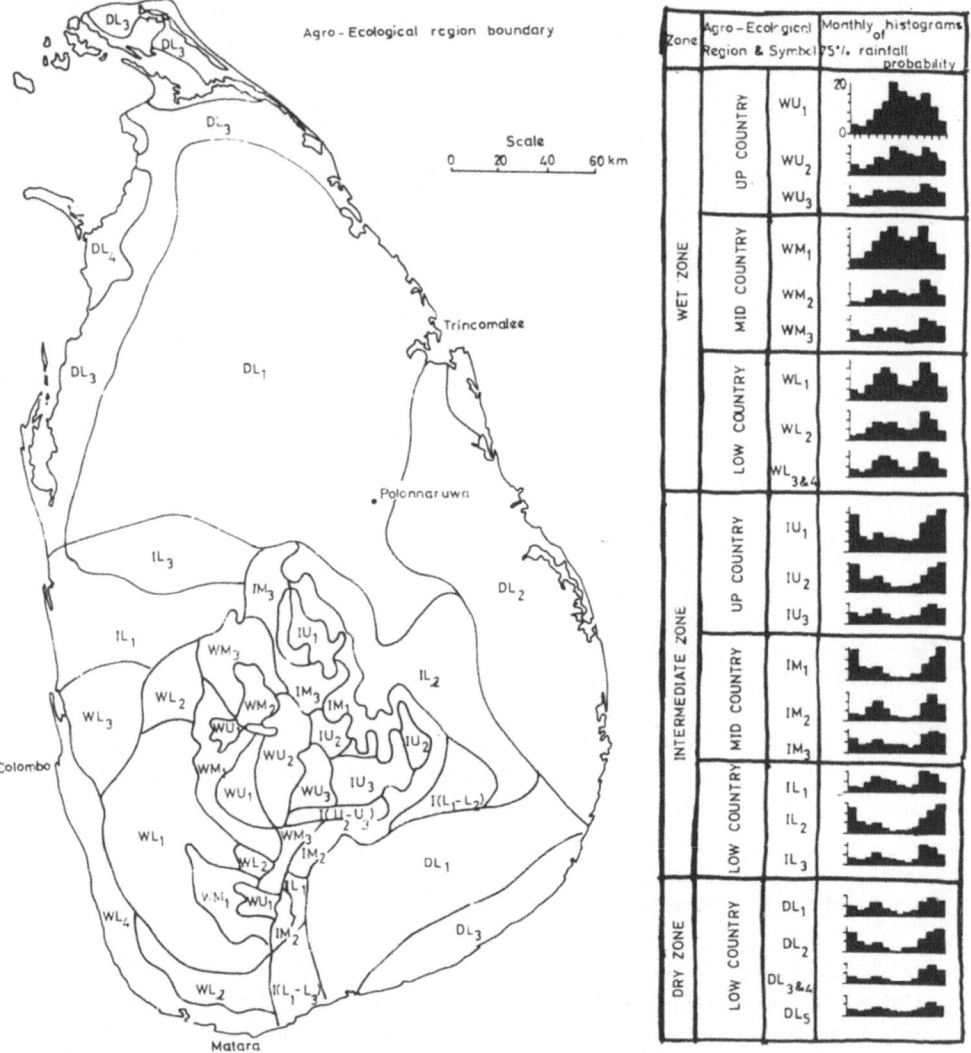

Fig. 2. Map showing the agro-ecological regions of Sri Lanka.

It should be noted that in general, areas covered by basic rocks (the charnock-ites, hornblendes and biotite gneisses) provide essential nutrient elements to the soil. These rocks are found mainly in the areas covered by the Highland Series and the Southwest Group. The Vijayan Series consists of granites and granitic gneisses which are deficient in some of the essential trace elements. The wet zone is nearly completely underlain by rocks of the Highland Series and Southwest Group as a result of which more of the essential elements are given over to the soils of the wet zone. This enhances the agricultural productivity of the wet zone.

Certain parameters are of vital importance in such a study. Fyfe and Kronberg (1980) in their detailed work on soils from the Amazon Basin outlined the major geological processes that produce the biomass nutrient reservoirs. These are:

1. Volcanism, particularly basaltic volcanism which produces the best spectrum of major and minor essential elements, often in an easily weathered form (glass or ash - Java and Hawaii as examples).
2. River processes, producing new sediments from fresh rock, and;
3. Glacial processes, producing fine and coarse rock debris.

In the processes associated with rock-soil-biomass interaction, Fyfe and Kronberg (1980) considered the following rates to be of importance.

Element X, Y, Z ----------- in rock rate K_1
Element X, Y, Z ----------- in soil rate K_2
Element X, Y, Z ----------- in biomass product

According to the laws of conservation, if rate K_2 is greater than the rate K_1, low fertility prevails. Thus the least fertile regions are associated with lands of low elevation, tropical climates with marked drought periods and great tectonic stability. In such a case, step K_1 is virtually non-existent and hence an extreme depletion of essential nutrients in overlying soils leads to reduced biomass production.

Sri Lanka provides an interesting situation for investigating the geochemical mass balance in the assessment of land for long range agricultural productivity. Sri Lanka has a number of features worthy of note in such a study; its tropical equatorial belt location with well marked, closely located dry and wet zones, its three clearly demarcated physiographic regions corresponding to the dry, wet and intermediate climatic zones, and its contrasting geological features that influence the rates K_1 and K_2 and result in contrasts in the biomass production. The influence of these factors has resulted in different social structures and population densities in the dry and wet zones.

Patterns of element migration

The patterns of element migration and secondary mineral formation depend very heavily on the climatic conditions. Keller (1962) has summarised the influence of climatic conditions on leaching of elements and these, as outlined below, are highly applicable to the dry and wet zones.

Where rainfall is profuse, the combined or average effect of pH from rain water (relatively unmodified in pH because of its profuseness) and the high pH of hydrolyzing silicates like nepheline and amphiboles produces an intermediate but alkaline pH in the weathering system. This results in separation of Al_2O_3 and SiO_2 as bauxite and soluble silica, or secondary silica in some places.

Where rainfall is scanty and annual evaporation exceeds precipitation, water is insufficient in quantity to carry away in solution all the SiO_2, Al_2O_3 and alkali and alkaline earths even though the pH is high. Products remaining from weathering of silicates include alkali and alkaline earths, Al_2O_3 and SiO_2 which combine to form montmorillonite, illite and zeolites.

Where rainfall is moderate, the precipitation is sufficient to remove most of the alkali and alkaline earths, but not profuse enough to flush away the H ions from H_2CO_3, organic clay and plant acids. The pH is then low enough to hold both Al_2O_3 and SiO_2 relatively insoluble. The combination of H, Al_2O_3 and SiO_2 results in the formation of the kaolin group of clay minerals.

Fig. 3 illustrates the clay mineral provinces of Sri Lanka (Herath & Grimshaw 1971). There is a marked correlation between the type of clay minerals and the climatic zones. The occurrence of these clay zones has a very significant bearing on the retention and soil availability of the essential nutrient elements and hence on agriculture. It was only recently that water from large irrigation schemes brought

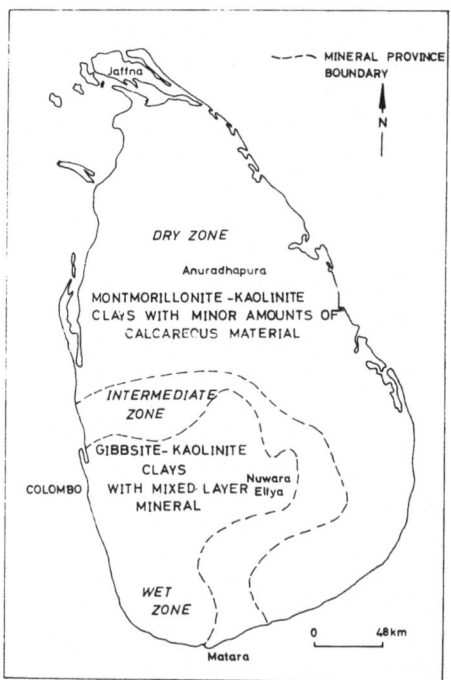

Fig. 3. The clay mineral provinces of Sri Lanka (After Herath & Grimshaw 1971).

vast areas of land in the dry zone again under rice cultivation. Prior to this, the dry zone was uncultivated for about 600 years, and consisted of jungle with grassland, deeply weathered soils being absent. Shifting cultivation was practised and all these features are a direct result of the type of weathering seen in the area and the deficiency of essential nutrients in the soil due to geochemical factors. An example of extreme depletion of major nutrient elements (Ca, Mg, K and Na) is seen in the Amazon soils, where the elements had been depleted to such an extent and not replenished that they reached levels of less than 0.1 % (Fyfe & Kronberg 1980).

Sri Lanka provides an excellent model to illustrate the relations of the geochemistry of weathering to soil and ecology since a traverse from north to south (an airline distance of 420 km (Fig. 2)) shows a complete transition from a dry to a wet climatic zone spreading out gradually. In the dry zone the rainfall is low and evaporation exceeds precipitation (Fig. 2). Further, being tectonically inactive, and having a flat topography, the rate of replenishment of the elements in the natural environment is very low. This means that rate K1 is extremely small, as a result of which low fertility prevails. This has resulted in the inevitable input of a very large amount of fertilizers.

Fig. 4 illustrates the variation of Mg, K, Fe_2O_3/MnO, base saturation (the measure of the saturation of the exchange complex with basic cations) and the pH of the soil in the area under consideration covering the three climatic zones. A low base saturation indicates depletion whereas a high base saturation indicates lesser depletion or replenishment. In the dry zone the base saturation is low, indicating the depletion of the essential elements. In the intermediate zone, the base saturation increases considerably, pointing to a rapid replenishment of the elements concerned. However, in the wet zone, the base saturation unexpectedly decreases and this is attributed to the extensive laterite formation in the region. Laterites present special geochemical environments and some elements in the surface soils are seriously depleted. Laterites, the result of residual concentrations of Al and Fe, are caused by extensive weathering and leaching and the surface is blanketed by the least soluble mineral phases, Al_2O_3-Fe_2O_3, diminishing the nutrient capacity. Phosphate in particular, another essential nutrient, is retained by the extremely insoluble Al and Fe oxide phases and hence the fertility of the lateritic terrains is reduced even further.

Southwards from the position of the high point of the bell-shaped curve (Fig. 4), weathering leads to a net effect of soil destruction, instead of soil formation. The optimum balance between the energy of the weathering agents and the energy of minerals to resist weathering has been exceeded. Towards the centre of the mountainous area, the excessive rainfall exceeding the rate of evaporation tends to leach nutrients from the soils. Towards Matara in the southern part, high rainfall, high temperature and destruction of organic residue left from macroflora accelerates the hydrolysis and and solution of silicates in the soil. This forms two-layer clays with cation exchange capacities and lateritic aluminium and iron oxides. At the northwestern end of the largest tropical rainforest, the Sinharaja forest, Ca and

74

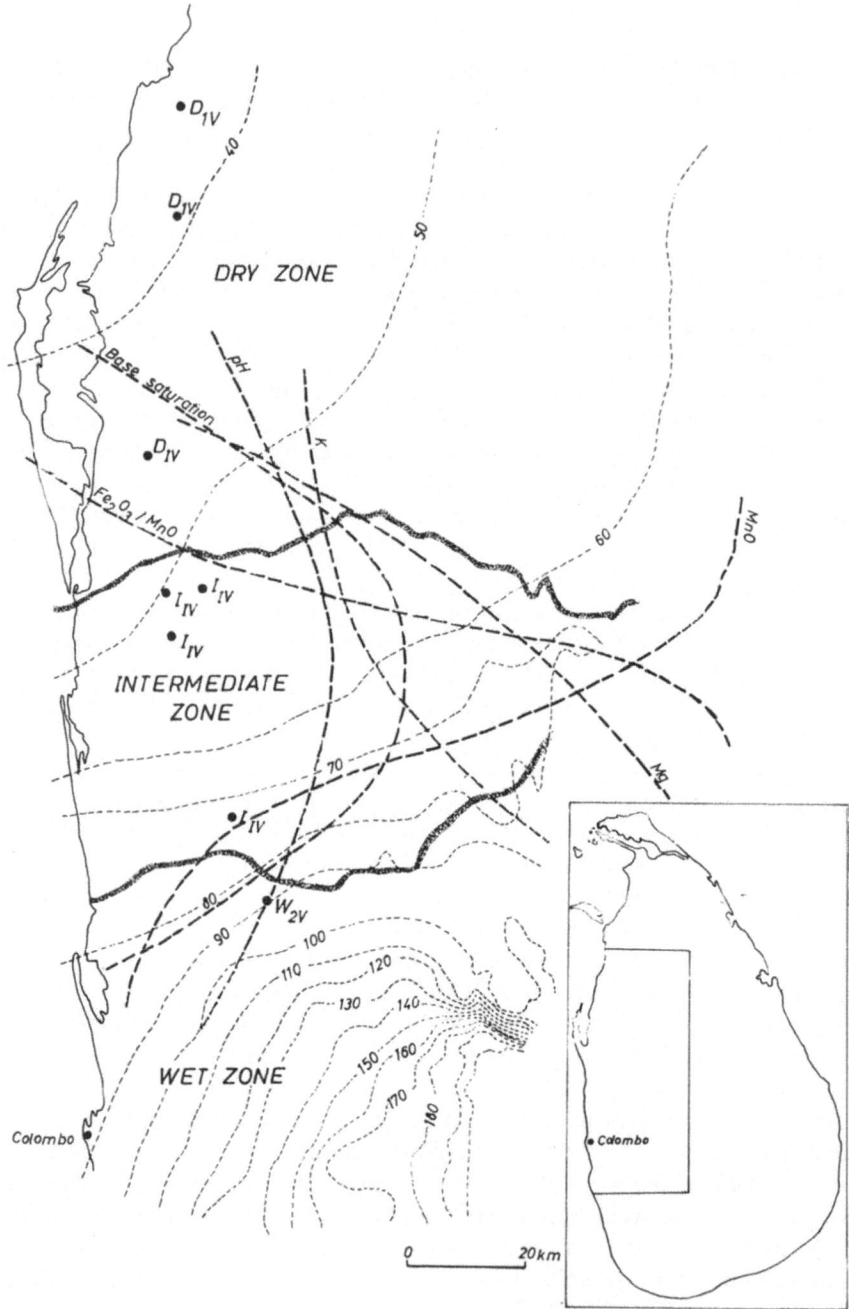

Fig. 4. Variation of some elements in the three climatic zones of Sri Lanka, with increasing concentration of nutrient elements Mg, K towards the wet zone. Contour lines show the average annual rainfall in cm. (Inset shows the area studied).

Mg are leached progressively to form acidic soil conditions, where H^+ions are substituted for Ca/Mg, giving rise to H-clay and H-organic colloids.

As shown in Figs. 4 and 5, the total MnO concentration decreases towards the wet zone and due to the high evaporation rate of the dry zone, Mn migrates into the upper layers as Mn^{2+}ions resulting in the formation of concretions. Due to the presence of Mn in the soil, Fe becomes unavailable (particularly in calcareous soils) because of the oxidizing action of Mn on Fe. The apparent increase of MnO in the upper 5 cm of wet zone soils is attributed to the trapping of Mn in the top humus layers.

The chemistry of Fe in the soils is different from that of Mn, as in the biosphere generally. For this reason, the Fe/Mn ratio (normally about 50 in rocks), decreases to about 30 in the soils, Fe being mobile and more easily removed from the soils.

The geochemical behaviour of Mg in the soils of the dry and wet zones of Sri Lanka is shown in Figs 4 and 6. Mg is depleted in the dry zone and appears to be replenished in the wet zone even though leaching is faster in the wet zone.

Environmental geochemistry of nitrates

Even though pollution due to industrial effluents has still not caused serious concern, pollution of ground water and drinking well waters due to biological sources has reached significantly high levels during the last decade. The bulk of the country's health problems are deeply rooted in the environment. It is estimated that only between 15-25 % of the people have access to safe water; or less than 10 % have access to piped water. The majority of the people use smaller unprotected

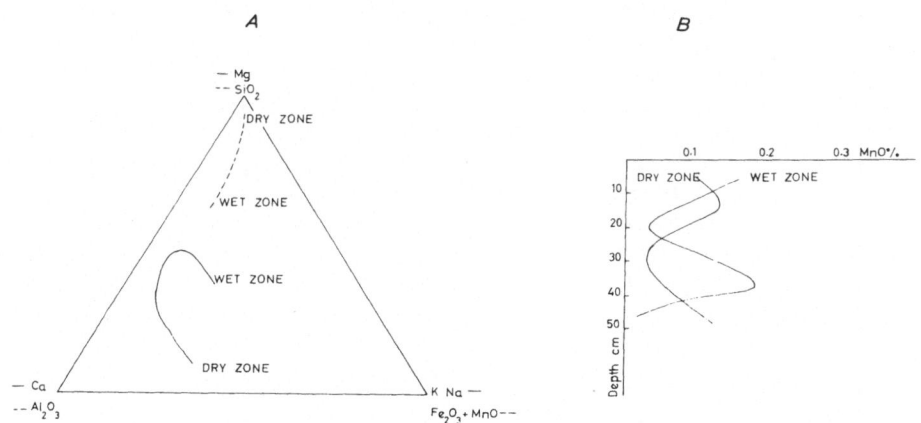

Fig. 5. (a) Ternary variation diagram for some elements in the contrasting climatic zones; (b) Variation of MnO with depth of soil in the dry and wet zones.

wells and in settlement areas tanks and water channels are the main source of drinking water. The proper disposal of human and other waste through sewage systems and latrines also is limited; less than one-third of the population has satisfactory sewage facilities. The poor water supply and excreta disposal programmes have resulted in 40% of the people seeking treatment for typhoid, amoebic and bacillary dysentry, infectious hepatitis, gastroenteritis, colitis and worm infections (Econ. Rev. 1980).

Monitoring of water pollution in Sri Lanka is an absolute necessity, but very few studies have been made. Among some of the studies carried out are those in by the Department of Geology of the University of Peradeniya (Sri Lanka), dealing with hydrogeochemistry of potable waters of Sri Lanka (Dissanayake & Hapugaskumbura 1980, Dissanayake & Ariyaratne 1980, Dissanayake & Jayatilaka 1980).

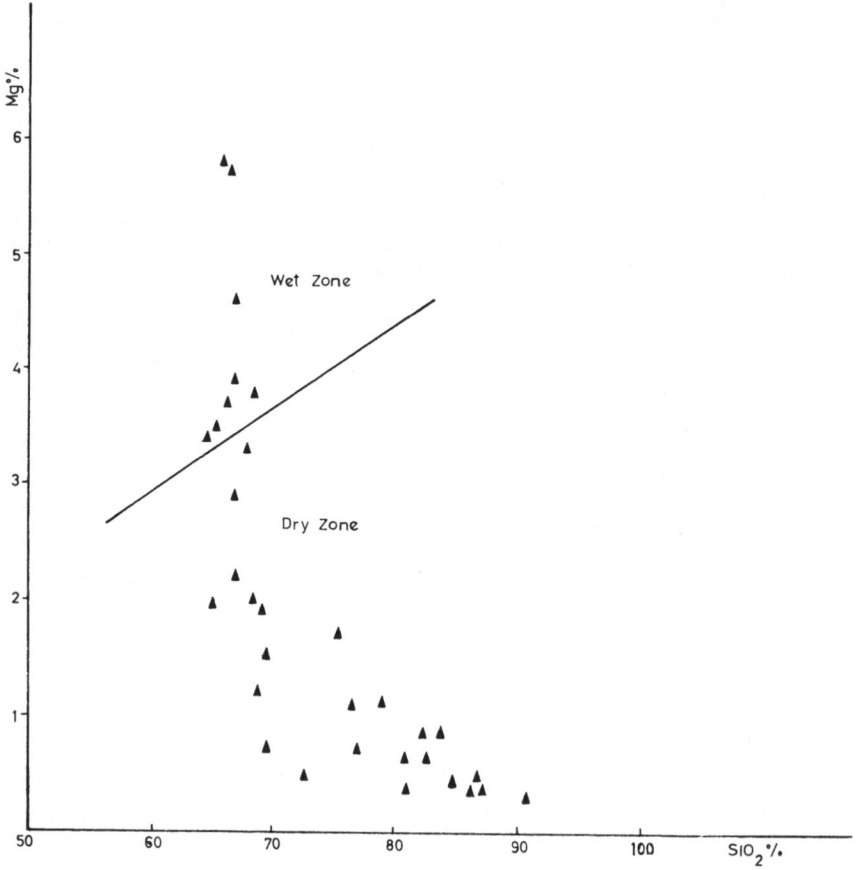

Fig. 6. Variation of Mg with Si O$_2$ in the dry and wet zones.

The potential health implication of the contamination of drinking water by nitrates has been the subject of scientific attention since 1945, in connection with infant methemoglobinaemia (Comly 1948, Weart 1948, Walton 1951). In the United States in particular, nitrate-rich waters have been studied intensely for their possible effects on human health (Miller 1971, Johns & Lawrence 1973, Gruener & Toplitz 1975, Brooks & Cech 1979). Further, nitrates in water are studied in relation to the formation of N-Nitroso compounds known to be carcinogenic and a voluminous literature is now available on this subject (Bogovski 1972, Bogovski & Walker 1975). The dangers of high nitrate concentrations in drinking water are well documented (W.H.O. 1978). These are essentially methemoglobinaemia and carcinogenesis. At present attention is sharply focussed on the problem of the intake of excessive nitrates because the compounds formed by nitrite which is the product of nitrate reduction and secondary amines are carcinogenic.

The nitrate distribution in Kandy Lake and Mid-canal

The author and his co-workers have studied the distribution of nitrates in Kandy Lake and its effluent canal which runs through a highly populated area. Fig. 7

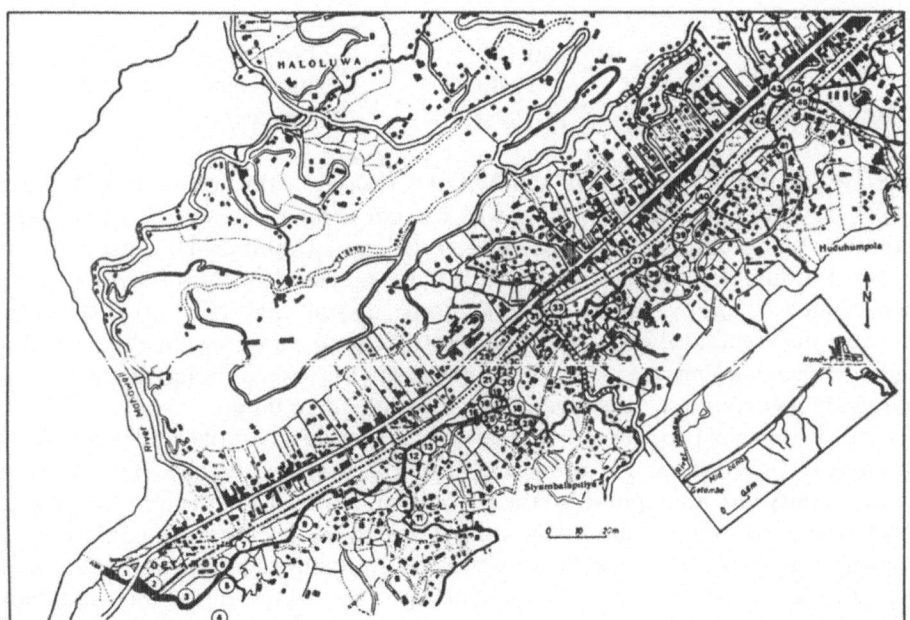

Fig. 7. Map showing the Mid-canal of Kandy. Circled numbers indicate the sampling points. (Inset shows the Kandy Lake and the origin of the Mid-canal).

illustrates the location of the Mid-canal and Kandy Lake. This study illustrates the possible detrimental effects due to nitrates of a sewage canal running through the city of Kandy (pop. 115,000).

Table 2 shows the kind of effluent inlets to the lake and also the maximum values obtained for some of the ions and elements.

Fig. 8 illustrates the distribution of nitrates in Kandy Lake. The monitoring of nitrate levels in the lake is of extreme importance since the water is used occasionally to supplement the town water supply. The maximum nitrate level detected was 70 mg l^{-1} at inlet No. 18, a drain carrying urine contaminated effluent from a school nearby. The improper disposal of sewage has resulted in most of the inlets carrying urine contaminated discharge and hence the presence of nitrates. As Fig. 8 shows, the maximum range of nitrates in the lake determined was between 5-10 mg l^{-1}. These levels were observed mainly in areas between inlets 1 and 9. Public lavatories, large hotels and guest houses in these parts of the area bordering the lake discharge some of their sewage effluents into the lake. Even though some of the inlets carry a high nitrate content, the high dilution has prevented the occurrence of unacceptable levels of nitrates in the lake water. The World Health Organization recommendation for nitrate levels in water are as follows:

	mg l^{-1} NO$_3^-$	category
general population	<50	acceptable
	50-100	borderline
	>100	unacceptable
infants under 6 months	>50	unacceptable

The level of nitrates observed in Kandy Lake, though within an acceptable level at present, could increase if municipal waste is not properly treated. With the rapid building construction and development of the area around the lake, proper waste disposal should necessarily take priority. The increasing tourist influx into Kandy has resulted in unprecedented building construction, and if the proper sanitary precautions are not taken, Kandy Lake could become a serious health hazard. A high nitrogen loading rate may not be reflected in the lake, if the incoming nitrogen is quickly incorporated into organic forms by the lake biota.

Nitrate is an end product of the decay of nitrogenous material, and this material could be fertilizers or animal and human excreta. The obvious lack of fertilizer use in the vicinity of Kandy points to the conclusion that most of the nitrate in Kandy Lake is due to contamination by animal and human excreta. Apart from the improper drainage of public lavatories and pit latrines, defaecation in secluded places also has contributed to the general input of nitrates. It is of interest to note that Hutton and Lewis (1980), in a study of nitrate pollution in Botswana, noted that there is a strong association between ground water pollution and population density, but the major towns served mainly by surface water were unaffected.

Table 2. The nature of effluents draining into the Kandy Lake.

Inlet	NO₃⁻ (mg l⁻¹)	Total P (mg l⁻¹)	Cu²⁺ (ppm)	F⁻ (ppm)	Coliform count	Remarks
1	7.35	0.46	—	—	—	Stream, densely populated valley, urine contaminated
2	4.60	0.38	0.016	0.801	275	As above
3	Very high	0.34	—	—	—	As above
4	1.10	—	0.029	—	30	Food stuff (decayed)
5	8.35	0.65	—	—	—	Densely populated valley
6	25.00	0.95	—	—	—	Dye stuff, contaminated
7	2.35	Very high	—	—	—	Food stuff, washouts
8	3.70	—	0.008	0.801	1800	As above
9	—	—	0.018	0.543	250	As above
10	1.55	0.65	0.008	0.632	175	Food stuff (decayed)
11	0.95	0.60	0.024	0.742	200	Solid washouts
12	1.79	0.25	—	—	900	Kitchen washouts
13	5.25	33.25	0.008	0.730	1800	Human excreta and urine contamination
14	1.34	0.60	0.002	0.721	1800	Foodstuff kitchen washouts (decayed)
15	30.00	1.90	0.002	0.721	250	Nursing home effluents, urine contamination
16	6.00	0.50	0.053	0.734	35	Urine contamination
17	5.50	2.55	0.053	0.734	14	Densely populated house washouts
18	70.00	2.00	0.053	0.732	—	Urine contaminated
19	3.50	—	—	—	—	As above

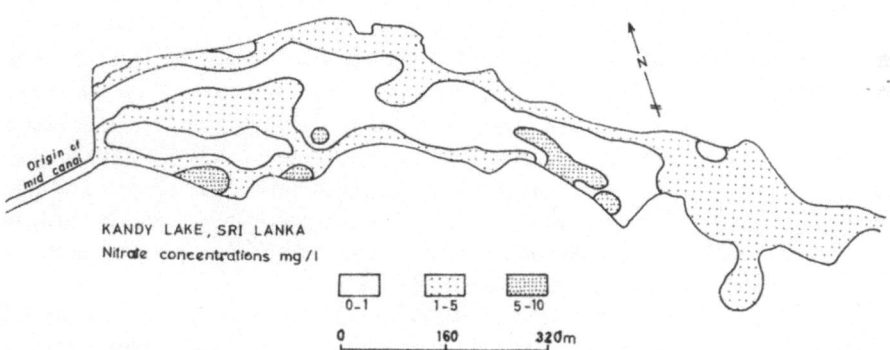

KANDY LAKE, SRI LANKA
Nitrate concentrations mg /l

0-1   1-5   5-10

0   160   320m

Fig. 8. Distribution of nitrates in the Kandy Lake.

The Mid-canal in Kandy runs through the city and many dwellings collecting massive quantities of effluent from side canals (Fig. 7) and drains into the River Mahaweli having traversed a distance of 8 km. The canal collects waste matter from the city and also from the run-off of other canals. Among the waste matter collected in the canal are human excreta, waste water from washing places and laundries, hospital effluents and general garbage from domestic dwellings. The dark colour of the canal water is due chiefly to the large quantities of diesel oil and other automobile waste liquids from garages and repair workshops for buses and other vehicles. Further, the slaughter house situated adjacent to the canal also discharges enormous quantities of sewage into the canal. The effluent from the Kandy Hospital adds a further load to the general sewage.

Kandy (pop. 115,000) has more than 3,000 houses with only bucket-type latrines, which are emptied into the open canal daily. A similar number of people are without any latrine facilities and the open canal is used by some of these people as latrines. The build-up of nitrogenous matter is obviously very high and the potential health hazards are enormous.

The canal is approximately 10-15 m in width and is stone paved only in some parts. In the other regions it is open and carries a very large load of waste matter, particularly during the rainy seasons. A large number of houses are found, particularly between the railway station and the point of discharge into the river at Getambe. In this area, many houses depend on unprotected well water for their domestic use and the close proximity of this sewage canal to the wells causes problems in environmental pollution.

The area lies in the wet zone of Sri Lanka and the temperature varies between 25-32 ° C throughout the year. Topographically, a hill and valley type of morphol-

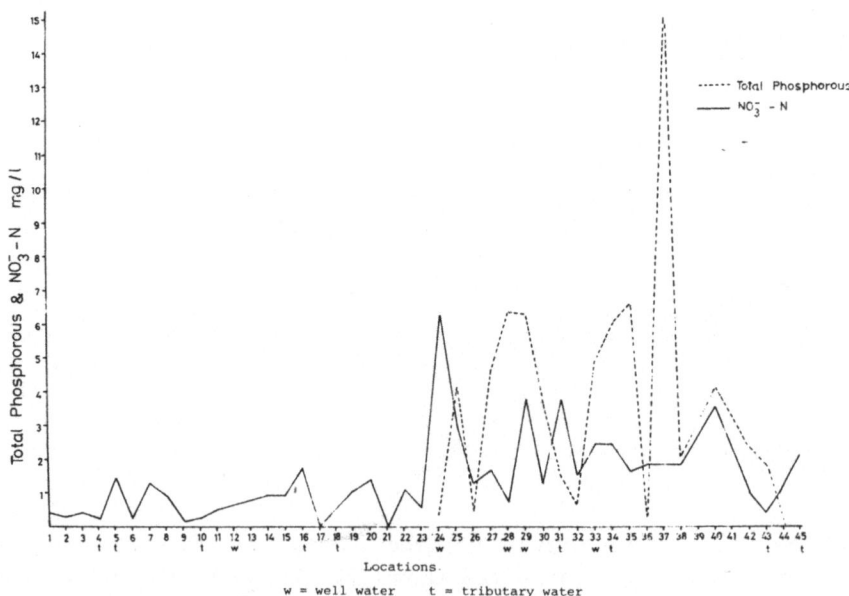

Fig. 9. Variation of total phosphorus and NO$_3$-N along the Mid-canal of Kandy.

ogy is seen, the lake and canal occupying a lower elevation, resulting in the draining of a large number of side-canals originating from the surrounding hilly region. Geologically, the area of the canal consists of highly weathered quartzites, marbles and a variety of gneisses. The alluvium is thick and the total amount of sediments is high, particularly closer to its discharge point.

Fig. 9 illustrates the variation of NO_3^--N and PO_4^{3-} along the canal, and Fig. 10 illustrates the relative variation of the NO_3^--N, NH_4^+-N and total phosphorus in a section of the canal between locations 25 and 43 where their concentrations reached higher levels. In general, NO_3^--N contents varied between 0.2-3.5 mg l-1 whereas the NH_4^+-N contents varied between 0.3-0.5 mg l-1. The nitrogen loading of the canal is chiefly of biological origin, the human and animal waste matter comprising a very large percentage of the total nitrogen loading. This is not reflected in the analytical data since some of the nitrogen is incorporated in organic forms, particularly in the bottom sediments. Further, the action of microbes such as denitrifying bacteria could release much of the nitrogen.

The relative variation of NO_3^--N with NH_4^+-N needs special emphasis. Fig. 11 illustrates the inverse relationship of the two species in the canal. Further, the NH_4^+-N content is enhanced in the canal whereas in the surrounding wells it

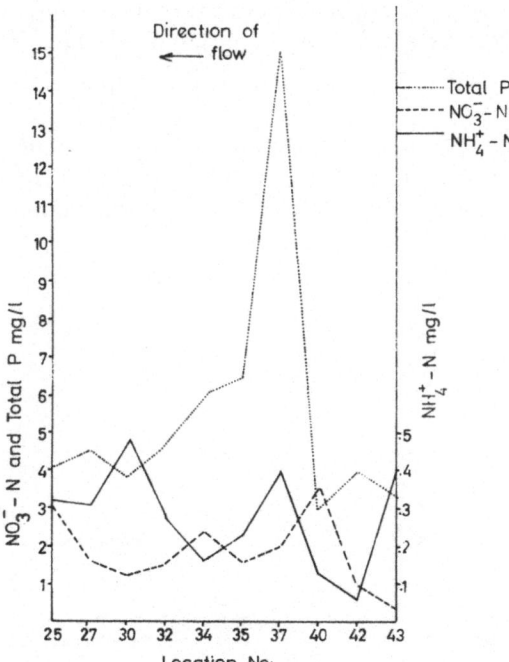

Fig. 10. Variation of total phosphorus, NO_3^--N and NH_4^+-N in a highly polluted section of the Mid-canal.

decreases rapidly. In the case of NO_3^--N the opposite effect is seen where the wells near the vicinity of the canal carry a higher concentration. Even within the canal, an elevated NH_4^+-N content did not show a proportionate increase of NO_3^--N concentration. Due to the presence of a large number of side canals carrying sewage, the instream nitrification (if any) of the Mid-canal cannot be assessed with any degree of certainty. Further, NO_3^--N unlike NH_4^+-N and organic nitrogen, is mobile in soil and it is likely to be the dominant form of nitrogen in surface run-off (non-point sources) (Cirello *et al.* 1979). Interpretations based on the relationship of NH_4^+-N and NO_3^--N could lead to erroneous conclusions due to the other forms of nitrogen transformations. Loss of NO_3^--N through denitrification could occur in organic-rich bottom sediments regardless of the oxygen status of the overlying water. Ammonification, the release of NH_4^+ from organic combination would tend to replace that consumed by nitrification.

Both the above mentioned processes lead to nitrification, which consequently could take place without leaving any trace in terms of the nitrogen species (Cirello *et al.* 1979). However, the elevated NH_4-N levels in the canal represent an unrealized potential for nitrification and this has had a very significant bearing on nitrate pollution of the large number of unprotected water wells in the vicinity, particularly in the areas around Mulgampola, Welate and Gatambe. Fig. 12 illustrates the potential danger of nitrate pollution of the wells. The highly permeable underlying alluvium aids the migration of nitrating species and the rate of reactions $NH_4^+ \rightarrow NO_2^- \rightarrow NO_3^-$ moves towards the nitrification step under these aerobic conditions.

A similar situation was studied in Botswana by Lewis *et al.* (1980) and Hutton and Lewis (1980), where gross faecal bacterial pollution of ground water was shown to result from pit latrines. A transit time of less than four hours had been

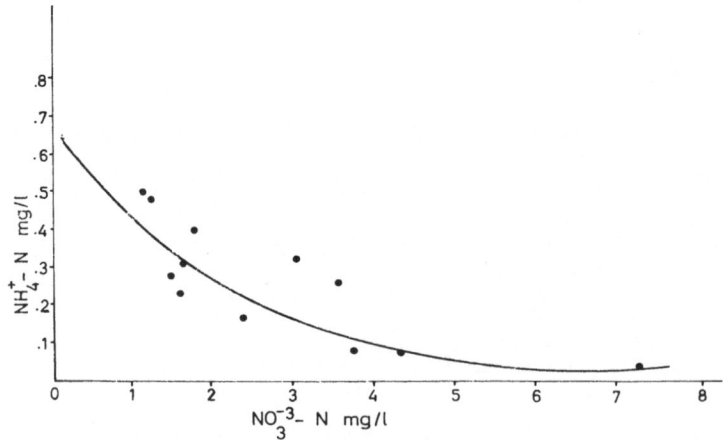

Fig. 11.   Relation of NO_3^--N with NH_4^+-N in the Mid-canal.

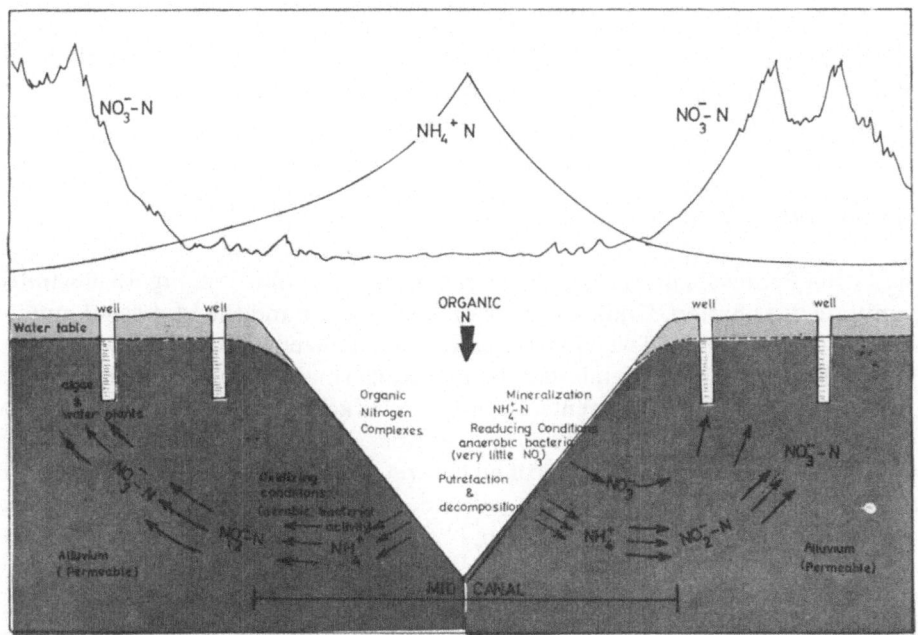

Fig. 12. Model showing the potential for nitrification of the well water close to the Mid-canal.

recorded between a pit latrine and the neighbouring bore-hole, at a distance of 25 metres. The pit latrines had caused a massive build-up of nitrogenous material in the surrounding soil and weathered rock. The nitrate leached from the contaminated soil clearly was the main cause of its extremely high concentration in the local ground water.

Brooks and Cech (1979), in a study of the rural water wells in eastern Texas, also observed that wills with the highest observed concentration of nitrates were those located in close proximity to septic tanks.

The Mid-canal of Kandy poses an extremely grave public health risk owing to its close proximity to the unprotected wells used by many people for domestic purposes. Indeed, some of the wells sampled for NO_3^--N reached concentrations of 7.2 mg l^{-1}. Even though this figure may appear to be within safe limits for drinking water, the very large amounts of human excreta discharged into the canal shows that the nitrification has not proceeded to completion and that the potential for extremely high nitrification of well water exists in the future. It appears very likely that most of the nitrogen loading into the canal exists at present in some other forms of nitrogen such as organic complexes, but these can eventually be converted into NO_3^--N.

This study clearly illustrates the serious ground water pollution hazard represented by pit latrines in hydrogeological environments. As shown by Hutton and

84

Lewis (1980), if pit latrines are close to water supply bore-holes or wells, a grave risk to public health arises. Bottle fed infants are at particularly high risk if nitrate contaminated waters are used in preparing their food, as boiling water will concentrate, not remove nitrogen.

Nitrates in the Jaffna Peninsula

The Jaffna Peninsula presents a special case of nitrate pollution. Fig. 13 illustrates the nitrate distribution. Unlike the case of Kandy Lake and the Mid-canal, nitrate abundance in the ground water is due mainly to fertilizers used in agriculture. The entire Jaffna Peninsula is underlain by limestone (mainly $CaCO_3$) and the soil is highly deficient in some nutrients. In order to make up for this deficiency, nitrogenous fertilizers are used in very large quantities, hence the abundance of nitrates in the ground water. Isolated spots of high nitrate levels, however, were observed in some areas due to sewage pollution, particularly within the urban limits of Jaffna and Point Pedro, ranging from 122 mg l^{-1} to 174 mg l^{-1}.

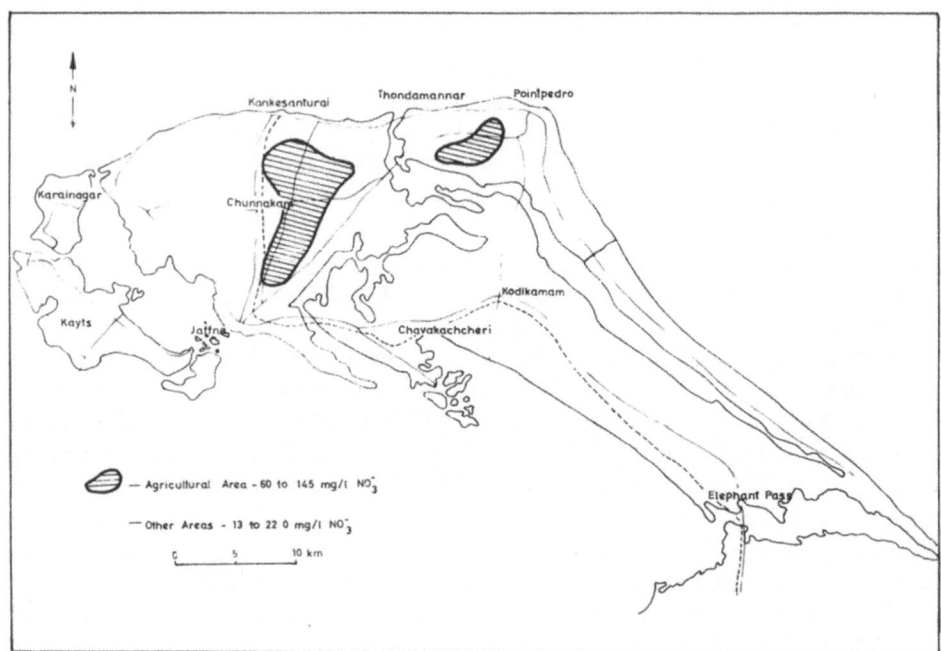

Fig. 13. Distribution of nitrates in the groundwater in the Jaffna Peninsula.

Flourine geochemistry and the incidence of dental diseases

Among the most common diseases prevalent in Sri Lanka are dental fluorosis and dental caries. In a developing country such as Sri Lanka, the need to study the distribution of dental diseases and the recognition of natural causes for the observed distribution can hardly be over-emphasized. Such a study is all the more important, since it has been found that a large number of school children in Sri Lanka suffer from dental fluorosis and dental caries. This fact has been established by the research work carried out by the Faculty of Medicine, University of Peradeniya (Senewiratne & Senewiratne 1975) and by the Department of Geology, University of Peradeniya (Dissanayake 1979). In a program oriented towards the prevention of such dental diseases, it is important to delineate the areas which show a high incidence of the disease and establish the geochemical causes which may be responsible.

The delineation of geochemical provinces pertaining to fluoride and dental diseases would help in programmes of defluoridation or fluoridation of water in Sri Lanka. A study of the geology is necessary, additional to climate, soil types, vegetation and the environmental geochemistry in relation to the distribution of diseases in an area.

Fig. 14 shows the distribution of fluoride containing water in Sri Lanka. The central hill country in which Kandy is located and the southwest coastal region are relatively free of fluoride. The lowlands dry zone contains a higher amount of fluoride in water with areas around Eppawala in the North Central Province, Maha Oya in the Eastern Province and Uda Walawe in the Southern Province showing a higher fluoride content.

It is of interest to correlate the fluoride-rich and fluoride-poor areas delineated with natural factors such as climate and geology. Fluoride-free areas are situated mainly in the wet zone, whereas the high fluoride bearing areas belong mainly to the dry zone (Fig. 14). It is likely that in the wet zone, where the average rainfall exceeds 200 inches in certain instances, leaching of soluble salts is high. In these areas, there is a tendency for the soluble ions to be efficiently leached and carried away in solution. Fluoride is easily leached from primary and secondary minerals (Hawkes & Webb 1982) and soils under the effect of high rainfall. In the dry regions evaporation tends to bring the soluble ions upwards due to capillary action in soils. This, although not the sole explanation for the observed distribution of the fluoride in well water in Sri Lanka, could nevertheless be an important factor. However, it is the geology of the areas that needs special consideration. The composition of the rocks in the area, particularly the easily leached constituents coupled with the climate are the key factors in the geochemical distribution of elements in a tropical region. The abundance of fluoride in the rocks and the ease with which it is leached under the effect of ground water has an important bearing on the abundance of fluoride in the areas concerned and hence the prevalence of dental diseases.

Table 3. The concentration of fluoride in well water in some cities in Sri Lanka.

	Anuradhapura	Polonnaruwa	Kandy
Dental fluorosis	77.5%	56.2%	13.0%
Dental caries	26.2%	26.5%	95.9%
Fluoride concentration	0.34-3.75 ppm	0.26-4.55 ppm	<0.2 ppm

Note: The maximum fluoride concentration in the Anuradhapura area was 9.0 ppm, 5.8 ppm in the Maha Oya and 4.8 ppm in Uda Waluwe (see Fig. 14 for localities).

Among the areas containing the highest fluoride concentrations in well water, the region around Eppawala and Anuradhapura (Fig. 14) is the largest. Senewiratne and Senewiratne (1975) reported fluoride concentrations is high as 9.0 ppm in this region. The abundance of fluoride which caused severe dental fluorosis among people of this area can be attributed to an abundance of fluorine in the rocks. It is significant that in this area there occurs an economically exploitable deposit of apatite (fluoro-hydroxy phosphate), classified as a carbonatite, and is known to contain reserves of 23 million tons. The apatite deposit is now being mined and analysis shows it to contain a fluorine concentration of 1.5-2.4% (Jayawardena 1976).

It is a well-established fact that the fluoride ion can take the place of the hydroxyl ion and that an equilibrium could be maintained. The substitution of fluoride for hydroxyl is to be expected from similarity of ionic radii and charges. Extensive research has been carried out on the fluoride-hydroxyl exchange in geological material (Gillberg 1964, Stormer & Carmichael 1971, Ekstrom 1972, Munoz & Ludington 1974). The presence of higher concentrations of fluoride in water in this area bearing fluorine-rich rocks is therefore explained on the basis of $FO = OH$ interchange between minerals and water. Apart from this, apatite which is present in abundance in this area also is known to exhibit this interchange of fluoride and hydroxyl (Nada & Ushio 1964).

The area in the southeast of Sri Lanka around Uda Walawe (Fig. 14), which contains high concentrations of fluoride, comprises different geological formations compared to the region around Anuradhapura. It has been discovered recently that the area around Uda Walawe consists of large deposits of serpentinites (Dissanayake & Van Riel 1978). These occur as long and narrow belts and a number of such deposits occur to the north and south of Uda Walawe. Serpentine, an iron-magnesium hydroxy silicate, also posseses the property of exchanging fluoride for the hydroxyl ion; it exhibits the property of taking up the fluoride ion into its structure from an aqueous solution and also releasing the fluoride into an aqueous medium. This release of exchangeable fluoride into ground water appears to be the most likely explanation for the abundance of fluoride in wells of the area lying in the dry zone. Serpentine generally contains 1,000-2,000 ppm of fluoride

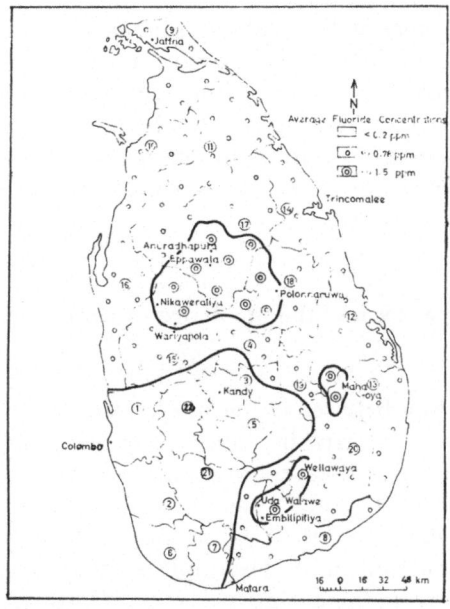

Fig. 14. Distribution of fluoride containing water in relation to the districts and population of Sri Lanka. Figures within brackets indicate the percentage of the total population of Sri Lanka living in that district. Faint lines show the district boundaries.

1. Colombo (21.8%); 2. Kalutara (6.0%); 3. Kandy (10.5%); 4. Matale (2.4%); 5. Nuwara Eliya (3.5%); 6. Galle (5.7%); 7. Matara (4.6%); 8. Hambantota (2.6%); 9. Jaffna (6.0%); 10. Manaar (0.6%); 11. Vavuniya (0.7%); 12. Batticaloa (2.0%); 13. Amparai (0.3%); 14. Trincomalee (1.5%); 15. Kurunegala (8.0%); 16. Puttalam (3.0%); 17. Anuradhapura (3.0%); 18. Polonnaruwa (1.3%); 19. Nadulla (4.8%); 20. Moneragala (1.5%); 21. Ratnapura (5.1%); 22. Kegalle (5.1%); .

and this is quite sufficient for an enrichment of fluoride in water in the vicinity, bearing in mind the large number of deposits present.

The third high fluoride zone lies around Maha Oya in the eastern part of Sri Lanka and also lies in an area of geological and geochemical significance. Around Maha Oya a number of hot springs, considered to be derived from thermally heated circulating ground water and gaseous emanations are present. Apart from the many dissolved ions, gases also are seen to bubble through the hot water. The abundance of fluorine among these gases in such terrains is well known. Further to the north and to the south of Maha Oya, there are other hot springs indicating a much larger area of thermally heated waters. If a more detailed survey of fluoride in water is carried out in the areas further north and south of Maha Oya, the boundary of the zone may have to be extended. The Maha Oya topographic region, however, contains the highest number of hot springs.

From the observations on the distribution of fluoride in drinking well-water in

Sri Lanka and the incidence of dental diseases in the areas concerned, it is apparent that a distinct correlation exists between this and the geology of the delineated areas. The geochemical distribution of fluoride in the rocks and minerals along with the climatic factors are responsible for the prevalence of dental diseases in certain parts of Sri Lanka. The areas so demarcated are basically geochemical provinces with neither an abundance or a lack of fluoride containing rocks and minerals.

Geochemistry of well water and the incidence of cancer and cardiovascular diseases

Whereas the relationships between iodine and goitre and also fluorine and dental diseases demonstrate very clearly the relationship between geochemical environment and human health and disease, other links are less convincing. Underwood (1971) pointed out that most of the associations of trace elements with diseases rest heavily upon correlation rather than causation, as in the case of cadmium and human hypertension and lead and and multiple sclerosis. The problem of cancer in humans certainly belongs to this category. Cancer is known to be caused by a wide variety of factors often acting in combination, over periods of many years. Even though research into the causes of cancer is now based on the hypothesis that all cancers are environmentally caused until the contrary is proved (Epstein 1974), a proper causative relationship, particularly a geochemical one, is yet to be established.

Cancer

Dissanayake and Senaratne (1981) have carried out a study to investigate the association of the geographical variation of total cancer in Sri Lanka with the geochemical status of potable waters. The fact that more than 300 compounds are known to be carcinogenic, as compiled by the International Agency for Research on Cancer (IARC), further complicates the isolation of natural environmental and geochemical factors as being associated with the incidence of cancer.

Fig. 15 illustrates the incidence of total cancer in Sri Lanka in the provinces, and Table 4 shows the type of cancer prevalent. Bearing in mind that only a very small percentage of the population uses piped water, the dissolved ions in the local water have an obvious effect on their general health. Fig. 16 illustrates the variation of total cancer rates with the total dissolved solids in water and the hardness for the water in the nine provinces of Sri Lanka. It is apparent from Fig. 16 that even though the curves for total solids and hardness against cancer incidence appear to be complementary, the relationship is not a simple one and there is an optimum level of hardness and total solids that corresponds to a minimum incidence of

89

cancer. Even though it cannot by any means be considered a causal relationship, an optimum amount of ions in the water appears to be conducive to a low incidence of cancer.

Cardiovascular diseases

Since 1957 studies have been made in many countries on the relationship between the hardness of drinking water, its content of bulk and trace elements and mortality and morbidity due to cardiovascular diseases. It is generally believed that mortality rates for cardiovascular diseases are higher in areas with soft water than in areas

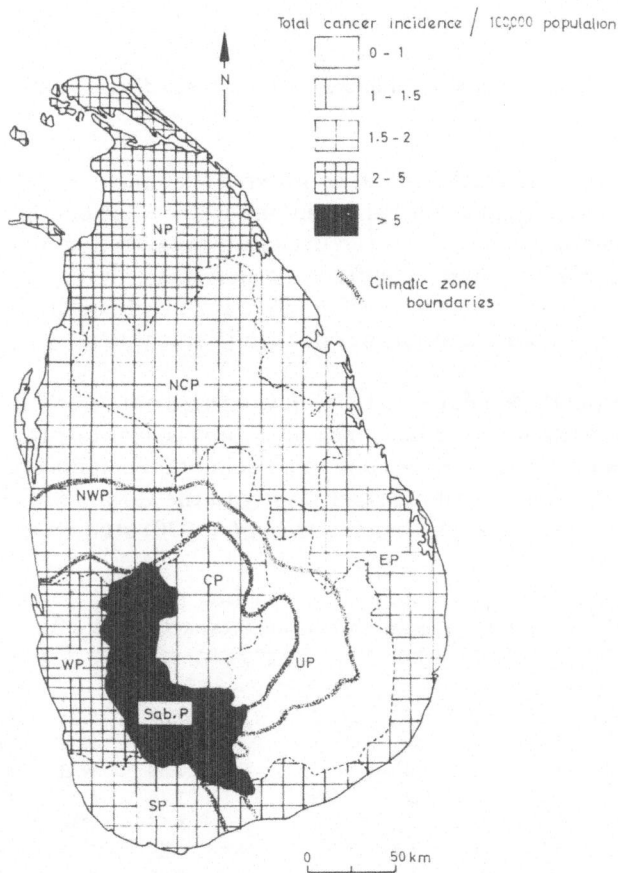

Fig. 15. The incidence of cancer in Sri Lanka. NP-Northern Province; NCP-North Central Province; NWP-Northwestern Province; CP-Central Province; UP-Uva Province; EP-Eastern Province; Sab.P-Sabaragamuwa Province; WP-Western Province; SP-Southern Province.

90

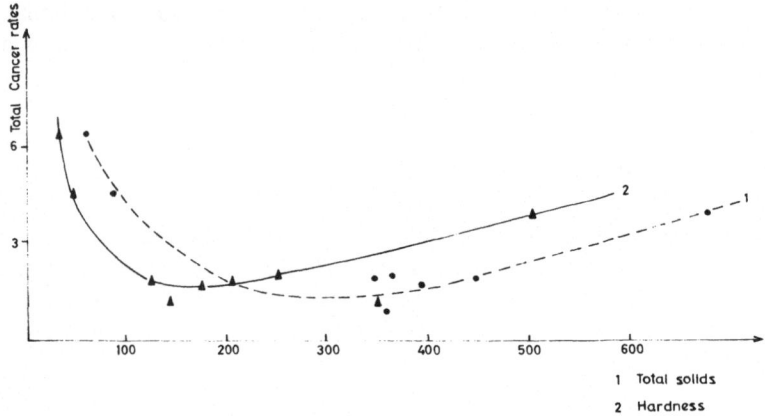

Fig. 16. Variation of total cancer with hardness of water and total dissolved solids.

with hard drinking water. Keil (1979) considered the following pathophysiologic mechanisms concerning soft and hard drinking water to be conceivable:
1. Soft water can be damaging to the organism because it contains fewer bulk elements (e.g. Ca ions) and/or more toxic elements.
2. Hard water can have a protective effect because it contains trace elements and bulk elements at higher concentrations than in soft water.

Gardner (1976) and Barker & Rose (1976) have carried out several epidemiological studies in Great Britain on the incidence of cardiovascular disease. They showed that drinking water properties besides socio-economic conditions, air pollution and precipitation rates contribute significantly to the variance of cardiovascular rates in the cities investigated. Shacklette et al. (1970) studied the geochemical

Table 4. The incidence of various types of cancer in the Provinces of Sri Lanka.

Province									
North	7.6	0.6	7.3	3.2	1.5	4.04	4734	372	1408
North Central	0.8	—	—	—	2.1	1.95	4433	293	408
North Western	3.0	—	0.78	—	—	1.89	5933	205	446
Central	—	—	—	0.9	1.35	1.12	4173	350	392
Western	—	—	4.6	4.6	—	4.55	5246	20	80
Sabaragamuwa	6.5	—	9.4	—	5.6	6.5	3863	30	58
Uva	—	0.9	—	—	—	0.9	3907	150	360
Eastern	3.8	—	0.13	—	—	1.99	9131	250	369
Southern	1.8	—	—	—	—	1.8	4394	175	392

* After Panabokke 1978; ** after Balendran 1970.

environments and cardiovascular mortality rates in Georgia, U.S.A. and concluded that if geochemical differences between the areas do in fact have a causal relationship with death from cardiovascular diseases, the cause would appear to be a dietary deficiency, rather than an excess of the elements.

Fig. 17 illustrates the incidence of hypertensive diseases per 10,000 population and Fig. 18 shows the variation of hypertensive diseases per 10,000 population with the hardness of water for some areas of Sri Lanka. It is interesting to note that the areas (e.g. Jaffna region) underlain by high Ca-bearing rocks have a lower rate of cardiovascular diseases. The North Central Province however appears to be a special case in that in spite of the high hardness of water, the cardiovascular disease rates also are high. The could well be accounted for by the adverse effects of the excess fluoride found in the water. As in the case of cancer, only a broad correlation rather than a causal relationship can be established between water hardness and cardiovasculr diseases. As pointed out by Keil (1979) in spite of doubts concerning a causal relationship, it seems likely that there is a factor which is closely associated with drinking water and has a damaging influence on the health of the population. It is not known whether this factor can be found in the water itself or only correlates with it. However, excess fluoride ions in drinking water could be such a factor and detailed investigations are warranted.

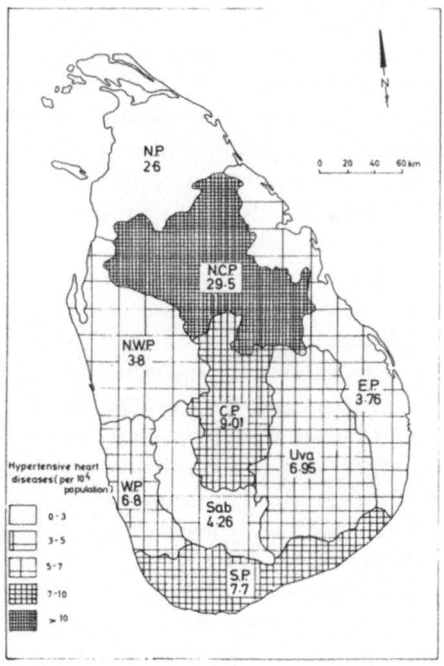

Fig. 17. Incidence of hypertensive diseases in Sri Lanka.

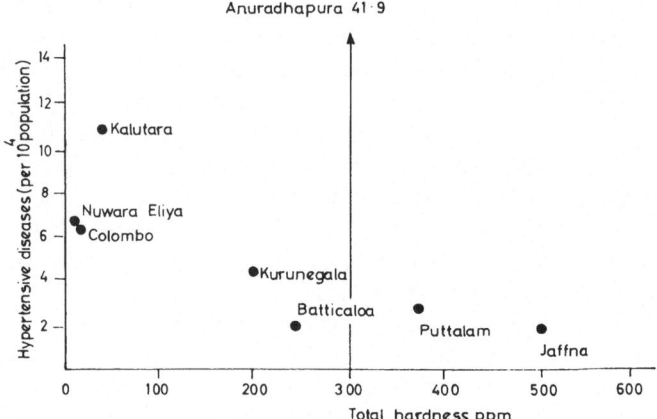

Fig. 18. Variation of hypertensive diseases with the hardness of water in some areas of Sri Lanka.

Environmental geochemistry of copper

Dissanayake and Ariyaratne (1980) studied the distribution of copper in the drinking well waters of Sri Lanka (Fig. 19). Unlike fluorine, copper is geochemically more mobile and in the aqueous phase occurs mainly in the ionic forms Cu^+ and Cu^{2+}, in floating organisms and soluble organic matter. In fresh water, copper occurs in the range of 0.001–0.01 ppm. Biologically, copper is an important plant nutrient, with deficiency symptoms evident when the copper content in soil falls below 10 ppm. Whereas a large number of copper based diseases are known in plants and animals (Todd 1969), the effect of a copper imbalance in human beings is not known with certainty.

Copper based industries are few and restricted to a few cities. The levels of copper encountered were those due to the natural geochemical levels of copper found in different areas. The wet zone (Zone 1) has lowest copper levels in ground water. Topographically, this same area includes only the highest parts of the central highlands in Sri Lanka, and the cumulative effect of topography and climate on the natural distribution of copper could be seen in this region.

The easily leached copper ions are carried away in solution aided by the topographical features of the terrain. The copper ions thus carried in solution are trapped and deposited in the lower lying areas (Zones 2A, 2B, 2C and 3). Some of the higher copper-bearing zones, particularly around 3B are composed of laterite terrain and this has a bearing on the accumulation of copper in the ground water, The alluvial deposits around the zones 3C also are likely material suitable for trapping copper in nature. Zone 3A, which covers a large part of the country, has about 15% of the population. This area lies in the dry zone and also in the flat

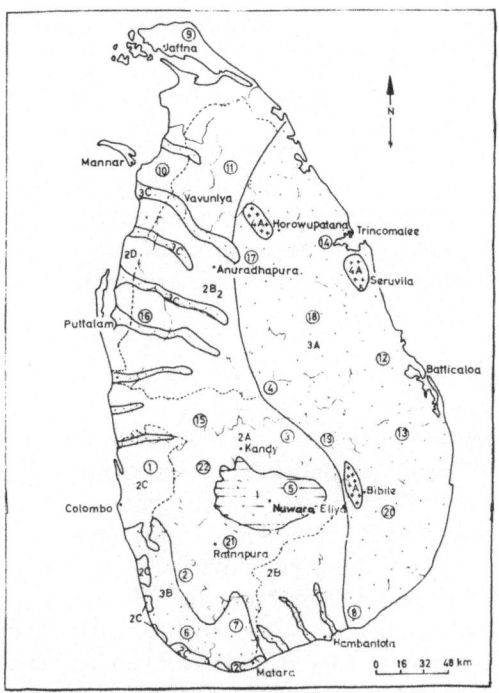

Fig. 19. Distribution of copper containing water in relation to the districts of Sri Lanka. Details regarding districts and population are the same for Fig. 14. Zone 1: Cu < 1 ppb; Zone 2: Cu 1-10 ppb; Zone 3: Cu 10-100 ppb; Zone 4: Cu > 100 ppb.

lowlands. The climatic factors play a very important role in the distribution of copper, in that due to seasonal drought and hence the fluctuating water table, copper becomes concentrated in surface waters more easily. The rapid evaporation of the surface waters brings up the deeper lying soluble salts due to the capillarity of the soil. The areas around Horowpatana, Seruwila and Bibile are possibly sites of copper mineralizations, and indeed a large Cu-Fe deposit has been discovered at Seruwila (Herath 1975).

With the establishment of a large industrial complex in a free trade zone around Colombo, the problem of industrial pollution is now receiving greater attention in Sri Lanka. Dissanayake and Jayatilake (1980) carried out a detailed study of the copper contents in waste and potable water in and around a few cities in Sri Lanka and their role in the possible contamination of ground water, Even though copper is not as toxic as Pb, Hg and Cd in the environment, it was selected as a case study since there are a few copper based industries in Sri Lanka and also due to its easy detectability. Table 5 shows the copper concentrations in the potable waters in the main cities located in different climatic zones. Except in and around Colombo, there appeared to be no significant correlation in the copper ion content between well water and the industrial wastes.

The natural factors responsible for the observed distribution of copper in Sri Lanka are essentially those due to geology, climate, topography and soil. The geochemical status of an element in a particular area must necessarily influence its uptake through food and drinking water. The understanding of the controls on the concentration of metals in natural waters and soil solutions, apart from helping in the study of areal incidence of certain diseases of man and animals, greatly facilitates the treatment of water for domestic and industrial use.

Conclusions

The case studies discussed above illustrate the importance of environmental geo-chemistry in health, disease, agriculture, nutrition and the general well-being of man and animals. Apart from serving the useful purpose of delineating areas with high levels of certain elements and hence in the geographical distribution of diseases, the geochemical data banks could also be of use in mineral exploration, farmland and plantation planning, and forestry. The above discussion shows that natural elemental levels indicated by geochemical data banks vary greatly in different areas due to climatic, topographic and geologic factors and such background levels of elements must necessarily be known before any epidemiological or agricultural studies are undertaken. The compilation of these data banks are of national importance in that 'contamination' can never be estimated if the 'natural' environmental levels of trace elements pertaining to that area are not known.

Sri Lanka in particular needs to be thoroughly investigated for its environmental geochemistry since a very large percentage of the population lives close to the soil and depend on it for their general living. As such the impact of the geochemical status of the soil and water on the humans can hardly be overestimated.

Table 5. Copper concentrations in potable waters in main cities located in different climatic zones.

	No. of samples	Lowest Cu conc. (ppb)	Highest Cu conc. (ppb)	Av. Cu conc. (ppb)	Climatic zone and mean annual rainfall
Anuradhapura	19	9	36	30	Dry zone (140 cm)
Kurunegala	22	8	36	29	Intermediate zone (200 cm)
Colombo	31	18	53	45	Wet zone (227 cm)
Kandy	19	10	20	12	Wet zone (215 cm)
Nuware Eliya	10	1	8	5	Wet zone (230 cm)

Acknowledgements

Grateful thanks are due to Messers A. Senaratne, S.V.R. Weerasooriya, K. Dunu-happawa and Miss S.J. Wijesekera for their assistance. Dr. Leslie Herath, Chairman of the Water Resources Board, Mr. Ti Gunasekaran, Resident Engineer, Jaffna, the Water Supply and Drainage Board and the Ministry of Health are thanked for the supply of some of the data discussed in this paper.

References

Barker, D.J.P. & G. Rose. 1976. Epidemiology in Medical Practice. Churchill Livingstone, New York. 79 pp.

Bogovski, P. 1972. The importance of the analysis of N-Nitroso compounds in international research. pp. 1-55 In: N-Nitroso Compounds Analysis and Formation (Ed. P. Bogovski, R. Preussman and E.J. Walker). IARC Sci. Publ. 3. Lyon.

Bogovski, P. & E.A. Walker. 1975. N-Nitroso compounds in the environment. IARC Sci. Publ. 9. Lyon.

Brooks, D. & I. Cech. 1979. Nitrates and bacterial distribution in rural domestic water supplies. Water Res. 13:33-41.

Cannon, H.L. & H.C. Hopps. 1971. (Ed.) Environmental geochemistry in health and disease. Geol. Soc. Am. Mem. 123.

Cirello, J., R.A. Rapaport, P.F. Strom, U.A. Natulewich, M.L. Morris, S. Goetz & M.S. Finstein. 1979. The question of nitrification in the Passaic River, New Jersey. Analysis of historical data and experimental investigation. Water Res. 13:525-537.

Comly, H. 1945. Cyanosis in infants caused by nitrates in well water. J. Am. Med. Assoc. 129:112-116.

Cooray, P.G. 1967. An Introduction to the Geology of Ceylon. Nat. Mus. Ceylon Publ. 319 pp.

Dissanayake, C.B. 1979. Geochemical provinces and the incidence of dental disease in Sri Lanka. Sci. Total Env. 13:47-53.

Dissanayake, C.B. 1980. Use of geochemical data banks in monitoring the natural environment. A case study from Sri Lanka. Env. Int. 3:293-296.

Dissanayake, C.B. & U.G.M. Ariyaratne. 1980. The significance of natural environmental factors in the distribution of copper in potable waters of Sri Lanka. Int. J. Env. Stud. 15:133-143.

Dissanayake, C.B. & A.K. Hapugaskumbura. 1980. The geochemistry of Na, K, Ca and F in well water around Kandy, Sri Lanka. Indian J. Earth Sc. 1:94-99.

Dissanayake, C.B. & G.M. Jayatilake. 1980. Distribution of copper ions in waste and potable waters in Sri Lanka cities. Wat. Air Soil Poll. 13:275-286.

Dissanayake, C.B. & A. Senaratne. 1981. Geochemical environments and the geographical distribution of some diseases in Sri Lanka. Wat. Air Soil Poll. 00,000-000.

Dissanayake, C.B. & B.J. Van Riel, B.J. 1978. Petrology and geochemistry of a recently discovered nickeliferous serpentinite in Sri Lanka. J. Geol. Soc. India 19:464-471.

Economic Review 1980. Medical Care and Public Health in Sri Lanka. People's Bank of Sri Lanka Publ. 5:3-14.

Epstein, S.S. 1974. Environmental determinants of human cancer. Cancer Res. 34:2425-2435.

Ekstrom, T.K. 1972. The distribution of fluorine among some coexisting minerals. Contr. Min. Petrol. 34:192-200.

Fortescue, J.A.C. 1972. The need for conceptual thinking in geoepidemiological research. pp. 333-339 In: Trace Substances in Environmental Health VIA. Symposium, Columbia University of Missouri.

Fortescue, J.A.C. 1973. Relationship between landscape geochemistry and exploration geochemistry. Report Ser. No.17. Dept. of Geological Sci., Brock Univ., St. Catherines, Ontario.

Fortescue, J.A.C. 1974. The environment and landscape geochemistry. Western Miner. 6 pp.

Fortescue, J.A.C. 1979. Role of major and minor elements in the nutrition of plants, animals and man. In: Review of Research on Modern Problems in Geochemistry. (Ed. F.R. Siegel). UNESCO, 290 pp.

Fyfe, W.S. & B.I. Kronberg. 1980. Nutrient conservation: the key to agricultural strategy. Mazingira 4:65-70.

Gardner, M.J. 1976. Soft water and heart disease? In: Health and the Environment Vol. 3. (Ed. J. Lenihan & W.W. Fletcher). Blackie, London, 121 pp.

Gillberg, M. 1964. Halogen and hydroxyl contents of micas and amphiboles in Swedish granite rocks. Geochim. Cosmochim. Acta 28:495-516.

Gruener, N. & R. Toeplitz. 1975. The effect of changes in nitrate concentration in drinking water on methemoglobin levels in infants. Int. J. Env. Stud. 7:161-163.

Hamilton, E.I. 1976. Review of the chemical elements and environmental chemistry - strategies and tactics. Sci. Total Env. 5:1-62.

Hawkes, H.E. & J.S. Webb. 1962. Geochemistry in Mineral Exploration. Harper & Row, N.Y., 415 pp.

Hemphill, D. (Ed.) 1967-1971. Trace substances in environmental health. Proc. 1st-5th Conf. Univ. Missouri Env. Centre and Extension Division.

Herath, J.W. 1975. Mineral resources of Sri Lanka. Econ. Bull. 2. Geol. Surv. Dept., Sri Lanka. 72 pp.

Herath, J.W. & Grimshaw, R.W. 1971. A general evaluation of the frequency distribution of clay and associated minerals in the alluvial soils of Ceylon. Geoderma 5:119-130.

Hutton, L.G. & Lewis, W.J. 1980. Nitrate pollution of groundwater in Botswana. pp. 1-4 In: 6th WEDC Conf. Water and waste engineering in Africa.

Jayawardena, D.E. de S. 1976. The Eppawala carbonatite complex in northwest Sri Lanka. Econ. Bull. 3. Geol. Surv. Dept., Sri Lanka. 41 pp.

Johns, M. & C. Lawrence. 1973. Nitrate-rich groundwater in Australia: a possible cause of methemo-globinemia in infants. Med. J. Aust. 2:925-926.

Keil, U. 1979. Hardness of drinking water (content of bulk and trace elements) and cardiovascular diseases. pp. 59-77 In: Geomedizin in Forschung und Lehre. (Ed. H.J. Jusatz). Steiner Verlag, Weisbaden.

Keller, W.D. 1962. The Principles of Chemical Weathering. Lucas, Columbia. 112 pp.

Kubota, J. & W.H. Allaway. 1972. Geographic distribution of trace element problems. Chap. 2 In: Micronutrients in Agriculture. Soil. Sci. Soc. Am. Madison.

Levinson, A.A. 1974. Introduction to Exploration Geochemistry. AppliedPubl. Calgary. 612 pp.

Lewis, W.J., J.L. Farr. & S.S.D. Foster. 1980. A detailed evaluation of the pollution hazard to village water-supply bore holes in eastern Botswana. Geol. Surv. of Botswana GS 10 Project. Evaluation of underground water resource.

Miller, L. 1971. Methemoglobinemia associated with well water. J. Am. Med. Assc. 216:1642-1643.

Munoz, J.L. & S.D. Ludington. 1974. Fluoride-hydroxyl exchange in biotite. Am. J. Sci. 274:396-413.

Nada, T. & T. Ushio. 1964. Hydrothermal synthesis of fluorine-hydroxyl phlogopite. II. Relationship between the fluorine content lattice constants and the conditions of synthesis of fluorine-hydroxyl phlogopite. Geochem Int. 1:96-104.

Senewiratne, B. & K. Senewiratne 1975. The distribution of fluoridecontaining water in Ceylon. Ind. J. Med. Res. 63:302-311. Shacklette, H.T., H.I. Sauer & A.T. Miesch. 1970. Geochemical environ-ments and cardiovascular mortality rates in Georgia. U.S. Geol. Surv. Prof. Pap. 574-C. 39 pp.

Shacklette, H.T., J.C. Hamilton., J.G. Boerngen & J.M. Bowles. 1971. Elemental composition of surficial materials in the conterminous United States. U.S. Geol. Surv. Prof. Pap. 574-D. 71 pp.

Stormer, J.C. & I.S.E. Carmichael. 1971. Fluorine-hydroxyl exchange in apatite and biotite. A potential geothermometer. Contr. Min. Pet. 31:189-197.

Todd, J.R. 1969. Chronic copper toxicity of ruminants. Proc. Nutr. Soc. 28:189-197.

Thornton, I. & J. Plant. 1980. Regional geochemical mapping and health in the United Kingdom. J. Geol. Soc. Lond. 137:575-586.

Walton, G. 1951. Survey of literature relating to infant methemoglobinemia due to nitrate contaminated water. Am. J. Publ. Health 41:986-995.

Weart, J. 1948. Effect of nitrates in rural water supplies on infant health. Ill. Med. J. 93:131-133.

Webb, J.S. 1975. Environmental problems and the exploration geochemist. pp. 5-17 In: Geochemical Exploration 1974. (Ed. I.L. Elliott & W.K. Fletcher). Sp. Publ. No. 2. Elsevier, Amsterdam.

World Health Organization 1978. Environmental Health Criteria No. 5. Nitrates, nitrites and N-Nitroso compounds. W.H.O. Geneva.

Underwood, E.J. 1971. Geographical and geochemical relationships of trace elements to health and disease. In: Man and his Physical Environment. (Ed. G.D. McKenzie & R.D. Utgard). Burgess, Minnesota.

Vitanage, P.W. 1970. A study of the geomorphology and morphotectonics of Ceylon. pp. 391-405 In: Proc. 2nd Seminar on geochemical prospecting methods and techniques. E. 72, II, F 2. U.N. New York.

4. Grasslands

M.A. Pemadasa

Introduction

Of the biotic resources of Sri Lanka, vegetation is most outstanding because of its wide variety, considerable species-richness, remarkably high degree of endemism, and great economic potential. Since industrial development is handicapped by the scarcity of local raw material, there must be more efficient and systematic exploitation of biotic resources. In this endeavour, grasslands are likely to play an economically vital and viable role, for they have a potential as pastures (Appadurai 1969, Pemadasa 1981a, Pemadasa & Amarasinghe 1982a). However, their exploitation has been rather unsystematic; increasing biotic interference by haphazard clearing for short-term cultivation, illegal burning, extensive removal of herbage for animal-fodder and over-grazing by cattle have caused considerable floristic and habitat, particularly edaphic, changes and severe erosion of many grasslands, with near complete destruction of some areas. In certain eroded habitats, particularly those on steep slopes, soil is highly truncated with protruding rock outcrops and severely impoverished (Pemadasa & Amarasinghe 1982a), and the native species are under threat of extinction as a result of continuous colonization by graminaceous and composite weeds (Mueller-Dombois & Perrera 1971, Pemadasa & Mueller-Dombois 1979, 1981; Amarasinghe & Pemadasa 1982a). The tragedy is that little attention has been focussed on the repercussion of unsystematic exploitation of these valuable ecosystems. They are ecologically extremely fascinating, and this, together with their economic potential, necessitates their management, improvement and conservation.

Grasslands of Sri Lanka can be divided into three general categories: montane (patana), savanna and lowland grasslands. Of these, patanas are the most extensive, and cover about 65,000 ha of the south-central highlands. The north-central, north-eastern and south-western lowlands, on the other hand, support a wide variety of climatically, edaphically and phytosociologically distinct pasture communities, the most noteworthy being the so-called damana, talawa and villu grasslands. At intermediate elevations in the eastern parts of the central highlands, patana merges into a savanna vegetation. The three types of grasslands,though,

100

share common features, are ecologically distinct, and have different economic potentials. They also have suffered different degrees of biotic interference.

In this chapter, the ecology and phytosociology of the more important of these grasslands are outlined, and problems of their management, improvement and conservation are discussed with special reference to the importance of human impact.

Patana grasslands

Patanas occur above 500 m altitude in topographically, altitudinally, climatically and edaphically diverse habitats in the south-central highlands of the humid and summer-dry zones of Sri Lanka (Mueller-Dombois 1968, Mueller-Dombois & Perera 1971) (Fig. 1). They have been the subject of repeated extensive surveys (Pearson 1899, Parkin & Pearson 1903, de Rosayro 1945, 1946a, b, c, 1969; Holmes 1951, Perera 1967). These have yielded comprehensive floristic accounts and detailed descriptions of the climate, geology and soils, but have caused considerable controversyregarding their origin, classification and ecology. This is not surprising, for there is a wide variety of phytosociologically and physiognomically different patanas, whose ecological differentiation is related to biotic as well

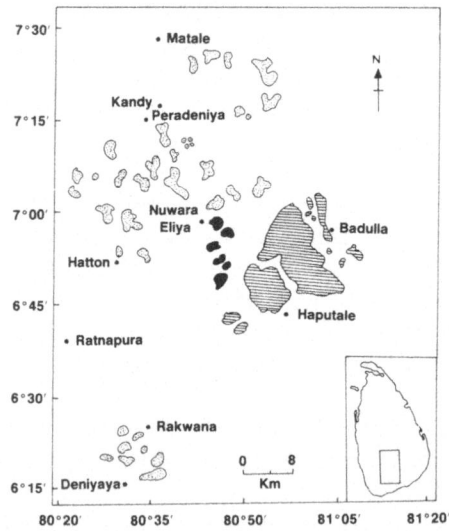

Fig. 1. Distribution of patana grasslands in the south-central highlands of Sri Lanka. Stippled, hatched and shaded areas represent humid zone dry patana, summer-dry zone patana and wet patana respectively. Inset: Sri Lanka showing the location of south central highlands (Modified from Holmes 1951).

as abiotic environmental diversity (Pemadasa & Mueller-Dombois 1979, 1981); apparently, the only feature common to all of them is their occurrence on mountains. Moreover, there is considerable heterogeneity within individual patanas (Pemadasa & Amarasinghe 1982a, Amarasinghe & Pemadasa 1982a). Knowledge of a complex of environmental and biological characteristics of patanas is essential to understand their origin, nature, dynamics and ecology more fully, and to provide a more reasonable classification.

Origin of patanas

A long-standing controversy exists about the origin of patanas, and the attempted explanations fall into two groups, the so-called disclimax and edaphic sub-climax hypotheses. Pearson (1899) and Holmes (1951) regarded patanas as plagioclimaxes resulting from frequent burning of woody vegetation; however, Pearson's presumption of a former savanna was disputed by Holmes. They argued that destruction of woody vegetation facilitated the establishment of grass-cover, and frequent burning, and perhaps clearing, prevented re-establishment of woody plants so maintaining the patanas in their present state. The scattered occurrence of charcoal in the soil and fire-tolerant trees of *Careya arborea* and *Phyllanthus emblica* (nomenclature of Abeywickrama 1959) is cited as evidence of this hypothesis. de Rosayro (1945, 1946a, b, c, 1961) repeatedly disputed this view, and postulated the patanas to be a natural edaphic sub-climax of a hydrosere which has never reached the woody vegetation stage; the occurrence of hygrophyllous sedges (e.g. *Fimbristylis nigrobrunnea*) and several endemic grasses in certain patanas (e.g. those in the Uva Basin (Figs 1 and 2)) is the basis of this postulation. The first hypothesis, originally mentioned by Broun (1898), received support from Chapman (1947) and Koelmeyer (1957) and the second was favoured by Szechowyez (1954). Perera (1967) was non-commital as to whether both these could be possible explanations.

Both hypotheses suffer many limitations, the most important being that neither can explain the origin of all the patanas. de Rosayro's view lacks support because: (a) Chronological evidence of the postulated hydrosere is not available; (b) Documentary evidence, that certain areas now covered by patanas were once under woody vegetation (Holmes 1951, Perera 1967), suggests that the former has replaced the latter and not vice versa as would be expected from de Rosayro's postulation. It is certain that wild animals such as elephant, buffalo, sambhur and axis deer played a critical role, well before man, in the ecological differentiation of natural communities in Sri Lanka, and their feeding activities could have facilitated establishment of grass-cover at the expense of woody communities; (c) In certain hills, e.g. those near Hakgala (Fig. 2), patana and montae forest occur side by side without prominant ecotones between them to indicate successional trends; (d) This hypothesis cannot explain the origin of patanas lacking hygrophyllous

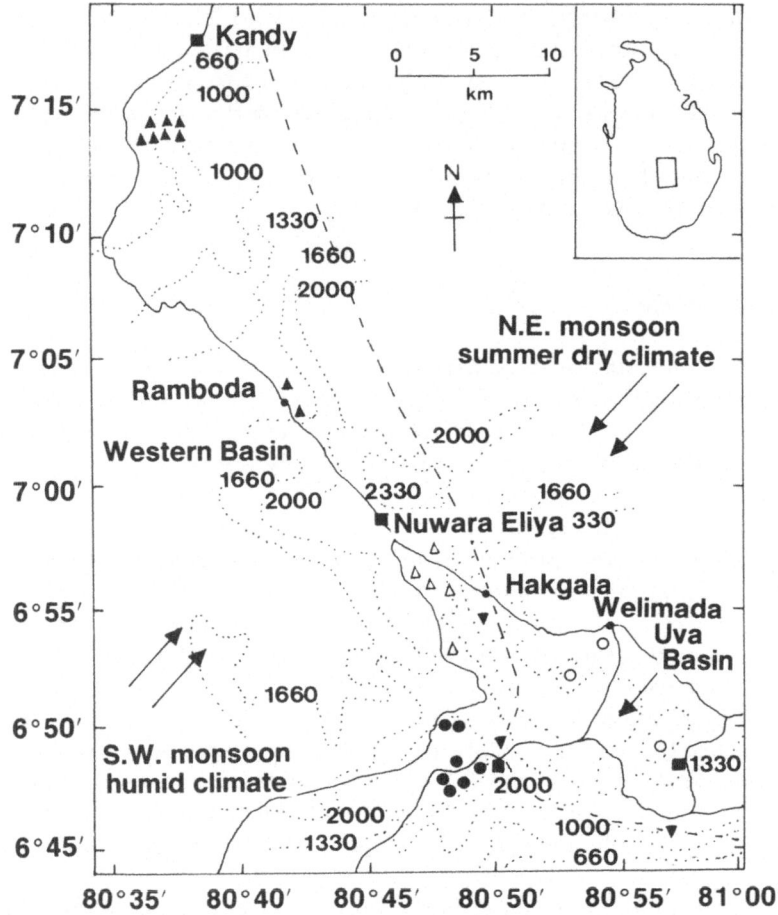

Fig. 2   Map of part of the south-central highlands of Sri Lanka indicating the approximate positions of twenty-seven sites sampled by Mueller-Dombois & Perera (1971) to represent the five patana types. ●, upper wet patana; △, lower wet patana; ▼, intermediate patana; ▲, humid zone dry patana; ○, summer-dry zone patana. Broken line separates humid zone from summer-dry zone. Contours are in metres. The position of the five stations for which climatological data are given in Fig. 5 also is shown (■). Inset: Sri Lanka showing the approximate location of the study area (from Pemadasa & Mueller-Dombois 1981).

sedges; in contrast, these sedges are common even in grass-covers of the dry-zone lowland, e.g. Ruhuna National Park (Mueller-Dombois & Perera 1971). Thus, their occurrence could be a reflection more of seasonal or periodic fluctuations of the soil moisture regimes than of an evolutionary relict of a former wet climate supporting swampy conditions.

Pearson's (1899) view that patanas are disclimaxes derived after fire-destruction of a savanna also is not satisfactory. The climatically- and biotically-controlled savannas occur in areas where the rainfall is much less, about 500 mm yr^{-1} (Walter 1964), than in some of the present patanas (e.g. those in the Western Basin (Figs. 1 and 2)). If patanas at >3,000 m altitude are disclimaxes, root remnants of woody plants must be present in the soil, but this is not so, at least in Horton Plains. Mueller-Dombois & Perera (1971) did not find charcoal in the soil of several patanas; however, if much of the soil was eroded, there is little chance of finding charcoal. It is unlikely that all the present patana-land was once under forest; pollen-analysis of soil may provide more decisive information.

It is hard to expect that the wide variety of patanas had a common origin. My own opinion is that some patanas are biotically-derived disclimaxes, while others are climatically-derived edaphic climaxes. Moreover, the considerable diversity within patanas (Pemadasa & Amarasinghe 1982a, Amarasinghe & Pemadasa 1982a) tempts one to suggest that in some of their habitats, such as those on slopes where soil dries out rapidly, establishment of grass-covers occurred following biotic destruction of former woody vegetation, while others, such as swampy depressions where woody plants are unlikely to flourish, supported edaphically-determined grass communities. Clearly, more careful exploration is necessary to elucidate the origin of different patanas.

Classification of patanas

Patanas are very diverse, yet most workers have divided them into only two (Pearson 1899, Holmes 1951) or three (de Rosayro 1945, 1946a, b, c, 1961; Perera 1967) general categories. Though Pearson and Holmes recognized two patana types, 'wet' and 'dry', their characterizing criteria were in sharp contrast. Pearson regarded patanas in the humid zone (e.g. Western Basin) as 'wet', because they receive a perennially high rainfall from both SW- and NE-monsoons, and those in the summer-dry zone (Uva Basin) as 'dry', because they experience a seasonal drought when the SW-monsoon acts as a desiccating wind; he set the altitudinal limit between them at 1,500 m. However, of the humid zone patanas, those around Kandy, Gampola, Pussellawa and Ramboda (Fig. 2) are altitudinally, edaphically and floristically quite different from the wet patanas in Nuwara Eliya, Sita Eliya and Hakgala, but resemble more the Uva Basin dry patanas; yet, the monthly distribution of rainfall is almost identical around Kandy and Nuwara Eliya with an annual total of 2,131 mm and 2,154 mm respectively (Fig. 5, see also Mueller-Dombois 1968). Evidently, Pearson's characterization of 'wet' and 'dry' patanas on the basis of rain climate is unsatisfactory.

Holmes (1951) recognized the importance of soil moisture in patana differentiation; he considered all grasslands between 500 m and 1,660 m altitude as 'dry' irrespective of their vast climatic diversity, because their soils dry out rapidly after rains, and those at higher altitudes with perennially moist soils as 'wet'. Thus, in his 'dry' patanas Holmes included a part of Pearson's 'wet' patanas. Moreover, there was some overlapping in their altitudinal limits to wet and dry patanas, indicating the difficulty of recognizing a distinct boundary and, therefore, the existence of an ecotone between them. de Rosayro resolved a part of the controversy by recognizing the ecotone between 1,500 m and 2,000 m altitude as a third type, an 'intermediate' patana, because it is floristically and ecologically somewhat intermediate between the other two, yet is topographically quite distinct for it occurs on steep slopes.

The altitudinal boundaries imply the possible significance of temperature variation on patana differentiation, as was emphasized by Perera (1967); he discarded the terms 'wet' and 'dry' as ambiguous, and renamed the three types as sub-montane (=dry), intermediate and upper montane (=wet).

Clearly, the main cause of the controversy is the differential emphasis on altitude, rain, climate, soil moisture or temperature as the sole factor responsible for patana differentiation, without considering their interactions. Moreover, these early classifications were very subjective.

In an attempt to rectify some of the discrepancies, Mueller-Dombois & Perera (1971) divided patanas into five macro-zones:
a) humid zone dry patana (mainly in the Western Basin),
b) summer-dry zone patana in the Uva Basin,
c) intermediate patana at 1,500-2,000 m altitude on the eastern escarpment,
d) lower wet patana at 2,000-2,330 m altitude, and
e) upper wet patana at >2,330 m altitude.

They based their classification on the individual and integrated effects of climatic, edaphic, topographic and altitudinal differences, and substantiated it by floristic comparisons, not only of higher plants but also of soil micro-fungi, based on So/renson's similarity coefficient and synthetic table technique (Mueller-Dombois & Ellenberg 1974). Thus, their approach was more objective and, therefore, more reasonable.

A subsequent more objective analysis of Mueller-Dombois & Perera's data (Pemadasa & Mueller-Dombois 1979, 1981) using reciprocal averaging ordination (Hill 1973) and association-analysis classification (Williams & Lambert 1959), two of the well-established multivariate techniques which reveal phytosociological interrelationships statistically, confirmed their tentative classification, and additional soil data substantiated its ecological justification. Some results of these analyses are outlined here to illustrate the affinities of the five patanas.

Despite their fundamental biological differences, higher plants and soil micro-fungi yielded remarkably similar results, so indicating their parallel ecological differentiation. The complete separation of wet patanas from the others along axis I of both ordinations (Fig. 3) and at the first division of association-analyses (Fig. 4) is convincing evidence of ecological justification of the separation of patanas into two major zones on the floristics not only of higher plants as repeatedly recognized by other works, but also, and more interestingly, of soil fungi; the grouping of intermediate patanas between wet and dry patana groups on axis I favours de Rosayro's recognition of this third type. The second axis separates wet patanas into lower and upper groups, and dry patanas into humid zone and summer-dry zone groups, and their distinctiveness is further evident from the association-analysis groupings. The floristic affinities of the five patanas were further tested by Sörenson's similarity coefficient (Greig-Smith 1964), which showed that two wet patanas are floristically more similar to each other than to the other three, which are themselves different from each other (Table 1). Thus, these statistical analyses have provided convincing evidence of the existence of at least five patana types in Sri Lanka; however, phytosociological heterogeneity within individual patanas is considerable, as is evident from the high degree of dispersion of sites within ordination-clusters (Fig. 3) and extreme fragmentation of sites in association-analyses (Fig. 4).

Ecology of five patanas

Geographic distribution

All five patanas occur in a radius of 15 km around Nuwara Eliya (Fig. 2). The summer-dry zone dry patana is confined to the Uva Basin, while the humid zone dry patanas are more scattered, the largest expanse being in the Western Basin (Kandy, Gampola, Pussellawa and Ramboda) with less extensive patches on the Knuckles ridge (NE of Kandy) and between Rakwana and Deniyaya (SSW of

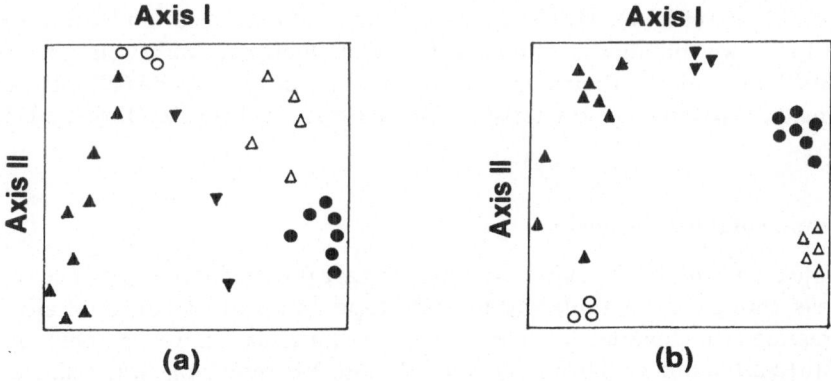

Fig. 3 Projection on the plane formed by the first two axes (I and II) of the reciprocal averaging ordination of the twenty-seven sites representing the five patana types of Sri Lanka, (a) and (b) represent ordinations based on higher plants and soil fungi respectively. Patana-type symbols as in Fig. 2 (from Pemadasa & Mueller-Dombois 1979).

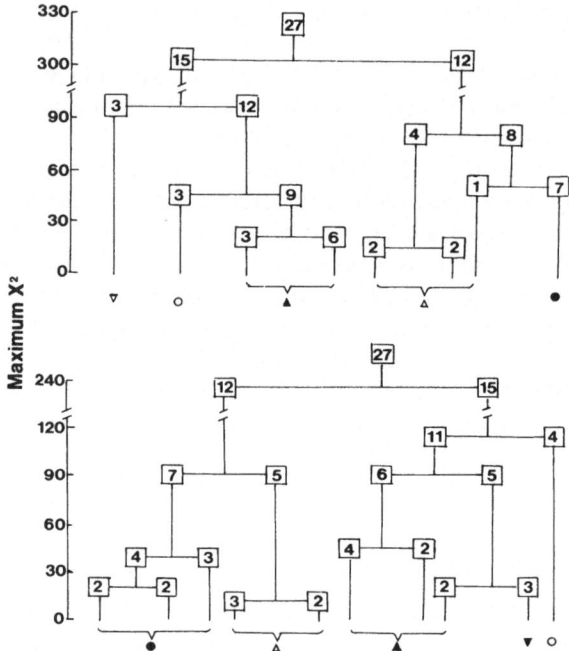

Fig. 4 Association analysis of twenty-seven sites showing the five condensed end-groups representing the five patana types of Sri Lanka. (a) and (b) represent analyses based on higher plants and soil fungi respectively. The numbers in boxes represent the number of sites at each division. Patana-type symbols as in Fig. 2 (from Pemadasa & Mueller-Dombois 1981).

Horton Plains) (Fig. 1). Much of the upper wet patanas occur on Horton Plains, while lower wet patanas are more widespread chiefly in Ambewela, Sita Eliya, Moon Plains and Elk Plains with a smaller tract in Madugoda (NE of Kandy). Intermediate patana is characteristic of steep slopes in Hakgala, Ohiya and Haputale.

Environmental heterogeneity

Climate diagrams for the major patana zones are illustrated in Fig. 5. The two dry patanas, though comparable altitudinally, topographically and edaphically, have contrasting rain climates. The humid zone dry patanas receive a perennially high rainfall ($>2,100$ mm yr^{-1}) from both the SW- and NE-monsoons with a single short dry spell in February. In contrast, summer-dry zone dry patanas receive a much lower rainfall ($1,450$-$1,750$ mm yr^{-1}) most of which occurs in October-November as fall-convectional-cyclonic thunderstorms, so that their rains are primarily inter-monsoonal and only secondarily N.E. monsoonal; they experience two dry peri-

Table 1. Sørenson's similarity quotients (%), calculated separately from higher plant and soil micro-fungal data, comparing five patana types (from Pemadasa & Mueller Dombois 1981).

Comparison	Higher plants	Micro-fungi
Upper wet and lower wet patana	81	88
Upper wet and intermediate patana	28	68
Upper wet and humid zone dry patana	15	62
Upper wet and summer-dry zone patana	12	50
Lower wet and intermediate patana	32	57
Lower wet and humid zone dry patana	32	75
Lower wet and summer-dry zone patana	18	72
Intermediate and humid zone dry patana	58	64
Intermediate and summer-dry zone patana	60	55
Humid zone and summer-dry zone patana	63	57

ods, a short spell in February and a longer more severe one in June-August (Mueller-Dombois 1968). Though similar in rainfall, the upper wet patana is 2 ° C cooler with more frequent mist and fog than the lower wet patana, a feature attributable to their altitudinal differences. The mean annual temperature in the

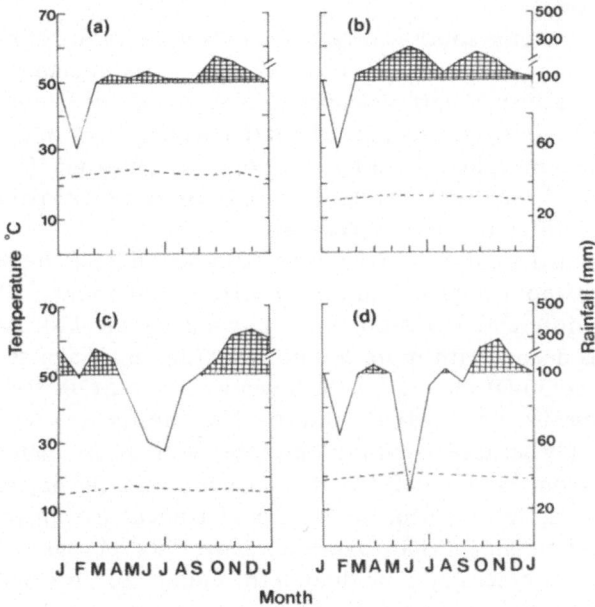

Fig. 5 Monthly variations in temperature (broken line) and rainfall (continuous line) in major patana zones. The data are averages for the period 1966-1975. Hatched areas indicate humid periods receiving a monthly rainfall in excess of 100 mm. (a) humid zone dry patana; (b) wet patana; (c) intermediate patana; (d) summer-dry zone dry patana (from Pemadasa & Mueller-Dombois 1981).

lower wet patana is 15.3 ° C with distinct diurnal and seasonal variations, the daily mean range being 7 ° C in February, 4 ° C in June-August, and 5 ° C during the rest of the year. In contrast, a mean day temperature of 13 ° C drops at night to >5 ° C in the upper wet patana; thus nocturnal ground frost is of common occurrence which could be a critical factor, as was stressed by de Rosayro (1945). The wet and humid zone dry patanas, though almost identical in rainfall pattern, differ altitudinally and edaphically, the former occurring above 2,000 m altitude and the latter below 1,660 m. These altitudinal differences are reflected in differential occurrence of cloud, mist, fog and frost, and in considerable temperature variations; the mean annual temperatures of 13-15.3 ° C in wet patanas are in sharp contrast to those in humid zone dry patanas where mean temperature varies from 18 ° C at 1,660 m to 24 ° C at 660 m altitude. The intermediate patana is somewhat intermediate between wet and dry patanas in altitude, edaphic conditions and mean annual temperature (18 ° C), but is topographically quite different from them being confined to steep slopes on the eastern escarpment. This characteristic topography has given it a unique rain climate; it lacks the February dry spell shared by other patanas because the NE-monsoon brings more than 100 mm rain to the eastern escarpment, and its dry period in June-August is less severe, though somewhat longer, than that in the summer-dry zone dry patana because it receives some rain from the SW-monsoon.

The parent rock in the south-central highlands is a mixture of khondalites and charnockites; being a mixture of metamorphosed sedimentary rocks, mainly quartzite, quartz schists, quartz-feldspar, gneisses and crystalline limestone, the former is more hgeterogeneous than the latter which constitutes both acid and basic rocks of metamorphosed sedimentary or volcanic origin (Cooray 1967, this volume). These rocks show no correlative zonal variations with patana differentiation; yet edaphic differences are considerable.

Both the wet and dry patanas have red-yellow podzolic soils, but the former with and the latter without a prominant A_1 horizon (Panabokke 1967), a feature atttributable to differential erosion. Though similarly blackish and organic, the topmost layer is deeper with more humic acids (James, Kalpage & Thenabadu 1963) in the upper (30-60 cm) than in the lower wet patana (20-50 cm); the amount of organic carbon also increases from intermediate through lower wet to upper wet patana, presumably because of reduced microbial activity associated with decrease in temperature from 18 ° C to 13 ° C. In sharp contrast, the dry patana soils are poor in organic matter, eroded, denuded and truncated with protruding rock outcrops; a highly compact, hardpan-like, quartz-gravel topmost layer is more prominant in the summer-dry zone than in the humid zone dry patana; the rough surface layer in the former shows little pedogenetic relation to the underlying rock rubble or finer soils and, therefore, appears to have resulted from lateral transport of quartz and gravel. The reddish to dark brown (air dry, 7.5 to 2.5 YR) dry patana soils in the humid zone are not very dissimilar to the grey to reddish brown (2.5 to 5 YR) soils in the summer-dry zone, but both these are distinctly different from the

Table 2. Edaphic characteristics of the five patana types (from Pemadasa & Mueller Dombois 1981).

	Upper wet patana	Lower wet patana	Intermediate patana	Humid zone dry patana	Summer-dry zone dry patana
pH	4.5-4.7	4.6-5.8	6.1-6.3	5.5-6.8	6.6-6.7
Percentage los-on-ignition	24-30	14-19	4-14	6-9	3-8
Percentage water-holding capacity	63-73	69-73	62-75	35-62	40-63
Potassium*	69-85	50-64	43-65	20-33	23-29
Sodium*	50-69	50-67	40-63	16-25	16-20
Calcium*	66-100	74-85	63-66	35-55	48-53
Nitrogen*	100-165	135-190	110-145	45-60	45-50
Phosphorus*	85-135	75-110	75-100	60-90	65-75

* Expressed as μg g^{-1} dry soil

dark to dark grey brown (air dry, 7.5 YR 3/2 to 2.5 YR 4/2) intermediate patana soils. This darkening of soil colour is not a result of change in amounts of organic carbon (Table 2) but its change in chemical nature with altitude; James *et al.* (1968) reported a predominance of humic acids over the yellowish or colourless fulvic acids in wet patana soils. The patana soils generally are poor in nutrients, but the concentrations of nitrogen, phosphorus and potassium are appreciably higher in wet patanas than in dry patanas, presumably because the degree of erosion is higher in the latter than in the former. The wet patana soils are slightly more acidic than dry patana soils. The water-retaining capacity is somewhat higher in wet than in dry patana soils, a feature related to differential amounts of organic matter.

Floristic heterogeneity

Although the five patanas have species in common, they differ in overall floristic composition, species-richness, species-predominance and physiognomy. The distribution of six of the chief higher plant species in different patanas is illustrated in Fig. 6. All six species occur in intermediate patanas. *Arundinella villosa*, though ubiquitous, is more abundant in wet than in dry patanas. Whereas *Chrysopogon zeylanicus* occurs in wet patanas but not in dry patanas, the reverse is true of *Cymbopogon nardus* and *Themeda tremula*. *Andropogon lividus* is very rare or absent in the humid zone dry patana. Several soil microfungi characteristic of different patanas also are known (Mueller-Dombois & Perera 1971, Pemadasa & Mueller-Dombois 1979, 1981). For example, *Circinella umbellata*, *Aspergillus candidus*, *Absidia blakesleana* and *Mucor genevensis* occur only in wet patanas, while *Absidia corymbosa*, *Penicillium brefeldianum*, *P. ehrlichii*, *P. javanicum*, *P. rugulosum* and *P. steckii* are more widespread in the humid zone for they extend from wet through intermediate to humid zone dry patanas. More restricted are

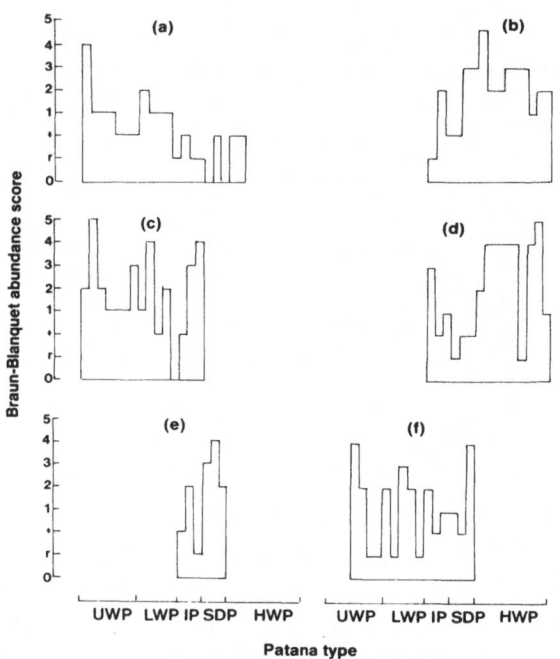

Fig. 6   The distribution of six of the chief higher plant species in different patanas. The histograms give the Braun-Blanquet cover-abundance scores in twenty-seven 10×10 m sites recorded by Mueller-Dombois & Perera (1971). (a) *Arundinella villosa*; (b) *Cymbopogon nardus*; (c) *Chrysopogon zeylanicus*; (d) *Themeda tremula*; (e) *Chrysopogon aciculatus* and (f) *Andropogon lividus*. UWP, LWP, IP, SDP and HDP represent upper wet, lower wet, intermediate, summer-dry zone and humid zone dry patana respectively (from Pemadasa & Mueller-Dombois 1981).

Aspergillus fumigatus, Helicostyllum pyriforme and *Penicilliumcitrinum*, which occur only in wet and humid zone dry patanas. The most restricted are *Aspergillis flavus* and *Circinella mucoides* which are found only in the intermediate patanas. Occuring in both dry patanas are *Absidia cylindrospora* and *Mucor microspora*, while *Penicillium herguei, P. wortmanii, Rhizopus nigricans* and *Sordaria fimicola* are confined to humid zone dry patanas. Though the habitat partitioning of soil micro-fungi is interesting, its ecological significance in relation to patana differentiation is not known, and merits investigation.

The humid zone dry patanas constitute tall grass (up to 1.5 m high) communities dominated by the dense-standing bunch-grass *Cymbopogon nardus* at lower altitudes and by the tussock-forming palatable grass *Themeda tremula* at higher altitudes, with the two species co-dominating at intermediate levels. The matrix-filling common associates include other grasses (e.g. *Dimeria gracilis, Eulalia trispicata, Imperata cyclindrica* and *Sacciolepsis indica*), tall composite herbs (e.g. *Blumea flexuosa, B. hieracifolia, Vernonia cinerea, V. setigera* and *V. zeylanica*), semi-woody shrubs (e.g. *Cassia mimosoides, C. kleini* and *Crotalaria calycina*) and ground-appressed herbs (e.g. *Desmodium heterocarpum, D. heterophyllum, D. triflorum, D. triquetrum* and *Oxalis corniculata*). The more important of the woody shrubs are *Lantana camara, Osbeckia octandra, Psidium guajava, P. guinense* and *Wikstroemia indica*. The predominant grasses, particularly *Cymbopogon nardus*, undergo a resting period from late-January to early-March when the aerial parts, especially older tillers and leaves, become yellow and dry, in response, presumably, to the annual dry spell in February. *Cymbopogon nardus, Eulalia trispicata* and *Themedia tremula* flower and set seeds during November-January, but much of their germination occurs in March-April synchronous with monsoonal rains (Pemadasa & Amarasinghe 1982b).

In contrast, the summer-dry zone dry patana is a heavily-grazed, closely-appressed, short-turf; composed predominantly of creeping perennial grasses such as *Alloteropsis cimicina, Bracharia distachya, Chrysopogon aciculatus, Digitaria longifolia, Eragrostiella secunda* and *Pseudanthistiria umbellata*, a short rosette-forming sedge, *Fimbristylis nigrobrunnea*, and small herbs such as *Elephantopus scaber, Evolvulus alsinoides* and *Leucas zeylanica*. Both *Cymbopogon nardus* and *Themeda tremula* are localized in places where grazing-pressure and human interference are less severe. Of the scattered woody plants, *Glochidion montanum, Microglossa zeylanica, Knoxia platycarpa* and *Vernonia wightiana* occur as shrubs and *Atylosia rugosa* as a creeper. In relatively moist hollows *Drosera* sp. is common, an indication of nitrogen deficiency.

There is some evidence that the predominant grass *Chrysopogon aciculatus* forms hummocks which follow the cycle of phases similar to that described by Watt (1947), Kershaw (1975) and Pemadasa (1981b) in several temperate and tropical communities. The progressive development of *Chrysopogon* clumps with upright aerial tillers and prostrate creeping shoots is paralleled by continued accumulation of soil particles resulting from erosion and carried by ants so forming hummocks of height not more than 10 cm. The growth of upright tillers maintain the depth of the hummock, and radiating creepers continue to expand the clump; the general stability of the hummock is ensured by the shelter provided by *Chrysopogon* shoots which restrict erosion. The creepers are progressively older towards the centre of the clump, and there is correlative reduction of density and growth-vigour of aerial shoots; the tillers die after flowering. The growth-vigour of plants increases with age to a maximum and then declines (Kershaw 1975). Thus, ageing prostrate shoots become less vigorous and eventually die with associated mortality of aerial tillers and reduced production of new ones, so that the centre of older hummocks becomes open enabling other species such as *Eragrostiella secunda, Evolvulus alsinoides* and *Leucas zeylanica* to colonize areas which they otherwise are unable to invade because of the intense competition of *Chrysopogon*. The death of the complete *Chrysopogon* clump eventually leads to erosion of the hummock, because other species cannot provide sufficient shelter. Accordingly, certain floristic changes in this patana appear to be related to the phasic development of the predominant *Chrysopogon aciculatus*.

The predominant grasses in the summer-dry zone dry patana undergo a short spell of resting in February, but a longer one in June-August synchronous with the more severe dry period when aerial parts become senescent and parched.

The two wet patanas are floristically not very dissimilar and are dominated by *Arundinella villosa* and *Chrysopogon zeylanicus*, with matrix-filling *Anaphalis brevifolia, A. subdecurrens, Centella asiatica, Cyanotis pilosa, Heracleum zeylanicum, Pedicularis zeylanicum, Pleiocraterum plantaginifolium* and *Satyrium neplense*, and scattered woody species such as *Hypericum mysorense, Osbeckia cupularis* and *Rhondodendron zeylanicum*. The most important differentiating species are *Ischaemum indicum, Justicia procumbens* and *Pseudanthistiria umbellata* which are absent in the upper wet patana. The grasses *Dichanthium polypticum* and *Tripogon bromoides*, the composite herb *Senecio gracilis* and the short woody shrub *Gaultheria rudis* occur mainly in the upper wet patana, while the short bamboo grass *Chimonobambusa densifolia* is restricted primarily to Horton Plains. Occurring in both wet patanas are several temperate species such as *Plantago lanceolata, P. major, P. minor, Leontodon* sp. and *Taraxacum* sp.

Physiognomically the two wet patanas are distinct. The cattle-grazed, stubble-tussock, short-turf (15-20 cm) of lower wet patana is co-dominated by *Arudinella villosa* and *Chrysopogon zeylanicus*. In contrast, in the upper wet patana, the vigorously growing, large *Chrysopogon* tussocks are more prominant and abundant than *Arundinella* clumps so giving its characteristic hummock-and-hollow physiognomy. These differences undoubtedly are related to differential grazing pressure.

Being co-dominated by *Arundinella villosa, Chrysopogon zeylanicus* and *Cymbopogon nardus* with *Themeda tremula* as a common associate and woody species such as *Knoxia platycarpa, Microglossa zeylanica, Rhondodendron zeylanicum* and *Vernonia wightiana*, the intermediate patana is floristically and physiognomically somewhat intermediate between wet and dry patanas.

Heterogeneity within individual patanas

The five macro-patana zones are ecologically distinct, but each shows considerable altitudinal, topographic, climatic and edaphic diversity which is manifested by the occurrence of floristically and phytosociologically different communities. A knowledge of the within-patana differentiation also is necessary to understand their ecology more fully and for their management and conservation.

An intensive study of the humid zone dry patana on the Hantane range near Kandy demonstrated its ecological diversity (Pemadasa & Amarasinghe 1982a, Amarasinghe & Pemadasa 1982a). It constitutes a series of hills of varying altitude (500-1,150 m), of which five hills (P-T on Fig. 7), selected to cover the entire altitudinal range, were sampled and the floristic data were ordinated by reciprocal averaging ordination (Hill 1973). Along axis I, sites from lower altitudes are

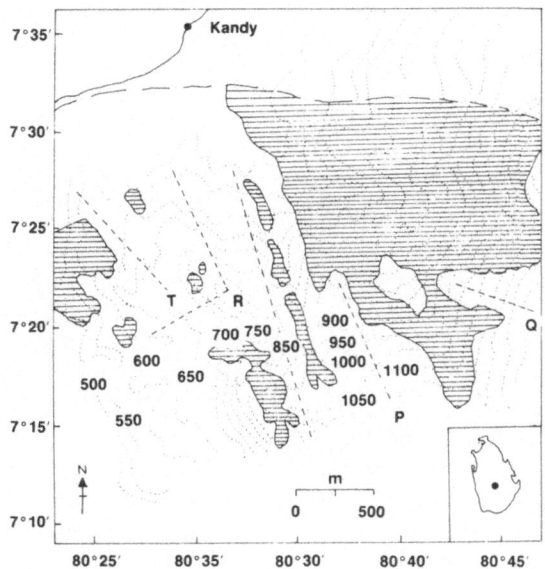

Fig. 7 Hantane range of the south-central highlands of Sri Lanka showing the position of transects (- - -) along which the five hills (P-T) were sampled by 2×2 m stands located at 15 m altitude intervals, contours (· · · · ·) with altitude (m), forest (hatched) and humid zone dry patana (unfilled). The nearest town (Kandy), the main road to Colombo (———) and the northern boundary of the Hantane range (-.-.-) also are indicated. Inset: Sri Lanka showing the position of the Hantane range (from Pemadasa & Amarasinghe 1982a).

separated from those from higher altitudes with intermediate ones in between (Fig. 8). Axis II separates hill P from Q at higher altitudes and hill T from R and S at other altitudes. A gradient of altitude and of soil particle-size occurs on axis I; the proportion of particles >0.8 mm is higher at higher than at lower altitudes, a feature related to the grading of particles by erosion. The loss-on-ignition and water-retaining capacity of soil are higher at higher altitudes, where the top layer is humus rich, than at lower altitudes where the soil is more eroded. *Cymbopogon nardus* is much less abundant at higher than at lower altitudes, while the reverse is true of *Themeda tremula* (Fig. 9). *Panicum maximum* and the weeds *Erigon sumatrensis* and *Pennisetum polystachyon* occur largely in disturbed habitats at lower altitudes.

The five hills are phytosociologically distinct, but each hill is still heterogeneous. The ordination results for hill R are illustrated as an example (Fig. 10). Axis I reflects and altitudinal gradient, while axis II separates northern from southern aspect. *Pennisetum polystachyon* is predominant at lower altitudes where herbage-removal is frequent; *Themeda tremula*, though predominant at higher altitudes, occurs more on the northern than on the southern aspect. Other hills also showed

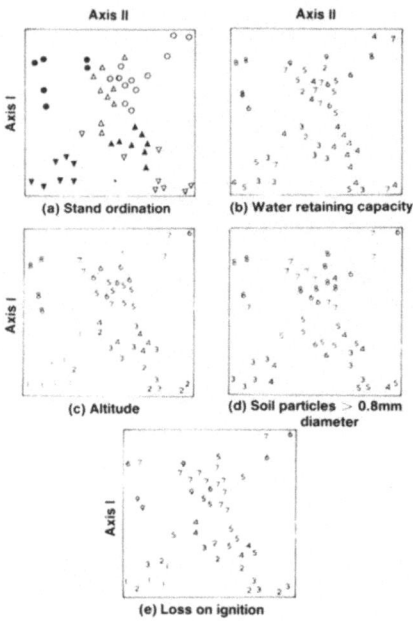

Axis II   **Axis II**

(a) Stand ordination   (b) Water retaining capacity

(c) Altitude   (d) Soil particles > 0.8mm diameter

(e) Loss on ignition

Fig. 8. (a) Projection on the plane formed by the first two axes (I and II) of reciprocal averaging ordination of fifty stands, selected at 45 m altitudinal intervals, from the five hills of the Hantane range. Symbols: ▲, stands from the 670-760 m altitudinal range of hills R and S; △, stands from steep slopes of hill S at 790-900 m altitudinal range; ●, stands from hill Q; ○, stands from hill P; △, stands from gentle slopes of hills R and S; △, stands from steep slopes of hill T; (b), (c), (d) and (e) represent the distribution on the stand ordination of water-retaining capacity, altitude (m), soil particles >0.8 mm diameter and los-on-ignition respectively. The environmental data are scored according to the following ranking method:

	Ranking score								
	1	2	3	4	5	6	7	8	9
Altitude (m)	550-610	611-670	671-730	731-790	791-850	851-910	911-970	971-1030	1031-1100
Soil particles >0.8 mm diameter (%)	0-10	11-20	21-30	31-40	41-50	51-60	61-70	71-80	81-90
Loss-on-ignition (%)	3.0-4.5	4.6-6.0	6.1-7.5	7.6-9.0	9.1-10.5	10.6-12.0	12.1-13.5	13.6-15.0	15.1-16.5
Water-retaining capacity	15-20	21-25	26-30	31-35	36-40	41-45	46-50	51-55	56-60

(From Pamadasa & Armasinghe 1982a)

similar phytosociological heterogeneity (for details see Pemadasa & Amarasinghe 1982a).

Evidently, most vegetational variation within Hantane range is related to altitude. The air temperature is about 2 °C less with greater diurnal and seasonal

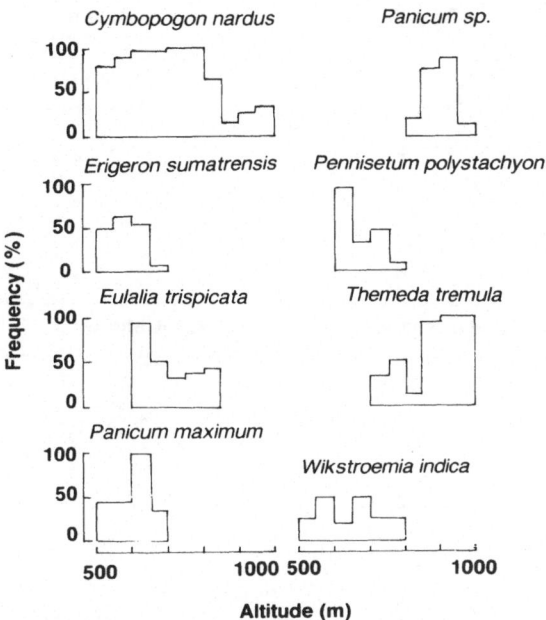

Fig. 9. The distribution of eight common species in the Hantane range in relation to altitude. Histograms give the percentage shoot frequency classes calculated from 164 2 × 2 m stands located at 15 m altitudinal intervals along transects on five hills (P-T) shown in Fig. 7 (from Pemadasa & Armasinghe 1982a).

fluctuations, more frequent mist, fog and cloud and less evaporation at higher than at lower altitudes, and the humus-rich black soil in the former contrasts sharply with the truncated soils in the latter. Thus, edaphically and microclimatically, the higher altitudes of the Hantane approximate to wet patanas. The occurrence of *Anaphalis brevifolia, A. subdecurrens* and *Polygala rosmarinifolia*, some of the species commonly found in wet patanas, is further evidence of this. The within-hill heterogeneity also is related partly to altitude, but mainly to topography, aspect and, especially, the degree of human interference. Selective herbage-removal for cattle fodder is most widespread at lower altitudes. The continued destruction of grass cover increases erosion and soil impoverishment, changes the soil's mechanical composition and water relations, and favours colonization by weeds; these changes are most severe on steep slopes. Moreover, selective forage-removal favours some species as a result of the elimination of others; for example, the non-palatable *Cymbopogon nardus* increases as the fodder grass *Themeda tremula* decreases, and this may partly explain their changing abundance with altitude (Fig. 9). The continued colonization of lower hills by *Pennisetum polystchyon* also is a consequence of destruction of natural turf which enables this species to flourish by suppressing the native species.

116

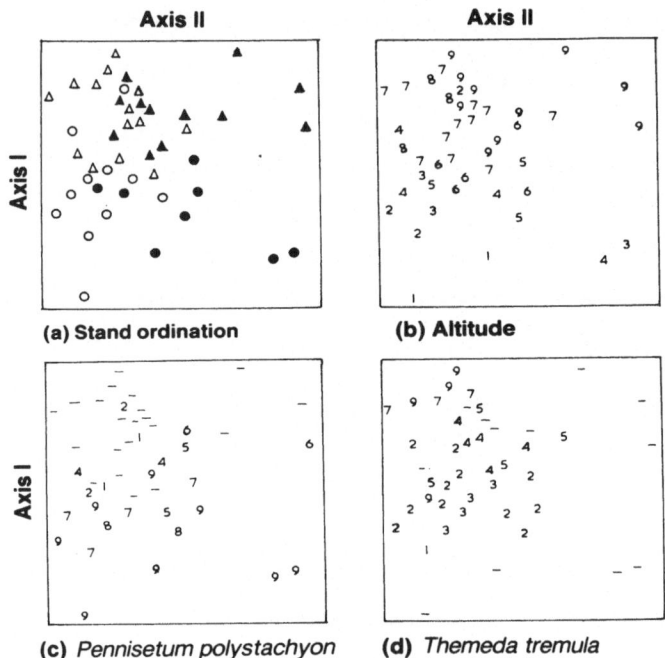

(a) Stand ordination

(b) Altitude

(c) *Pennisetum polystachyon*

(d) *Themeda tremula*

Fig. 10. (a) Projection on the plane formed by first two axes (I and II) of reciprocal averaging ordination of forty-five stands from hill R. △, ▲, stands from >750 m altitude; ○, ●, stands from ×750 m altitude; unfilled and filled symbols represent stands from northern and southern aspects respectively; (b) distribution of altitude in 25 m class on stand ordination (altitude ranges: 1 = 626-650; 2 = 651-675; 3 = 676-700; 4 = 701-725; 5 = 726-750; 6 = 751-775; 7 = 776-800; 8 = 801-825; 9 = 826-850); (c) and (d) abundance patterns of *Pennisetum polystachyon* and *Themeda tremula* on stand ordination. Abundance data for individual species are expressed as classes of frequency (%):- =absent; 1 = 1-10; 2 = 11-20; 3 = 21-30; 4 = 31-40; 5 = 41-50; 6 = 51-60; 7 = 61-70; 8 = 71-80; 9 = 81-90; 10 = 91-100 (from Pemadasa & Armasinghe 1982a).

The humid zone dry patanas constitute a variety of communities in addition to the typical patana covers. The lower slopes near streams, where the soil is wetter than usual, support a turf of luxuriantly growing *Axonopus compressus* and *Cynodon dactylon* with scattered *Cympogon nardus, S. urtifolia, Fimbristylis* sp., *Alocasia* sp. and *Commelina* sp. On steep slopes, where the soil is shallow, excessively and rapidly drained and highly truncated with protruding rock out-crops, patches of *Dicranopteris linearis* occur with occasional *Psidium guajava* and *Wikstroemia indica* shrubs and very few ground-appressed herbs. Valleys between hills support forest, and ecotones between them and patana have grass covers with sporadic trees and shrubs. The north-to-east slopes on the lee side relative to the SW-monsoonal winds simulate conditions and communities in the intermediate patana. Clearly, each humid zone dry patana is a complex of plant communities.

The summer-dry zone dry patana is somewhat less heterogeneous, yet several deviations from the normal pattern cover can be seen. For example, on lower seepage slopes shrubs (e.g. *Psidium* spp.) and ferns (e.g. *Dicranopteris linearis* and *Pteridium aquilinum*) predominate with grasses as subordinates. Relicts of perennially marshy, swampy depressions are dominated by sedges, especially *Fimbristylis nigrobrunnea*. The soils are more blackish with more organic matter in these habitats than in the patana. Unlike in the humid zone dry patana, steep slopes are absent, but some knolls have shallow soils with few protruding rock outcrops where the grass cover is less vigorous with more *Dicranopteris linearis* than elsewhere.

The slopes are more gentle and the elevational differences between knolls and depression are much smaller in wet patanas than in the summer-dry zone dry patana. The depressions are perennially marshy and give a unique bog-like character to wet patanas; frequently occurring are *Eriocaulon* sp. and *Juncus* sp. The ecotonal knolls in depressions are less wet and shallower with more wet patana species and isolated *Eriocaulon* and *Juncus* clumps; these are common particularly in Sita Eliya and Nuwara Eliya. On Horton Plains, the higher knolls and slopes are better drained and drier than normal, and are occupied predominantly by *Arundinella villosa* which can tolerate relatively dry conditions as indicated by its occurrence even in some parts of the summer-dry zone dry patana.

Evidently, each macro-patana zone constitutes a mosaic of floristically, physiognomically, ecologically and microclimatically distinct plant communities which occupy easily recognizable micro-habitats. More intensive exploration of habitat-partitioning and zonation of these communities is likely to indicate successional trends which may be useful in understanding the general ecology, the differentiation, and, perhaps, origin of patanas.

Savannas

Much of the grasslands occurring north and east of the Uva Basin at 300-500 m altitudes may be regarded as savannas, because they are characterized by a prominant tree component; trees play an ecologically more important role in savannas than in patanas and lowland grasslands. Savannas are presumed to have resulted from fire-destruction of dry mixed evergreen forest which is supposed to be the climax vegetation in the dry zone (de Rosayra 1950, Holmes 1951); the frequent occurrence of fire-toleranttree species such as *Careya arborea*, *Phyllanthus emblica*, *Terminalia belerica* and *T. chebula* favours this presumption. Accordingly, they may be regarded as disclimaxes maintained in their present state by frequent fires which enable establishment of grasses and fire-tolerant trees at the expense of forest species more vulnerable to burning. In certain areas, savannas merge into patanas at higher (>500 m) altitudes where the grass cover is short with fire-tolerant trees occurring sporadically, and into dry mixed evergreen forest at other altitudes. de Rosayro (1950) and Holmes (1951) presented extensive descriptive accounts of floristics and ecology of savannas.

Climatically savannas approximate to the Uva Basin dry patanas, but they receive a somewhat higher annual rainfall (1,450-2,000 mm) with a single distinct dry spell in June-August. On the basis of the topographic, edaphic, floristic and physiognomic differences, two general categories of savanna can be recognized; the so-called upland and lowland savannas.

Upland savannas occur on hillsides where soil is eroded and denuded as a result of frequent destruction of the vegetation; however, they are somewhat less truncated with sporadic rock outcrops than the dry patanas, and the prominant A_1 horizon is reddish to dark brown with low proportions of gravel and quartz. The tall (to 1.0 m high) grass cover is predominantly of *Cymbopogon polyneuros* and *Themeda triandra* with occasional clumps of *Cymbopogon nardus* and *Themeda tremula*. Between these predominant grasses occur other grasses (e.g. *Panicum maximum, Panicum* sp. and *Heteropogon triticeus*), woody shrubs (e.g. *Lantana camara* and *Osbeckia octandra*), semi-woody shrubs (e.g. *Cassia mimosoides*) and ground-appressed herbs (e.g. *Desmodium* spp., *Elephantopus scaber, Evolvulus alsinoides* and *Exacum trinerve*). Fire-tolerant tree species such as *Careya arborea, Phyllanthus emblica, Terminalia belerica* and *T. chebula* constitute a major component of the vegetation, particularly on the lower slopes between hillsides. Recently disturbed habitats are characterized by the luxuriantly growing *Imperata cylindrica* with *Cymbopogon polyneuros* and *Themeda triandra* occurring as subordinates and isolated herbs (e.g. *Elephantopus scaber* and *Exacum trinerve*). Weed species such as *Hedyotis* sp., *Mimosa pudica, Tridax procumbens* and *Vernonia cinerea* occur more abundantly in these habitats than elsewhere, and the soil is more eroded and denuded with a greater proportion of gravel and sand than fine particles. A typical tract of upland savanna occurs in Pethiyagoda (NW of Bibile).

In contrast, the lowland savannas occur in valleys and plains near hillsides, particularly in Bibile and Moneragala. The soils largely are alluvial with a more prominent, dark brown A_1 horizon containing a higher proportion of fine particles in these savannas than in the upland ones. Protruding rock outcrops occur only sporadically. The grass-cover is much taller (to 1.5 m high) and is dominated by *Aristida setacea, Cymbopogon polyneuros* and *Panicum* sp. with *Cymbopogon nardus, Themeda tremula* and *T. triandra* occurring sporadically. Common herbs include *Desmodium* spp., *Elephantopus scaber, Evolvulus alsinoides* and *Exacum trinerve*, but they are of low growth-vigour, presumably because of intense competitive pressure of the almost complete, thick grass cover. Fire-tolerant trees are rather sporadic.

The floristic and physiognomic differences between the upland and lowland savannas are undoubtedly related to edaphic differences. The more eroded residual soils of the former are comparatively poor in nutrients, water-retaining capacity and organic matter with a greater proportion of coarse particles than the alluvial soils of the latter. Such subtle differences are exaggerated if the soil tends to be dry and nutrient-poor, as in the dry zone. It is likely that the degree of human interference was more in the lowland valleys and plains than in the upland hillsides, and this may be one of the reasons for the rarity of trees in the former; the more frequent occurrence of root-remnants and fire-destroyed logs in the lower savannas than in the upland ones supports this view.

In both types of savannas, the predominant grasses undergo a resting period during June-August synchronous with the annual dry spell when leaves and older tillers become yellow, senescent, and sometimes even parched; during this time wilfully set fires are frequent. The fruits of *Phyllanthus emblica, Terminalia belerica* and *T. chebula* are used widely in Ayurvedic medicine, and are, therefore, economically valuable. The grass covers around these trees are cleared periodically in order to enable easy collection of fruits, and this is likely to play an important role in ecological differentiation within savannas.

Very little quantitative information is available regarding the floristics, phytosociology, dynamics and productivity of savannas and the factors responsible for their ecological differentiation; these aspects merit intensive investigation, for savannas are ecologically extremely fascinating.

Lowland grasslands

The lowland (<300 m in altitude) surrounding the central highlands and covering about four-fifths of Sri Lanka is climatically highly variable, and, therefore, is characterized by a wide variety of vegetation types of which grasslands are of particular importance for their potential as pastures. Included in lowland grasslands are a multitude of grass-dominated communities which provide forage for animal fodder and grazing-land; accordingly, they may be collectively termed pastures, though some are not useful as pastures. On the basis of climatic differences, four major zones of pasture can be recognized: wet-, intermediate-, dry-, and arid-zone pastures. Each zone, though climatically distinct, comprises a mosaic of edaphically and biotically different habitats supporting floristically different grass covers.

Wet-zone pastures

Occurring extensively in the south-western lowlands, the wet-zone pastures are characterized by a perennially high rainfall of over 2,300 mm yr^{-1} which is bimodally distributed with maxima in April-May and October-November due to SW and NE monsoons (Fig. 11a); even the driest month, February, has more than 100 mm

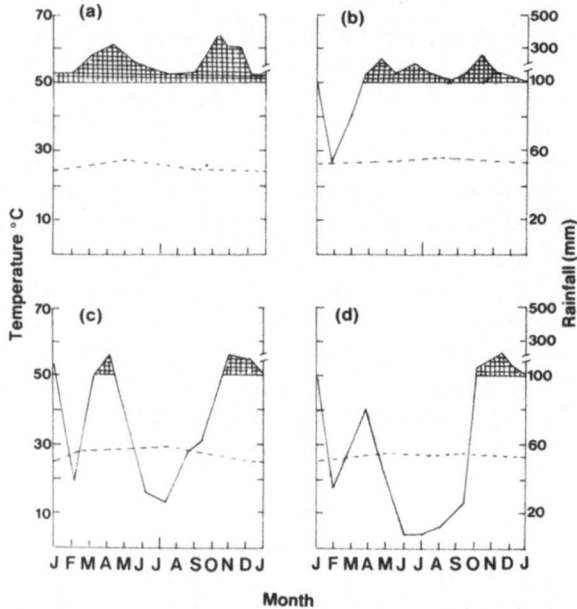

Fig. 11. Monthly variations in temperature (broken line) and rainfall (continuous line) in the four lowland pasture zones. The data represent averages for the period 1966-1975. Hatched areas indicate humid periods receiving monthly rainfalls in excess of 100 mm. (a) wet zone pasture; (b) intermediate zone pasture; (c) dry zone pasture; (d) arid zone pasture.

rainfall. Thus, the wet-zone pastures are the wettest grasslands in Sri Lanka. The mean air temperature of about 27 °C shows very little diurnal and seasonal fluctuations.

Much of the wet-zone pastures occurs under coconut cultivation, except certain maritime wild grass covers which differ from them edaphically and floristically. The wet-zone lowland is believed to have once been under forest and/or maritime vegetation which subsequently has been cleared to introduce permanent crops. Especially under coconut much of the ground is bare and, therefore, provides suitable niches for grasses because continued maintenance of the crop prevents re-establishment of other wild communities. Thus, wet-zone pastures are presumed to have arisen in this way.

The maritime pastures, dominated by *Cynodon dactylon* with occasional clumps of *Axonopus compressus* and *Chrysopogon aciculatus*, are generally of low vigour and productivity, presumably because of the low moisture and nutrient conditions of the sandy soil and frequent salt spray. Here the turf is short and sparse. Even during a relatively short dry spell of 2-3 weeks, the turf becomes dry and parched. *Crotalaria verrucisa, Ocimum* sp., *Stachytarpheta indica* and *S. urticifolia* occur as occasional shrubs. This grass cover is of limited value as pasture.

Pastures under coconut cultivation are the most widespread. The short turf (to 15 cm high) is predominantly of the carpet-grass *Axonopus compressus* and the love-grass *Chrysopogon aciculatus*; both these are low, creeping, sward-forming perennials producing animal fodder, but the former is more palatable than the latter. The soil is rather sandy, but moister and more fertile than the soils of maritime pastures. In alluvial soils are communities dominated by the perennial creeping grasses *Bracharia miliiformis* and *B. subquadripara* which are very palatable and, therefore, most valuable pasture species. Occurring in exposed inland habitats with less fertile soils are extensive communities of *Ischaemum timorense*, a creeping grass of low productivity. On lateritic and rocky soils *Apluda aristata* forms a short turf (10-15 cm high) of low vigour. These last two communities are of limited use as pasture.

Herbs common to all these communities include *Desmodium heterophyllum, D. triflorum* and *Alysicarpus vaginalis*, but they are more abundant in lateritic soils and road banks. The species of more herbage value are *Asystasia gangetica, A. variabilis, Ipomoea angustifolia* and *Ruellia ringens*. Frequent grazing and forage removal create suitable niches for many weed species of which most widespread are *Acanthospermum hispidum, Eleutheranthera ruderalis, Euphorbia hirta, Mimosa pudica, Synedrella nodiflora, Tridax procumbens* and *Vernonia cinerea*. Much of the grass covers under neglected coconut cultivation has now been replaced by thick populations of the troublesome weed *Eupatorium odoratum*.

Unlike in patanas, the predominant grasses in the wet-zone pastures do not have a prominant resting period presumably because the rainfall is perennially high; most of them flower, set seeds and germinate throughout the year.

The floristic heterogeneity and consequent ecological differentiation within wet-zone pastures may be related chiefly to localized differences in edaphic, particularly moisture and nutrient, conditions.

Intermediate-zone pastures

The intermediate-zone pastures are widespread in the northwestern lowlands (<200 m altitude), mostly under coconut cultivation. This region must originally have been under intermediate evergreen forest (de Rosayro 1950) most of which was cleared for agricultural purposes. Much of the present pastures are presumed to have established after coconut cultivation, as a result of continued maintenance of the crop which prevented re-establishment of natural communities; thus they are regarded as disclimaxes maintained in the present state by biotic interference such as forage removal for animal fodder and grazing by cattle. The only notable exceptions are wild grass covers occurring in habitats wetter or drier than usual, which may be considered as edaphic climaxes.

Climatically, intermediate-zone pastures are somewhat intermediate between wet-zone and dry-zone pastures. A mean annual rainfall of over 2,000 mm is bimodally distributed (Fig. 11b) as in the wet-zone pastures, but intermediate-zone pastures also approximate to dry-zone ones because they share a common dry spell in February. This intermediate climate is manifested by the occurrence of several grass species characteristic of wet and dry-zone pasture, e.g. *Aristida setacea* and *Axonopus compressus.*

The short-turf (20-30 cm) is dominated by *Aristida setacea, Axonopus compresses, Chrysopogon aciculatus, Cyrtococcum trigonum* and *Digitaria* sp. and frequently is grazed by cattle. Soils are rather sandy with a prominant A_1 horizon, and are more fertile than some of the dry-zone soils. Disturbed areas are characterized by isolated populations of *Imperata cylindrica* and *Themeda triandra.* Occasionally sporadic clumps of *Cymbopogon nardus* and *Themeda tremula* occur. *Panicum maximum* grows luxuriantly in relatively dry habitats. In more fertile soils are communities of *Bracharia mutica.* Shrubs and herbs common to all these grass covers include *Achyranthes aspera, Alysicarpus vaginalis, Cassia occidentalis, Ocimum* sp., *Stachytarpheta indica* and *S. urticifolia.* Frequent biotic interference creates favourable microsites for weed species such as *Eleutherantha ruderalis, Euphorbia hirta, Mikania scandens, Mimosa pudica, Tridax procumbens* and *Vernonia cinerea.*

All the predominant grasses provide forage for animal-fodder. They flower, set seeds and germinate throughout the year, and show a brief spell of resting during the February dry period when leaves and old tillers become yellow and senescent.

Dry-zone pastures

The dry-zone pastures, occurring in the northern and northeastern plains, are the most extensive and diverse of all lowland grasslands. Climatically, they approximate to the Uva Basin patanas, with a mean annual rainfall of about 1,400 mm which is seasonal and NE-monsoonal. The February dry spell is shorter and less severe than the dry period in June-September (Fig. 11c).

According to Senaratna (1956), the dry-zone grasslands have resulted in one of two ways: (a) tank-border grasslands specifically developed for pasturage of animals used for cultivation and other purposes, and (b) grass covers established in land abandoned after shifting cultivation; thus the former are man-made pastures while the latter are natural but resulting from human destruction of woody vegetation. However, this postulation cannot explain the origin of some of the inland pastures, e.g. villu grasslands characteristic of perenially moist depressions, which must be edaphically-determined climaxes. Before man started destroying his environment, much of the dry zone must have been under forest and/or thorn scrub. The great density of irrigation tanks (Abeywickrama 1955, Fernando & de Silva, this volume) is clear evidence that the dry zone was highly populous and under intensive cultivation in the past. It is very likely that man developed pastures on tank borders to serve a dual purpose: to protect the tank borders from erosion and to provide pasturage for animals. Historical evidence indicates that most of these tanks subsequently were abandoned and neglected for centuries (Fernando & de Silva this volume). This should have allowed the natural succession to progress towards forest and/or thorn scrub. The continued existence of grass covers is indicative of frequent human interference which led to a deflected succession culminating in grasslands. Similarly, shifting cultivation, where the land is cleared, burned, cultivated for several seasons and then abandoned, would have lead to the establishment of grass covers. Accordingly, I believe that much of the dry zone pastures are disclimaxes resulting from continued human interference; the possible exceptions are edaphically determined villus and similar pastures.

Holmes (1951) described three major types of dry zone grasslands, the so-called damanas, talawas and villus, and Senaratna (1956) supplemented these with a variety of other grass-dominated communities specific to certain habitats which are edaphically different from the normal dry zone conditions.

Extensive patches of damanas occur in Amparai and Batticaloa districts. The dense, tufted, coarse grass cover is dominated by *Cymbopogon nardus* with sporadic clumps of *Aristida setacea, Imperata cylindrica, Themeda tremula* and *T. triandra*. Recently-disturbed habitats are characterized by thick populations of *Imperata cylindrica* with occasional tussocks of *Cymbopogon nardus* and *Themeda tremula*. Subordinate grasses common to both communities include *Bracharia distachya, B. repatans, Cynodon dactylon, Echinochloa colonum, Eragrostis tenella, E. viscosa* and *Sporobolus diandrus* . Occurring particularly in the *Cymbopogon* -dominated community are ground-appressed herbs, semiwoody shrubs and sporadic trees such as *Acanthospermum hispidum, Alysicarpus vaginalis, Asparagus racemosus, Calotropis gigantea, Cassia auriculata, Desmodium heterophyllum, D. triflorum, Lantana camara* and *Pterocarpus marsupium*; the latex of the last species is used in Ayurvedic medicine for diabetes. When compared with the neighbouring forest soils, the damana soils have lower carbon and nitrogen contents and higher Mg/Ca ratios (Holmes 1951), features attributable to denudation resulting from frequent destruction of ground vegetation. Wild elephants and buffalos are two of the most decisive biotic factors controlling the ecology of damanas. During the dry season the soils bake hard and crack, and the predominant grasses become yellow, senescent and very often parched.

An extensive patch of talawa grassland occurs in Haldummulla. The dense, tall (to 1.5 m high) turf is dominated by *Cymbopogon nardus* with *Andropogon lividus, Arundinella villosa, Chrysopogon aciculatus* and *Themeda tremula* as common subordinates. The soils are less eroded, truncated and denuded with comparatively more nitrogen, phosphorus and potassium than those of the Uva Basin patanas; unlike the latter, the former lacks a hardpan-like top layer. Dispersed in the grass cover are herbs and shrubs such as *Desmodium triflorum, Elephantopus scaber, Evolvulus alsinoides, Lantana camara* and *Psidium guajava*.

Contrasting sharply with the disclimaxes damanas and talawas are villus, which are edaphically-characterized climaxes occurring in perennially moist depressions in Polonnaruwa and the surrounding areas. This floristically and phytosociologically unique community is characterized by the succulent, creeping grasses *Iseilema laxum* and *Paspalidium flavum* with sporadic tussocks of *Fimbristylis* spp. The relatively humus-rich soils are a dark brown with more fine particles than coarse sand, gravel and quartz, and have a higher carbon, nitrogen, phosphorus and potassium content when compared with other dry zone pasture soils. The predominant grasses flower, seed and germinate throughout the year and do not have a predominant resting period, presumably because the soil moisture is perennially favourable.

Of the grass-dominated communities described by Senaratne (1956), the following are the most noteworthy. Occurring in large tracts of fertile land, particularly alluvial soils near rivers and forest margins, are communities of *Imperata cylindrica* growing luxuriantly to 1.2 m high. It occurs also on road banks and roadsides often in less luxuriant form with culms rarely exceeding 20 cm in height. A common associate of this grass cover is *Paninum maximum*. Long-neglected land often supports a vigorous turf of the densely-tufted, coarse grass *Aristida setacea* with spiny shrubs *Asparagus racemosus, Carrisa spinaru,* and *Flueggea leucopyrus* as common associates. This is of limited use as a pasture because of the unpalatability of *Aristida* and the presence of spiny shrubs. Senaratna (1956) believes it to be the climax stage of the deflected grassland succession. Particularly plentiful in recently-opened land with favourably moist soils is a turf of *Chloris inflata*, a grass that flowers and seeds within 2-3 months and then becomes fibrous and unpalatable. The quick-growing, creeping, much-branched, palatable, annual *Brachiara repatans* forms a dense sward up to 30 cm high in arable lands immediately after harvesting crops. If the land is abandoned, this seral turf gives way to the *Aristida* -dominated climax. Along riverine tracts with moist alluvial soils is a grass cover of *Stenotaphrum dimidiatum*, a creeping, carpet-grass producing a main pasturage to animals. Large areas of the floor of sparse forest support almost pure populations of *Cyrtococcum trigonum*. This forest is main grazing-land of cattle during the dry season.

Subordinate grasses occurring in varying frequencies in most of these communities include *Cynodon dactylon, Echinochloa colonum, Paspalum conjugatum, P. metzii, Paspalidium flavidum* and *Zoysia matrella. Alysicarpus vaginalis, Desmodium triflorum, D. triquetrum, Indigofera chinata, Impomoea ungustifolia, Phaeleolus trilobus, Teramnus labialis* and *Zornia diphylla* are some of the useful dicotyledonous herbage species. Frequent biotic interference has enabled a variety of weed species to invade many of these pastures, the most noteworthy being *Achyranthus aspera, Alternathera pungens, Croton glandulosus, C. sparsiflorus, Euphorbia geniculata, Synedrella nodiflora* and *Tridax procumbens. Pennisetum polystachyon* is one of the recently established troublesome graminaceous weeds (Amaratunga & Lazardes 1970) which is spreading extensively and rapidly in the dry zone.

Thus, most ecological differentiation in dry zone pastures can be attributed to local edaphic differences associated with and resulting from topographic variations and differential degrees of biotic interference.

Arid-zone pastures

The grass covers occurring on certain sandy and muddy flats of the arid northwestern and southeastern coasts, mainly in Hambantota, Mannar and Puttalam, are regarded as arid zone pastures. Of the mean annual rainfall of less than 1,200 mm, about 75% is NE-monsoonal (October-March) (Fig. 11d). In some years the period from June to August is entirely without rain. The air temperature remains high throughout the year, and during the dry SW-monsoonal season it may exceed 35 °C, thus giving rise to a high rate of evaporation and a low relative humidity. The desiccating southwesterly winds are most severe from April to August.

Included in the arid zone pastures is a variety of grass-dominated communities (Pemadasa *et al.* 1979), which are the driest grasslands in Sri Lanka. Sandy habitats on relatively high ground (about 10 m altitude), adjoining thornscrubs which are believed to be the climax vegetation in the arid zone, support a dwarf turf with about 30-40% bare ground. Of the predominant grasses, *Eragrostis tenella* and *Zoysia matrella* contribute as much as 60% of the total cover, and *Chloris* sp. and *Cynodon dactylon* about 10-15%. Sporadic individuals of *Acacia planifrons, Aloe barbadensis, Asparagus recemosus, Coccinia grandis, Calotropis gigantea, Dactyloctenium aegyptium, Epalis divaricata, Euphorbia antiquorum, Grewia tenax* and *Sida retusa* are scattered in the grassy carpet, along with the herbs *Enicostema verticillare, Hybanthus enneaspermus, Portulaca tuberosa* and *P. wightiana.* The soil contains more clay, fine sand and coarse sand. The pH and calcium, carbonate, nitrogen, phosphorus and potassium contents generally are low. This community, though exposed to occasional shallow floods during heavy rains, rarely is influenced by sea or groundwater. During prolonged droughts the soils bake hard and crack, and most plants appear parched.

Relatively low-lying damp habitats in saltmarshes are covered by a closed turf of *Cynodon dactylon* which contributes more than 80% of the total cover. *Arthrocnemum indicum, Cressa cretica, Cyperus pumilus, C. rotundus* and *Salicornia brachiata* occur very sporadically. During wet months, bare areas between grassy patches are colonized by the non-heterocystous blue-green alga *Microcoleus* sp. The community is flooded in wet months, and may also undergo occasional inundation during high tides. The soil is blackish, with considerable amounts of humus, clay, fine sand and very little coarse sand.

The ecotones between these grass covers and thornscrubs are covered by a mixed sward of *Chloris* sp., *Cynodon dactylon, Eragrostis tenella* and *Zoysia matrella,* with frequent individuals of *Acacia eburnea, A. planifrons, Aloe barbadensis, Asparagus racemosus, Ocimum canum, Portaluca tuberosa, P. wightiana* and *Tylophora* sp.

All these communities may be regared as edaphic climaxes. In contrast, localized swards of *Zoysia matrella* occurring in open places among thornscrubs appear to be disclimaxes resulting from human-destruction of woody vegetation. Scattered in this grass cover are sporadic individuals of thornscrub species such as *Cissus quadrangularis, Dyschoriste madurensis, Evolvulus alsinoides, Portulaca tuberosa* and *P. wightiana*, along with tree species such as *Acacia eburnea* and *A. planifrons*. Very often thornscrub trees are cut for firewood, and the resulting gaps provide suitable niches for grasses to establish.

Localized sandy hillocks on the coast support an open turf of *Cynodon dactylon*, with isolated individuals of *Salicornia brachiata* and *Suaeda maritima*. Arthrocnemum indicum, Portaluca tuberosa and *Evolvulus alsinoides* also occur very occasionally. The soil is composed mainly of coarse sand, with plentiful amounts of shell fragments and very little clay, fine sand and humus.

Disturbed habitats on roadsides support a dense sward of *Cynodon dactylon* with *Zoysia matrella* as a frequent subordinate. Isolated individuals of *Salicornia brachiata* and *Evolvulus alsinoides, Ocimum canum, Portulaca tuberosa* and *P. wightiana* occur along with occasional trees of *Acacia eburnea* and *A. planifrons*. The soil generally is dry with very little humus, clay and fine sand.

The arid zone pastures generally are of low vigour, especially during the dry season, when the predominant grasses become yellow, senescent and very often parched; however, with the onset of NE-monsoonal rains, the grasses quickly spring to active growth and in a short time the ground is covered with a green carpet. Though most of the predominant grasses are palatable, they are of limited value as pasturage because their productivity is very low; yet very often cattle roam these grass covers for food. The low productivity is attributed chiefly to hostile arid conditions and partly to deficiency of nitrogen, phosphorus and potassium.

Most of these grass covers are open, and therefore many suitable microsites are available for weeds such as *Acanthospermum hispidum, Alternathera pungens, Martynia diandra, Pedalium murax* and *Tribulus terrestris*. Grazing by cattle also favours establishment of weeds, particularly *Tribulus terrestris*. Cattle avoid this species because of its spiny fruits.

Grassland productivity

The economic viability of grasslands as pastures depends partly on their productivity and food value of constituent species. These aspects have received very little attention in Sri Lanka, though some information is available on the effects of fertilizer application on biomass production of pasture species established in the wet zone (Appadurai & Goonawardene 1974, Appadurai & Arsaratnam 1969, Sivasupramaniam *et al.* 1974). Amarasinghe (1979) followed the natural productivity of two communities of the humid zone dry patana in the Hantane Hills; one of these was at 900 m altitude (upper site) and the other at 560 m altitude (lower site). The vegetation was harvested at ground level at monthly intervals from 1×1 m plots, and the weight of live and dead material of each species was determined after oven-drying at 85 °C for 72 h; the experiment lasted one year. The total productivity was calculated as $\Sigma(H_i\text{-}H_j)$ where H_i and H_j are standing crop biomass values at successive harvests. The estimates for the upper and lower sites were 911 and 680 g m^{-2} respectively. The standing crop of the upper site was predominantly of *Cymbopogon nardus, Eulalia trispicata* and *Themeda tremula*, their contribution being 65%, 10% and 25% respectively. At the lower site the major constituents were *Cymbopogon nardus, Panicum maximum* and *Pennisetum polystachyon* which respectively contributed 65%, 19% and 8% of the total biomass.

Grassland productivity depends on many factors, including annual rainfall and its seasonal variations, soil fertility, species composition and grazing pressure (Murphy 1975). Ambasht *et al.* (1972) estimated net productivity of 3,810 g m^{-2} yr^{-1} in Indian grasslands where the rainfall was about 1,000 mm yr^{-1}. The values in the humid zone dry patana were much lower despite precipitation of about 2,200 mm yr^{-1}, and this is attributed to low nutrient status of the soil (Pemadasa & Amarasinghe 1982a). There is evidence that nutrient enrichment increases the growth of predominant grasses *Cymbopogon nardus, Eulalia trispicata, Pennisetum polystachyon* and *Themeda tremula* substantially (Amarasinghe & Pemadasa 1982b). Addition of nutrients also has been found to increase biomass production in a lower-wet patana (Pemadasa 1981a). Amarasinghe's data show a considerable difference in productivity between the two sites, which may be attributed to differences in moisture conditions and floristic composition; the soil is drier and the turf is more open in the lower than in the upper site.

Little is known of the productivity of other patanas, savannas and lowland grasslands, but the general view is that they are of low productivity. This is supposed to be a reflection of deficiency of nutrients, particularly nitrogen, phosphorus and potassium (Pemadasa 1981a, Amarasinghe & Pemadasa 1982b). Why the grassland soils are deficient in nutrients is not known; the rarity of legumes, loss of nutrients by leaching and erosion and supposedly slow rate of microbial activity (Mueller-Dombois & Perera 1971) are some of the possible causes.

The low productivity of Sri Lanka grasslands is a critical limitation to their potential as pastures, and fertilizer application and introduction of legumes and/or more productive grasses with a low nutrient demand are two possible approaches to alleviating this problem. As already mentioned, addition of nutrients can improve productivity, but this approach has several limitations. It may be practicable in lowland pastures but not in patanas and savannas because they occur on mountain slopes which makes it difficult to prevent loss of fertilizer by leaching and erosion. It is arguable whether Sri Lanka can afford to spend vast amounts of money to import prohibitively expensive fertilizer and, consequently, whether this approch is economically profitable. Thus, introduction of legumes and more productive pasture species appears to be the more promising alternative. However, such introductions undoubtedly modify the floristics and natural ecological balance of native grasslands. On the other hand, productivity and soil fertility would be increased and erosion prevented; these are essential prerequisites of conservation and improvement of these ecosystems. Much of the ground in coconut plantations is open, which could profitably be used for pasture development without seriously interfering with the main crop; thus, coconut plantations can be converted to multi-purpose agricultural ecosystems. Clearly, careful experimentation is necessary to develop methods which could be employed to improve productivity of Sri Lanka grasslands without seriously disrupting their natural balance. This area of research is likely to yield considerable rewards.

Biotic impact on grasslands

Grasslands probably are the most disturbed terrestrial ecosystems in Sri Lanka, and the most severe deleterious impact has been that of domesticated animals and man, although wild animals also play a critical role in the ecological balance of certain communities. The continued destruction of grass covers by clearing for agricultural purposes, burning, removing herbage for animal fodder and grazing by cattle has caused increased erosion and soil impoverishment, changed the soil's mechanical composition and water relations, and facilitated colonization by weeds. These floristic and habitat changes are most severe in patanas and savannas.

The majority of steep slopes in the Western and Uva Basins (Fig. 2) which are known to have once been under dry patanas but subsequently cleared for agricultural and other purposes are now highly eroded, denuded and truncated with protruding rock outcrops and very shallow soils lacking a prominant A_1 horizon, and are, therefore, unable to support native grass covers; instead, they are often colonized by many composite and graminaceous weeds, the most noteworthy being *Eupatorium odoratum, Pennisetum polystachon* and *Tithoniadiversifolia.* Amarasinghe & Pemadasa (1982a) reported evidence of the effect of human impact in modifying the micro-distribution of native species within humid zone dry patana communities in Hantane hills. The micro-distribution of predominant grasses was found to be related to variations in the soil-surface height. In undisturbed communities, *Cymbopogon nardus* occupied hollows and *Eulalia trispicata* and *Themeda tremula* hummocks. In contrast, in disturbed habitats, *Pennisetum polystachon* occurred more abundantly and luxuriantly in hollows suppressing the native species. Thus, *Cymbopogon nardus* was less abundant in hollows than on hummocks; clearly, its habitat reversal is a result of human interference.

Most of the wet patanas around Nawura Eliya have already been put under cultivation as cropland, plantation forest or grazing range. Moreover, much of the grazing range has been vastly altered from its original conditions. Certain areas abandoned after short-term cultivation already are covered with weeds such as *Eupatorium riparium.*

Selective herbage removal is common to almost all grasslands in Sri Lanka. This favours some species as a result of continued destruction of others. For example, the non-palatable grass *Cymbopogon nardus* increases as the fodder grass *Themeda tremula* decreases. Amarasinghe (1979) provided experimental evidence that frequent clipping favoured colonization of *Panicum maximum* at the expense of the predominant grass *Cymbopogon nardus*.

Grazing pressure is particularly severe in lower-wet and summer-dry zone dry patanas and lowland grasslands. In fact, the general physiognomy of the stubble-tussock short turf of lower wet patanas is a result of continued cattle grazing. Particularly in the dry zone and arid zone pastures, over-grazing during the dry season causes severe destruction of the turf, which hampers the native species and favours ready establishment of weeds. Large areas of wet zone and intermediate zone pastures under neglected coconut plantation are increasingly being replaced by the fast spreading weed *Eupatorium odoratum* which is harmful to both the pasture and the main crop.

The available evidence is that unsystematic exploitation of grassland ecosystems in Sri Lanka leads to their inevitable destruction. There is a danger that 'virgin' grass covers will have disappeared before long unless remedial steps are taken.

Conclusions

Grasslands of Sri Lanka are ecologically fascinating for their wide floristic, phyto-sociological and environmental diversity, and are economically important for their great potential as pastures. They have been the subject of repeated extensive surveys, but many of their ecological aspects are yet to be understood. Though comprehensive floristic lists and descriptions of climate, geology and soils are available, little is known of the dynamics, energy relations, productivity, phenology, seasonality and general biology of various communities in general and individual species in particular. Thus, more intensive experimental investigations are necessary to elucidate the ecology of graslands more fully.

Presumably because of the limited ecological knowledge, the exploitation of grasslands for agricultural purposes has been rather unsystematic. Agricultural expansion inevitably leads to reduction of natural vegetation. Arguably, natural and agricultural ecosystems are equally indispensable rivals, and man has to resolve his basic dilemma of opposing priorities. Natural ecosystems must be conserved for many reasons. Systematic agriculture and associated land management must be based on scientific research which requires natural vegetation to be employed as controls for otherwise it is impossible to assess the repercussions of various practices. Destruction of native species already adapted to local environments essentially means extinction of their genetic potential, and the only way to avoid this catastrophe is to preserve native species which would provide a gene pool for future potential combinations. Since grassland ecosystems are little understood, it is a crime to allow their destruction, thereby destroying their future uses as a source of knowledge. Thus, conservation of Sri Lanka grasslands is a prime necessity. On the other hand, continued demographic explosion and consequent aggravation of energy burden necessitates agricultural expansion. Thus, scientist, agriculturalists, economists and politicians must get together to formulate a sound basis for systematic exploitation of these valuable ecosystems.

References

Abeywickrama, B.A. 1955. The origin and affinities of the flora of Ceylon. Proc. Ceylon. Assoc. Adv. Sci. Part 2:1-23.

Abeywickrama, B.A. 1959. A provisional check list of the flowering plants of Ceylon. Ceylon. J. Sci. 2:119-240.

Amarasinghe, L. 1979. Ecology of grassland vegetation of Hatane Hills. M.Sc. thesis, University of Peradeniya, Sri Lanka.

Amarasinghe, L. & M.A. Pemadasa. 1982a. The ecology of a montane grassland in Sri Lanka. II. The pattern of four major species. J. Ecol. 70:000-000.

Amarasinghe, L. & M.A. Pemadasa. 1982b. The ecology of a montane grassland in Sri Lanka. IV. Vegetative growth of four major grasses. J. Ecol. 70:000-000.

Amaratunga, K.L.D. & M. Lazardes. 1970. A noteworthy new grass record from Ceylon. Trop. Agric. 76:195-198.

Ambasht, R.S., A.N. Maurya & U.N. Singh. 1972. Primary production and turnover in certain protected grasslands of Varanasi, India. pp. 43-50 In: Tropical ecology with an emphasis on organic production (Ed. P.M. Golley & F.B. Golley). Institute of Ecology, Univ. of Georgia, Athens.

Appadurai, R.R. 1969. Grassland farming in Ceylon. Gunasena, Colombo, Sri Lanka.

Appadurai, R.R. & R. Arsaratnam. 1969. The effects of large applications of urea nitrogen on the growth and yield of an established pasture of Brachiaria brizantha (Hoscht. and Stapf.). Trop. Agric. 133:153-158.

Appadurai, R.R. & L.A. Goonawardane. 1974. Effect of nitrogen and defoliation on the productivity and feeding value of Bracharia ruziziensis (Germain and Everard) in the mid-country wet zone of Sri Lanka. Trop. Agric. 138:1-9.

Broun, A.F. 1898. Administration report of the Forest Department. Ceylon Govt Press.

Chapman, V.J. 1947. The application of aerial photography to ecology as exemplified by the natural vegetation of Ceylon. Indian Forester 73:287-314.

Cooray, P.G. 1967. The geology of Ceylon. Spolia. Zeylan 31(1). Ceylon Govt Press, Colombo, Sri Lanka. 324 pp.

Greig-Smith, P. 1964. Quantitative Plant Ecology, 2nd edit. Butterworth, London.

Hill, M.O. 1973. Reciprocal averaging: an eigenvector method of ordination. J. Ecol. 61:327-249.

Holmes, G.H. 1951. The grass, fern and savannah lands of Ceylon, their nature and ecological significance. Imp. For. Inst. Lond. Pap. 28.

James, H., F.S.C.P. Kalpage & M.W. Thenabadu. 1968. The nature of organic matter in soils of the hill country wet zone. Ceylon Assoc. Adv. Sci. part 1:22-23.

Kershaw, K.A. 1975. Quantitative and dynamic plant ecology, 2nd edit. Arnold, London.

Koelmeyer, K.O. 1957. Climatic classification and the distribution of vegetation in Ceylon. Ceylon Forester 3:144-163.

Mueller-Dombois, D. 1968. Ecogeographic analysis of a climate map of Ceylon with special reference to vegetation. Ceylon Forester 8:39-58.

Mueller-Dombois, D. & H. Ellenberg. 1974. Aims and methods of vegetation ecology. Wiley, N.Y.

Mueller-Dombois, D. & M. Perera. 1971. Ecological differentiation and soil fungal distribution in the montane grasslands of Ceylon. Ceylon J. Sci. 9:1-41.

Murphy, P.G. 1975. Net primary productivity in tropical terrestrial ecocystems. pp. 217-231 In: Primary productivity of the biosphere (ed. H. Leith & R.H.Wittaker). Springer-Verlag, New York.

Panabokke, C.R. 1967. Ceylon General Soil Map. Ceylon Survey Dept, Colombo, Sri Lanka.

Parkin, J. & H.H.W. Pearson. 1903. The botany of the Ceylon patanas. J. Linn. Soc. Lond. 85:430-463.

Pearson, H.H.W. 1899. The botany of the Ceylon patanas. J. Linn. Soc. Lond. 34:300-365.

Pemadasa, M.A. 1981a. The mineral nutrition of the vegetation of a montane grassland in Sri Lanka. J. Ecol. 69:125-134.

Pemadasa, M.A. 1981b. Cyclic changeand pattern in an *Arthrocnemum* community in Sri Lanka. J. Ecol. 69: 565-574.

Pemadasa, M.A. & L. Amarasinghe 1982a. The ecology of a montane grassland in Sri Lanka. I. Quantitative description of the vegetation. J. Ecol. 70:000-000.

Pemadasa, M.A. & L. Amarasinghe. 1982b. The ecology of a montane grassland in Sri Lanka. III. Germination of three major grasses. J. Ecol. 70: 000-000.

Pemadasa, M.A., S. Balasubramaniam, H.G. Wijewansa & L. Amarasinghe. 1979. The ecology of a saltmarsh in Sri Lanka. J. Ecol. 67:41-63.

Pemadasa, M.A. & D. Mueller-Dombois. 1979. An ordination study of montane grasslands in Sri Lnka. J. Ecol. 67:1009-1023.

Pemadasa, M.A. & Mueller-Dombois, D. 1981. An association-analysis of montane grasslands in Sri Lanka. Aust. J. Ecol. 6:111-121.

Perera, N.P. 1967. An evaluation of the human impact on the nature and distribution of wild plant communities in the Ceylon highlands. Ph.D. thesis, University of Leeds.

de Rosayro, R.A. 1945. The montane grasslands (patanas) of Ceylon. I. Trop. Agric. 101:206-216.

de Rosayro, R.A. 1946a. The montane grasslands (patanas) of Ceylon. 2. Trop. Agric. 102:4-16.

de Rosayro, R.A.1946b. The montane grasslands (patanas) of Ceylon. 3. Trop. Agric. 102:81-94.

de Rosayro, R.A. 1946c. The montane grasslands (patanas) of Ceylon. 4. Trop. Agric. 102:139-148.

de Rosayro, R.A. 1950. Ecological conceptions and vegetational types with special reference to Ceylon. Trop. Agric. 151:108-121.

de Rosayro, R.A. 1961. The nature and origin of secondary vegetation communities in Ceylon. Ceylon Forester 5:23-49.

Senaratna, S.D.J.E. 1956. Regional survey of the grasslands of Ceylon. pp. 175-180 In: Study of tropical vegetaion. Proc. Kandy Symp. UNESCO, Paris.

Sivasupramaniam, S., J.S. Siththamparanathan & R.R. Appadurai. 1974. Studies on pasture improvement in the hill country dry zone (patanas) of Ceylon (Sri Lanka). III. Effects of high nitrogen application on the productivity and feeding value of *Paspalum dilatatum* . Trop. Agric. 138:35-44.

Szechowyez, R.W. 1954. Some observations on climate, soil and forest climac. Ceylon Forester 1:131-141.

Walter, H. 1964. Die Vegetation der Erde in öko-physiologischer Betrachtung. I. De tropischen und subtropischen Zonen. 2nd edit. VEB Fischer Verlag, Jena.

Watt, A.S. 1947. Pattern and process in the plant community. J. Ecol. 35:1-22.

Williams, W.T. & J.M. Lambert. 1959. Multivariate methods in plant ecology. I. Association-analysis in plant communities. J. Ecol. 47:83-101.

5. Some aspects of the ecology of ricefields

S. Weeraratna and C.H. Fernando

Introduction

Rice is the second largest cereal crop in the world after wheat. Many monographs and books have been devoted to the scientific aspects of rice cultivation. A few of these are: Angladette (1966), Anon (1975a, b), Bertin *et al.* (1976), Blankenberg (1933), Denishko (1954), Eregin (1950) and Grist (1975). There is, however, relatively little published on the ecology of ricefields, and of these most are very recent publications. The most recent and comprehensive studies on ricefield ecology are the monograph by Heckman (1979) and the proceedings of a symposium (Furtado 1980). Fish culture in ricefields has attracted considerable interest because of the economic incentives. The literature on this subject and a discussion of the methods is available in the studies of Coche (1967) and Meijen (1940) and the bibliographies of Fernando *et al.* (1979) and Temprosa & Shehadah (1980). The agronomic aspects of rice cultivation have a very extensive literature. Some of the major works are referred to above.

Studies on the ricefield ecosystem of a broad nature include the works of Moroni (1961) and Heckman (1979). Koch (1952) found tropical plants and Fox (1965) found tropical ostracods in Italian ricefields. Pashitnova (1935) and Mukhamediev (1960) studied the hydrology of Central Asian ricefields. Fernando (1956a) listed the fish found in ricefields in Sri Lanka and gave a bibliography of the occurrence of fish in ricefields. The ecology of the aquatic mesofauna and macrofauna of South East Asian ricefields was studied by Fernando (1977) and the Ostracoda of South East Asian ricefields were studied by Victor & Fernando (1980). Fernando *et al.* (1979) gave an extensive bibliography on the aquatic fauna of ricefields listing over 900 references.

In Sri Lanka, extensive rice cultivation was the backbone of a flourishing civilization which lasted from 200 B.C.-1200 A.D. in the dry zone (Brohier 1975). *Orzya sativa* L., the most common species of rice, was first cultivated about 6,000 years ago below the foothills of the Himalayas (Roschievicz 1931, Ramiah 1937, Chatterjee 1951, Chang 1964, 1976a, 1976b). Hogan (1970) reported that four terraces built for rice cultivation on the banks of the Ravi river, India, date back about 3,500 years. Excavation in Mohenjadaro indicates that rice was an important staple in the region around 2,500 B.C. (Andris & Mohammed 1958).

Rice cultivation in Sri Lanka was started by immigrants from the Indian mainland sometime before 600 B.C., when a settled civilization developed and an elaborate irrigation system was established (Brohier 1975). In the heyday of the Sinhalese kingdoms (100-1,200 A.D.) Sri Lanka served as a 'granary of the East' (Paul 1945).

In this chapter we attempt to provide an account of rice cultivation in Sri Lanka at the present time. The methods of cultivation, soils, physical and chemical properties and their seasonal changes are documented. Special attention is paid to the ecology of the flora and fauna based on the meagre data available.

Ricefields and their soils

The total area of ricefields in Sri Lanka is about 580,000 ha, distributed over a wide range of climatic, geomorphic and edaphic conditions (Panabokke 1978). Fig. 1 gives the extent of ricefields in different parts of Sri Lanka, and Fig. 2 a diagram of the ricefield biotope.

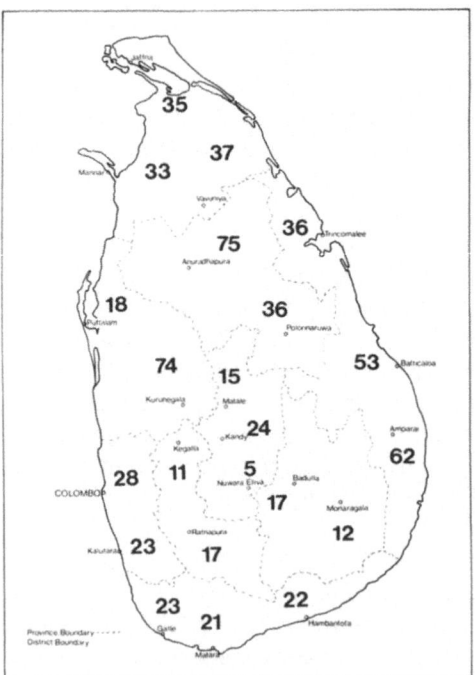

Fig. 1. Areas of ricefields (1,000 ha) in the districts of Sri Lanka.

Ricefields are distributed over three broad climatic zones: wet, intermediate and dry. These zones differ mainly in the annual rainfall. The average annual rainfall in the wet zone is about 320 cm while the figures for the intermediate and dry zones are 200-320 and 200 cm respectively.

The moisture content of ricefields is influenced by their geomorphology, hence a short description of this factor will be useful in understanding the ecology of ricefields.

The dry zone

Nearly 275,000 ha of ricefields are found in the dry zone. The geomorphology of most of these ricefields is confined to moderately broad to broad, gentle sloping smooth valleys with concave side slopes and gradients ranging from 0.5-1.0%.

Surface drainage is satisfactory in individual ricefield tracts within a valley, but vertical drainage is poor due to the underlying impervious basement rocks. This results in a rising ground water level in the land located in the valley bottoms during the rainy season. Thus, most of the rice lands in the dry zone are for the most part, poorly or inadequately drained (Panabokke 1978).

Soils in the valleys consists of a catenary sequence, viz., well drained, reddish brown earths (Rhodustalfs and Haplustalfs), inadequately drained reddish brown earths and low humic gley soils (Tropaqualfs). Ricefields are located on the imperfectly drained reddish brown earths and low humic gley soils.

Imperfectly drained reddish brown earths have sandy clay loams to sandy loam, dark brown (10 YR 3/3) to dark yellowish brown (10 YR 3/4) colour in the upper horizons. Soils in the lower horizons are yellowish brown (10 YR 5/6) to brown (7-5 YR 4/4) and are sandy clay loam in texture. These soils have a pH ranging from 6-7.

Soils of the upper horizon of the low humic gley soils are sandy loam to sandy clay loam and dark grey brown (10 YR 3/2). Soils of the lower horizon are lighter coloured (10 YR 4/3) sandy clay loam to sandy clay. Soil pH for the surface and subsurface are 6-7.5 and 7-8.5 respectively.

The intermediate zone

About 125,000 ha of ricefields are found in this zone. Most of these lands are located in moderately broad, gently sloping valleys, which show a weak gradient.

The rice soils of this zone may be grouped into four suborders: Aqualfs, Aquults, Udaijs and Ustutls (Panabokke 1978). Rice soils of Aqualfs are more or less similar to Rhodustalfs in the dry zone. However, these soils have a slightly lower pH as compared with the Rhodustalfs. Aquults, the rice soils of the low country, are the most abundant soils of this zone. These soils are situated within moderately broad,

136

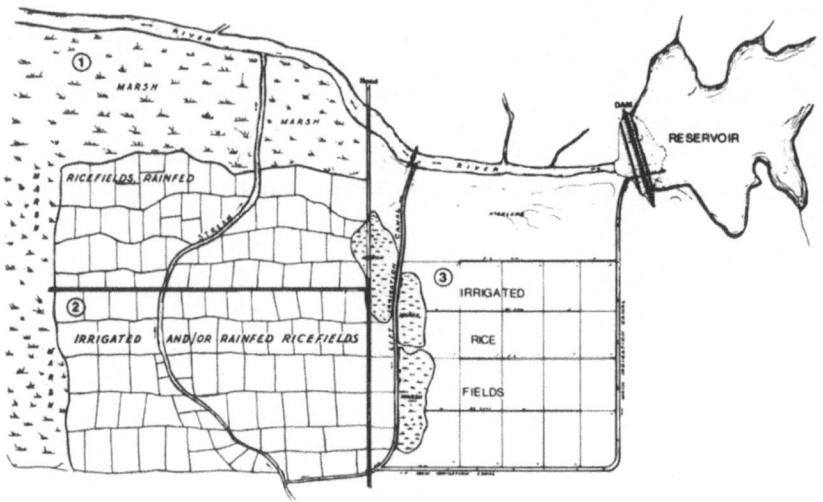

Fig. 2. The ricefield biotope (after Fernando 1977).

gently sloping valleys and are distributed in a caternary sequence consisting of red-yellow Podzolic soils (Tropudults) and their drainage associates. Most of the rice lands are located in inadequately or poorly drained soil types, corresponding to the low humic gleys. These soils are dark brown to dark yellowish brown in the upper layers and are loamy to sandy loam in texture. Mottling is observed in the lower layers. Soil pH is between 5.5 and 6.0 in the surface layers.

The upper horizons of these inadequately drained low humic gley soils are dark brown to dark yellowish brown in colour and are loamy sand to sandy loam. Subsurface layers are mottled clay to sandy clay loam in texture and have pH values of 5.2-5.7.

The wet zone

Nearly 180,000 ha of rice lands are found in the wet zone and are distributed in widely different geomorphological regions.

Rice soils in the lower peneplains of the wet zone are found in broad inland valley systems. Drainage conditions vary from moderately well drained to poorly drained soils. These Aquult soils have a pH of 4.5-5.5 in the surface layers and 5.0-5.5 in the subsurface layers. Rice also is cultivated in the west and southwest coastal plains of the wet zone and associated flood plains. These soils are classified as Aquepts and Hemists. Most of these soils are located in the poorly drained lands of the low country wet zone. These lands are confined to the filled lagoons, tidal marshes and back swamps of rivers, where a considerably different ecological pattern exists.

In the mid country (middle peneplain, see Cooray, this volume), rice is grown in narrow sloping valleys and on terraced slopes. Drainage generally is good in these soils and their pH varies from 5.0-5.5 in the surface layers and 5.5-6.0 in the subsurface layers.

Methods of cultivation

Before actual seeding or transplanting commences, water is impounded in the ricefield for two to three weeks, killing most of the weeds. The well soaked soil is then drained and tilled, inverting and mixing the soil and burying the remaining weeds. The process of impoundment and tillage is repeated, bringing about complete weed burial and pulverization and homogenization of soil clods. Normally, the depth of ploughing is about 25 cm. During preparation the soil is thus kept completely saturated with water for a good portion of this period the land is kept flooded.

Seeds are sown in the moist drained soil, and about 7 to 10 days after germination the soil is again flooded and remains flooded with 5-8 cm water. The moisture conditions of the soil promote the growth of weeds and other organisms, such as algae. Instead of sowing seeds, rice often is transplanted from nurseries. Here soils prepared in the same manner are flooded within 2-3 days after transplanting and hence there is little time for weeds to develop. In irrigated ricefields, the soil is kept flooded most of the time, enabling aquatic flora and fauna to develop. In rain fed areas, moisture conditions vary, making the aquatic habitat unstable. However, many pf the aquatic inhabitants burrow into the moist soil or retreat into the irrigation channels during dry periods orlow water levels.

When the crop is mature, the water supply is cut off, marking the onset of dry conditions. This state exists until preparation for the cultivation of the next crop begins. During the intervening period, grasses and other weeds grow, except in areas where farmers grow vegetable crops under upland conditions. Ricefields in the coastal areas of the southwestern region of the country are poorly drained throughout the year and fields in this region remain wet. Some of these lands tend also towards salinity, due to the inflow of sea water during high tides.

Thus the ecological factors effecting ricefields tend to vary considerably depending on the method of cultivation, water availability and soil type. A number of physical, chemical and biological changes take place in the soil with flooding (Ponnamperuma 1965). These changes can be great enough to alter the ecological community of the field.

Physico-chemical changes

With flooding, ricefield soils tend to become anaerobic. Oxygen movement by diffusion through flood water and water filled pores in the soil is much slower compared with diffusion rates through air filled pores. This slow oxygen supply rate and rapid consumption by the aerobic flora and fauna in the soil result in anaerobic soils.

Pearsall (1938) reported on the development of such anaerobic zones in submerged soils and studies conducted by Ponnamperuma (1955), Yamane (1974) and others confirm the existence of such anaerobic zones. This change in the aerobic nature of the soil leads to changes in the organisms inhabiting the soil. With the development of an anaerobic zone, the populations of obligate aerobic flora and fauna gradually decrease, while the facultative aerobes and anaerobes tend to increase (Takai 1969). A reduction in the number of nitrifiers (Schreven & Sieben 1972) and nitrification capacity (Weeraratna 1981) has been reported. The changes in the type of organisms also is influenced by the changes in pH. The pH of most soils tend towards neutrality after submergence (Ponnamperuma 1965). Associated with these changes is the alteration in the availability of nutrients such as phosphate, iron, manganese, copper and molybdenum (Patrick & Reddy 1978). In addition, a number of other soil characters also tend to change with flooding (Ponnamperuma 1977).

The net efect of these physical and chemical changes in the soil is to change the fauna and flora. These changes are made more complex by variations in the moisture content of soils, usually in rainfed rice lands and the application of nitrogenous fertilizers, such as urea or ammonium sulphate, and biocides, at different stages of cultivation

The pattern of changes in the kind of the organisms inhabiting ricefields is of ecological importance as well as being important from the point of view of plant growth, since the activity of these organisms influences considerably the development and production of the rice plants. In irrigated wetland rice, weed growth has been found to reduce yields by as much as 50% (Jayasekara & Velmurugn 1967).

Flora

Because of the inherent high moisture content of most ricefields, the type of flora associated with them varies considerably. Unfortunately, very few studies have been carried out on the succession of plant forms occurring in ricefields. The recent papers of Akobundu & Fagage (1978) and Merlier (1978) deal with weeds and their control in African ricefields. DeDatta (1978) deals with the problem of weeds in tropical Asia.

Weeds

Many varieties of weeds are found in ricefields in Sri Lanka. This can be attributed to either different ecological conditions existing over the cultivation season or differences in climatic factors (Velmurugu 1980). Table 1 lists common ricefield weeds in Sri Lanka.

Cyperus irria and *C. dehiscens* appear soon after the crop is sown, and flower in about a month, or may produce a second crop. Water hyacinth, *Eichornia crassipes* and *Salvinia auriculata* are common in waterlogged ricefields.

Many agrochemicals are used to control these weeds, in addition to various cultural practices. Of these chemicals, chlorophenoxy compounds such as 2,4-dichlorophenoxy acetix acid (2,4-D), 2,4,5-trichlorophenoxy acetic acid (2,4,5-T) and 2-methyl, 4-chloropheroxy acetic acid (MCPA) and propanil (3,4-dichloro-proprionanilide) are the most common. Addition of these compounds to the ricefields has a profound effect on the ecology, depending on their degree of persistence. 2,4-D and MCPA are rapidly broken down in the soil (Weeraratna 1977), but 2,4,5-T persists for some time (Audus 1951). In addition to herbicides, a large number of insecticides and fungicides also are applied to control harmful insects and pathogenic fungi, respectively. The toxicity of these compounds tends to disrupt the natural ecological balance by their harmful effects on predators and pests of insects that hitherto have not reached pest population levels (Metcalf 1971).

Microflora

Few ecological studies have been carried out on the microflora in rice soils. A number of diseases in the rice are caused by bacteria, fungi and nematodes found in ricefield soils. Among the fungi present in rice soil environments are *Pricularia oryzae, Helminthosporium oryzae, Sclerotium oryzae, Rhizoctonia solani, Nigrospora oryzae* and *Ustilaginoidea virens*. In addition to the above species, Seneviratna (1980) recently listed *Thanatephorus cucmeris, Leptosphaeria salvinii, Acrocylindrium oryzae* and *Rhyneosporium oryzae*.

A large number of algae occur in ricefields except during the latter stage of crop development when there is little sunlight received at the water surface. Thirukkanasas *et al.* (1977) reported the occurrence of heterocystous blue greens such as *Nostoc, Anabaena, Cylindrosperum, Gloeotrichia, Rivularia, Aulosira, Scytoaema* and nonheterocystous forms such as *Oscillatoria, Aphanotheca* and *Microcystis* in some paddy fields in Jaffna during the Northeast monsoon period. They found that some species of *Nostoc, Anabaena* and *Oscillatoria* appeared during the early stages of cultivation but disappeared at later stages. Other species of *Nostoc, Anabaena, Aulosira, Aphamotheca* and *Microcystis* existed throughout the monsoon period.

Fauna

At different times of the year the ricefield fauna is dominated by aquatic or terrrestrial fauna depending on whether the ricefield is in the cultivation stage (usually wet) or the fallow phase (usually dry). Wet ricefields may remain fallow however, and support a rich aquatic flora and fauna for considerable periods.

The ecology of the aquatic fauna of ricefields has not been studied to any great extent. The scattered literature on the fauna has been listed in the bibliography of Fernando *et al.* (1979). The recent symposium on ricefield ecology (Furtado 1980) dealt with some of the aquatic invertebrates (Lim 1980) and the Ostracoda (Victor & Fernando 1980) in some detail. Previous studies of a broad nature include Pashitnova (1935), Meijen (1940), Mukhamediev (1960), Moroni (1961), Fox (1965), Heckman (1974, 1979) and Fernando (1977, in press). The aquatic fauna of Sri Lanka ricefields has been referred to in many publications dealing with both invertebrates and vertebrates. Studies which deal specifically with the ricefield fauna or refer to them exclusively include Fernando (1956a, 1956b, 1959, 1960, 1977, 1980b), Lord (1927), Mendis & Fernando (1962), Neale (1977) and Weerekoon & Samarasinghe (1958).

The aquatic invertebrates found in ricefields include plankton, benthos and nekton. Fernando (1980a) recorded 71 species of Rotifera, 41 species of Cladocera and 18 species of Copepoda in an intensive and extensive study of zooplankton in Sri Lanka. The ricefields had a high diversity of Cladocera and cyclopoid and harpacticoid Copepoda and a slightly lower diversity of Rotifera and Calanoida than ponds and small reservoirs, which had the highest zooplankton diversity. This same relative diversity pattern was shown also for the Protozoa, Ectoprocta, Nematoda, Oligochaeta, Mollusca, Conchostraca, Anostraca, *Caridina* and aquatic insect adults and larvae (Fernando 1980a). Weerekoon (1950) pointed out the diversity of the aquatic fauna of Sri Lankan ricefields. Weerekoon & Samarasinghe (1958) showed the high densities of insect larvae and the diversity of soil fauna in a Sri Lankan ricefield. Fernando (1960) gave an account of the freshwater crabs in ricefields and their ecology and Lord (1927) dealt with the damage caused by crabs to rice seedlings.

The fish species inhabiting ricefields include almost all the species found in Sri Lanka. Many fish use the ricefield habitat as nursery (Fernando 1956b). Ricefields are used for fish culture in many countries. Coche (1967) and Temprosa & Shehadeh (1980) have listed the extensive literature dealing with organized fish culture in ricefields. However, many rice farmers obtain considerable quantities of fish from natural stocks of fry which enter the ricefields at each cultivation. Sometimes fallow ricefields which contain water also serve as fish nurseries.

Although there is a varied terrestrial fauna in ricefields, there are little published data on its occurrence and ecology. Pests of rice, though, especially insects, have been studied extensively. The literature is summarized in a number of monographs, textbooks and symposia, for example Dammerman (1929), Grist (1975),

Table 1. Common ricefield weeds.

Family	Weeds	Areas commonly found		
		Dry	Intermediate	Wet
Acanthaceae	Justica betonica	+	—	—
	Asteracantha longifolia	+	—	—
Amaranthaceae	Amaranthus spinosus	+	+	—
	Amaranthus viridis	+	+	—
	Achyranthes aspera	+	+	—
Compositae	Spilanthis calva	+	+	+
	Acanthospermum hispidum	+	+	—
	Epasltes divaricata	+	+	+
	Eupatorium odoratum	+	+	—
	Vernonia zeylanica	+	—	—
Convolvulaceae	Ipomea sp.	+	—	—
Cyperaceae	Cyperus flavidus	+	+	+
	Cyperus deformis	+	+	+
	Cyperus irria	+	—	—
	Cyperus rotundus	+	—	—
	Fimbristylis miliacea	+	+	—
	Fimbristylis tetragona	—	—	+
	Fimbristylis acuminata	—	+	+
	Lipocarpa argentae	+	+	+
Euphorbiaceae	Euphorbia hirta	+	—	—
	Euphorbia indica	+	—	—
	Eragrostis unioloides	—	—	—
	Echinochloa crusgalli	+	+	+
	Echinochloa stagenina	+	+	+
	Echinochloa frumentacea	+	+	+
	Echinichloa colonum	+	+	+
	Panicum repens	+	—	—
	Imperata cylindrica	+	—	—
	Digiteria ascendenus	+	—	—
	Dactyloctenium aegyptium	+	—	—
	Cryza perennius	+	—	+
	Elusine corocane	+	—	—
Leguminosae	Aeschynomene americana	—	—	+
	Aeschynomene indica	—	—	+
	Crotoloria verusosa	+	—	—
	Cassia tora	+	—	—
	Desmodium heterocarpon	+	—	—
	Mimosa pudica	+	—	—
	Phaseolus lathyroides	+	—	—
Malvaceae	Sida acuta	+	—	—
Oxalidaceae	Oxalis corneculata	+	+	—
Pontederiaceae	Eichornia crassipes	—	—	+
	Monochoria vaginalis	—	—	+
	Pistia stratiotes	—	+	+
Salvinia auriculata	Salvinia auriculata	—	—	+

Nishida & Torii (1970), Buddenhagen & Perstey (1978) and Furtado (1980).

Birds are among the ubiquitous visitors to ricefields. The small, common pond heron *Ardeola grayi* (Sykes) is referred to as the 'paddybird' because it is so common (Henry 1955). A wide range of birds visit the ricefields in its 'aquatic phase'. These include egrets, herons, snipe, plover, sandpipers and storks. Less common visitors are cormorant, teal and ducks. Bitterns and waterterns are permanent residents, nesting in the shrubs at the margins. Henry (1955) recorded a large number of Sri Lankan birds in ricefields, while Glennister (1971) provided an equally long list of Malaysian ricefield birds.

Fernando (in press) has suggested the use of ricefields as sites for basing instruction in environmental ecology to schoolchildren.

Acknowledgements

The paper by Dr C.R. Panabokke on rice soils of Sri Lanka (Panabokke 1968) was used extensively in writing this chapter. Thanks are due to Miss M. Kaluarachchi for typing the manuscript.

References

Akobundu, I.O. & S.O. Fagage 1978. Weed problems of African rice lands. pp. 181-192 In: I.W. Buddenhagen & G.J. Pearsley (Eds.) Rice in Africa. Acad. Press, London.
Andiris, J.R. & A.F. Mohammed. 1958. The economy of Pakistan. Oxford Univ. Press.
Angladette, A. 1966. Le Riz. Paris.
Anon. 1975a Report of the International Seminar on deep water rice, August 21-26, Dacca, Bangladesh. 225 pp.
Anon. 1975b. Arroz. Inst. Cubano Libr., Habana. 502 pp.
Audus, L.J. 1951.The biological breakdown of hormone herbicides in soil. Pl. Soil 3:170-192.
Berten, J., J.J. Hermardinguer, M. Kaul & W.C.L. Randles. 1976. Atlas of food crops. Publ. Ecole Prat. Hague. 41 pp.
Blankenburg, P. 1933. Der Reis: Eine wisschaftgeographische untersuchungen. Berlin.
Brohier, ?. 1975. Food and the people. Lake House, Colombo. 200 pp.
Buddenhagen, I.W. & G.J. Persley (Eds.). 1978. Rice in Africa. Acad. Press, London. 365 pp.
Chang, T.T. 1964. Present knowledge of rice genetics and cytogenetics. Int. Rice Res. Inst. Tech. Bull. 1:1-96.
Chang, T.T. 1976a. Rice. pp. 98-104 In: Evolution of crop plants. N.W. Simmonds (Ed.). Longmans, London.
Chang, T.T. 1976b. The rice culture in the early history of agriculture. Phil. Trans. R. Soc. (B) 275:143-157.
Chatterjee, D. 1951. Note on the origin and distribution of wild and cultivated rices. Ind. J. Genet. Pl. Breed. 11:18-22.
Coche, A.G. 1967. Fish culture in ricefields: a worldwide synthesis. Hydrobiologia 30:1-44.
Dammerman, K.W. 1929. Agricultural study of the Malay Archipelago. The animals injurious and beneficial to agriculture, horticulture and forestry in the Malay Peninsula, the Dutch East Indies and the Philippines. De Bussey, Arustidien. 473 pp.

de Batta, S.K. 1978. Weed control and soil management in rainfed rice at IRRI and other locations in tropical Asia. pp. 202-211 In: I.W. Buddenhagen & G.J. Persley (Eds.). Rice in Africa. Acad. Press, London.

Denischenko, M. (Ed.) 1954. Rice cultivation. (Slovak). Bratislava. 173 pp.

Eregin, P.S. 1950. [Physiological basis of rice cultivation] (in Russian). Ak. Nauk USSR Physiol. Inst. Moscow. 208 pp.

Fernando, C.H. 1956a. On the food of four common freshwater fish of Ceylon. Ceylon. J. Sci. 7:201-217.

Fernando, C.H. 1956b. The fish fauna of paddy-fields and small irrigation ditches in the western lowlands of Ceylon and a bibliography of references to fish in paddy fields. Ceylon J. Sci.(C) 7:223-227.

Fernando, C.H. 1959. Colonization of freshwater habitats with special reference to aquatic insects. pp. 182-186 In: R.D. Purchon (Ed.) Proc. Cent. Bicent. Congr. Singapore.

Fernando, C.H. 1960. The Ceylonese freshwater crabs. Ceylon J. Sci. (Bio. Sci.) 3:191-222.

Fernando, C.H. 1977. The ecology of the aquatic fauna of ricefields with special reference to South East Asia. Geo-Eco-Trop. 1:169-188.

Fernando, C.H. 1980a. The freshwater zooplankton of Sri Lanka with a discussion of tropical freshwater zooplankton composition. Int. Rev. ges. Hydrobiol. 65:85-125.

Fernando, C.H. 1980b. The freshwater invertebrate fauna of Sri Lanka. Cent. Vol. Colombo Mus. Spol. Zeylan 35:15-42.

Fernando, C.H. 1981. The regionality and topicality of environmental education. Sinet, Addis Ababa 4: 101-113.

Fernando, C.H., J.I. Furtado & R.P. Lim. 1979. The aquatic fauna of the world's ricefields. Wallaceana Suppl. Kuala Lumpur 2. 105 pp.

Fox, H.M. 1965. Ostracod Crustacea from ricefields in Italy. Mem. Ist. Ital. Idrobiol. Dott. Marco Marchi. 18:207-215.

Furtado, J.I. (Ed.). 1980. Ricefield ecosystems. pp. 939-1035 In: Tropical ecology and development. Proc. 5th Int. Symp. Trop. Ecol. Kuala Lumpur.

Glennister, A.G. 1971. The birds of the Malay Peninsula, Singapore and Penang. Oxford Univ. Press. 219 pp.

Grist, D.H. 1975. Rice. Longmans, London. 601 pp.

Heckman, C.W. 1974. The seasonal succession of species in a rice paddy in Vientiane, Laos. Int. Rev. ges. Hydrobiol. 59:489-507.

Heckman, C.W. 1979. Ecology of the fauna of a ricefield in northeastern Thailand. Monogr. Biol. 34. 228 pp.

Henry, G.M. 1955. A guide to the birds of Ceylon. Oxfor Univ. Press. 432 pp.

Hogan, J.T. 1970 Footnote facts on rice. Rice. J. 73:5.

Jayasekara, E.H.W. & V. Velmurugu. 1976. Weed control in rice. pp. 132-138 In: Proc. Symp. Res. Prod. Rice in Ceyl.

Koch, W. 1952. Zur Flora der oberitalienischen Reisfelder. Ber. Schweiz. bot. Ges. 62:628-670.

Lim, R.P. 1980. Population changes in some aquatic invertebrates in ricefields. pp. 971-980 In: J.I. Furtado (Ed.). Tropical ecology and development. Proc. 5th Int. Symp. Trop. Ecol., Kuala Lumpur. Lord, L. 1927. Land crabs in paddy fields. Trop. Agric. 69:141-144.

Meijen, V.A. 1940. [Fish culture in ricefields]. (In Russian). Moscow. 96 pp.

Mendis, A.S. & C.H. Fenando. 1962. A guide to the freshwater fauna of Ceylon. Bull. Fish. Res. Stn. Ceylon 12:1-162.

Merlier, H. 1978. Weed problems in pluvial rice cultivation in Ivory Coast. pp. 191-200 In: I.W. Buddenhagen & G.J. Persley (Eds.). Rice in Africa. Acad. Press, London.

Metcalf, R.L. 1971. Pesticides. J. Soil Wat. Cons. 26:44-65.

Moroni, A. 1961. L'ecosistema di risaia (Monografia). Ann. Fac. Agr. dall'U.C.S.C. 31:8-53.

Mukhamediev, A.M. 1960. [Contribution to the hydrobiology of the ricefields of the Fergana Valley]. (In Russian) Sci. Notes. Uzbeqk State Pedagog. Inst. Fergana (Biol.) 6:1-82.

Neale, J.W. 1977. Ostracods from the ricefields of Sri Lanka (Ceylon). pp. 271-283 In: H. Löffler & D.C. Danielopol (Eds.). 6th Int. Ostr. Symp., Saarfelden, Austria.

Nishida, T. & T. Torii. 1970. A handbook of field methods for research on rice stem-borers and their natural enemies. I.B.P. Handbook 14:1-132. Blackwell, Oxford.

Panabokke, C.R. 1978. Rice soils of Sri Lanka. pp. 19-33 In: Soils and rice. Int. Rice Res. Inst. Los Banos, Philippines.

Pashitonova, Z.A. 1935. [Microfauna of a typical rice field in a Samarkand Oasis]. (In Russian) Acta Univ. Asiae Mediae 18:1-64.

Patrick, W.H. & C.N. Reddy. 1978. Chemical changes in rice soils. pp. 361-379 In: Soils and rice. Int. Rice Res. Inst.

Paul, W.R.C. 1945. Paddy cultivation. Ceylon Govt Press. 86 pp.

Pearsall, W.H. 1938. The soil complex in relation to plant communities. 1. Oxidation reduction potentials in soil. J. Ecol. 26:180-193.

Ponnamperuma, F.N. 1955. The chemistry of submerged soils in relation to the growth and yield of rice. Ph.D. thesis, Cornell University. 208 pp.

Ponnamperuma, F.N. 1965. Dynamic aspects of flooded soil. pp. 295-328 In: The mineral nutrition of the rice plant. Johns Hopkins Press, Baltimore.

Ponnamperuma, F.N. 1977. Physicochemical properties of submerged soils in relation to fertility. IRRI Research Pap. 5:32 pp.

Ramiah, K. 1937. Rice in Madras, a popular handbook. Govt Pres, Madras. 88 pp.

Roschevicz, R.J. 1931. [A contribution to the knowledge of rice]. (In Russian). J. Appl. Bot. Gent. Plant Breeding 27:1-133.

Schreven, D.A. van & W.H. Sieben. 1972. The effect of storage of soils under waterlogged conditions upon subsequent mineralization of nitrogen, nitrification and fixation of ammonia. Pl. Soil 37:245-253.

Seneviratna, S.N. de S. 1980. Research on rice diseases in Sri Lanka. A review. In: Rice Symp. Dept Agric. Sri Lanka. 30 pp.

Takai, Y. 1969. The mechanism of reduction in paddy soil. Jpn Agric. Res. Q. (JARQ) 4:20-23.

Temprosa, R.M. & Z. Sgebadah. 1980. Preliminary bibliograohy of rice-fish culture. ICLARM Bibliog. (Manila) 1:1-20.

Thirukkanasan, A., S.A. Kulasooriya & K. Theirvendrarajan. 1977. A periodic survey of blue greens found abundantly during the monsoon period in a paddy field near Jaffna campus at Vaddukoddai and a general survey of blue greens in the Jaffna Peninsula during the same period. Proc. Assoc. Adv. Sci. Sri Lanka 33, part 1:50-51.

Velmurugu, V. 1980. A review of weed control in rice. In: Rice Symp. Dept Agric. Sri Lanka. 43 pp.

Victor, R. & C.H. Fernando. 1980. Freshwater Ostracoda from the ricefields of South East Asia. pp. 957-970 In: J.I. Furtado (Ed.). Tropical ecology and development. Proc. 5th Symp. Int. Soc. Trop. Ecol., Kuala Lumpur.

Weereratna, C.S. 1977. Residual toxicity of some herbicides. J. Nat. Sci. Counc. Sri Lanka 5:147-155.

Weereratna, C.S. 1981. Effect of submergence on subsequent nitrification in a wetland cultivated rice soil. Pl. Soil 59:407-414.

Weerekoon, A.C.J. 1950. Some animals in the paddy field. Loris (Ceylon) 7:1-8.

Weerekoon, A.C.J. & E.L. Samarasinghe. 1958. Mesofauna of the soil of a paddy field in Ceylon. Prelim. Surv. Ceylon. J. Sci. (Bio. Sci.) 1:155-170.

Yamane, I. 1974. Solution pH of submerged soils. Rep. Inst. Agric. Res. Tohoku Univ. 25:1-11.

6. Freshwater invertebrates: some comments

C.H. Fernando

The publication of a general chapter on freshwater invertebrates of Sri Lanka by Fernando (1980c) in the Colombo Museum Centenary volume makes it unnecessary to deal with this subject fully again. I wish, however, to give a brief presentation in order to provide an updated perspective of studies to date. To accomplish this I shall refer specifically to studies made after those referred to in Fernando (1980c). Also some interesting features of the faunal composition which have been noted previously are repeated here. I think it is apt to make a few comments on the historical aspects of the study of the freshwater fauna of Sri Lanka because some of the earlier work has an important bearing on subsequent, present and future work in the field in Sri Lanka and elsewhere too.

The freshwater fauna of Sri Lanka is better known systematically than that of any other tropical country. The documentation of the substantial literature on this subject is contained in 'A guide to the freshwater fauna of Ceylon' by Mendis & Fernando (1962) and four supplements by Fernando (1963, 1964, 1969, 1974). Fernando (1974) categorized the level of systematic knowledge of the 50 groups of freshwater fauna as well known (A), fairly well known (B) and poorly known (C). Five groups were placed in the A category, while 20 and 25 groups were placed in the B and C categories respectively. In Fernando (1980c), now expanded to 52 groups, 9 were listed in A, 21 in B and 22 in C.

At the present time, taking into account literature published subsequently and the chapters in this volume, the Temnocephalida, Ostracoda, Macrura and Plecoptera should be moved into category A and the Simuliidae should be now in category B and not C.

In the present volume six of the 22 chapters deal largely with the systematics of freshwater invertebrates. This bias towards the freshwater fauna is merely a reflection of the relatively high emphasis given in systematics research to the freshwater biotope as compared to the terrestrial and marine biotopes. Three of these six chapters have contributed to the upgrading of systematic knowledge, namely the chapters on Ostracoda, Meso and Macrocrustacea and the Simuliidae. Also the work of Kawai (1973-1975) and Zwick (1980, 1982) has made the Plecoptera well known.

Compositional and zoogeographic features of the freshwater invertebrate fauna have been discussed by Fernando (1980c). Additional comments on these two features are given in the relevant chapters in the present volume. I shall mention a few compositional features. Fernando (1971), Wroblewski (1972) and Fernando (1980c) have commented on the high endemicity of the wet zone (southern) stream fauna based on Micronectinae, Parastenocarididae and fishes. More recently Senanayake *et al.* (1977), basing their studies on fishes, amphibians and reptiles, stated that 'the southern forests and highlands of Sri Lanka are a unique biological treasure trove of evolutionary relicts'. These comments also are substantiated by Crusz (this volume).

Sri Lanka received relatively early attention to the fauna as compared to other tropical regions of the world. An indigenous zoologist, Kelaart (1852), published a relatively detailed faunal study. Two eminent naturalists, Emerson Tennent and Ernst Haeckel, travelled in Sri Lanka and wrote treatises. Tennent (1861) lists a number of aquatic animals including insects and leeches and very surprisingly, Rotifera. Haeckel (1883) included natural history in his travelogue.

A major landmark in the study of the freshwater fauna of Sri Lanka was the work of the outstanding Hungarian biologist Daday (1898) on the microscopic freshwater fauna of Ceylon. He summarised earlier work on the zooplankton of Sri Lanka and described a number of crustacean species, mainly Cladocera. These species, like many small freshwater invertebrates, are widely distributed.

Recent sophisticated studies on the systematics of the Cladocera (e.g. Frey 1982) has necessitated the accurate characterization of Daday's species. Fortunately the types of Daday's (1898) material are available in Budapest. Rajapaksa & Fernando (1982a, 1982b) have studied this material as part of an ongoing systematic study of Daday's types to provide up-to-date, accurate descriptions.

Perhaps the earliest ecological study of tropical lakes is the work of Apstein (1907, 1910) on two man-made lakes in Sri Lanka. There was, however, a long hiatus in subsequent studies. Ecological studies in Sri Lanka were reviewed by Fernando & Indrasena (1969), Costa (1980) and Fernando (1980c). As elsewhere in the tropics ecological studies lag behind those of the temperate zone and depend heavily on sound systematic studies. In recent years there has been an upsurge of ecological work on tropical freshwaters mainly because of the concern on a worldwide scale of the threat of widespread and intensive aquatic pollution and the economic value of freshwaters, especially in the tropics, for fisheries and fish culture.

Sri Lanka has benefitted from two recent expeditions sponsored by Lund University (Brinck *et al.* 1971) and the University of Vienna (Starmühlner 1972) and a more recent joint venture with an international team to study the ecology of a reservoir (Schiemer 1980, 1983). These have provided a great deal of sophisticated systematic work and important insights on ecological aspects of tropical fresh-waters.

Fernando (1980c) made a detailed study of the freshwater zooplankton of Sri

Lanka. This work has provided the data for some important generalization on the composition of tropical freshwater zooplankton. Fernando (1980b) combined the data from this study with published data from tropical and non-tropical regions and showed that (a) the limnetic zooplankton was less diverse (especially Cladocera) in the tropics than in the temperate region, a reversal of the usual situation, (b) the genus *Daphnia*, common in temperate zooplankton, was poorly represented in species diversity and rare in the tropical region and (c) the common limnetic zooplankton in tropical regions consists exclusively of eurytopic species common in the full range of standing freshwater habitats in the region, i.e. there are no typical limnetic species as opposed to littoral species as in higher latitudes.

In summary, the Sri Lankan freshwater fauna is relatively well known taxonomically. The availability of reliable and extensive documentation of the invertebrate fauna of freshwaters makes Sri Lanka a very suitable place for ecological studies on freshwaters. The range of freshwaters in Sri Lanka is wide and the climatic patterns diverse. The only types of habitats which are conspicuous by their absence are very large lakes and large rivers.

In the past, co-operative work with foreign institutes has been fruitful in providing a great deal of valuable systematics and ecological work. There is now a rapidly growing group of indigenous researchers on freshwater flora, fauna and the ecology of freshwaters.

References

Apstein, C. 1907. Das plancton im Colombo-See auf Ceylon. Zool. Jb. (Abt. Syst.) 25:201-244.

Apstein, C. 1910. Das plancton der Gregory-Sees auf Ceylon. Zool. Jb. (Abt. Syst.) 29:661-680.

Brinck, P., H. Andersson & L. Cederholm. 1971. Reports from the Lund University Ceylon expedition in 1962. Report No. 1, Introduction. Ent. Scand. Suppl. 1:4-36.

Costa, H.H. 1980. The physical, chemical and biological characteristics of freshwater bodies in the lowlands of Sri Lanka. Spol. Zeylan. 35:43-99.

Daday, E. 1898. Microscopische Süsswasserthiere aus Ceylon. Termesz. Füz. (Budapest) 21:1-123.

Fernando, C.H. 1963. A guide to the freshwater fauna of Ceylon, Suppl. 1. Bull. Fish. Res. Stn Ceylon 16:29-38.

Fernando, C.H. 1964. A guide to the freshwater fauna of Ceylon, Suppl. 2. Bull. Fish. Res. Stn Ceylon 17:177-211.

Fernando, C.H. 1969. A guide to the freshwater fauna of Ceylon, Suppl. 3. Bull. Fish. Res. Stn Ceylon 20:101-134.

Fernando, C.H. 1971. The role of introduced fish in fish production in Ceylon's freshwaters. pp. 295-310 In: E. Duffey & A.S. Watt (Eds.). The scientific management of animal and plant communities for conservation. Blackwell, Oxford.

Fernando, C.H. 1974. A guide to the freshwater fauna of Ceylon (Sri Lanka), Suppl. 4. Bull. Fish. Res. Stn Sri Lanka (Ceylon) 25:27-81.

Fernando, C.H. 1980a. The freshwater zooplankton of Sri Lanka with a discussion of tropical freshwater zooplankton composition. Int. Rev. ges. Hydrobiol. 65:85-125.

Fernando, C.H. 1980b. The species and size composition of tropical freshwater zooplankton with special reference to the Oriental Region (South East Asia). Int. Rev. ges. Hydrobiol. 65:411-426.

Fernando, C.H. 1980c. The freshwater invertebrate fauna of Sri Lanka, Spol. Zeylan. 35:14-42.

Frey, D.G. 1982. The reticulated species of *Chydorus* (Cladocera,Chydoridae): two new species with suggestions of convergence. Hydrobiologia 93:255-279.

Haeckel, E. 1883. A visit to Ceylon. Kegan Paul, French & Co., London. 337 pp.

Kawai, T. 1975. Plecoptera from Ceylon. Ent. Scand. Suppl. 4:65-78.

Kelaart, E.F. 1852. Prodromous faunae Zeylanicae. Ceylon Govt. Press, Colombo. 250 pp.

Mendis, A.S. & C.H. Fernando. 1962. A guide to the freshwater fauna of Ceylon. Bull. Fish. Res. Stn Ceylon. 12:1-160.

Rajapaksa, R. & C. H. Fernando. 1982a The Cladocera of Sri Lanka (Ceylon), with remarks on some species. Hydrobiologia 94:49-69.

Rajapaksa, R. & C.H. Fernando. 1982b. The first description of the male and ephippial female of *Dadaya macrops* (Daday, 1898) (Cladocera,Chydoridae), with additional note on this common tropical species. Can. J. Zool. 60:1841-1850.

Schiemer, F. (Ed.) 1980. Parakrama Samudra, Sri Lanka limnology project. Interim Rept Inst. Int. Coop. Vienna. 112 pp.

Schiemer, F. (Ed.) 1983. Parakrama Samudra, Sri Lanka project. Junk, The Hague. 236 pp.

Senanayake, F.R., M. Soule & J.W. Sennar. 1977. Habitat values and endemicity in the vanishing rain forests of Sri Lanka. Nature 265:351-354.

Starmühlner, F. 1972. The Austrian-Ceylonese Hydrobiological Mission 1970. Part 1. Introduction and description of stations. Bull. Fish. Res. Stn Sri Lanka (Ceylon) 32:43-76.

Tennant, E. 1861. Sketches of the natural history of Ceylon. Longman, Green, Longman & Roberts, London. 500 pp.

Wroblewski, A. 1972. Further notes on the Micronectinae from Ceylon. Bull. Ent. Polon. 52:5-52.

Zwick, P. 1980. The genus *Neoperla* (Plecoptera:Perlidae) from Sri Lanka. Oriental Insects 14:263-269.

Zwick, P. 1982. A revision of the Oriental stonefly genus *Phanoperla* (Plecoptera:Perlidae). Syst. Ent. 7:87-126.

7. Monogenea from freshwater fishes

A.V. Gussev & C.H. Fernando

Introduction

The Monogenea of the whole Oriental region are relatively poorly known. Yamaguti (1963) lists few species from the region. More recently some major contributions have been made to the systematics of Monogenea of freshwater fishes of Sri Lanka (Gussev 1963), India (Gussev 1976) and Malaysia (Lim 1979). The present account is a brief report on the Monogenea recorded from freshwater fishes in Sri Lanka.

The recent work of Gussev (1978) on Indian freshwater fish monogeneans has enabled a listing of probable species on the 55 species of freshwater fishes found in Sri Lanka. Only 12 of these species have been examined.

There is certainly a rich and interesting fauna of freshwater fishmonogeneans awaiting study in the varied and numerous fishes found in the Oriental Region. The recent study of Lim (1979) of fish from a single habitat provided a rich Monogenea fauna.

Species recorded

At present 23 species of Monogenea are known from freshwater fishes of Sri Lanka. Material on 21 of them was obtained from 12 species of formalin preserved fishes presented to AVG by CHF during the stay of the research vessel 'Ob' (Soviet Marine Antarctic Expedition) in Colombo in 1957. Two species of ancyrocephalins from thestomach of fish were collected later (Gussev & Fernando 1973).

All the species of Monogenea recorded in Sri Lanka belong to the suborder Dactylogyrinea of the order Dactylogyridea. A list is given below indicating the region where their hosts were found. All the species were found on gill filaments of fishes except for members of the genus *Enterogyrus* which live in the folds of the stomach walls.

List of species, localities and fish hosts

Fam. Dactylogyridae

1. *Dactylogyrus aequipinnati* - Meegoda - *Danio aequipinnatus*
2. *D. curiosus* - neighbourhood of Colombo - *Rasboro daniconius*
3. *D. daniconii* (Syn. *D. pedunculatus*) - Nugegoda - *Rasbora daniconius*
4. *D. saranae* - Nugegoda, Meegoda - *Puntius sarana*
*5. *Dactylogyroides bimaculati* - Meegoda - *Puntius bimaculatus*
6. *D. dorsalis* - Meegoda - *Puntius dorsalis*
7. *D. fernandoi* - Meegoda - *Puntius dorsalis*
8. *D. tripathii* - Nugegoda - Puntius filamentosus, P.melanamphys sinhala, P. sarana
9. *D. vittati* - neighbourhood of Colombo - *Puntius vittatus*

Fam. Ancyrocephalidae

Subfam. Ancyrocephalinae

10. *Ancyrocephalus* (s.l.) *aequalis* - Sri Lanka - *Rasbora daniconius*
11. *A. daniconii* - Sri Lanka - *Rasbora daniconius*
12. *A. danionis* - Meegoda - *Danio aequipinnatus*
13. *A. esomi* - Nugegoda - *Esomus danrica*
14. *A. etropli* - neighbourhood of Colombo - *Etroplus suratensis*
15. *A. heteranchoris* - Nugegoda - *Rasbora daniconius*
16. *A. kirtisinghei* - Sri Lanka - *Rasbora daniconius*
17. *A. rasborae* - Sri Lanka - *Rasbora daniconius*
18. *A. tripathii* - Sri Lanka - *Rasbora daniconius*
19. *Ceylonotrema colombensis* (Syn. *Onchiodiscus pterodiscoides*) - neighbourhood of Colombo - *Etroplus suratensis*
20. *Enterogyrus globodiscus* - Polonnaruwa - from stomach of *Etroplus suratensis*
21. *E. papernai* - Polonnaruwa - from stomach of *Etroplus suratensis*

Subfam. Ancylodiscoidinae

22. *Cornudiscoides jaini* - Nugegoda - *Macrones keletius*
23. *Bifurcohaptor lanki* - Nugegoda - *Macrones keletius*

* Record of three testes in the original description of the genus apparently is erroneous.

Comments

In the paper by Venkatanarsaiah & Kulkarni (1980) *Ceylonotrema colombiensis* was identified as *O. pterodiscoides* with mistakes in the redescription (with regard to the number of marginal hooks, treating their anterior pair as 'onchium'). The first description of *C. colombensis* could possibly contain mistakes in the data or anatomical structure and body shape, since they could be obtained only from material that was badly fixed. However features that confirm the correctness of the indentification of Gussev (1962) of *C. colombensis* and *O. pterodiscoides* were revealed with sufficient exactness (the number of marginal hooks, nature of 'onchium', shape of other chitinoid structures, host). The authors of the above paper may not agree to this view, but their disregard of Gussev's work (1963, 1976, 1978) and also of the presence of no less than 7 pairs of marginal hooks in all members of Dactylogyridea seems at least puzzling.

The descriptions of all the Sri Lanka species with figures and measurements were given in Gussev (1963, 1976) and Gussev & Fernando (1973). Gussev (1976) also discussed critically and in detail the morphological features of the Indian genera and some aspects of the zoogeography of Oriental species.

To judge by the list of freshwater fishes of Sri Lanka (Munro 1955), it is probable that at least 31 other monogenean species described from fishes of India and Pakistan will be found on further investigation. These species are as follows:

Dactylogyrus barbi - Lucknow - *Puntius sarana*

D. barbusi - (*D. sarani* = Neodactylogyrus indicus) - Rivers of India and Pakistan - *Puntius stigma, P. sarana* D. bucinus - Bhavanisager water reservoir - *Puntius dorsalis*

D. catlarius - Lucknow - *Catla catla*

D. hyderabadensis - Hyderabad - *Puntius sarana*

D. kalyanensis - Fish farm, Kalyani - *Catla catla*

D. moorthyi - Mysore state - *Puntius ticto*

D. rizvii - Indus river - *Puntius sarana*

D. sarani (= *Neodactylogyrus indicus* = D. barbusi?) - Rivers of India and Pakistan - *Puntius sarana, P. stigma*

D. sphyrnoides - Bhavanisagar reservoir - *Puntius sarana*

D. spinitubus - Bhavanisagar reservoir - *Catla catla*

Dogielius catlaius - Lucknow - *Catla catla*

Dactylogyroides longicirrus - Indian ponds - *Puntius ticto*

D. zulfikari - Indus river - *Puntius sarana*

Ancylodiscoides micracanthus - Hyderabad - *Wallaga attu*

Silurodiscoides gomtius - Lucknow - *Wallago attu*

S. indicus - Hyderabad, Bhavanisagar reservoir - *Wallago attu*

S. longicirrus - Indian rivers - *Wallago attu*

S. parvulus - Lucknow - *Macrones vittatus*

S. sudhakari - Bhavanisagar reservoir - *Wallago attu*
S. tengra - Indian rivers - *Macrones gulio*
S. wallagonius - Lucknow, Hyderabad - *Wallago attu*
Cornudiscoides geminus - Lucknow - *Macrones vittatus*
C. proximus - Lucknow - *Macrones vittatus*
Bychowskyella wallagonia - Lucknow - *Wallago attu*
Mizelleus indicus - Lucknow - Wallago attu
M. linorchis - Hyderabad - *Wallago attu*
Bifurcohaptor - Lucknow - *Macrones vittatus, M. tengara*
Gyrodactylus ('elegans') *indicus* - Ponds in Barrackpore - *Catla catla* and some
    other Indian fishes.
Metagyrodactylus indicus - West Benghal - *Ophicephalus marulius*
Diplozoon indicum - Lucknow - *Puntius sarana*

Notes on zoogeography

The monogenean fauna of Sri Lanka as well as the ichthyofauna is most closely related to the Indian. Sri Lanka has in fact a considerably impoverished Indian fauna. This refers both to common species and to endemic fishes that are obviously genetically connected with continental components. The depauperate fish fauna of Sri Lanka freshwaters of only 55 species means fewer Monogenea.

Therefore the description of Indian monogeneans given earlier by Gussev (1976, 1978) in general remains valid for the fauna of Sri Lanka. Its main features are:
1) considerable percentage of dactylogyrids, ancyrocephalins and ancylodiscoidins (in accordance with prevalence of their hosts in the ichthyofauna);
2) presence of only one bar (dorsal) of haptor in most dactylogyrine;
3) absence of dactylogyrine species with I and ⊥-shaped ventral bar;
4) presence of many species (from different genera) with attachment of parasites by means of embracing of gill filaments by anchors (*Dactylogyroides, Bifurcohaptor, Cornudiscoides*).

Unfortunately, it is impossible now to characterize the composition or differences in the fauna of monogeneans of the different rivers: Mahaveliganga, Kelaniganga, Aruviaru, Kumbukanoya, etc. The main material was found in the vicinity of Colombo and some fragmentary material from Polonnaruwa. Therefore it isimpossible to perform zoogeographic analysis of the parasites within theisland before research is done in different rivers.

A recent study on Malaysian freshwater fish Monogenea (Lim 1979) shows that the Malaysian species are quite different from those of the Indian subcontinent. Three new genera also were described from 15 species of fish mainly from one habitat, a swamp lake, Tasek Berah. Ha Ky (1971) found 23 monogeneans in only 16 species of freshwater fishes in Vietnam.

Only a few species, and relatively few specimens, of fish in the Oriental region have been examined for Monogenea. It is therefore likely that the present list of Monogenea in freshwater fishes of Sri Lanka and the Oriental region will be greatly increased with more intensive and extensive studies of Monogenean systematics.

References

Gussev, A.V. 1963. New species of Monogenoidea from fishes of Ceylon. Bull. Fish. Res. Stn Ceylon 16:53-93.
Gussev, A.V. 1973-74. Freshwater Indian Monogenoidea. Principles of systematics, analysis of the faunas and their evolution. Ind. J. Helm. 25-26:1-241.
Gussev, A.V. 1978. [Monogenoidea of freshwater fishes. Principles of systematics, analysis of world fauna and its evolution]. Parasitol. Sbornik. 28:96-198 (in Russian).
Gussev, A.V. & C.H. Fernando. 1973. Dactylogyridae (Monogenoidea) from the stomach of fishes. Folia Parasitol. (Prague) 20:207-212.
Ha Ky. 1971. New species of monogeneans from freshwater fishes of North Vietnam Part 2. Parasitologia 5:429-440.
Lim, Lee-Hong. 1979. Monogenea and Digenea of freshwater fishes mainly from Tasek Bera, Malaysia. M.Sc.thesis, University of Malaya, Kuala Lumpur. 201 pp.
Munro, I.S.R. 1955. The marine and freshwater fishes of Ceylon. Dept Ext. Affairs, Canberra. 351 pp.
Venkatanarsaiah, J., T. Kulkarni. 1980. Studies on the monogenetic trematodes of Andhra Pradesh: contribution to our knowledge of *Tympanocirrus* Tripathi, 1959 with a description of *T. trigoni* n. sp. and a rediscription of *Onchiodiscus pterodiscoides* Kulkarni, 1969. Revista Parasitol. 41:85-91.
Yamaguti, S. 1963. Systema Helminthum. Vol. 4. Monogenea and Aspidocotylea. Interscience, N.Y. 699 pp.

8. Freshwater Zooplankton

R. Rajapaksa & C.H. Fernando

Introduction

The animal components of the freshwater plankton are important in the production of freshwater ecosystems yet little is known of their role in tropical freshwater habitats.

The freshwater zooplankton in Sri Lanka has been studied sporadically by taxonomists and ecologists since the mid nineteenth century. Perhaps the earliest ecological study of tropical zooplankton was made in Sri Lanka by Apstein (1907, 1910). The accessibility of the island and its attractiveness to freshwater biologists travelling in the region has resulted in the zooplankton being studied by a series of scientists residing in many parts of the world.

Fernando (1980a) provided a detailed analysis of the zooplankton species found in all types of freshwater habitats in the country and has used these results as a basis for discussing the composition of tropical freshwater zooplankton.

The taxonomy of the zooplankton is well documented and almost all the species present have been recorded except perhaps in the Rotifera. The freshwater limnetic zooplankton in Sri Lanka, as in most tropical regions, is less diverse than that of the temperate zone. There is also a paucity of larger zooplankton species (Fernando 1980a), and predatory zooplankton are rare. The uniform high temperature, historical factors and fish predation seem to be the most probable explanations (Fernando 1980a, Duncan & Gulati 1981). The broad characteristics of composition of zooplankton in Sri Lanka is shared by most tropical regions.

The lack of endemicity and occurrence of many tropicopolitan or even cosmopolitan species are characteristic to the assemblage of Sri Lanka freshwater zooplankton.

In this chapter we discuss the species composition and distribution of the major zooplanktonic groups (Rotifera, Copepoda and Cladocera) and other faunal groups which contribute representatives to the plankton. The data on ecology and production of zooplankton is very meagre. However, available data relating to each group are discussed. The zoogeographical relationships of the zooplankton will be considered separately.

Composition and distribution

Protozoa

Of all the zooplankton groups, Protozoa are the most neglected group in Sri Lanka as elsewhere, and there are few published records. Most of these are based on the examination of a small number of samples (e.g. Daday 1898, Apstein 1907).

Over 40 species of free-living freshwater protozoans are known from Sri Lanka (Fernando 1974, 1981). Most of them are planktonic. Members of the thecate sarcodines, particularly *Difflugia, Arcella* and *Centropyxis* (Figs. 1-3) were frequently found, with *Arcella* spp. most common (Daday 1898, Apstein, 1910, Fernando 1980a). Duncan (1981) recorded *Lionotus* spp. (Ciliata) abundant in the limnetic zone of Pakrama Samudra, a man made lake, in 1980.

Rotifera

About 140 species belonging to 42 genera are known from Sri Lanka. The major papers dealing with Sri Lanka Rotifera are those of Daday (1898), in which he listed 46 species, Apstein (1907, 1910), Mendis & Fernando (1962), Fernando (1969, 1980a), Mendis (1964, 1965), Costa & De Silva (1969,1978), Chengalath & Fernando (1973a) and Chengalath *et al.* (1973, 1974). A comprehensive study of the taxonomy of Sri Lanka rotifers has been provided in the last three papers mentioned. These studies were based on material collected from over 300 habitats throughout the island.

In this account, species of both limnetic and littoral forms are included. Almost all the known species of Sri Lanka belong to the class Monogononta.

The planktonic rotifers are represented mainly by members of the families Brachionidae, Asplanchnidae, Synchaetidae, Hexarthridae, Conochilidae, Filinidae and Collothecidae. Species diversity is lower in the limnetic than in the littoral region. *Brachionus calyciflorus, B. falcatus, B. forficula, Keratella tropica, Asplanchna brightwelli* and *Hexarthra intermedia* are some of the most common and abundant species (Chengalath *et al.* 1974). Some of these species are illustrated in Figs. 4-8.

The genus *Brachionus* is represented by 26 species (Fernando 1980a). Green (1972) has recorded many *Brachionus* species in the tropics together with the absence of the temperate genus *Notholca*.

Sri Lanka has no natural lakes. Most of the limnetic species are eurytopic and commonly found in man-made reservoirs and other habitats. Only a few species (*Hexartha intermedia, Filinia opoliensis, Brachionus falcatus* and the rare *H. mira, Horaella brehmi* and *Kellicottia longispina* seem to be predominantly limnetic.

Of the littoral fauna, the diversity is greatest in the genus *Lecane* with 35 species;

of which four, *L. ceylonensis, L. lankae, L. kohouteki* and *L. plesiades* were recorded as new species recently (Chengalath *et al.* 1974, Chengalath & Fernando 1973). *Lecane bulla*, a cosmopolitan form, is the commonest species. The rotifers so far recorded have been included (except *L. kohouteki* and *L. scutali*) in Fernando (1980a).

In Sri Lanka, rotifers occur in a wide range of freshwater habitat types, namely, rivers, ponds, man-made reservoirs, rice fields, villus (similar to ox-bows or Reiss's (1977) Varzea type lakes), and small (temporary or permanent) water collections. More than 40 species were recorded as common species, since they occurred in most of the habitat types (Fernando 1980a). Of these *Lecane bulla* and *Euchlanis dilatata* were the commonest species occurring in all the habitat types (Chengalath *et al.* 1974, Fernando 1980a). Although *E. dilatata* is littoral, it is often found in open water specially in eutrophic conditions (Ruttner-Kolisko 1974).

No endemic genera are known for Sri Lanka. Of the new species recorded, *L. ceylonensis* has been recorded from the Philippines (Mammaril & Fernando 1978). Most of the recorded species are cosmopolitan or tropicopolitan.

A few ecological investigations have been done on Sri Lanka rotifers. Apstein studied the zooplankton in two different lakes (Lake Colombo, 1907 and Lake Gregory, 1910). He recorded the seasonal variations in numbers of some rotifer species. Also he mentioned an endo-parasitic protozoan (*Plistophora* sp.) from one of the *Brachionus* species.

According to the studies of Costa & De Silva (1978), there is no clear seasonal pattern in rotifer occurence in Lake Colombo (Beira). The maximum number of rotifers has been recorded in the drier months (especially June and July) (Apstein 1907, 1910) or before the monsoon rains (Costa & De Silva 1978). Fernando & Rajapaksa (1983) studied the zooplankton of a man-made reservoir (Parakrama Samudra) in the dry zone of Sri Lanka, over a period of 3 years. They recorded rotifers as the dominant zooplankton group, during most of the study period. There were no clear seasonal fluctuation. The dominance of rotifers in this reservoir is also recorded by Duncan & Gulati (1981) and Schiemer (1981). It appears that the dominant zooplankters were Rotifera following a cyclone, after which the water level was reduced. However, crustaceans have become more prominent after high water levels have been restored (Fernando & Rajapaksa 1983).

Cladocera

Thanks to the work of both earlier and recent authors (Brady 1886, Poppe & Mrázek 1895, Daday 1898, Bär 1924, Apstein 1907, 1910, Gurney 1916, Fernando & Ellepola 1969, Fernando 1974, 1980, Rajapaksa & Fernando 1982), the taxonomic knowledge of freshwater Cladocera of Sri Lanka is better than that of other tropical countries. Daday (1898) recorded 30 species, of which 17 were described as new. There are some doubts attached to Poppe & Mrázek's (1895) findings. They recorded five species, four of which (*Leptodora kindtii, Bosmina japonica, Daphnia galeata* and *Chydorus ovalis*) have not since been recorded. These are temperate species. Fernando (1974, 1980a) considered *L. kindtii* as an introduced species. Unfortunately, none of Poppe or Mrázek's material is available for re-examination. The record of *D. galeata* probably refers to *D. lumholtzi*, a common small *Daphnia* species found in man-made reservoirs and ponds in Sri Lanka.

A total of 68 species of Cladocera belonging to 7 families (Sididae, Daphniidae, Bosminidae, Moinidae, Macrothricidae, Chydoridae and Leptodoridae) have been recorded. A summary of the recorded cladoceran genera is given below (numbers in parentheses refer to the number of species).

Sididae: *Diaphanosoma* (4), *Pseudosida* (1), *Latonopsis* (1)
Daphniidae: *Ceriodaphnia* (2), *Daphnia* (3), *Scapholeberis* (1), *Simocephalus* (3)
Moinidae: *Moinodaphnia* (1), *Moina* (2)
Bosminidae: *Bosmina* (2), *Bosminopsis* (1)
Macrothricidae: *Macrothrix* (5), *Grimaldina* (1), *Guernella* (1), *Ilyocryptus* (1)
Chydoridae: *Pleuroxus* (1), *Alonella* (1), *Chydorus* (8), *Pseudochydorus* (1), *Dunhevedia* (2), *Daday* (1), *Alona* (11), *Leydigia* (2), *Graptoleberis* (1), *Kurzia* (1), *Oxyurella* (1), *Euryalona* (1), *Indialona* (1)
Leptodoridae*: *Leptodora* (1).

The composition of the Cladocera in Sri Lanka is typically tropical with few *Daphnia* and *Ceriodaphnia* species which are the dominant elements of the zooplankton in the temperate zone (Fernando 1980a). Some of the factors influencing the species and size composition of zooplankton in the tropics are discussed in Fernando (1980b).

Of the species originally described from Sri Lanka by various authors, 12 (*Macrothrix triserialis (Brady 1886), Chydorus reticulatus, C. ventricosus, C. parvus, Dunhevedia serrata, Dadaya macrops, Kurzia longirostris, Oxyurella singalensis, Euryalona orientalis, Indialona globulosa, Indialona (Alona) macronyx* (Daday 1898) and *C. invaginatus* (Frey in press) are valid species.

Except for some members of the genera *Diaphanosoma, Daphnia, Moina* and the family Bosminidae, all the others are littoral species. However, some of these (e.g. *C. (Ephemeroporus) barroisi, C. sphaericus, M. triserialis*) often are found in open waters. *Diaphanosoma excisum, Ceriodaphnia cornuta* and *Moina micrura* (Figs 9-15) are the dominant limnetic species in Sri Lanka (Apstein 1907, Fernando 1980a, Rajapaksa & Fernando in press). The abundance of these three species or related ones in the limnetic zone in the tropics has been recorded by various

Leptodora kindtii probably is introduced.

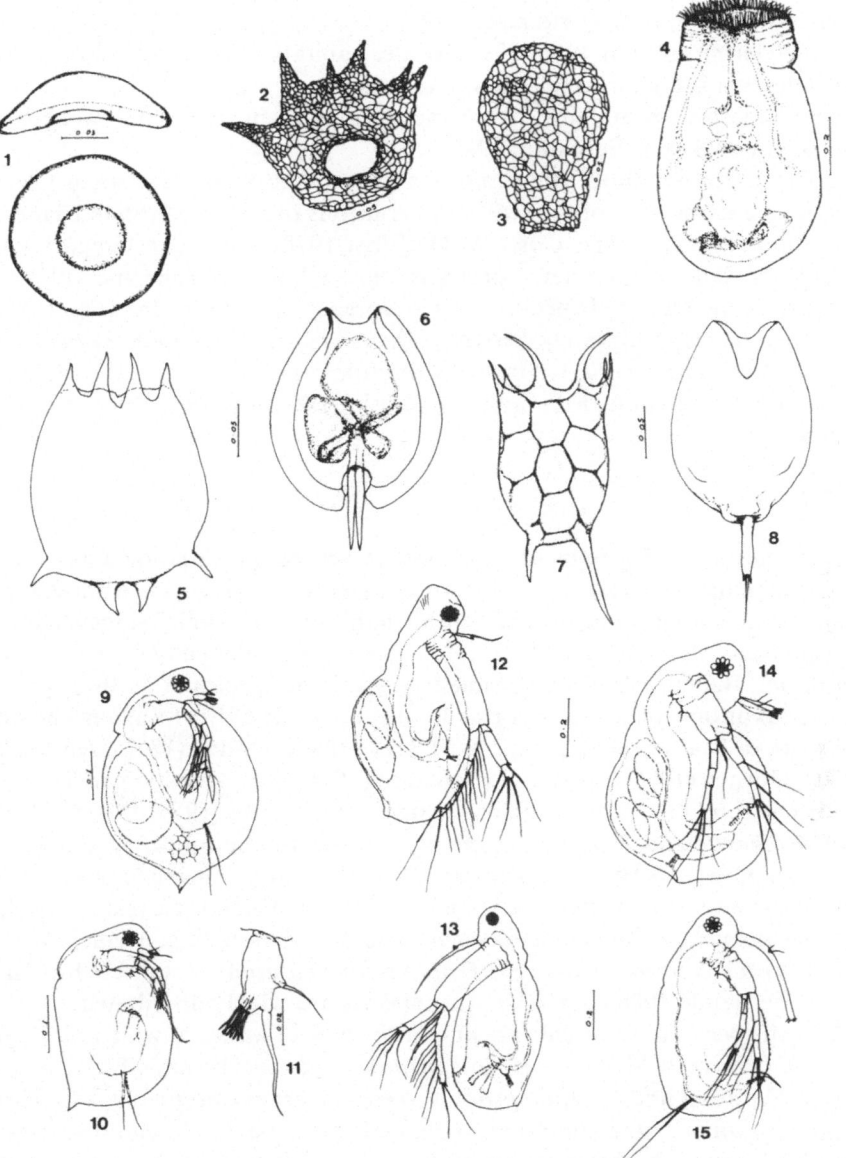

Fig. 1. *Arcella* sp. lateral and ventral view; 2. *Centropyxis* sp.; 3. *Difflugia* sp.; 4.*Asplanchna brightwelli*; 5. *Brachionus calyciflorus*; 6. *Euchlanis dilatata*; 7. *Keratella tropica*; 8. *Lecane bulla*.

Fig. 9-11. *Ceriodaphnia cornuta:* 9. Parthenogenetic female; 10. Male; 11. Antennule of male.

Fig. 12-13. *Diapohanosoma excisum:* 12. Perthenogenetic female; 13. Male.

Fig. 14-15. *Moina micrura:* 14. Parthenogenetic female; 15. Male.

160

authors (Brook & Rzoska 1954, Green 1962, Brehm 1933, Burgis *et al.* 1975, Fernando 1980b, Idris & Fernando 1981).

Considering different types of freshwater habitats in Sri Lanka, species diversity of Cladocera is high in small man-made reservoirs and in ponds. The number of species recorded from villus, rivers and streams is considerably lower (Fernando 1980a, Rajapaksa & Fernando 1982).

Little is known about the ecology and production of Sri Lanka Cladocera. According to the data of the seasonal variations of zooplankton in 2 lakes (Lake Colombo (Apstein 1907, Costa & De Silva 1978) and Lake Gregory (Apstein 1910)), the density of limnetic species is high in January, June and July.

Most of the recorded species are widespread throughout the island. Many of them are tropicopolitan species (e.g. *D. macrops, D. serrata, Ephemeroporus barroisi* etc.). The genera *Acroperus, Camptocercus* and some of the species found in neighbouring countries (India, Malaysia, Indonesia, etc.) are absent in Sri Lanka.

Copepoda

The species of the 3 groups of Copepoda (Cyclopoida, Calanoida and Harpacticoida) are found in a variety of freshwater habitats in Sri Lanka. Although most of the copepods occur in littoral or benthic habitats, their larval stages (nauplii and copepodites) represent a considerable percentage numerically of the pelagic zooplankton. Therefore it is worthwhile to discuss the species of all three groups.

The taxonomy of freshwater free-living copepods of Sri Lanka is known from the work of Brady (1886), Poppe & Mrázek (1895), Daday (1898), Apstein (1907, 1910), Gurney (1906, 1916) and Fernando (1974). The free-living cyclopoids are represented by 13 species, which are listed in Fernando (1974, 1980a). *Mesocyclops leuckarti* [1]and *Thermocyclops crassus* are the common cyclopoids in the pelagic region (Figs 16-24). Other species are very rare.[1] This species occurs only in Europe according to Kiefer (1981). In Sri Lanka *M. leuckarti* is really three species; *M. thermocyclopoides, M. aspericornis* and *M. ruttneri*. (e.g. *Metacyclops minutus, Microcyclops moghulensis*). *Tropocyclops confinus* is restricted to high altitudes (Fernando 1980a). All the other species are widely distributed.

Eleven species of calanoids are known from Sri Lanka, of which *Eudiaptomus drieschi* (Poppe & Mrázek), *Heliodiaptomus viduus* (Gurney), *Phyllodiaptomus annae* (Apstein) and *Paradiaptomus greeni* (Gurney) originally were described from Sri Lanka. *Megadiaptomus hebes* (Kiefer), a giant calanoid species (body length 4.0 mm) was recorded recently from Sri Lanka by Fernando & Hanek (1976). In the Calanoids, only *Phyllodiaptomus annae* is a common and eurytopic species (Fernando 1980a).

Three species of harpacticoids are known. Only *Attheyella cingalensis (Brady)* seems to be an endemic species. *Elaphoidella grandidieri* occurs in almost all the habitat types (Fernando 1980a).

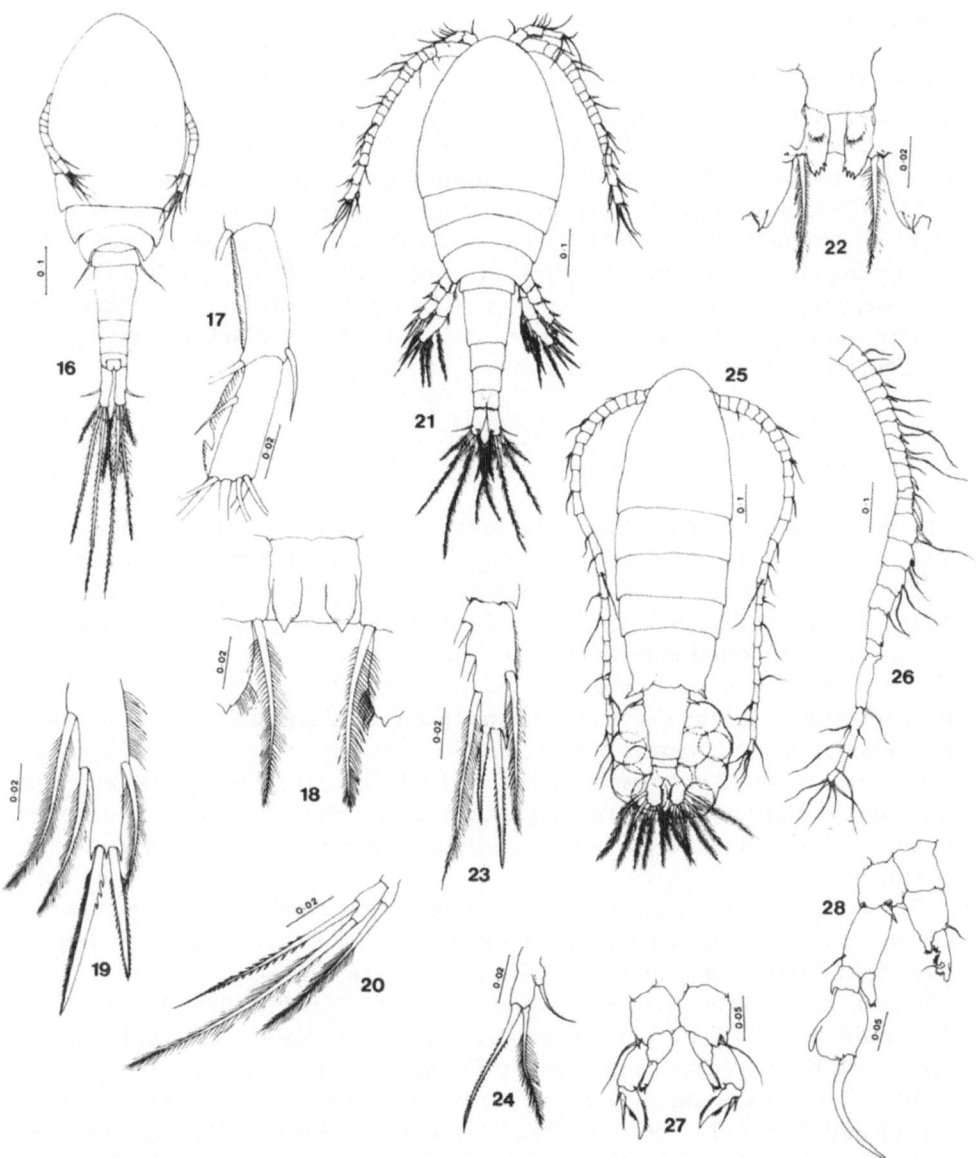

Fig. 16-20. Mesocyclops thermocyclopoides female: 16. Dorsal view; 17. Terminal segments of antennule; 18. Leg P₄, connecting membrane; 19. Terminal segment of endopodite of leg P₄; 20. Leg P₅.
Fig. 21-24. Thermocyclops crassus female: 21. Dorsal view; 22. Leg P₄ connecting membrane; 23. Terminal segment of endopodite of Leg P₄; 24. Leg P₅.

Fig. 25-28. Phyllodiaptomus annae: 25. Female, dorsal view; 26. Prehensile antennule of male; 27. Leg 5 of female; 28. Leg 5 of male.

Ostracoda

A few species of Ostracoda occur in zooplankton. Taxonomic studies of Sri Lankan Ostracoda have been done by Brady (1886), Daday (1898), Apstein (1907, 1910), Gurney (1916), Neale (1977) and Neale & Victor (1978a, 1978b). Most of the recorded species are benthic. Some occur among the aquatic vegetation.

Strandesia purpurascens is a planktonic species which was originally described from Sri Lanka by Brady (1886) as *Cypris purpurascens*. The occurrence of this species in plankton was noted by Brady (1886), Apstein (1907) and Neale (1977).

The ecology of freshwater ostracods in Sri Lanka is poorly known but they form an important portion by weight of the benthic biomass of shallow habitats and can have high densities in the zooplankton.

Other fauna

Members of other faunal groups occasionally are found in zooplankton samples. They are mainly the larval and adult stages of some freshwater invertebrates.

The floating statoblasts of bryozoans were recorded by many authors (Apstein 1907, 1910, Fernando 1974, 1980a, b). The statoblasts of the ectoproct *Plumatella repens* often are found in a variety of habitats.

The planktonic stages of the parasitic cyclopoid copepods were recorded from the zooplankton samples of man-made reservoirs (Fernando 1980a). Four genera of parasitic copepods were recorded from Sri Lanka, namely *Ergasilus, Paraergasilus, Lernaea* and *Lamproglaena* (Fernando & Hanek 1973a, b). *Paraergasilus brevidigitus* and *Lernaea cyprinacea* have been introduced with imported species of Chinese carp and trout (Mendis & Fernando 1962, Kirtisinghe 1964, Fernando & Hanek 1973a).

Branchiura (fish-lice) is another group found in the zooplankton samples. A few species of branchiuran have been recorded from Sri Lanka. Species of *Argulus* have been introduced with fish (Fernando & Hanek 1973a).

Hydracarines often appear in plankton samples (Daday 1898, Apstein 1910, Fernando 1974, 1980a). Some of the Sri Lankan species are given in Fernando (1974) and their occurrence recorded in Fernando (1980a).

Juveniles of atyid and paleomonid prawns were recorded by Costa & De Silva (1978) and Fernando (1977, 1980a). Costa & De Silva (1978) reported their presence in Lake Colombo throughout the year and peaks of abundance during July and August. Larval forms of some aquatic insects (especially emphemeropterans and dipterans) as well as free-living nematodes also are recorded (Fernando 1980a). However, these groups are not well studied in Sri Lanka.

The true planktonic dipteran *Chaoborus* has been recorded in small and large reservoirs but is rare. Perhaps in most habitats it is benthic, especially during the day.

General comments

In general it can be said that the freshwater zooplankton of the limnetic zone is relatively poor in species as compared with temperate regions (Fernando 1980a). Also there are no typically limnetic crustaceans in the zooplankton in Sri Lanka freshwaters. All the species found in the limnetic region are eurytopic (Fernando 1980a). In the Rotifera a few species seem to be predominantly limnetic (Fernando 1980a).

The size of the crustacean zooplankton in Sri Lanka is smaller than in temperate regions. This is to be expected. However the rarity of *Daphnia* spp. is unexpected. Also the predaceous cladocerans *Leptodora* and *Polyphemus* are absent. Ostracoda occur commonly in tropical freshwater zooplankton. Often they go unrecorded because of the lack of expertise on this group of animals.

Seasonality, long term changes and production

The data available on seasonality of tropical freshwater zooplankton is meagre. Fernando & Rajapaksa (1983) found that high densities of Rotifera, Cladocera and Copepoda occurred both in the dry and wet seasons in a shallow lake, Parakrama Samudra in Sri Lanka. Duncan & Gulati (1981) recorded very low densities of crustacean zooplankton in the same lake during August-September 1979 (dry period-beginning of rainy season). The zooplankton consisted mainly of smaller Rotifera species. They considered that temperature, flushing rate of water and fish predation were the major factors affecting rotifer abundance. High densities of zooplankton, both in the dry and wet seasons, were recorded for Lake Colombo and Lake Gregory by Apstein (1907, 1910).

Egborge (1981) claims that in a Nigerian reservoir crustacean zooplankton maxima were associated with high turbidity (rainy season) while Rotifera maxima were coincident with high transparency (dry season). In some other tropical lakes in Africa (cf. Robinson & Robinson 1971, Burgis 1974), high densities of zooplankton were recorded only in the dry season. Lewis (1979) found that the zooplankton biomass declines during the seasonal mixing in Lake Lanao, a much deeper tropical lake than any others mentioned above. The presently available data on seasonality of zooplankton in tropical lakes and reservoirs does not permit any generalizations. However, the data comes from widely different types of lakes and reservoirs. The density of the zooplankton could also be markedly influenced by fish predation in some of the shallower lakes and reservoirs especially at low water levels as suggested by Fernando and Rajapaksa (1983) for Parakrama Samudra.

Tropical standing waters largely are composed of reservoirs. The zooplankton colonizing these must come from river systems. We know very little about the changes that take place in the formation of a stable zooplankton community in a reservoir after impoundment. In Sri Lanka Fernando & Rajapaksa (1983) found that at low water levels the zooplankton composition is typically littoral with Rotifera dominating probably as a result of high predation by fish. When high water levels were restored in the aftermath of cyclone damage to the dam (of Parakrama Samudra), a more typical limnetic zooplankton was restored. They attribute this to the unsuitability of shallow water for limnetic species and also the higher predation by fish at low water levels.

Zooplankton production in tropical lakes seems to be dominated by cyclopoid copepods. The herbivorous cyclopoid *Thermocyclops crassus* accounts for 46% of the production in Lake Lanao, Philippine (Lewis 1979) and >80% of the crustacean zooplankton in Lake George, Uganda (Burgis & Walker 1972). In Sri Lanka *Thermocyclops crassus* is relatively common in larger reservoirs (Fernando 1980a). This species, and the carnivorous *Mesocyclops leuckarti* s.l. (see Kiefer 1981), dominates the zooplankton throughout S.E. Asia (Fernando 1980b). There is some indication from the study of Burgis (1974) that zooplankton production in the highly productive (in fish) Lake George, Uganda, is not high in comparison with temperate lakes. Shallow tropical lakes like Lake George often are dominated by blue green algae. The presence of blue green algal feeders in the fish community (Cichlidae) accounts for the high fish yields. Perhaps under these circumstances the zooplankton production remains low and contributes relatively little to the fish food especially for the young fish.

Notes on zoogeography

It can be said in general that many zooplankton species found in Sri Lanka are either cosmopolitan or tropicopolitan. The range of distribution of some rarer species may well be extended as more intensive and extensive studies are made. Recently, however, Pejler (1977a), Frey (1982) and Kiefer (1981) have indicated that the extent of cosmopolitanism in zooplankton has been exaggerated. Many species on closer study have proved to be more restricted in their distribution.

The Rotifera have long been considered as composed largely of cosmopolitan species. Early workers (Rousselet 1909) considered cosmopolitanism as widespread in the Rotifera. Pejler (1977a, b) cited evidence to show that some Rotifera species had restricted distributions, although no geographic barriers exist for members of this group. However, the actual distribution of species claimed by Pejler (1977) as being restricted to the Nearctic (*K. earlinae*) and the temperate zones (*Kellicottia longispina*) was extended to Sri Lanka by Chengalath *et al.* (1973). It seems very likely that more species will have this range extended as more studies are made. De Ridder (1981) found 141 species and forms (78%) of Rotifera in Zaire to be cosmopolitan.

In Sri Lanka some cosmopolitan species recorded are: *B. angularis*, *B. calyciflorus*, *B. quadridentatus*, *B. rubens*, *E. dilatata*, *F. longiseta*, *L. bulla*, *L. luna*, *L. lunaris* and *P. vulgaris*. Some common tropicopolitan species found are *Brachio-*

nus caudatus, B. falcatus, B. patulus, Keratella tropica and *Tetramastix opoliensis.*

The predominantly temperate genera *Synchaeta* and *Notholca* have not been recorded in Sri Lanka. *Brachionus donneri* , endemic to the Indian region has, however, been recorded. Also the genus *Brachionus* is represented by a large number of species. This feature has been noted by Green (1972) and Pejler (1977b). Pejler (1977a) stated that the latitudinal distribution is associated, at least partly, with the distribution of suitable food. The dominant food organisms in temperate regions are the chrysophyceans, cryptomonads and green algae, whereas in the tropics blue green algae usually are dominant.

In all, 63 species of Cladocera are known from Sri Lanka. This number probably represents the total fauna barring taxonomic revision. Some of these species are cosmopolitan, e.g. *Moina micrura* and *Graptoleberis testudinaria. Chydorus sphaericus*, another cosmopolitan species, also has been recorded, but according to Frey (1980) this species consists of a number of species (species complex). A number of pantropical species have been found, namely *Dadaya macrops, Dunhevedia serrata* and *Chydorus eurynotus.*

Indialona (Alona) macronyx and *Chydorus ceylonicus* originally were described from Sri Lanka (Daday 1898). They have not been found in Sri Lanka since 1898 (a few specimen slides are available at the National Museum of Hungary, Budapest). However, *I. macronyx* has been recorded from Celebes (Brehm 1938), Malaysia (Idris & Fernando 1981) and the Philippines (Mamaril & Fernando 1978).

Species of *Disparalona, Camptocercus* and *Acroperus* recorded from the neighbouring countries (India, Malaysia, Indonesia) are not reported in Sri Lanka. The genus *Daphnia* is rare in Sri Lanka; as in other tropical regions (Fernando 1980a). Also *Pleuroxus*, a commonly occurring genus in temperate regions, is very rare in Sri Lanka.

In discussing the Copepoda we must realize that considerable alterations are taking place in the specific status of some common species found in Sri Lanka, hence the present comments must remain very tentative.

Eucyclops serrulatus, though it is considered a cosmopolitan species, seems more restricted (Dumont 1981). *Tropocyclops confinus* is found in the tropics but only at high altitudes. This species was recorded from Sri Lanka (Fernando 1980a), the Philippines (Mammaril & Fernando 1978), Nepal (Dumont & Van de Velde 1977) and India (Lindberg 1939). *Mesocyclops leuckarti* (s.l.) is the commonest tropical species. However, according to Kiefer (1981) this species does not occur in the paleotropics or Australia. It is restricted to Europe and the western part of northern Asia. *Mesocyclops* occurring in Africa, Australia and southern and eastern Asia belong to other species than *M.leuckarti*. The common Sri Lanka species probably is *M. aspericornis* (Daday) found in South East Asia (Kiefer 1981).

In the calanoids, in spite of extensive studies recently by Lai & Fernando (1980) and Kiefer (1982), the systematics of many species remains confused. Eleven species of calanoids are recorded (Fernando 1980a). They belong to the genera *Tropodiaptomus* widely distributed in Australia, Asia and Africa; *Neodiaptomus* and *Phyllodiaptomus*, found in Asia and Europe; *Eudiaptomus* recorded from Asia, Australia, Europe and Africa. *Paradiaptomus*, recorded from the Ethiopian and Oriental region, like *Megadiaptomus*, a very large diaptomid, is found in astatic waters. *Heliodiaptomus* and *Rhinediaptomus* are common in the Indian subcontinent.

Ostracoda occur in the plankton in tropical freshwaters unlike in temperate regions. In Sri Lanka, the true limnetic species *Strandesia purpurascens* is widely distributed throughout the island. This is a common species in the Oriental regions. It has been recorded from Sumatra, Java and Thailand under its original name (Cypris purpurascens) (Sars 1903, Vavra 1906, Müller 1906, Grochmalicki 1915). From India (Victor 1976, Victor & Fernando 1979), Sri Lanka (Neale 1977, Neale & Victor 1978), Phillipines and Malaysia (Victor 1979) recorded it under its amended name.

References

Apstein, C. 1907. Das plancton im Colombo See auf Ceylon. Zool.Jb. (Abt. Syst.) 25:201-244.

Apstein, C. 1910. Das plancton des Gregory Sees auf Ceylon. Zool. Jb. (Abt. Syst.) 29:661-680.

Bär, G. 1924. Über Cladoceren von der Insel Ceylon. Jena Z. Naturw. 60:83-126.

Brady, G.S. 1886. Notes on Entomostraca collected by Mr A. Haly in Ceylon. J. Linn. Soc. Lond. (Zool) 10:293-317.

Brehm, V. 1933. Die cladoceren der Deutschen Limnologischen Sunda-Expedition. Arch. Hydrobiol. Suppl. 11:631-771.

Brehm, V. 1938. Die Cladocera der Wallacea Expedition. Int. Rev. ges. Hydrobiol. 38:99-124.

Brook, A.J. & J. Rzoska. 1954. The influence of the Gebel Aulyia dam on the development of Nile plankton. J. Anim. Ecol. 23:101-114.

Burgis, M. 1974. Revised estimates for the biomass and production in Lake George, Uganda. Freshwat. Biol. 4:535-541.

Burgis, M. & A.F. Walker. 1972. A preliminary comparison of the zooplankton in a tropical and a temperate lake (Lake George, Uganda and Loch Leven, Scotland). Verh. Internat. Verein. Limnol. 18:647-655.

Burgis, M., J.P.E. Darlington, I.G. Dunn, G.G. Ganf, J.J. Gwahaba & L.M. McGowen. 1973. The biomass and distribution of organisms in Lake George, Uganda. Proc. R. Soc. Lond. (B) 184:271-298.

Chengalath, R. & C.H. Fernando. 1973a. Rotifers from Sri Lanka (Ceylon). 1. The genus *Lecane* including the description of two new species. Bull. Fish. Res. Stn Sri Lanka (Ceylon 24:13-27.

Chengalath, R., C.H. Fernando & W. Koste. 1973. Rotifera of Sri Lanka. 2. Further studies on the Eurotatoria including new records. Bull. Fish. Res. Stn Sri Lanka (Ceylon). 24:29-62.

Chengalath, R., C.H. Fernando & W. Koste. 1974. The Rotifera of Sri Lanka. 3. Descriptions of new species and new records with a list of species in Sri Lanka and discussion of distribution. Bull. Fish. Res. Stn Sri Lanka (Ceylon) 25:83-96.

Costa, H.H. & S.S. De Silva. 1969. Hydrobiology of Colombo (Beira) Lake. 1. Diurnal variations in temperature; hydrochemical factors and zooplankton. Bull. Fish. Res. Stn Ceylon 20:141-149.

Costa, H.H. & S.S. De Silva 1978. The Hydrology of Colombo (Beira) Lake. III. Seasonal fluctuations of plankton. Spol. Zeylan. 32:35-53.

Daday, E. 1898. Microscopische susswasserthiere aus Ceylon. Termesz. Füz. (Budapest) 21:1-123.

De Ridder, M. 1981. Hydrobiological Survey of the lake Bangweulu Luapula river basin. Vol. 11, part 4. 191 pp. Ed. J.J. Symoens, Brussels.

Dumont, H.J. 1980. Zooplankton and the science of Biogeography: The example of Africa. pp. 685-696 In: Evolution and ecology of zooplankton communities. Ed. W.C. Kerfoot. Univ. Press of New England, New Hampshire. 793 pp.

Dumont, H.J. & I. Van de Velde. 1977. Report on a collection of Cladocera and Copepoda from Nepal. Hydrobiologia 53:55-65.

Duncan, A. 1983. The composition, density and distribution of the zooplankton of Parakrama Samudra 1980. In: F. Schiemer (Ed.) Limnolgoy of Parakrama Samudra, Sri Lanka: A case study of an ancient man-made lake in the tropics. Developments in Hydrobiology. Dr W. Junk, The Hague.

Duncan, A. & R.D. Gulati. 1981. Parakrama Samudra (Sri Lanka) Project, a study of a tropical lake ecosystem. III. Composition, density and distribution of the zooplankton in 1979. Verh. Internat. Verein. Limnol. 21:1007-1014.

Egborge, A.B.M. 1981. The composition, seasonal variation and distribution of zooplankton in Lake Asejire, Nigeria. Rev. Zool. Afr., 95:137-180.

Fernando, C.H. 1969. A guide to the freshwater fauna of Sri Lanka (Ceylon). Suppl. 3. Bull. Fish. Res. Stn Ceylon 20:15-25.

Fernando, C.H. 1974. A guide to the freshwater fauna of Sri Lanka (Ceylon). Suppl. 4. Bull. Fish. Res. Stn Sri Lanka (Ceylon) 25:27-81.

Fernando, C.H. 1977. Investigations on the aquatic fauna of tropical ricefields with special reference to South East Asia. Geo. Eco. Trop. 1:169-188.

Fernando 1980a. The freshwater zooplankton of Sri Lanka with a discussion of tropical freshwater zooplankton composition. Int. Rev. ges. Hydrobiol. 65:85-125.

Fernando, C.H. 1980b. The species and size composition of tropical freshwater zooplankton with special reference to the Oriental Region (South East ASia). Int. Rev. ges. Hydrobiol. 65:411-426.

Fernando, C.H. 1981. The freshwater invertebrate fauna of Sri Lanka. Spol. Zeylan. 35:15-42.

Fernando, C.H. & W.B. Ellepola. 1969. A preliminary study of two village tanks (reservoirs) in the Polonnaruwa area with biological notes on these reservoirs in Ceylon. Bull. Res. Stn Ceylon 20:3-13.

Fernando, C.H. & G. Hanek. 1973a. Some parasitic Coepoda from Sri Lanka (Ceylon) with a synopsis of parasitic Crustacea from Ceylonese freshwater fishes. Bull. Fish. Res. Stn Sri Lanka (Ceylon) 24:63-67.

Fernando, C.H. & G. Hanek. 1973b. Two species of the genus *Ergasilus* Nordmann (Copepoda,Ergasilidae) from Ceylon. Crustaceana 25:13-20.

Fernando, C.H. & G. Hanek. 1976. A new genus of calanoid copepod from Sri Lanka (Ceylon). Crustaceana 30:82-88.

Fernando, C.H. & R. Rajapaksa. 1982. Some remarks on long-term and seasonal changes in the zooplankton of Parakrama Samudra. Developments in Hydrobiology 12: 77-84.

Frey, D.G. 1980. On the plurality of *Chydorus sphaericus* (Cladocera:Chydoridae) and a designation of a neotype from Sjaelso, Denmark. Hydrobiologia 69:83-123.

Frey, D.G. (in press). Questions concerning cosmopolitanism in Cladocera. Arch. Hydrobiol.

Green, J. 1962. Zooplankton of the River Sokoto. The Crustacea. Proc. Zool. Soc. Lond. 138:415-453.

Green, J. 1972. Latitudinal variation in associations of planktonic Rotifera. J. Zool Lond. 167:31-39.

Grochmalicki, J. 1915. Beitrage zue Kenntnis der susswasser fauna Javas. Phyllopoda, Copepoda und Ostracoda. Anz. Akad. Wiss. Krakau Math. nat. Kl. Reise B. 217-242.

Gurney, R. 1906. On two new Entomostraca from Ceylon. Spol. Zeylan. 4:126-134.

Gurney, R. 1916. On some freshwater Entomostraca from Ceylon. Proc. Zool. Soc. Lond. (1916):333-343.

168

Idris, B.A.G. & C.H. Fernando. 1981. Cladocera of Malaysia and Singapore with new records, redescriptions and remarks on some species. Hydrobiologia 77:233-256.

Kiefer, F. 1981. Beitrag zur kenntnis und geographischen verbreitung von *Mesocyclops leuckarti* auctorum. Arch. Hydrobiol. Suppl. 62:148-190.

Kiefer, F. (in press). Vergleichende Untersuchungen uber Morphologie, Taxonomie und geographische Verbreitung der Arten der Gattung *Tropodiaptomus* Kiefer (Copepoda, Calanoida) aus asiatischen Binnengewasser. Hydrobiologia.

Kirtisinghe, P. 1964. A review of the parasitic copepods of fish recorded from Ceylon with descriptions of additional forms. Bull. Fish. Res. Stn Ceylon 17:45-132.

Lai, H.C. & C.H. Fernando. 1980. Geographical distribution of Southeast Asian freshwater Calanoida. Hydrobiologia 74:53-66.

Lewis, W.M. 1979. Zooplankton community analysis: studies on a tropical system. Springer-Verlag, N.Y. 163 pp.

Lindberg, K. 1939. Etude de representants Indien du sous genre *Tropocyclops* (Crustaces, Copepodes, Cyclopoides). Rec. Ind. Mus. 41:1-15.

Mammaril, A.C. & C.H. Fernando. 1978. Freshwater zooplankton of the Philippines (Rotifera, Cladocera and Copepoda). Nat. App. Sci. Bull. 30:109-221.

Mendis, A.S. 1964. A contribution to the limnology of Colombo Lake. Bull. Fish. Res. Stn Ceylon 17:213-220.

Mendis, A.S. 1965. A preliminary survey of 21 Ceylon lakes. 2. Limnology and food production potential. Bull. Fish. Res. Stn Ceylon 18:7-16.

Mendis, A.S. & C.H. Fernando. 1962. A guide to the freshwater fauna of Ceylon. Bull. Fish. Res. Stn Ceylon 12:1-160.

Müller, G.W. 1906. Ostracoden aus Java. Mitt. Hamb. Zool. Mus. Inst. 23:139-142.

Neale, J.W. 1977. Ostracoda from the ricefields of Sri Lanka. pp. 271-283 In: H. Löffler & D.L. Danielopol (Eds.) Aspects of Ecology and Zoogeography of recent and fossil Ostracods. Dr. W. Junk, The Hague.

Neale, J.W. & R. Victor. 1978a. On *Indiacypris luxata* a freshwater ostracod (Crustacea, Entomostraca) from Sri Lanka. Zool. J. Linn. Soc. 64:71-77.

Neale, J.W. & R. Victor. 1978b. The Lund University Expedition freshwater Ostracod from Sri Lanka. Can. J. Zool. 56:1081-1087.

Pejler, B. 1977a. General problems on rotifer taxonomy and global distribution. Arch. Hydrobiol. Beih. Ergebn. Limnol. 8:212-220.

Pejler, B. 1977b. On the global distribution of the family Brachionidae (Rotatoria). Arch. Hydrobiol. Suppl. 53(2):255-306.

Poppe, S.A. & Mrázek, S. 1895. Die von Herrn Dr. H. Driesch auf Ceylon gesammelten Susswasser-Entomostraken. Beih. Jahrb. Hamb. Wiss. Anst. 12:139-142.

Rajapaksa, R. & C.H. Fernando. 1982. Cladocera of Sri Lanka with remarks on some species. Hydrobiologia 94:49-69.

Rajapaksa, R. & C.H. Fernando (1982). The first description of the male and ephippial female of *Dadaya macrops* (Daday, 1898)(Cladocera,Chydoridae) with additional notes on this common tropical species. Can. J. Zool. 60:1841-1850.

Reiss, F. 1977. The benthic zoocoenoses of Central Amazon Varzea lakes and their adaptations to annual water level fluctuations. Geo. Eco. Trop. 1:65-75.

Robinson, A.H. & P.K. Robinson. 1971. Seasonal distribution of zooplankton in the Northern basin of Lake Chad. J. Zool. Lond. 163:25-61.

Rousselet, C.F. 1909. On the geographical distribution of the Rotifera. J. Queckett Microsc. Cl. Ser 2 10:465-470.

Ruttner-Kolisko, A. 1974. Plankton rotifers. Biology and Taxonomy (Die Binnengewasser vol. XXVI; Part 1: Chapter 'Die Rotatorien') Stuttgart. (English translation). pp.99-234.

Sars, G.O. 1903. Freshwater Entomostraca from China and Sumatra. Arch. Math. Naturv. Kristiana 25:3-44.

Schiemer, F. 1981. Parakrama Samudra (Sri Lanka) project, a study of a tropical lake ecosystem. I. An interim review. Verh. Internat. Verein. Limnol. 21:987-993.

Vavra, W. 1906. Ostracoden von Sumatra, Java, Siam, Sandwich inseln and Japan. Zool. Jb. (Syst.) 23:413-438.

Victor, R. 1976. A taxonomic study of freshwater ostracods (Crustacea:Ostracoda) of the Indian Subcontinent. M.Sc. thesis, University of Waterloo. 180 pp.

Victor, R. 1979. The taxonomy and Distribution of freshwater Ostracods (Crustacea,Ostracoda) of Malaysia, Indonesia and the Philippines. Ph.D. thesis, University of Waterloo. 348 pp.

Victor, R. & C.H. Fernando. 1979. The freshwater ostracods (Crustacea,Ostracoda) of India. Rec. Zool. Surv. India 74:147-242.

9. The freshwater Ostracoda

J. W. Neale

Introduction

The Ostracoda, a group of small entomostracan Crustacea, form an important element in the Sri Lankan fauna. The adults range from 0.3 to 5 mm in length. Occupying a wide variety of habitats, the rather rare terrestrial forms have never been found there and the marine species are regarded as lying outside the terms of reference of this account. With the exception of the euryhaline brackish water species *Ghardaglaia ambigua* (Neale 1979c), which belongs in the Paracypridinae, all the ostracods in this account are inhabitants of freshwater and found in lakes, tanks, ponds, ditches, reservoirs and rice fields.

Work on the ostracods of Sri Lanka has been mainly taxonomic and the first publication, that of Brady (1886), described 11 freshwater species of which 9 were new. Later papers by Daday (1989), Apstein (1907) and Gurney (1916) described small collections and/or new species from the island. More recently Ferguson (1969) published on some of the stenocyprines and a resurgence of work in the mid-1970's (Neale 1976a-f, 1977a, b, 1979 a-c, and Neale & Victor 1978 a, b) has added to our knowledge of the fauna. These were the result of extensive collecting carried out by Professor C.H. Fernando during the last ten years, and by the University of Lund and the Joint Universities of Aberdeen and Colombo Expeditions and the Parakrama Samudra Project of the Vienna Limnological Institute. In addition, many other publications on the freshwater ostracods of India and S.E. Asia have a bearing on the Sri Lankan fauna. Except for a single species belonging in the Darwinulinacea characterised by its rosette shaped muscle scar pattern on the valves, the ostracods all fall within the Superfamily Cypridacea of the Suborder Podocopina and covered here are 31 species beloning in 19 genera.

A number of records appearing in the older literature are doubtful or difficult to interpret and thus not included here in the text and distribution maps. These include *Cypris monilifera, Ilyocypris australiensis* (in Daday 1989 and Apstein 1907), *Cypricercus reticulatus* (and ? in Gurney 1916), *Cypridopsis minna* (in Daday 1898) *Cypridopsis assimilis* (in Daday 1898), *Candonella albida* (in Apstein 1907), *Cypridopsis newtoni* (in Gurney 1916) (now referred to *Plesiocypridopsis*).

In addition there are a number of new species that need describing belonging to the genera *Strandesia* and *Cypridopsis* and the latter genus still awaits detailed investigation.

Superfamily Darwinulinacea

Comprising a single family this group of ostracods is poorly represented on the island. The smooth, weakly-calcified, elongate-ovate valves with narrowly rounded anterior end have a characteristic adont hinge and rosette muscle scar pattern. The only species *Darwinula lundi* (Figs. 1,2) belong in the *D. pagliolii* group of Danielopol (1980), whose epigean species stretch in a band across the tropics from Indonesia through Sri Lanka and Africa to subtropical Brazil. Collected by the 1962 Lund Expedition and described by Neale & Victor (1978b), *D. lundi* is some 0.4 mm long and, as is general in this group, no males have been found. Its sole recorded occurrence is in Sabaragamuwa Province, 9 miles N.N.W. of Ratnapura ca. 6°49N, 80°22W where it was collected in running water at a sandy river bank with coarse gravel (Fig. 11).

Subfamily Paracypridinae

The brackish to marine *Ghardaglaia ambigua* is the only representative of this subfamily (Fig. 3) and was found by the 1978 Aberdeen-Colombo University Expedition at four localities in the Yala National Park at the south east end of the island. This species is particularly interesting since it shows a melange of characters found in *Aglaia* and *Ghardaglaia*. It has the setal brush at the top of limb 7 so characteristic of *Ghardaglaia*. The 6 segmented antennule is more typical of *Aglaia* than *Ghardaglaia* which has 5 segments although the proportions of the second segment are more in accordance with a division of this segment in the type species of *Ghardaglaia* than in the type species of *Aglaia*. For these reasons the present species is placed in the former. Hitherto the genus has only been found in the Red Sea area and particular interest attaches to its occurrence here (Fig. 11). It is obviously euryhaline and at Gonnalabba, where it was the only species, salinity was 44-72% with pH of 7.5. At Yala 1, where similarly it was the sole species, salinity was 15-30% and pH 7.5-8.5. It was again the only species at Yala 2 but at Mahasilawa it occurred in conjunction with *Indiacypris luxata, Strandesia elongata* and *?Cytheridea pusilla* which again underlines its wide accommodation to variations in salinity. The remaining species, all freshwater Cypridacea, may be roughly divided according to size.

Large species

Subfamily Herpetocypridinae

In the tropics the stenocyprines are the analogue of the candonids of the more temperate realms and in *Stenocypris* are characterised by the smooth, elongate shells with prominant marginal septa and pore canal zones and the marked asymmetry of the left and right furcal rami. They are the largest and most conspicuous of the Sri Lanka ostracods, *S. fernandoi* attaining a length of as much as 5 mm. Six species are recorded, of which *S. major* (Baird 1859) (Figs. 4, 6) is the commonest and is distributed more or less throughout the island (Fig. 13). *Cypris malcolmsi*, generally reaching 2.0 to 2.2 mm in length, was regarded by Daday (1898) as a synonym, a view upheld by Ferguson (1969). Found in a wide range of habitats, it is a freshwater species like other members of the subfamily. It has been noted in water temperatures ranging from 27.5 to 33 °C and pH 6.8 to 9.4 on varying substrates.

S. hislopi (Fig. 8) formerly was confused with *S. major* but is smaller (length 1.38-1.50 mm) and also differs in other respects, notably the termination of the seventh pair of appendages (the cleaning limb). The original material was collected by the Rev. S. Hislop in 1845 from ponds in Nagpur. A single specimen has been found in northern Sri Lanka at Kudattanai (Neale & Victor 1978b) by the Lund Expedition of 1962, and at a number of localities in Yala National Park in the south (Fig. 14) where it was collected by the Aberdeen-Colombo University Expedition of 1978 (Neale 1979a). Like other species it is a denizen of freshwater but can withstand low salinities for short periods and has been found in waters whose temperatures varied between 26 and 34 °C adn pH 6.8 to 9.4. The substrate with which is has been found associated varied widely from those with an organic content as little as 0.75% to those with as much as 11.09%.

The distributions of the other species are limited and sporadic as known at present. *S. fernandoi* (Fig. 7), the largest species on the island, has been found at only one locality, in a pond at Yala-Palatupana approx. 6° 19N, 81°27E (Fig. 12). *S. derupta*, first described from W. Java and also known from Sumatra and India, is a medium-sized stenocyprine species 2.5 mm long and has been found only in Habarana Tank in Sri Lanka (Figs. 5, 12).

Chrissia lacks the radial septa of *Stenocypris* and is represented by two species. *C. halyi*, about 1.4 to 1.5 mm long (Fig. 9), has been found at three localities (Fig. 12) and is probably commoner than this would suggest. *C. ceylonica* was caught in February 1898 in wells in the Kalawewa area (Fig. 12). This medium sized stenocyprine species (length 1.7 mm) had never been seen since, until this year, when it was found in material from the Parakrama Samudra Project collected by the Vienna Limnological Institute. Part of Daday's original figure is given here as Fig. 10.

174

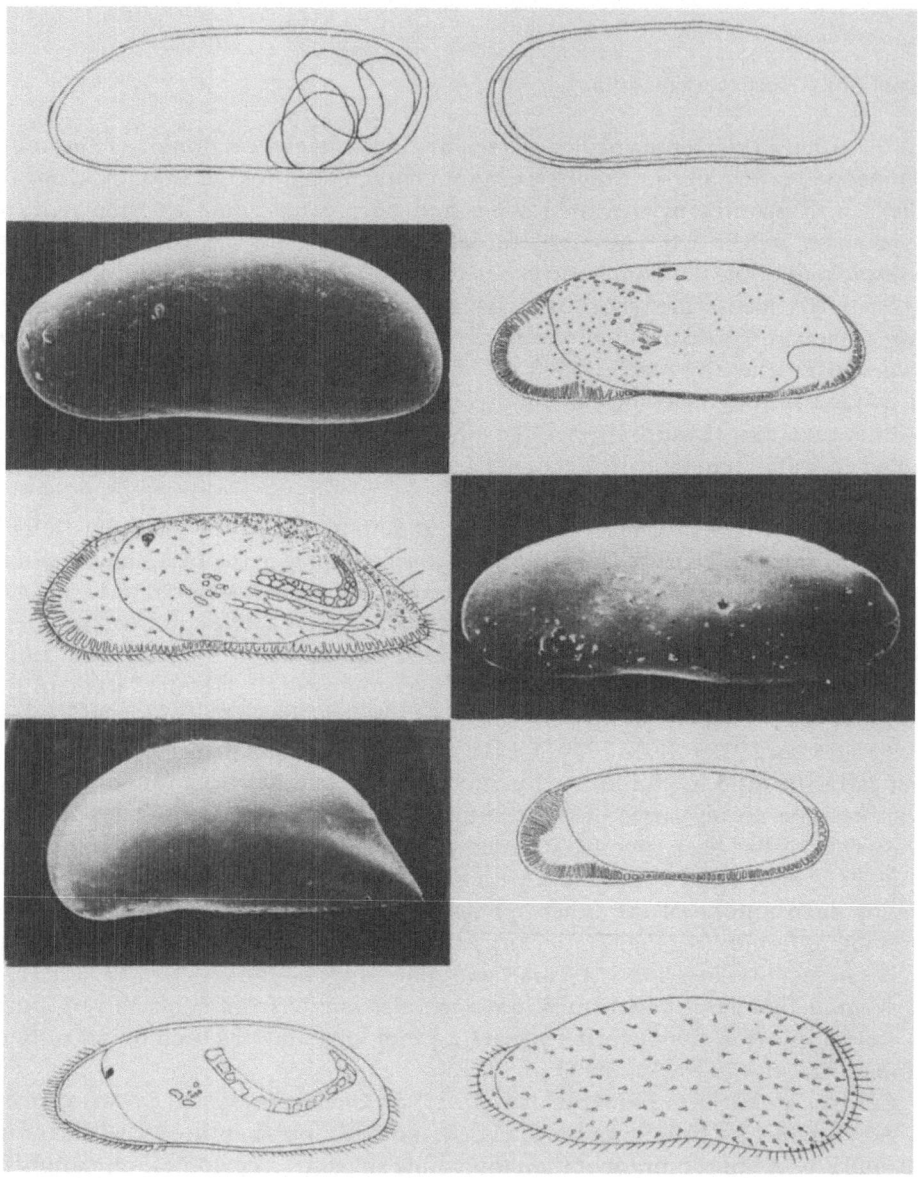

Explanation of Fig. 1-10.

1, 2. *Darwinula lundi* x 128, female carapace: 1 from left, 2 from right; 3. *Ghardaglaia ambigua* x 78, female carapace from left; 4, 6. *Stenocypris major*: 4. internal view of right valve x 24, 6. external view of left valve x 29; 5. *Stenocypris derupta* x 21, external view of female left valve; 7. *Stenocypris fernandoi* x 10, external view of female left valve; 8. *Stenocypris hislopi* x 29, internal view of female right valve; 9. *Chrissia hayli* x 31, internal view of female right valve; 10. *Chrissia ceylonica* x 29, female carapace from right.

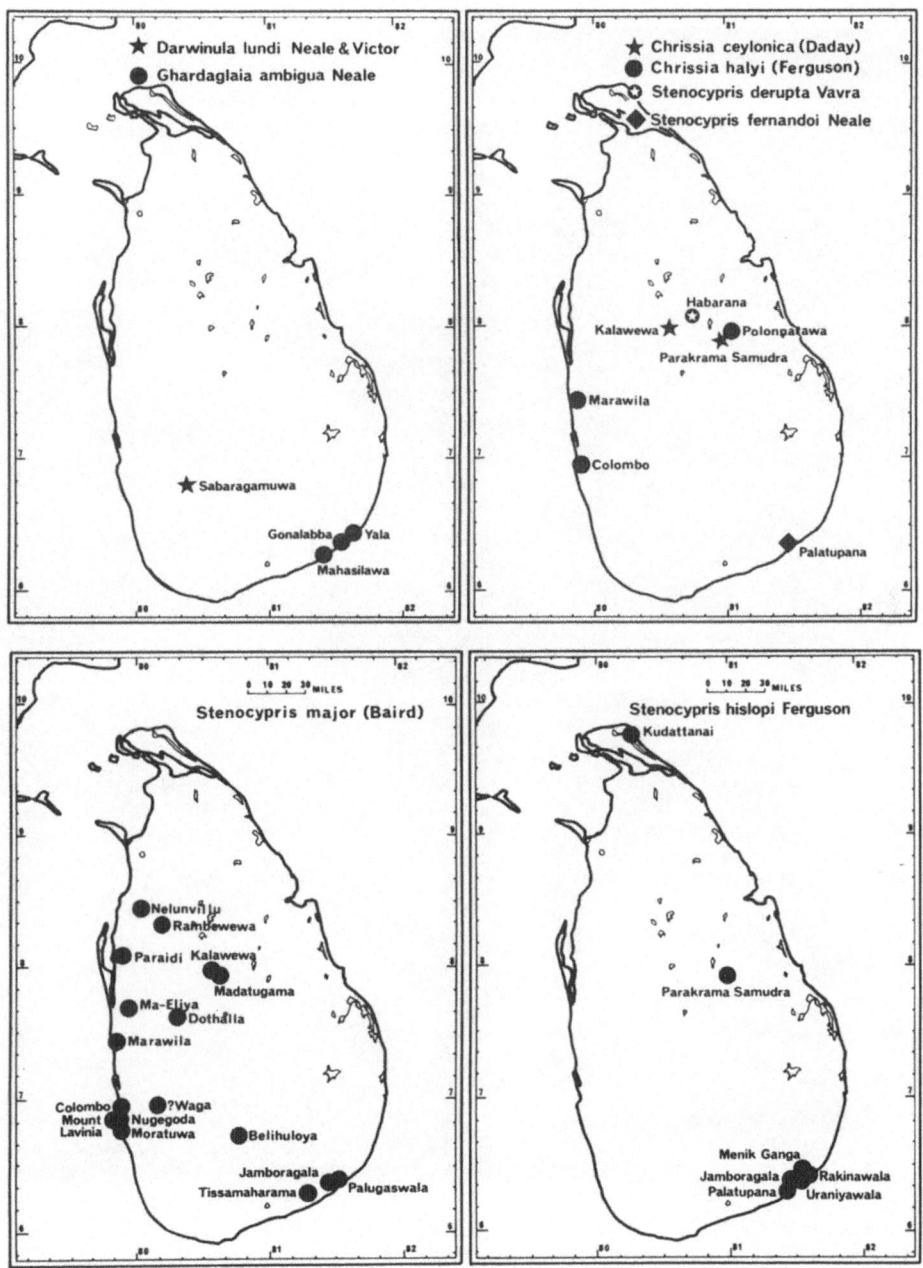

Explanation of Fig. 11-14.
11. Distribution of *Darwinula lundi* and *Ghardaglaia ambigua*; 12. Distribution of *Chrissia ceylonica*, *Chrissia halyi*, *Stenocypris derupta* and *Stenocypris fernandoi*; 13. Distribution of *Stenocypris major*; 14. Distribution of *Stenocypris hislopi*.

176

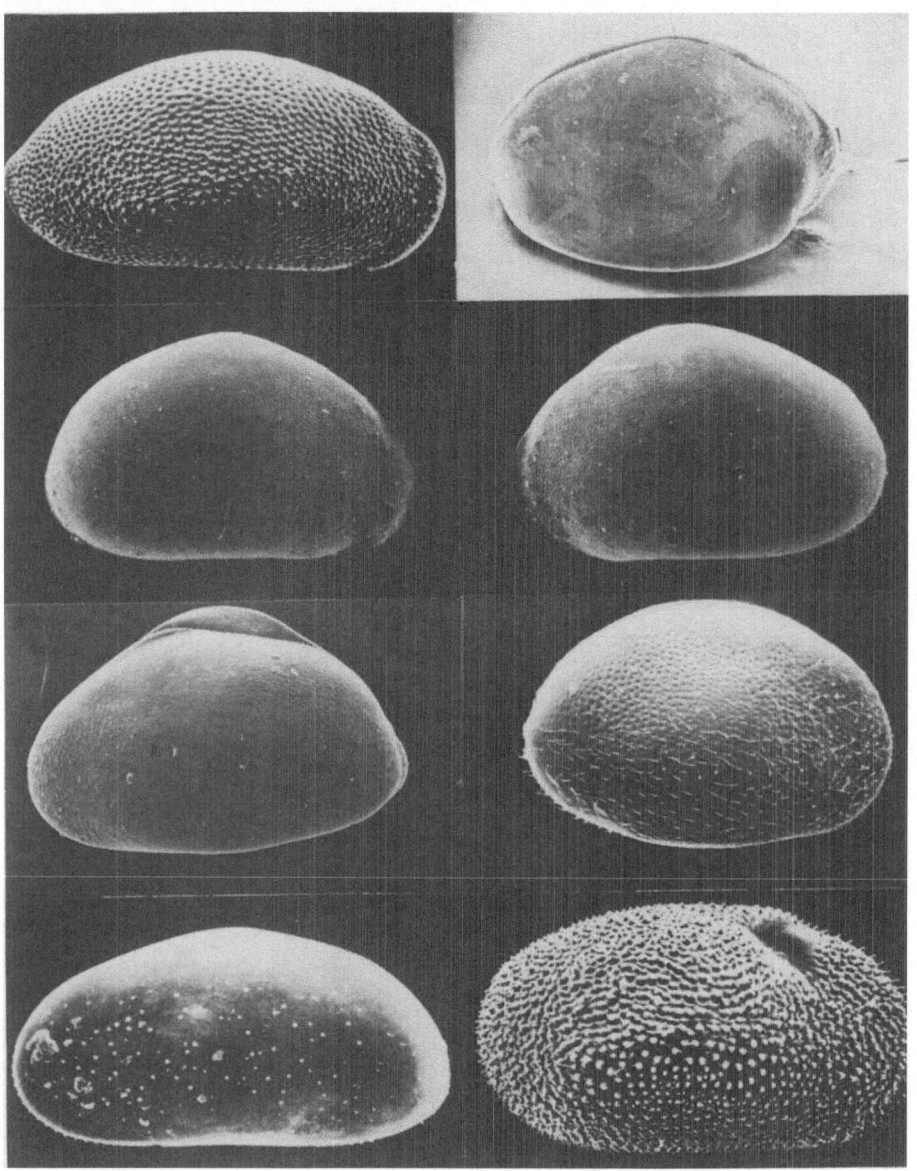

Explanation of Fig. 15-22.
15. *Cypris subglobosa* x 42, external view of right valve; 16. *Cypris latissima* x 31, dorso-lateral view of
female carapace from the right; 17, 18. *Cypris decaryi* x 30: 17. external view of female right valve, 18.
external view of female left valve; 19. *Cyprinotus cingalensis* x 45, carapace from left; 20. *Hemicypris
pyxidata* x 51, female carapace from right; 21. *Heterocypris dentatomarginatus* x 64, external view of
female left valve; 22. *Centrocypris viridis* x 60, external view of female right valve.

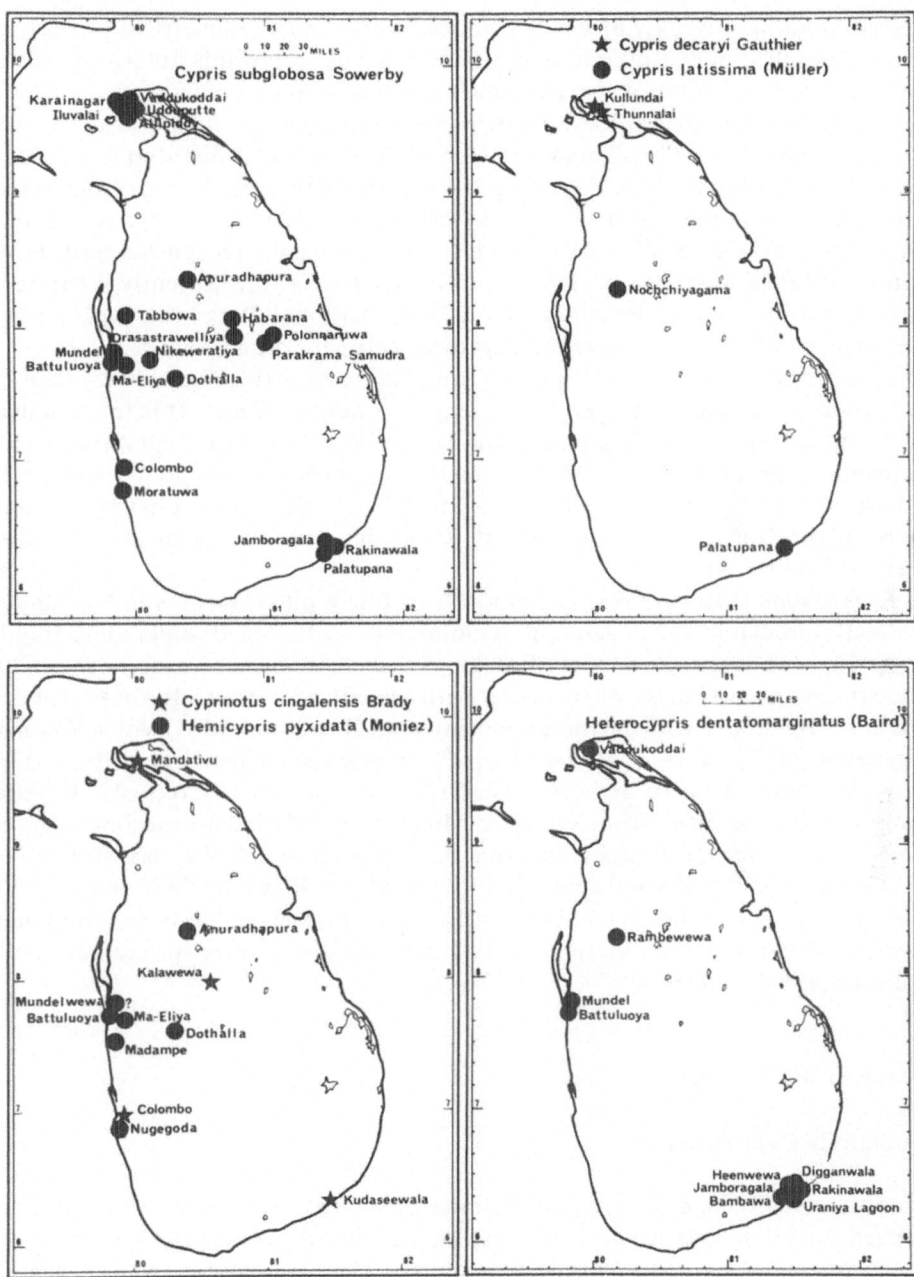

Explanation of Fig. 23-26.
23. Distribution of *Cypris subglobosa*; 24. Distribution of *Cypris decaryi*; 25. Distribution of *Cyprinotus cingalensis* and *Hemicypris pyxidata*; 26. Distribution of *Heterocypris dentatomarginatus*.

Large, smooth or weakly pitted Cypridacea with almost symmetrical valves with strongly arched dorsal margin and selvages displaced inwards anteriorly with a secondarily fused anterior margin which projects as a lip. Only the genus *Cypris* has been found in Sri Lanka, with three species present. *C. subglobosa* first was described from Indian fossil material. One of the most widely distributed species, it has been recorded world wide in tropical and warm temperate areas from Japan and Indonesia in the east to Venezuela and the U.S.A. in the west, and is found throughout Sri Lanka (Fig. 23). Some 1.4-1.5 mm long, it is recognized easily by its pitted surface and serrate postero-ventral margin (Fig. 15). Recently it has been found at three further localities in the Yala National Park where pH varied between 7.4 and 9.6 and water temperatures between 25 and 33°C.

C. decaryi (Gauthier 1933) is more rounded posteriorly (Figs. 17, 18) then *C. subglobosa* and slightly larger with a length of about 1.7 mm. It is fairly widely distributed from Dutch Guiana and the Dutch West Indies through Africa to Sri Lanka but appears to be rare. McKenzie (1971) recorded it from pools and lakes in Aldabra, where salinity ranged from 1.3 to 8.9%. In Sri Lanka it has been found only in the North where it occurs at Kallundai and Thunnalai in the Jaffna Peninsula (Fig. 24).

C. latissima (Müller 1898) is a smooth or finely pitted form which is almost perfectly circular in dorsal view and is globose with a flattened venter and a fragile shell (Fig. 16). It is known from Madagascar where it reaches 2 mm in length but the Sri Lanka material is rather smaller with a length of 1.48 mm. In shape and size the latter agrees well with the Indian specimens of Hartmann (1964) which Victor & Fernando (1979) have now placed in *C. dravidensis*, a new name for *Cypris globulosa* (preoccupied). However, the Sri Lanka specimens agree with the African material and differ from Victor & Michael's (1975) Indian specimens in the nature of the second thoracic leg and the furca on which they lay stress, and dissection of material kindly supplied by the Musée Royal de l'Afrique Central, Tervuren, confirms this. Apparently rather rare, this species has been found only near Nochiyagama and Palatupana (Fig. 24). The above three species have been fully described in Neale (1976e, f, 1977b).

Medium sized species

Subfamily Cyprinotinae

Robust, usually smooth, equidimensional, asymmetrical valves with strongly arched dorsal margin. *Cyprinotus cingalensis* (Fig. 19) is the type species of the genus which has a widely recorded distribution. The right valve shows a characteristic expansion dorsally and this valve also is denticulate anteriorly and ventrally. The left valve margin overlaps the right valve all round the shell, which is 1.0 mm to 1.1 mm long. the Zenker's Organ has 11-12 rosettes and the posterior seta lies just beyond the mid-length of the furcal ramus. The type locality is Lake Colombo, but it is no longer found there, probably because of pollution. Daday (1898) found it in a marsh in the Lake Kalawewa area but did not figure it. Victor (*in litt.*) recently has recorded it from Mandativu in the north of the island. The only other specimens known from Sri Lanka come from a brackish/saline environment at Kudaseewala in Yala National Park, where well preserved carapaces, but devoid of appendages or soft parts, were found in 1978 (Fig. 25). The species was fully described by Neale (1979b) and a further revision of the type material has been published recently (Pinto & Purper 1980). The reasons for its rarity are not apparent. It is unlikely that the species actually lived in this environment where the salinity at times reached 45%. The specimens probably were washed in from elsewhere.

Hemicypris differs in lacking the dorsal expansion of the right valve and in the opposite overlap relationships between the valves. In Sri Lanka the genus is represented by *H. pyxidata* (Fig. 20); about 1 mm long, marked with dark brown patches and with the left valve considerably smaller than the right. Found in the Nugegoda ricefields and in other water bodies in central and southern Sri Lanka it has not yet been found north of Anuradhapura (Fig. 25). However, it does occur in Indonesia, and it also has been recorded from India. The absence in the north of the island has been suggested as due to limited migration into the island (Neale 1977a) but it may be due to collection failure and recently Fernando (*in litt.*) has suggested that it may be due to the drier regime of the northern part of the island although it should not affect its occurrence in permanent and semi-permanent water bodies.

The third genus of this subfamily represented in Sri Lanka is *Heterocypris* which, while showing the same type of valve overlap as *Cyprinotus*, lacks the marked dorsal expansion in the right valve. *H. dentatomarginatus*, brown in colour when freshly collected, is about 0.9 mm long and in the left valve has serrate antero-ventral and postero-ventral margins (Fig. 21). It is widely distributed geographically (Fig. 26) and in the island has been collected in waters of pH varying between 7.4 and 9.0 and temperatures from 25 to 33 °C.

Subfamily Eucypridinae

The eucyprine genera *Strandesia* and *Eucypris* both occur in Sri Lanka and are of medium size. The former is represented by a number of species of which four are common and widely distributed. The furca generally is slender and very narrow in proportion to the length.

S. purpurascens , Lectotype 0.972 mm long, is a relatively elongate, elliptical species (Fig. 27) suffused with purple when freshly caught and widely distributed in the central and southern parts of the island, although is not known north of the latitude of Anuradhapura (Fig. 40). It is, however, found in India (Victor & Fernando 1979) and is well known from Indonesia and Thailand.

S. marmorata is a smaller species with lectotype 0.576 mm long (Fig. 28) and with white valves distinctively patterned in black or dark blue-purple. It is less elongate than *S. purpurascens* and shows a similar distribution in the island although not recorded from India or S.E. Asia. It is not known from the Jaffna Peninsula and northern Sri Lanka (Fig. 39).

S. wierzejskii , with elliptical valves showing characteristic fingerprint-like ornamentation was described first from Sitoe Sampora and Tjitajam in Java (Fig. 30). The taxonomic problems concerning this species and *S. striatoreticulata* have been outlined in Neale (1977a). The Sri Lanka material is about 0.75 mm long and shows a distribution similar to the two foregoing species in that it has not been found in the north of the island (Fig. 41). Nor has it been found in India.

S. elongata is an elongate species some 0.7 to 0.8 mm long (Fig. 29) which was described first from India and occurs fairly widely in Sri Lanka.

S. vittata, originally described from Puching, China (Fig. 31) is recorded as common in a pond at Peradeniya by Gurney (1916) (Fig. 38).

A number of other species of *Strandesia* occur which still need description including a large species which has been found in four ricefield localities in the Jaffna Peninsula and a Nugegoda.

Eucypris tenuicauda is 0.88 mm long and we still are dependent on Brady's (1886) figures for our knowledge of the shell. The furcal attachment and nature of the sixth pair of appendages suggest that it belongs in *Eucypris*. It is known only from the original material collected in Lake Colombo (Fig. 44).

Subfamily Centrocypridinae

Containing only the genus *Centrocypris* this typically tropical and subtropical family is characterised by a short, stout, tuberculate or spinose shell with a narrow fused zone, slender furca and Zenker's Organ with a considerable number, usually 24-32 rosettes. The only species in Sri Lanka is *C. viridis* (Fig. 22), which is moderately large (length 0.92-1.0 mm long), and although not particularly common, has been found at three localities (Fig. 42). Bright green when freshly caught it occurs in ricefields in the Jaffna Peninsula and a reservoir at Ma-Eliya. Recently McKenzie (in Victor & Fernando 1979) has suggested that *C. viridis* is a synonym of *C. matthai*, which originally was placed in the genus *Eurycypris* (Arora 1931). Quite apart from the fact that Arora's material came from Lahore, 2,000 miles (3,000 km) to the north, which may or may not be significant, one is at a loss to reconcile McKenzie's statement with Arora's figures. In side-view the sub-parallel dorsal and ventral margins, the less strongly rounded posterior margin and the more elongate rounded anterior margin in *C. matthai* contrast sharply with the posteriorly tapering dorsal and ventral margins, and the strongly, evenly curved posterior anterior margins of *C. viridis*. In dorsal view Arora's *C. matthai* shows a most distinctive 'pinch-in' in the anterior part of the shell. Again in *C. matthai* the distal part of the furcal ramus is strikingly narrow compared with the same

181

structure in *C. viridis*. The proportions and details of the other two appendages figured by Arora also differ from the comparable appendages in *C. viridis*. Either we must postulate that Arora could not draw and that his figures bear no relation to the species they purport to illustrate or *C. viridis* is a separate and distinct species. The latter seems the most logical assumption. At the same time it would be useful to re-collect from Arora's type locality and re-figure his species in accordance with modern standards of illustration and description. Michael & Victor (1975) re-figured and re-described what they considered to be Arora's species although their material came from Madurai in southern India and not from the type locality. There is considerable difference in the valve shape between the Madurai material and the type Lahore material as figured by Arora so some doubt must still attach to their interpretation until the type locality has been re-investigated. Nevertheless, Michael & Victor's appendage figures show a reasonable approximation to those of Arora, given the quality of the drawing and, since the modern figures are much better, can be compared more closely with those of *C. viridis*. They all differ, without exception. The difference is most strongly noticeable in the proximal endite of the sixth appendage and in the width of the furcal ramus. There can be little doubt that the Indian and Sri Lankan species are different and distinct.

Subfamily Indiacypridinae

In the only genus *Indiacypris* the larger left valve overlaps the smaller right valve all round and the antero-ventral and postero-ventral margins show well-developed pore canals which form distinctive areas. the seventh limb differs from all other cypridinids in carrying three bristles and Zenker's Organ has 16 rosettes. *Iniacypris luxata* (Fig. 33) is a greenish-brown species from 0.79-0.98 mm long first described by Brady (1886) from the Colombo area. Apstein (1907) also found it in Lake Colombo. More recently Neale & Victor (1978a) found it at Kalpitiya and Belihuloya and the Aberdeen-Colombo University Expedition (1978) found it at seven localities in the Yala National Park (Fig. 43). Essentially a freshwater species and generally in waters of less than 2‰ salinity, nevertheless it seems to be an adaptable species and has been found in environments where the salinity could rise to as much as 6‰ for short periods. Water temperatures ranged between 25 and 34 °C and pH from 7.4 to 10.1 though generally towards the higher end of the scale. Substrate varied from poorly sorted grey-green silt and moderately sorted red-brown sand-clay to rich organic mud and it seems generally associated with substrates containing a relatively high organic content in the region of 6-12%.

I. dispar, the other species of *Indiacypris*, which occurs in India, also is apparently present at Parakrama Samudra (Figs. 34, 43). It was not possible to confirm this on the valves which were somewhat decalcified but the seventh limb showed the typical structure and proportions characteristic of the Indian specimens of *I. dispar*. Males were present, which enabled the Zenker's Organ to be examined and shown to contain 16 rosettes.

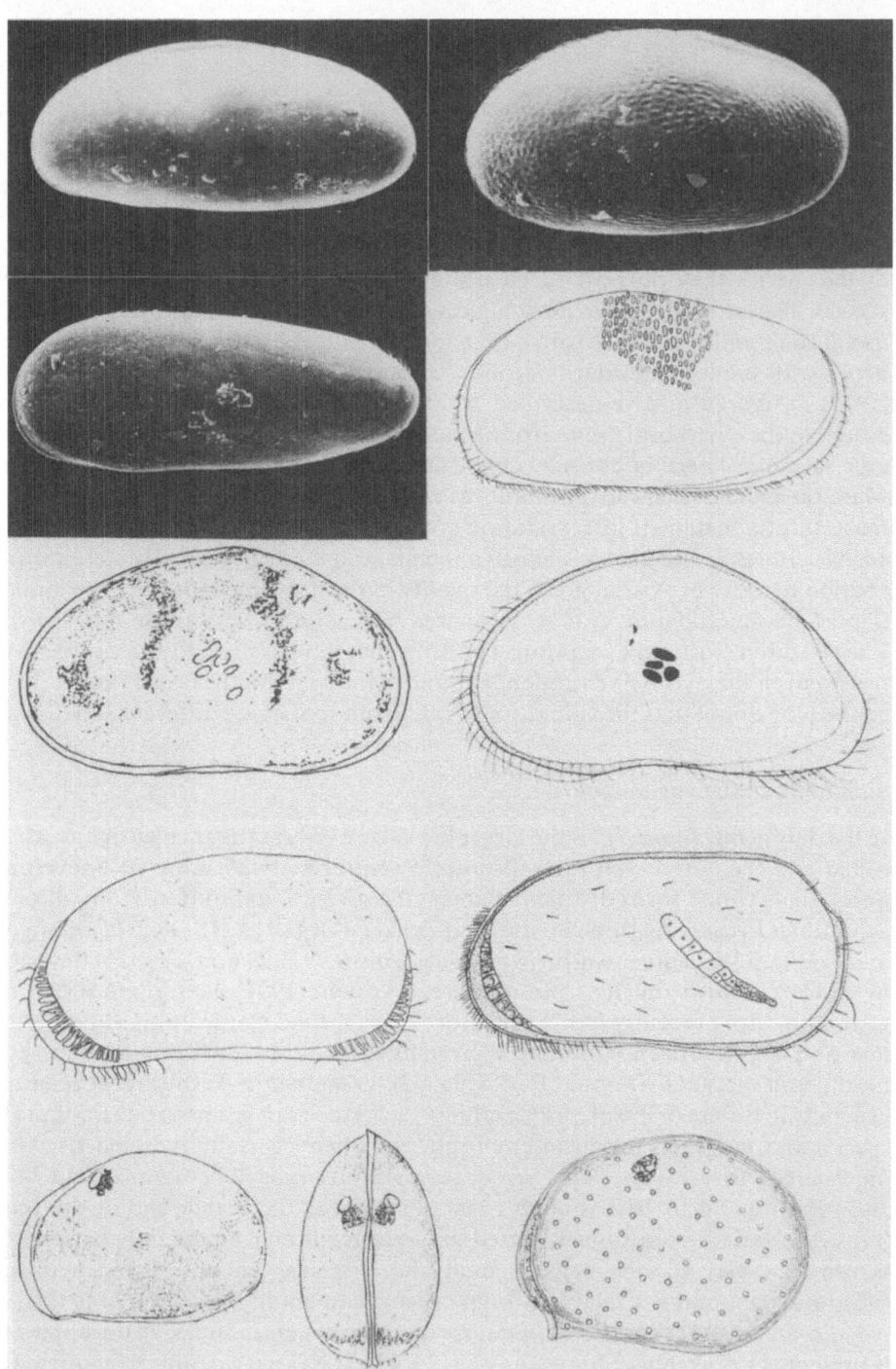

Explanation of Fig. 27-31.

27. *Strandesia purpurascens* x 55, external view of left valve; 28. *Strandesia marmorata* x 55, external view of left valve; 29. *Strandesia elongata* x 74, external view of left valve; 30. *Strandesia wierzejskii* x 70, external view of left valve; 31. *Strandesia vittata* x 61, female carapace from right; 32. *Cypridopsis dispar* x 86, external view of female left valve; 33. *Indiacypris luxata* x 55, external view of female left valve; 34. *Indiacypris dispar* x 68, external view of female left valve; 35, 36. *Notodromas oculata* x 58: 35. female carapace from left; 36. dorsal view of carapace; 37. *Notodromos entzi* x 49, female carapace.

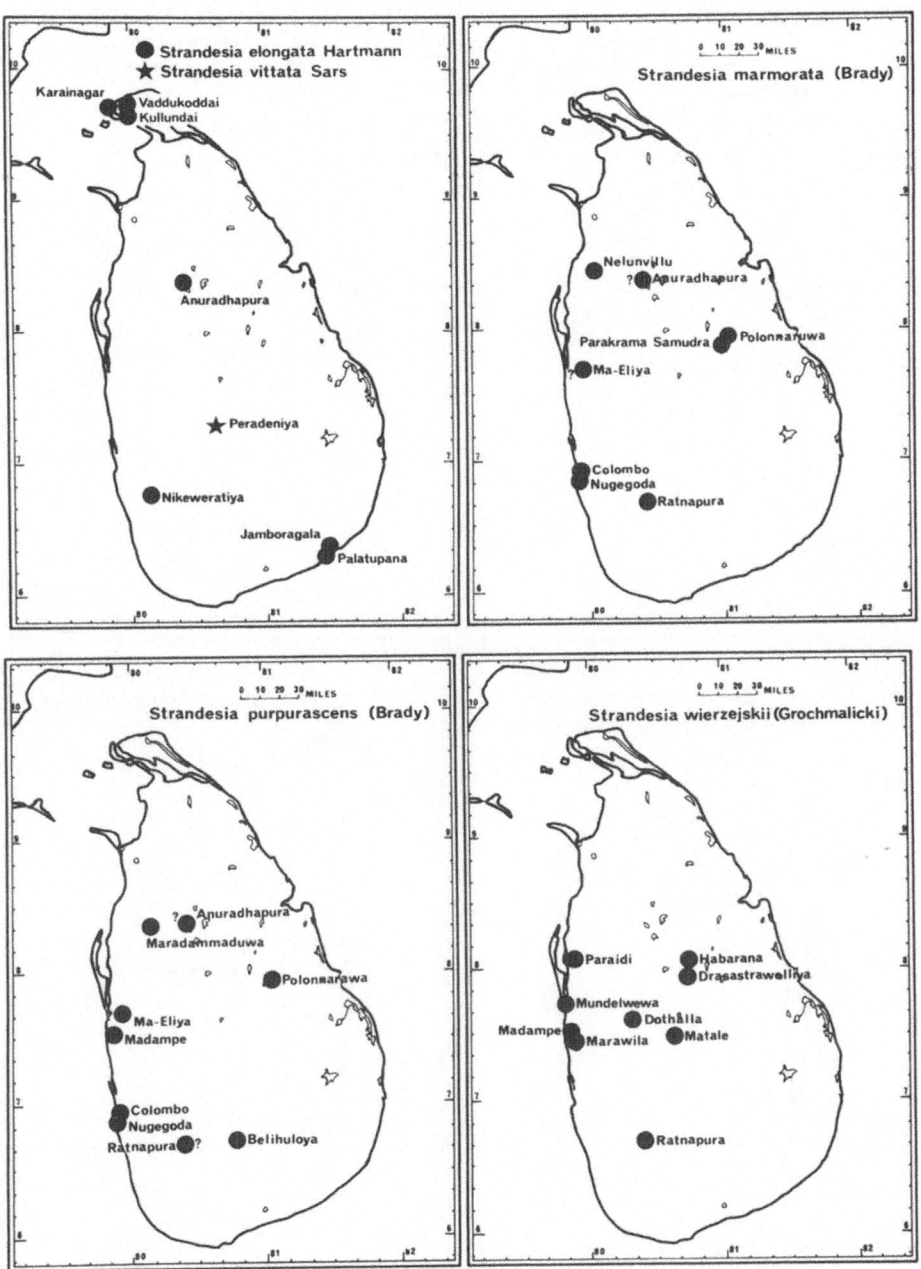

Explanation of Fig. 38-41.
38. *Distribution of Strandesia elongata* and *Strandesia vittata*; 39. Distribution of *Strandesia marmorata*; 40. Distribution of *Strandesia purpurascens*; 41. Distribution of *Strandesia wierzejskii*.

184

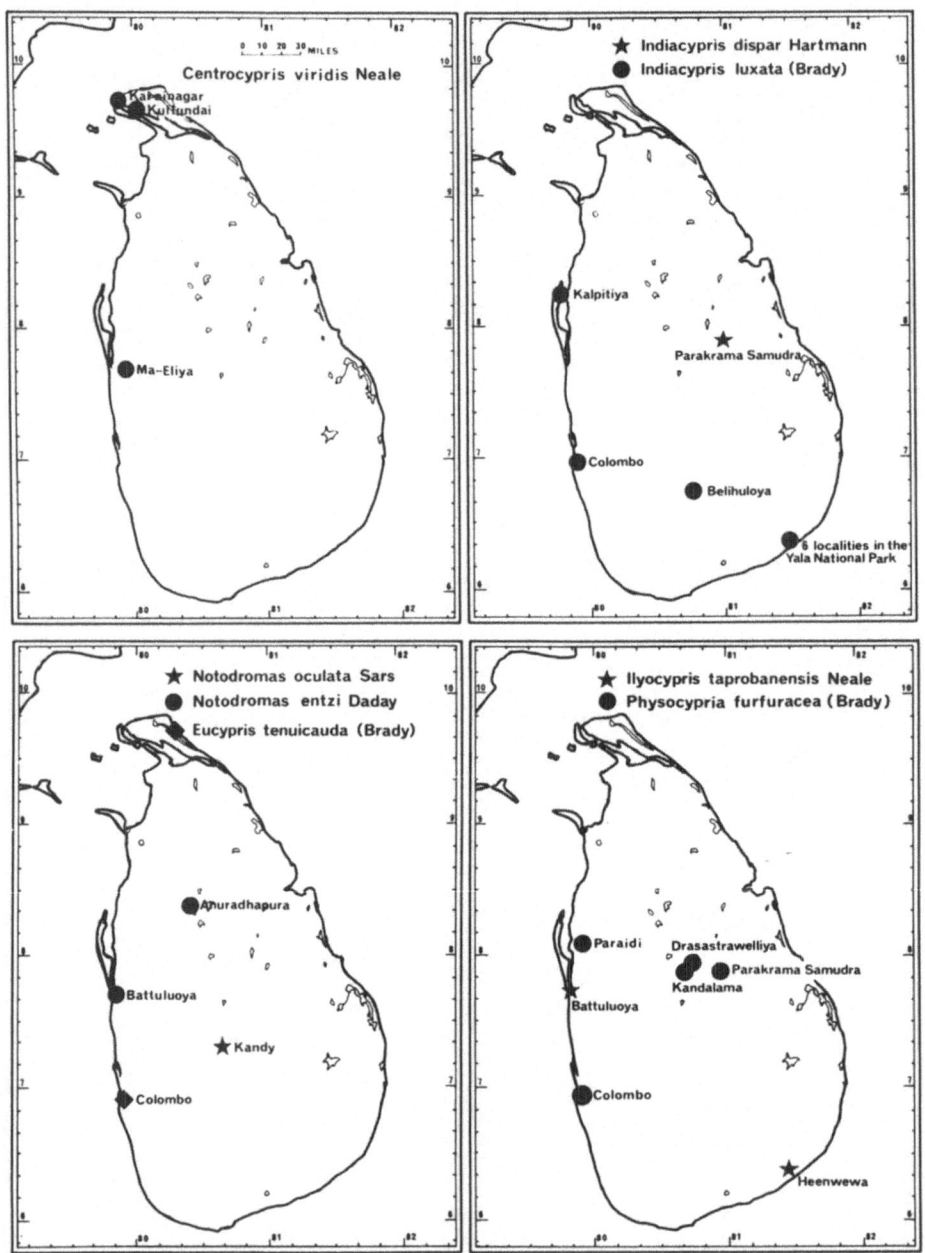

Explanation of Fig. 42-45.
42. Distribution of *Centrocypris viridis*; 43. Distribution of *Indiacypris dispar* and *Indiacypris luxata*; 44. Distribution of *Notodromas oculata, Notodromas entzi* and *Eucypris tenuicauda*; 45. Distribution of *Ilyocypris taprobanensis* and *Physocypria furfuracea*.

Subfamily Notodromatinae

Notodromas entzi has small, stout, equal sized valves with a flattened venter, narrow zone of concrescence and few marginal pore canals. The eyes are well developed and there is marked sexual dimorphism and a marked ventral projection in the female (Fig. 37). It has been found at Battuluoya and Anuradhapura (Fig. 44). The related and somewhat similar *N. oculata*, which differs principally in being more regularly oval in dorsal view (Figs. 35, 36) was found by Gurney (1916) in a tank at Kandy (Fig. 44). The subfamily occurs only rarely in Sri Lanka.

Family Ilyocypridinae

Ilyocyprids with their distinctive elongate, quadrangular, sulcate shells occur at at a number of localities in Sri Lanka but in many cases prove to be juveniles which probably are *Ilyocypris taprobanensis* but cannot be placed with certainty. *I. taprobanensis*, some 0.73 mm long, shows distinctive ornamentation with its strongly pitted surface and variable spination with about twelve spines anteriorly and ten posteriorly (Fig. 46, 47) and definitely occurs at Battuluoya and Heenwewa (Fig. 45). *Ilyocypris australiensis* also has been thought to occur in the island where Apstein (1907) recorded one specimen from Lake Colombo variously as *Ilycypris australis, Iliocryptus australis* and *Ilicypris australis*, and Daday (1898) recorded it as *Iliocypris australiensis* Sars from the Lake Kalawewa area. Neither of these authors figured it, the species has not been found since, and these records are regarded as doubtful since it may easily be confused with other species.

Small species

Family Candoniidae

Subfamily Cyclocypridinae

The only cyclocyprinid found in Sri Lanka is *Physocypria furfuracea*, a small species (0.44-).55 mm long) with fairly tumid, sub-reniform shell with arched dorsal margin which slopes more gradually anteriorly (Fig. 50) and with denticulate anterior and posterior right valve margins. Originally found in Lake Colombo, whence it was also described by Gurney (1916) as *Physocypria tuberata* (a synonym) it has been found recently at four other localities (Fig. 45) in a belt across the centre of the island. Because of its small size it is easily overlooked and is probably widespread throughout the island.

Subfamily Cypridopsinae

Generally small ostracods in which the body ends in a flagelliform, not leg-like furca. The genus *Oncocypris* is widely distributed in the tropics and is small, equidimensional, heavily pitted and with marked eye tubercles and sexual dimorphism. *Oncocypris pustulosa* is a distinctive tuberculate form (Figs. 48, 49) in which the females (length ca 0.55 mm) are more elongate than the males (length ca. 0.49 mm). It has been found at five widely distributed localities on the island, but not north of Nochiyagama (Fig. 54). It is known also from S. India (Michael & Victor 1975). The species is fully described in Neale (1976b).

Pseudocypretta maculata is only about 0.3 mm long (Fig. 51) but is distinctive and easily recognized by its three cuticular, bright violet spots on each valve. Found originally in Sumatra and Java, it has been found more recently in S. India (Victor & Fernando 1979). In Sri Lanka it occurs in the central and southern parts of the island but not north of Marawila (Fig. 55).

Cypridopsis is a genus which generally reaches a length of about 0.5 mm and which is fairly common and widespread in Sri Lanka where there are a number of species. These still need careful analysis since the genus is nodescript and not as distinctive as most. *C. dispar* recently has been recognized at Madampe by Neale & Victor (1978b) (Figs. 32, 54).

Family Cypridopsidae

Subfamily Cyprettinae

A subfamily of small cypridopsids in which the furca is weak, but leg-like, with two claws distally, and the shell has septa in both valves. In Sri Lanka the subfamily is represented by two species of *Cypretta*, both about 0.5 to 0.55 mm long. The species generally are pilose but in addition *C. globosa*, which is pale brown to gold in colour, has short strong bristles on the carapace (Fig. 52) whilst *C. globula* is a pale yellowish-green and lacks the spines (Fig. 53). *C. globosa* has a wide distribution in Sri Lanka (Fig. 56) and is known from the northern part of the island but has not been recorded in India or S.E. Asia. On the other hand *C. globula* is widely distributed in the centre and south of the island but is not known in the north (Fig. 57). However, *C. globula* has been found in Java (Grochmalicki 1915) and Celebes (Tressler 1937) in Indonesia and in India (Arora 1931).

Discussion

Freshwater ostracods show considerable adaptability and their general production of drought resistant eggs means that they can colonise temporary bodies of fresh water often as easily as more permanent habitats. Generally they are some of the first organisms to appear, although this is at variance with the work of Selvarajah & Costa (1979) in the Jaffna Peninsula, who found that they were the last of the microcrustaceans to appear, being first discovered in a paddyfield at Vaddukoddai 42 days after it filled with water, and in a pond at Vaddukoddai 84 days after is filled with water. Nevertheless, the results here may be unusual since the widespread occurrence of large *Cypris subglobosa* in the paddyfields of the Jaffna Peninsula suggest that there is no lack of time for this species to run through its full life cycle of nine instars and reproduce its kind so that an early time of appearance is not critical.

Whilst ostracods are ubiquitous, they prefer quiet waters and find rivers and streams with even moderate current velocities uncongenial; in these situations they are rare. With this reservation, other waterbodies such as ponds, lakes (both natural and artificial) and paddyfields yield a well-developed and often abundant fauna except where local factors such as high pollution come into play (such as Lake Colombo). On the data available it is not possible to recognise any significant differences between the species colonising these habitats.

Endemic species

Some species clearly are rare and of these present knowledge suggests that eleven are endemic. *Ghardaglaia ambigua* is known only from the south-eastern Sri Lanka, although one must bear in mind that this is a brackish water species and may occur at other suitable localities round the island and in adjacent regions. *Stenocypris fernandoi*, the largest species in the island, also is known from but a single locality in the south. *Chrissia ceylonica*, recently found in the Parakrama Samudra fauna and discovered in the island for the first time since Daday's (1898) original description, similarly is rare and apparently endemic. To this may be added *Chrissia halyi* known from three localities on the island, whence came the type designated by Ferguson in 1969. *Centrocypris viridis*, also known from three localities, appears to be endemic. *Indicypris luxata*, known only from three localities as recently as 1978, is known now to be widespread and common in the Yala National Park in the south, but is not known from the north of the island and certainly is not recorded outside Sri Lanka. *Ilyocypris taprobanensis* is another rare but distinctive endemic species. It may, however, be less rare than appears at first sight, since a number of juvenile *Ilyocypris*, which were too immature to assign to a species with certainty, have been found at various localities. Thus collection failure may be an important factor here, and one that may apply equally in the case of the other species. In contrast *Cypretta globosa* is common and widely distributed throughout the island, including the north, but even so is not known further afield. The only Darwinulinacean, *Darwinula lundi* is known only at a single locality, where it was collected in running water, and has not been found outside Sri Lanka.

To these must be added an undescribed species of *Strandesia* from northern Sri Lanka and an undescribed cytherid recently recorded in material from Parakrama Samudra supplied by the Institute of Limnology of the University of Vienna. Cytherids are rare in freshwater, and although the shells are decalcified it is possible to make out that this one had dorsal tubercles and a ventro-lateral ridge,

188

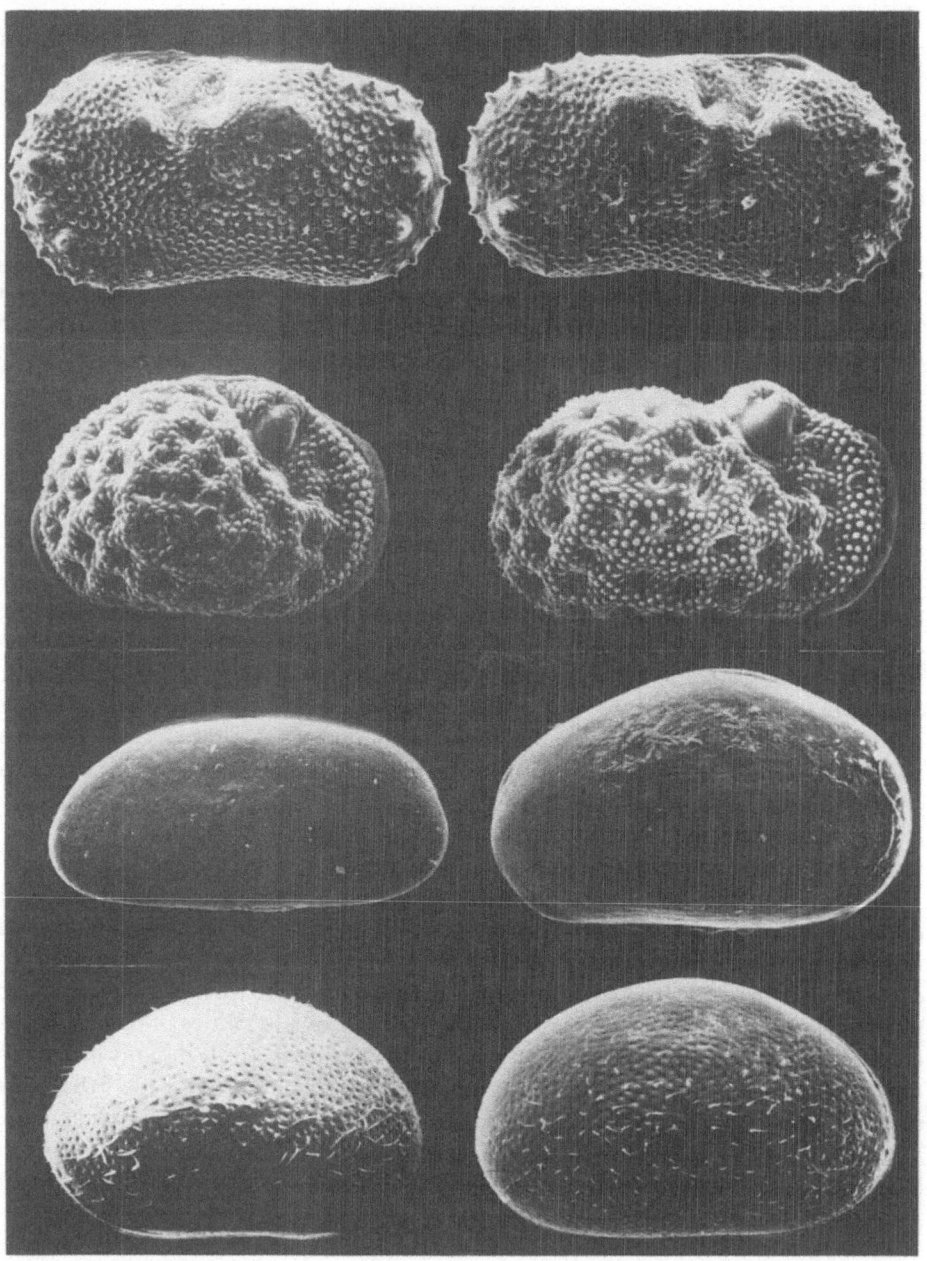

Explanation of Fig. 46-53.
46,47. *Ilyocypris taprobanensis* x 81: 46. external view of female left valve, 47. external view of female
right valve; 48,49. *Oncocypris pustulosa* x 96: 48. external view of male right valve, 49. external view
of female right valve; 50. *Physocypria furfuracea* x 106, external view of male right valve; 51. *Pseudocypretta maculata* x 226, carapace from right; 52. *Cypretta globulosa* x 102, carapace from right; 53.
Cypretta globula x 98, external view of right valve.

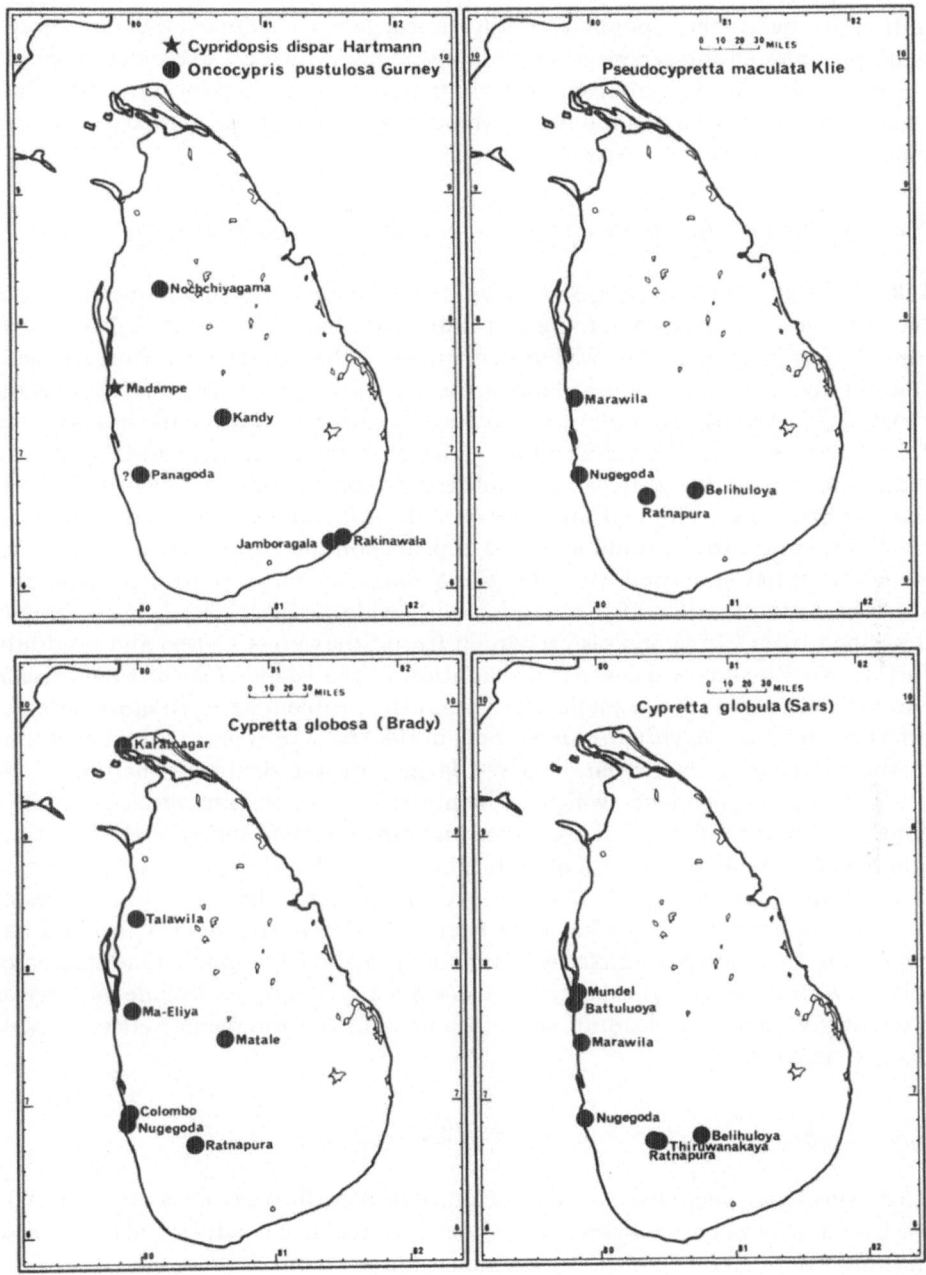

Explanation of Fig. 54-57.
54. Distribution of *Cypridopsis dispar* and *Oncocypris pustulosa*; 55. Distribution of *Pseudocypretta maculata*; 56. Distribution of *Cypretta globulosa*; 57. Distribution of *Cypretta globula*.

although it has not been possible to assign it to a genus. No doubt other new species will appear when the genera *Strandesia* and *Cypridopsis* are examined in detail. Further collecting will almost certainly produce additional localities for these species in Sri Lanka and a number of these species currently regarded as endemic may eventually prove to extend further afield.

Species which are rare but comparatively widespread geographically

Like the larger stenocyprines, *S. derupta* is rare, being found only at one Sri Lanka locality, but it occurs also in India and Indonesia. The two cyprids *Cypris decaryi* and *C. latissima* are similar. Widely distributed in the tropics, the former has been found only in the north of the island, a situation for which collection failure is the most likely explanation. *C. latissima* is equally rare in Sri Lanka and has not been found in the north of the island, although it occurs in Madagascar and the African mainland. *Cyprinotus cingalensis*, another rare species, has been recorded widely outside Sri Lanka, although most records do not stand up to scrutiny. It does, however, appear to occur in India, and appears confined to these two areas, being absent from Indonesia and Australia. For *Strandesia vittata* we are dependent on a single record by Gurney (1916) but it is a fairly widespread species, being originally described from China and also recorded from Sulawesi (Celebes) and no doubt further work will reveal a wider distribution in the island. *Eucypris tenuicauda* similarly is known from a single locality but the problem here probably is one of interpretation. *Indiacypris dispar* has only in 1982 been recognised for the first time in the island in a determination based largely on the distinctive nature of the seventh pair of appendages, which are identical with Hartmann's species and differ considerably from Brady's. This extends its range outside India for the first time and it will no doubt be found at other localities. Whether the reverse will prove true in the case of the endemic *Indiacypris luxata* remains to be seen. Both species of *Notodromas* are rare. They have not been recorded in India but *N. oculata* has been found outside Sri Lanka in Indonesia and the Philippines. One species of *Cypridopsis* has been named in recent work but any attempt to include this here or elsewhere would be misleading, since the current situation merely reflects lack of taxonomic work.

Species which are fairly common and widespread geographically

Three species are included here which cannot be described as rare in occurrence on the island. *Physocypria furfuracea* has been found at five localities and occurs also in India, while *Stenocypris hislopi*, which is known from both north and south of the island, probably is much more common than present records suggest and also is present in India. *Strandesia elongata*, which occurs also in India, has been recorded from seven localities in the island.

Species which are very common and widely distributed

Species which are abundant and very widely distributed belong here, namely *Cypris subglobosa* and *Heterocypris dentatomarginatus*. The former is a particularly widespread species as noted earlier, and obviously a successful coloniser. *H. dentatomarginatus* also is highly successful, and has been reported from India and Australia, although the Australian specimens differ in a number of features and belong in another species.

Species often common and geographically widespread but not known from northern Sri Lanka

These include *Oncocypris pustulosa, Stenocypris major, Cypretta globula, Strandesia marmorata, S. purpurascens, S. wierzejskii, Hemicypris pyxidata* and *Pseudocypretta maculata*.

In many ways these are the most interesting group of species and some attempt already has been made to account for the present distribution of some of them (Neale 1977a). Any such attempt must be regarded as highly speculative, and the following remarks seen in this light. New evidence may render the suggestions made here obsolete at any time.

One of the ever-present problems in the case of freshwater ostracods is how they colonise discrete and isolated bodies of water. The principal agencies seem to be the transport of eggs by animals (birds and insects being the groups generally involved) and on which a considerable literature exists, by humans as in the transport of rice seed for which there is good evidence in the case of the Italian rice fields and by the wind transport of resistant eggs which laboratory studies indicate is possible as regards viability but which is difficult to prove in the field.

The failure of species to appear in the northern part of the island may be due to three factors:

1. Collection failure, 2. Unsuitable conditions, 3. Lack of movement, or transport in, of relatively late arrivals in the island. In general unsuitable conditions may include a wide variety of parameters from temperature, salinity and pH to food supply, water chemistry and current strength. In the present situation the most probable effective parameter is the duration and season of occurrence of ephemeral waters such as paddyfields which are the source of most collections from the area, compared with the breeding cycle of the individual species. Which of these three factors seems the most likely to be the controlling one varies from species to species.

Oncocypris pustulosa is the least common of these species, having been noted at five localities in Sri Lanka and only one in southern India. It seems to be generally fairly rare, but also is small and in consequence easily overlooked and so here collection failure probably is the answer.

Stenocypris major and *Cypretta globula* are in a different category since they are widespread and well-established in Sri Lanka, and occur as far away as Europe, where their occurrence is believed to be due to the importation of rice seed. The northern areas of Sri Lanka have been fairly well sampled and in the case of these common species collection failure seems the least likely answer. In the case of ephemeral water bodies the presence of water and the breeding cycle may be out of phase and irreconcilable and such unfavourable conditions would provide a reasonable explanation. On the other hand, conditions appear to present no problem in the case of the closely related *Cypretta globosa* and *Stenocypris hislopi* which both occur in the north and in this respect detailed studies on the breeding cycle of the various species are needed urgently. In the case of permanent water bodies one would need to invoke some degree of collection failure as well. The distribution could be equally explained by the suggestion for the group below.

The remaining five species have strong links with Indonesia and in 1977 were unknown outside that area except for a record of *S. purpurascens* in Thailand. Collection failure seemed the least likely explanation and there seemed no reason to postulate inimical environmental factors in view of closely allied species living in the north (rather as in the case of *S. major* and *C. globula* noted above). Since there was no obvious animal or physical agency it was suggested that the importation of rice from Indonesia into the south and centre of the island with failure to colonise the north was the most probable explanation. Since 1977 *Hemicypris pyxidata, Strandesia purpurascens* and *Pseudocypretta maculata* have been recorded in S. India (Victor & Fernando 1979). Transfer of rice seed would still explain the distribution of these three species although the Indian records increase somewhat the likelihood of unsuitable conditions being the controlling parameter.

Future work

Whilst great progress has been made there is still some taxonomic work needed on the Sri Lanka fauna. It is estimated that there probably are another fifteen or so species in addition to these listed here. These include both new and undescribed species as well as older records. Among the latter it is doubtful if the true nature of *Cypris monilifera* Brady will ever be satisfactorily ascertained and it seems best regarded as a *nomen nudem*. It is suspected that the records of *Ilyocypris australiensis* by Daday and Apstein refer to one of the species of *Indiacypris*-probably *I. luxata*, detailed study of the cypridopsids will establish whether the true *C. assimilis* and *Plesiocypridopsis newtoni* occur or whether these names have been appended to other species, whilst detailed work may still yield a satisfactory answer to the identity of *Cypricercus reticulatus* Daday and the *Candonella albida* Vavra of Apstein.

On this sound taxonomic basis a large and interesting field opens up, and every effort should be made to solve the problem of the absence of certain species in the north. A carefully controlled collecting programme with detailed physical and chemical measurements at the time of collection of the individual samples would yield much valuable information. Coupled with studies of the position of this group in the food chain, for they are known to form an element, albeit a variable one, in the diet of some fish, this might give interesting results and even possible commercial implications.

References

Apstein, C. 1907. Das Plancton im Colombo-See auf Ceylon. Zool. Jahrb. 25:201-244.

Arora, G.L. 1931. Fauna of Lahore 2. Entomostraca (Water fleas) of Lahore. Bull. Dept. Zool. Panjab. Univ. 1:62-100.

Baird, W. 1859. Description of some new Entomostraca from Nagpur, collected by the Rev. S. Hislop. Proc. Zool. Soc. Lond. 389:231-234.

Brady, G.S. 1886. Notes on Entomostraca collected by Mr. A. Haly in Ceylon. J. Linn. Soc. Zool. 19:293-317.

Daday, E. von. 1898. Mikroskopische Süsswasserthiere aus Ceylon. Termesz. Füz. 21:69-85.

Danielopol, D.L. 1980. An essay to asses the age of the freshwater interstitial ostracods of Europe. Bijdragen Tot de Dierkunde 50:243-291.

Ferguson, E. 1964. Stenocyprinae, a new subfamily of freshwater ostracods (Crustacea) with description of a new species from California. Proc. Biol. Soc. Wash. 77:17-24.

Ferguson, E. 1969. The type species of the genus *Stenocypris* Sars 1889 with descriptions of two new species. pp. 67-75 In: J.W. Neale (ED.) The Taxonomy, Morphology and Ecology of Recent Ostracoda. Oliver & Boyd.

Gauthier, H. 1933. Entomostracés de Madagascar 1re Note. Description d'une nouvelle *Cypris* (Ostracodes). Bull. Soc. Zool. Fr. 58:209-216.

Grochmalicki, J. 1915. Beiträge zur Kenntnis der Süsswasserfauna Javas. Phyllopoda, Copepoda und Ostracoda. Bull. Int. Akad. Sci. Cracovie Ser. B. Sci. Nat.:217-242.

Gurney, R. 1916. On some fresh-water Entomostraca from Ceylon. Proc. Zool. Soc. Lond:333-343.

Hartmann, G. 1964. Asiatische Ostracoden. Systematische und zoogeographische Untersuchungen. Int. Rev. ges. Hydrobiol. Hydrogr. Syst. Beihefte 3:1-155.

Klie, W. 1932. Die Ostracoden der Deutschen Limnologischen Sunda-Expedition. Arch. Hydrobiol. 11:447-502.

McKenzie, K.G. 1971. Entomostraca of Aldabra, with special reference to the genus *Heterocypris* (Crustacea,Ostracoda). Phil. Trans. Roy. Soc. Lond. B. 260:257-297.

Michael, R.G. & R. Victor. 1975. Redescription of two little known Indian freshwater ostracods *Oncocypris pustulosa* Gurney 1916 and *Cypris matthaii* Arora 1931. J. nat. Hist. 9:509-512.

Moniez, R. 1892. Entomostracés d'eau douce de Sumatra et de Celebes. 2. Ostracodes. pp. 129-135 In: M. Weber (Ed.) Zoologisches Ergebnisse einer Reise in Niederlandisch Ost-Indien, 2.

Müller, G.W. 1898. Die Ostracoden. pp. 257-296 In: Wissenschaftliche ergebnisse der Reisen in Madagaskar und Ostafrika in den Jahren 1889-95 von Dr. Voelzkow. Abh. senckenb. Naturforsch. Ges. 21.

Neale, J.W. 1976a. On *Centrocypris viridis* Neale sp. nov. Stereo-Atlas of Ostracod Shells 3:13-20.

Neale, J.W. 1976b. On *Oncocypris pustulosa* Gurney. Stereo-Atlas of Ostracod Shells 3:21-28.

Neale, J.W. 1976c. On *Stenocypris fernandoi* Neale sp. nov. Stereo-Atlas of Ostracod Shells 3:29-36.

Neale, J.W. 1976d. On *Ilyocypris taprobanensis* Neale sp. nov. Stereo-Atlas of Ostracod Shells 3:37-40.

Neale, J.W. 1976e. On *Cypris subglobosa* J. de C. Sowerby. Stereo-Atlas of Ostracod Shells 3:125-132.

Neale, J.W. 1976f. On *Cypris decaryi* Gauthier. Stereo -Atlas of Ostracod Shells 3:133-140.

194

Neale, J.W. 1977a. Ostracods from the rice-fields of Sri Lanka (Ceylon). pp. 271-283 In: H. Löffler & D. Danielopol (Eds.) Aspects of Ecology and Zoogeography of recent and Fossil Ostracoda. Proc. 6th Int. Symp. Ostracods, Saalfelden (Salzburg) 1976.

Neale, J.W. 1977b. On *Cypris latissima* (G.W. Müller). Stereo-Atlas of Ostracod Shells 4:139-144.

Neale, J.W. 1979a. Ostracod Taxonomy. pp. 89-114 In: Dunnet *et al.* Report of the Joint Aberdeen and Colombo Universities Expedition in Sri Lanka, 1978.

Neale, J.W. 1979b. On the genus *Cyprinotus* and its interpretation. pp. 77-85 In: N. Krstic (Ed.) Taxonomy, Biostratigraphy and distribution of Ostracodes. Proc. 7th Int. Symp. Ostracodes, Belgrade, Jugoslavia, 1979.

Neale, J.W. 1979c. On *Ghardaglaia ambigua* Neale sp. nov. Stereo-Atlas of Ostracod Shells 6:99-106.

Neale, J.W. & R. Victor. 1978a. On *Indiacypris luxata* (Brady), a freshwater ostracod (Crustacea: Entomostraca) from Sri Lanka. Zool. J. Linn. Soc. 64:71-77.

Neale, J.W. & R. Victor. 1978b. The Lund University Expedition freshwater Ostracoda from Sri Lanka (Ceylon) Can. J. Zool. 56:1081-1087.

Pinto, I.D. & I. Purper. 1980. Redescription and new data about the soft parts of the holotype of *Cyprinotus cingalensis* Brady, 1886. Perquisas, Porto Alegre (Brasil) 3:43-61.

Sars, G.O. 1889. On some Ostracoda and Copepoda raised from dried Australian mud. Arch. Math. Naturv. Kristiana 8:3-79.

Sars, G.O. 1903. Freshwater Entomostraca from China and Sumatra. Arch. Math. Naturv. Kristiana 25:3-44.

Selvarajah, N. & H.H. Costa. 1979. The distribution of Anostraca and Conchostraca (Crustacea) in the Jaffna Peninsula. Bull. Fish. Res. Stn Sri Lanka 29:79-88.

Tressler, W.L. 1937. Ostracoda. In Mitt. 19 von der Wallacea Expedition. Int. Rev. ges. Hydrobiol. 34:188-207.

Vavra, W. 1906. Ostracoden from Sumatra, Java, Siam, den Sandwich-Inseln und Japan. Zool. Jahrb. 23:413-438.

Victor, R. & C.H. Fernando. 1979. The freshwater Ostracods (Crustacea: Ostracoda) of India, Rec. Zool. Surv. India 74:147-242.

Victor, R. & R.G. Michael. 1975. Nine new species of freshwater Ostracoda from Madurai area in Southern India. J. nat. Hist. 9:361-376.

10. The ecology and distribution of free-living Meso and Macrocrustacea of inland waters

H.H. Costa

Introduction

Five classes of Crustacea are represented in Sri Lankan freshwaters, namely Branchiopoda, Copepoda, Branchiura, Ostracoda and Malacostraca. Most of the published information on them is in the form of taxonomic descriptions. Many details regarding the biology and the ecology of these animals are unknown.

The crustacean fauna is circumtropical and an important component of the freshwater fauna, and consists of species occurring also in Africa, India and the Malayan Archipelago. The freshwater Crustacea recorded from Sri Lanka comprise about 10% of all animal species inhabiting freshwaters (Fig. 1). They range in size from microscopic to large (>10 cm long). The present work deals mainly with the larger Branchiopoda and Malacostraca.

Subclass Branchiopoda

The subclass Branchiopoda consists of semi-microscopic cladocerans and larger forms such as the anostracans (fairy shrimps) and conchostracans. The latter groups are restricted mostly to the Arid Zone. They produce resting eggs which can withstand high temperatures, pH differences and seasonal desiccation. Under favourable conditions, the eggs hatch and the young swarm in lentic waters. In these forms, dry conditions are a necessary stimulus for the hatching of eggs.

The representatives of the Anostraca and the Conchostraca in Sri Lanka are mostly freshwater; till the discovery of *Artemia* sp. in salterns in the Northern Province, no marine representatives were known.

The periodicity of rainfall in the Dry Zone produces temporary pools, immediately after the rains. These are ponds, rain pools and ricefields. In the dry zone, the anostracans and conchostracans appear in temporary waters after rains around October and disappear in most places by January or February.

Observations made on the succession of microcrustacea and macrocrustacea in the ponds and ricefields at Vaddukoddai (Selvarajah & Costa 1979) have indicated that the first Crustacea to appear in these habitats are copepods, which appear on day 2 and last until day 7. These are followed by cladocerans (which appear on day 21 and lasted till day 42). The anostracans and conchostracans appeared on day 28 and were observed till day 56. The last to appear were the Ostracoda, which appeared on day 42 and lasted till the pond dried up on day 82. The times of appearance of the different crustaceans, however, tend to differ in the different habitats.

Observations made by Selvarajah & Costa (1979) on the anostracan *Streptocephalus spinifer* have indicated the appearance of successive generations with changing physical and chemical conditions. With the onset of rains and the first filling with water, a generation of small individuals appear in the water pools. These disappear on drying. With further rains and with the second filling of water a generation of larger individuals appear. Their size and number of eggs are given in Table 1.

The anostracans and conchostracans described from Sri Lanka are given in Table 2, their known distribution is as follows:

Streptocephalus spinifer is distributed widely in the Jaffna Peninsula and occurs more often in ricefields than in other habitats. This species also has been recorded in areas further south such as Puttalam, Maho and in Nikaweratiya (unpublished

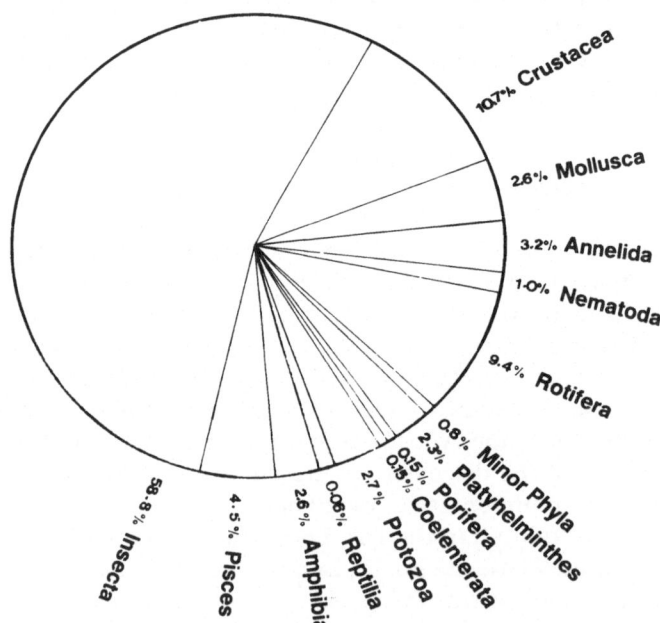

Fig. 1. The faunal spectrum (%) in Sri Lankan freshwaters (total recorded ca. 1350 spp.) (after Costa 1980).

data). *S. dichotomus* has been collected only from one locality, Chavakachcheri, while *Branchinella kugunamaensis* also has been collected from nearby Manipay in the Jaffna Peninsula.

The anostracans are very local in occurrence; one pool may have large numbers while adjoining pools have few individuals or none. Their duration varies with duration of water in the pools.

The Conchostraca are bivalve-like animals. They swarm in temporary waters after rain, occur also in perennial ponds in the southwest sector of Sri Lanka. The Conchostraca are encountered more often in rain pools, ricefields and flood waters than are the Anostraca. As a group they are more local. *Eulamnadia michaeli* has been recorded only from the Jaffna Peninsula where it is widespread. *Eocyzicus plumosus* and *Lynceus serratus* also are restricted in distribution and have been recorded only from a few localities in the Jaffna Peninsula.

The pantropical conchostracan *Cyclestheria hislopi*, an inhabitant of the wet tropics, occurs in the littoral areas of ponds, drains, ricefields and other lentic waters in the south of Sri Lanka. It has not been found in the northern parts of the island. *C. hyslopi* has been recorded from Chilaw, Bingiriya, Wattala, Colombo and Panadura (unpublished data). In the Colombo (Beira) Lake they form an important component of the fauna living among the roots of *Eichhornia crassipes*

Table 1. The succession of *Streptocephalus spinifer* in a pond at Vaddukoddai (from Selvarajah & Costa 1979).

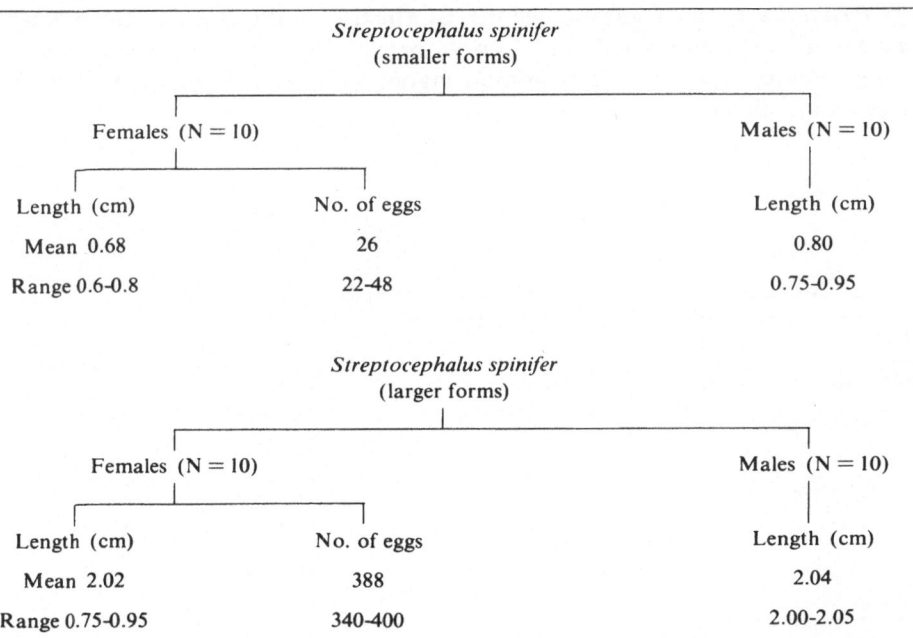

Streptocephalus spinifer (smaller forms)		
Females (N = 10)		Males (N = 10)
Length (cm)	No. of eggs	Length (cm)
Mean 0.68	26	0.80
Range 0.6-0.8	22-48	0.75-0.95
Streptocephalus spinifer (larger forms)		
Females (N = 10)		Males (N = 10)
Length (cm)	No. of eggs	Length (cm)
Mean 2.02	388	2.04
Range 0.75-0.95	340-400	2.00-2.05

(Costa & De Silva 1978b). *C. hislopi* is a typical inhabitant of floating vegetation in Africa, South America and Asia and does not occur in water with a high concentration of suspended solids (Junk 1977).

Caenestheriella indica has an extensive distribution. This species has been collected from rain pools in several localities in the southern and northern parts of the island.

All of the crustaceans living in temporary waters can withstand fairly high temperatures. *S. spinifer* has been collected from waters ranging from 25-36 ° C and *E. michaeli* from waters ranging from 23-37 ° C. These two species also show tolerances to pH 6.4-7.8.

Subclass Malacostraca

This Subclass is represented in fresh and brackish waters by members of all three orders: Amphipoda, Isopoda and Decapoda.

Amphipoda

Unlike temperate regions, the Amphipoda are very poorly represented in Sri Lankan freshwaters. This order, however, is richly represented in the lagoons and coastal habitats. Only two species have been reported from inland waters; *Paracallipe fernandoi*, the exact locality records of which are unknown, and *Grandidierella megnae* from Colombo (Beira) Lake. A species of *Parochistia* has been reported living among oysters in the Negombo lagoon (Pinto & Wignarajah 1980).The isopoda are all parasitic.

Table 2. Sri Lankan freshwater Anostraca and Conchostraca.

Order	Species	Distribution
Anostraca	*Streptocephalus spinifer*	Dry zone
	Streptocephalus dichotomus	Dry zone
	Branchinella kugunamaensis	Dry zone
Conchostraca	*Cyclestheria hislopi*	Wet zone
	Caenestheriella indica	Wet zone & Dry zone
	Eulamnadia michaeli	Dry zone
	Eocyzicus plumosa	Dry zone
	Lynceus serrata	Dry zone

Decapoda

The Decapoda are represented in freshwaters by shrimps, prawns and crabs. The atyid shrimps characteristically are freshwater forms, while the palaemonid prawns are found in both fresh and brackish waters. The distribution of the Decapoda is given in Table 3.

Atyidae

The island is rich in atyid species, some of them abundant. Their distribution is shown in Fig. 2. They seem to occupy the same niche as the Amphipoda and Isopoda in temperate freshwaters. They are more or less absent in the fast flowing streams, e.g. in the Maskeliya area north of Adam's Peak wilderness. Streams in the Nuwara Eliya area, except those in the Moon Plains and the Horton Plains, do not have atyids. Two of the endemic species, *Caridina singhalensis* and *C. pristis* are confined to cold water mountain streams in the central hilly regions. The distribution of *C. singhalensis* is limited to the region above 2,000 m. They are found in the slow flowing streams in the Moon Plains and Horton Plains at Nuwara Eliya. *C. pristis* is confined to hill streams situated at altitudes between 400-1,000 m. They have been recorded from streams in both the northern and the southern flank of the central mountains. *C. pristis* has been collected from Singharaja forest, Kitugala and in the Ratnapura district, Hunnasgiriya and the Knuckles mountains. Of the other species, *C. fernandoi* is found up to the same elevations as *C. pristis*. However, *C. fernandoi* is more widely distributed and has been

Table 3. The geographical distribution of Sri Lankan decapod Crustacean genera living in inland waters.

Family	Geographical distribution	Sri Lankan genera
Atyidae	Circumtropical	*Caridina* *Atya*
Palaemonidae	Circumtropical	*Macrobrachium*
Penaeidae	Circumtropical	*Penaeus* *Metapenaeus*
Parathelphusidae	Asia, Australia	*Ceylonthelphusa* *Spirothelphusa* *Oziothelphusa*
Sundathelphusidea	Asia	*Perbrinckia*
Portunidae	Circumtropical	*Scylla* *Portunus*

200

collected from several reservoirs and rivers in the coastal plains (Arudpragasam & Costa 1962). *C. typus* and *C. costai* are limited in distribution and have been collected from the lower reaches of the Menik Ganga near Yala and Singharaja forest, respectively. All of these species are heavily built. The most widely distributed shrimp in Sri Lanka *C. simoni*, extending even to brackish waters. This 'shrimp' is abundant in the man-made reservoirs in the dry zone, in the coastal areas of the wet zone and in many parts of the hill country. The atyid species of the lowland areas include *C. zeylanica, C. gracilirostris* an *C. propinqua. C. zeylanica* is found in smaller numbers in ricefields and small streams in Nawala, Attidiya, Vaddukoddai and Wellawaya and in brackish waters in Colombo and Moratuwa, especially in Bolgoda Lake where they are abundant.

  C. singhalensis forms an important food item of the rainbow trout, *Salmo gairdneri*, in the streams of the Horton Plains (Costa 1974). The presence of these shrimps in such large numbers must be due, *inter alia*, to food availability. The dense growth of aquatic plants at the edges affords protection, especially the trailing grasses which prevent the shrimp from being washed downstream during torrential conditions.

  Atya spinipes is the sole Sri Lankan representative of its genus. Unlike the other atyids it is large and lives exclusively in rivers. It has not been found in any of the lentic waters. Specimens have been collected from Deduru Oya, Maguru Oya, at Kurunegala, in the Maha Oya at Mawanella and Alawwa, in the Menik Ganga at Kataragama and Ruhunu National Park and in the lower reaches of the Kumbukkan Oya.

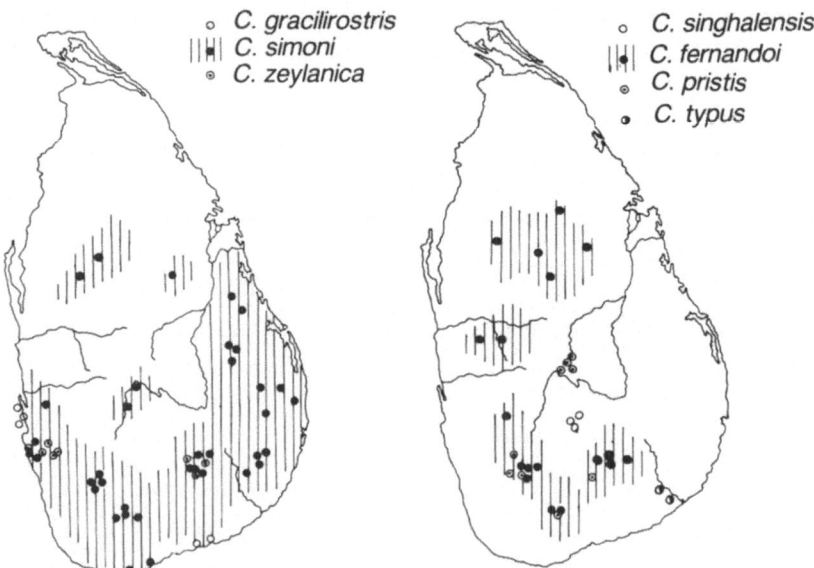

Fig. 2.   The distribution of atyid shrimps.

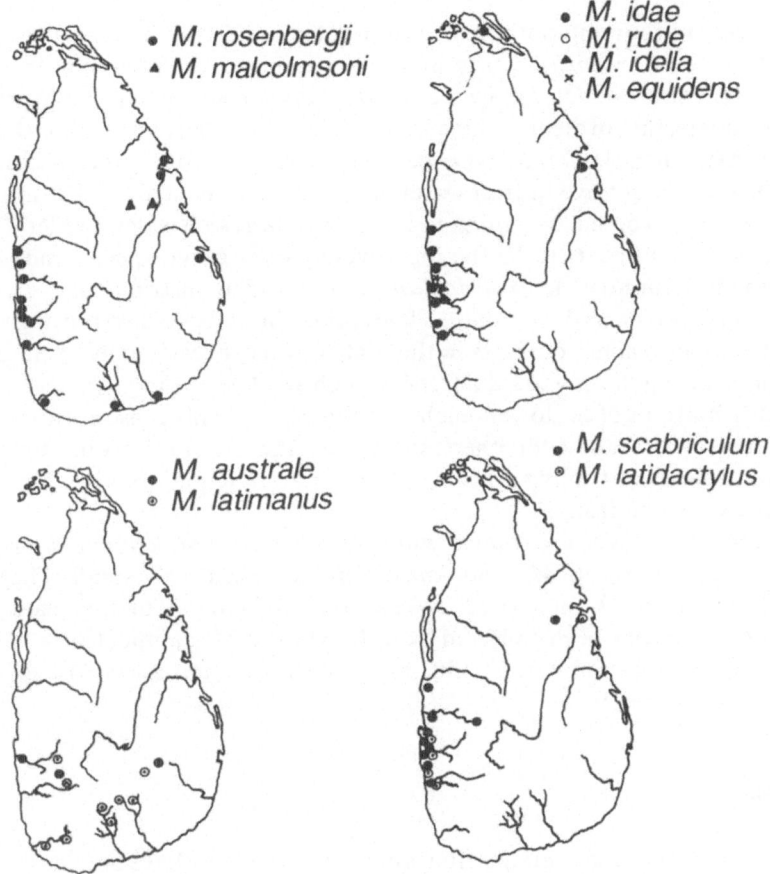

Fig. 3. The distribution of Palaemonid shrimps (after Costa 1978).

Palaemonidae

The Palaemonidae are represented by a single genus: *Macrobrachium*. Species of *Macrobrachium* are found in both the low and mid country rivers, streams, reservoirs, irrigation channels, marshes, estuaries, lagoons and even in temporary ponds formed by flood waters. However, they are absent from streams at high elevations of the Nuwara Eliya district. There are twelve species of *Macrobrachium* in Sri Lanka. The physical and chemical conditions of the different habitats of the species, and their distribution are given in Table 4 and Fig. 3.

In Sri Lanka, *M. latimanus* and *M. australis* are confined to the Southwest quarter of the island, including the hilly regions, while *M. kistnensis* and *M. malcolmsoni* are found in the north central plains. *M. idae, M. idella* and *M.*

scabriculum are found in both fresh and brackish water in the coastal areas. The distribution of *M. scabriculum* is much wider, however, extending to the mid country. *M. rude, M. latidactylus* and *M. srilankanse* tolerate both fresh and brackish waters; the former two species have been collected from the southwestern coastal areas, while *M. latidactylus* has been collected only from the east coast. Adult *M. rosenbergii* live in freshwater close to the coastal areas, but migrate to brackish waters to breed. *M. equidens* is mainly a brackish water dweller. Of these prawns, *M. idae* appears to be the most widely distributed species and is found even in the low country. *M. scabriculum* also is widely distributed.

Some prawns such as *M. latimanus* lives under boulders in fast flowing streams and in the upper reaches of rivers in the mid country, occasionally coming out of the crevices during the day to feed. Others, such as *M. rosenbergii* prefer the sandy and muddy bottoms of the lower reaches of rivers, especially those portions shaded by large trees and shrubs. Still others, such as *M. idae* occur mostly in habitats with luxuriant growths of trailing grasses and other aquatic macrophytes and among roots close to the banks.

Most of these prawns are robust, while a few species (*M. kistnensis, M. equidens, M. latidactylus* and *M. srilankanse*) are comparatively small. The largest species found in Sri Lanka is *M. rosenbergii*. Specimens of this giant prawn measuring 30 cm have been collected from Ja-Ela and Madampe (Costa 1979). All of these species are omnivorous and many species exhibit allometric growth in their larger chelipeds.

Penaeidae

The penaeid prawns and their distribution in inland waters have been described by Bruin (1965, 1971) (Table 5). Of the 31 species present in Sri Lankan waters, 13 occur in the low and high salinity lagoons and estuaries.

The most numerous are species of the genus *Metapenaeus*, found in the inland waters of the southern, southwestern and southeastern sectors of the island. *M. dobsoni* and *M. elegans* are abundant, while *M. burkenroadi, M. ensis* and *M. mutatus* are scarce in these lagoons and when present, are restricted to the mouths of estuaries where the salinity is higher than 10% throughout the year, except during the Southwest Monsoon. In northern and northeast lagoons, *M. dobsoni* and *M. elegans* are either absent or very scarce, while *M. burkenroadi, M. ensis* and *M.mutatus* are abundant.

All but two species of these penaeid prawns appear to be migratory. These are *M. elegans* and *M. monoceros*, which appear to complete their life cycles within the estuarine system.

In the Negombo Lagoon, where an important prawn fishery exists, Bruin (1971) reported that the most abundant species are *Penaeus indicus, P. semisulcatus, M. dobsoni* and *M. elegans*. Because of the migratory nature of the first three species,

Table 4. The ecological conditions of the habitats of the palaemonid prawns of Sri Lanka.

	M. rosenbergi	M. malcolmsi	M. australe	M. kistnense	M. srilankense	M. idae	M. idella	M. rude	M. latimanus	M. scrabiculum	M. latidactylus
altitude	0-20 m	0-20 m	15-60 m	0-20 m	0-5 m	0-10 m	0-20 m	0-3 m	15-700 m	0.5-150 m	0.5-15 m
depth	0.3-3 m	0.3-3 m	0.3-3 m	0.34 m	0.3-3 m	0.3-3 m	0.3-3 m	0.3-3 m	0.3-3 m	0.3-2 m	0.3-2 m
temperature (°C)	27-31	28-31	25-31	28-31	28-31	28-31	38-30	28-31	20-26	25-31	28-31
pH	6.4-8.0	7.8-8.0	6.0-6.4	7.8	6.6-7.6	6.4-7.6	6.5	7.1-7.4	5.5-6.6	6.4-7.2	6.5-7.3
total hardness (dH)	0-36-1.83	—	0.36-0.70	—	—	103	0.51	21	25-46	0.36-20	13-22
conductivity (µS cm⁻¹)	27-288 20°C	221 25°C	27-340 20°C	506 25°C	690-4,680 —	288 20°C	42,6 20°C	2,000 20°C	0.5-36 20°C	27-2,200 20°C	27-280 20°C
Ca ppm	1.6-140	22.4	1.81-3.20	32.1	2-140	1.15	2.09	101.0	1.19-6.0	1.81-107.2	1.81-2.96
Mg ppm	0.50-40.0	9.45	0.48-1.10	19.3	2-43.0	5.81	2.96	21.0	1.45-4.03	0.48-21.6	0.48-5.81
Na ppm	2.0-180.0	8.50	1.80-2.00	4.16	2-180.0	1.00	4.80	225.2	1.90-2.30	2-222	2.0-4.80
K ppm	3.0-141.0	3.55	0.95-1.10	10.3	3.5-14.0	3.5	5.0	31.0	0.80	0.95-31.0	0.95-50
Cl ppm	2.84-240.0	8.65	2.41-2.84	121.6	65-240.0	85.2	7.38	745.5	1.99-2.41	2.84-745.5	2.84-7.38
SiO₂ ppm	3.50-6.95	9.55	3.50-6.40	4.0	3.50-6.40	6.95	3.00	7.50	9.7-105	3.0-7.50	3.0-3.50
NO₃ ppm	0.12-2.60	0.7	0.13	—	1.02-2.40	—	—	3.0	0.01	0.25-3.0	0.25-1.75
P₂O₅ ppm	0.02-0.19	0.15	0.18	—	0.4-1.54	0.08	0.04	0.19	0.1	0.02-0.19	0.02-0.04
NH₄ ppm	0.0-0.01	1.0	0.08	—	0.16-0.25	0.04	0.01	0.25	0.1	0.025	0.01

large fluctuations in relative abundance occur. In the Negombo Lagoon a great increase of *P. indicus* was observed in November and of *P. semisulcatus* in March and September. These prawns appear to migrate into the sea during intermonsoonal periods.

There are several economically unimportant macruran species living in the lagoons. Among them is the snapping shrimp, *Alpheus edwardsii* which is common in Negombo Lagoon and lives among the oyster shells (Pinto & Wignarajah 1980).

Potamonidae

The family Potamonidae is represented by seven species belonging to four genera (Bott 1970). They occur throughout the country (Fig. 4), extending from the sea coast to streams situated at elevations of 2,300 m. Their habitats are varied: ponds, ditches, ricefields, irrigation channels, man-made reservoirs, slow flowing and torrential streams.

Some species are found mainly in the water, while others often inhabit burrows close to the water. In the low country, their favoured habitat is the ricefield

Table 5. Genera and species of penaeid prawns in Sri Lankan lagoons and estuaries (from De Bruin 1965, 1971).

Genus	Species	Occurrence	Distribution	Abundance
Penaeus	*latisulcatus*	Negombo	mouths of estuaries	not abundant
	canaliculatus	S.W. coast	mouths of estuaries	not abundant
	indicus	S.W. & S. coast	lagoons	abundant
	semisulcatus	Negombo	lagoons	abundant
	monodon	Negombo Panadura Koggala Arugam Bay		
Metapenaeus	*burkenroadi*	S.W. coast	mouths of lagoons	not abundant
	monoceros	Mullaitivu	mouths of lagoons	not abundant
	affinis	Panadura Negombo	mouths of lagoons	not abundant
	ensis	Mullaitivu	lagoons	abundant
	dobsoni	S.W., S.E. & S. coast	lagoons	abundant
	elegans	S.W., S.E. & S. coast	lagoons	abundant
	rysianassa	Jaffna	lagoons	rare
Parapeneopsis	*cornuta*	Negombo	mouths of estuaries	rare
	nana	Mutwal	mouths of estuaries	rare

(Fernando 1960). Here they burrow into the bunds or the sides of irrigation channels. In the hilly regions, they are found in fast flowing streams sheltering under rocks or burrowing into the banks of streams.

The young crabs, carried by the mother, are dispersed with the flood waters. With the recession of floods they construct burrows in the soft mud close to the water. These burrows serve as shelters from which they emerge to feed periodically. The burrows constructed vary with the different species. *Oziothelphus senex* generally burrows into the bunds of ricefields. *Spiralothelphusa hydrodroma* often is found burrowing into the bed of the ricefield itself, while *Ceylonthelphusa rugosa* makes shallow, more or less horizontal burrows into the sides of streams (Fernando 1960).

The freshwater crabs are widely distributed in Sri Lanka (Fernando 1960, Bott 1970, Pretzmann 1973). Their general distribution pattern is given in Fig. 4. Of the species recorded, *O. senex, O. minneryensis, S. hydrodroma* and *C. inflatissima* are found in the low country. *O. senex* is the most widely distributed as well as the most abundant species, occurring extensively in coastal areas and extending even to the valleys in the mid country. *O. senex* is absent from the Jaffna Peninsula, however, where it is replaced by *S. hydrodroma*. This species extends from the north along the eastern coast to Yala, and along the western coast to Puttalam. *O. minneryensis* has a restricted distribution and is confined to areas around Minne-

     ə *Spiralothelphusa hydrodroma*
     ☰ *Oziothelphusa senex*
   |H| *Ceylonthelphusa rugosa*
    × *Ceylonthelphusa inflatissima*
    ◼ *Ceylonthelphusa soror*
    + *Perbrinckia enodis* ,

Fig. 4. The distribution of Potomonidae (after Bott 1969).

riya, Polonnaruwa, Habarana and Ritigala. *C. inflatissima* is restricted to the Sabaragamuwa Province.

Three other species: *C. rugosa, C. soror* and *Perbrinckia enodis* are found in the hilly areas of the island. Although *C. rugosa* also has been recorded from the lowland areas such as Trincomalee and Anuradhapura, its main distribution pattern begins with the Singharaja Range in the south. It is found throughout the Deniyaya Region to Horton Plains, to the north towards Knuckles group and to the east towards Badulla. *C. soror* is found in streams above 300 m in the hilly areas of Southern Province, Sabaragamuwa Province, Uva Province and the Central Province.

Parasites and associated organisms

There are still no records of parasites or associated organisms on Branchiopoda from Sri Lanka.

An interesting group of commensals which are small (<3 mm in length), are the Temnocephalida (Rhabdocoela, Platyhelminthes), which live in gill chambers of some atyids. These commensals have been observed only on *C. simnoi* and *C. fernandoi* (unpublished data). Three species from two genera have been recorded; *Caridinicola platei, Monodiscus parvus* and *M. macbridei. C. platei*, which has a wide distribution, is found in Kandy Lake and in water bodies in the Central Province, where it is found living with *M. parvus*. In the low country and in the dry zone, *M. platei* and *M. macbridei* are found living together (Fernando 1952).

A macrocrustacean commensal belonging to the genus *Probopyrus* (Family Bopyridae) lives in the gill chamber of *M. idella*, a palaemonid prawn, and causes a bump on the side of the carapace.

An ectoparasite and several endoparasites have been described from freshwater crabs. The rhyncobdellid leech, *Paraclepsis vulnifera* is a common parasite in the branchial chambers of *O. senex* and *C. rugosa* (Fernando 1960, Costa 1978). This parasite is sometimes also found living on the limbs and undersides of these crabs.

The freshwater crabs also are the known intermediate hosts for the infective larval stages of the trematodes *Paragonimus westermani* and *Pleurogenoides sitapuri*. The metacercariae of *P. sitapuri*, which is progenetic and protandrous, occur in cysts, which appear as whitish or brownish spherical bodies in the muscles of the cephalothorax and limbs of *O. senex* (Dissanaike & Fernando 1960). It appears that this infection is very common in wet zone crabs. The adults live in *Rana cyanophlyctis*.

The human lung fluke *P. westermani* is present in wild carnivores; the civet cat, *Vivericula indica* and the fishing cat, *Felis viverrina*, in the Eastern Province. The cercariae encyst in freshwater crabs.

Several parasitic nematodes have been reported in freshwater crabs. Adult *Rhabdochona* and juvenile *Proleptus* and *Rhabdochona* have been recovered from the hepatopancreas of *O. senex, C. soror* and *C. rugosa*. The juveniles mature in the crab's hepatopancreas. It is likely that their entire life cycle is completed within the host, a specialised type of behaviour for a nematode belonging to the Spiruridae (Poinar & Kannangara 1962). The juveniles of *Proleptus* have been recorded in high numbers in *C. rugosa* and *O. senex* found in Sabaragamuwa, Northcentral, Northwestern and Northern Provinces.

Economics of Crustacea in Sri Lanka

Although few of the microcrustaceans are of direct economic importance, several meso and macrocrustaceans in inland waters are. They form important food items for man and economically important animals.

The atyid shrimps and palaemonid prawns form important items in the diet of several carnivorous and omnivorous fish. They have been found in the stomachs of several species of reservoir, riverine, estuarine and lagoon fish. Many amateur and commercial fisherman use these shrimps as bait.

None of the atyids have yet been commercially exploited. *C. simoni* is worthy of investigation for this purpose because of its occurrence in large numbers in both fresh and slightly brackish waters. Very recently, Weerakkody & Costa (in press) have studied the development of this shrimp under laboratory conditions with a view to culturing it. In some localities in the southwest coastal area, *C. simoni* is collected and dried. This dried shrimp is considered a delicacy.

For many reasons, marine crustaceans are still preferred to those of rivers. Of the palaemonid prawns, only *M. rosenbergii* and *M. idae* are commercially important. Presently, there is a flourishing fishery (10,000 kg yr^{-1}) at Madampe, Katunayake, Colombo, Moratuwa and Batticaloa for *M. rosenbergii* (Wanninayake & Costa in press, unpublished data). *M. latismanus* occurs in exploitable numbers in the hill streams of the Deniyaya region (Costa 1974), but has not been commercially exploited. However, in some localities, the villagers catch these prawns for domestic consumption. This species also shows promise for further investigation and exploitation.

Several projects for the mass culture of the giant river prawn, *M. rosenbergii* have been initiated. These are not yet as commercial success. Prawn larvae have been successfully grown to juvenile stage and experiments have been conducted to culture the prawn using locally available artificial diets (Costa 1980b). Extensive natural habitats exist for stocking, especially those from which clay has been extracted for brick, tile and pottery industries.

Some potamonid crabs, especially *C. rugosa, C. soror* and *P.enodis* are usedas food by the poor. They also are eaten because of their supposed medicinal properties (Fernando 1960). *O. senex*, found in large numbers in rice fields, damage the bunds by burrowing and causing considerable water loss. They may also cause damage to rice plants (Fernando 1960). Crabs also burrow into earth

208

dams of large and small man-made reservoirs and bunds of irrigation channels. These burrows cause seepage and result in considerable water loss in the dry zone.

Freshwater crabs also harbour larval stages of parasites which, as adults, could cause diseases in higher vertebrates and man.

Juvenile or small crabs are important items in the diets of fish and other aquatic vertebrates. *C. rugosa* and *P. ènodis* form the most important components of the diet of the introduced rainbow trout stock in the Horton Plains streams (Costa 1974) (Fig. 5).

Of the penaeid prawns in the inland waters, only a few species, belonging to the genera *Penaeus* and *Metapenaeus* are economically important. Although catch statistics are not available, important fisheries exist for *P. indicus, P. monodon, P. semisulcatus, M. dobsoni* and *M. elegans* in estuaries and lagoons (primarily Negombo, Batticaloa and Mullaitivu) in both western and eastern coastal areas. Most of this catch is sold to tourist hotels. A subsistence fishery exists for the crab *Scylla serrata* in Negombo lagoon, where the catch is marketed locally, as well as to Colombo.

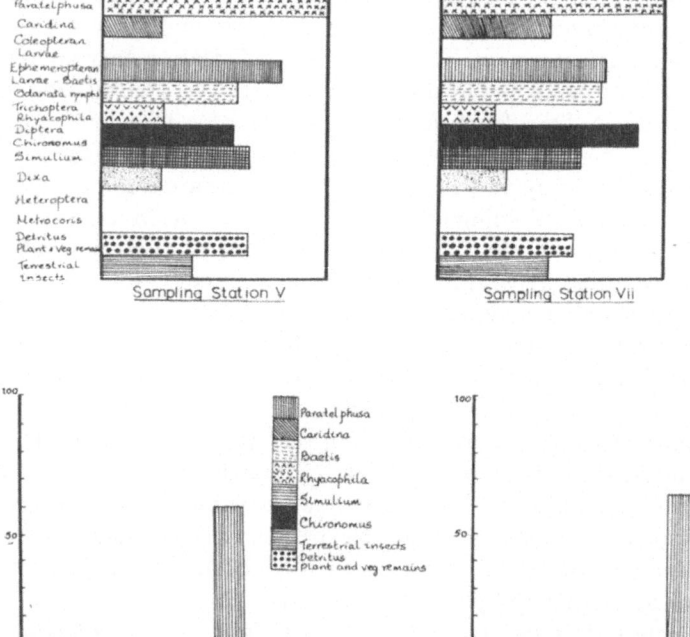

Fig. 5. The food of the introduced fish *Salmo gairdneri* in Horton Plains. Crabs and shrimp form a major component in the diet.

Zoogeography

Classical zoogeography is based principally on the distribution of vertebrates, particularly mammals. The zoogeographical regions have been characterized by faunal assemblages which have become distinctive because of the past geographical history of the region.

Because of the interest shown by zoogeographers in the higher vertebrate groups, freshwater fish and crustaceans have received scant attention in regional faunistic studies. Zoogeographers contend that these organisms disperse too readily and that many of these groups still are too poorly known systematically to be of use in zoogeographical studies.

Johnson (1960) contended that the invertebrate groups, more specifically the freshwater groups, were as important as objects of zoogeographical study as the higher terrestrial vertebrates. The generally held view that freshwater animals, being members of ancient groups, could only provide a confused picture of former land connections, and thus would probably not be useful in zoogeographical studies, was refuted by Johnson. He argued that the present distribution patterns of freshwater fish and prawns, because of their aquatic lifestyle and limited powers of dispersion, could be considered to be determined at least partly by historical factors.

The study of the distribution of freshwater shrimps, prawns and crabs in the Indo-Pacific region reveals distribution patterns which differ from those generally accepted as characterizing the distribution of higher vertebrates, and which sometimes conflict with the classical concept of an Oriental Region. *Macrobrachium* and *Caridina* extend from Africa through Asia to Amurland. *Atya*, on the other hand, is a pantropical genus which occurs at the southern fringe of eastern Asia (Table 3).

The degree of tolerance of salt water is important in determining the distributional patterns of shrimps and prawns. Those that are not tolerant of salt water show a restricted range, while those that show a slight tolerance could withstand short sea passages. An example of the former is *M. malcolmsoni*, which is restricted to the Indian region, while examples of the latter include *M. rosenbergii*, extending from Pakistan through India, Sri Lanka to Papua-New Guinea and *Atya spinipes* extending from Travancore through Sri Lanka to New Caledonia. *Caridina typus*, which is present in Sri Lanka, is absent from continental Asia but shows a peripheral distribution from Africa to Asia. On the other hand, the distribution of marine animals is not primarily determined by the past distribution of land masses, as the ocean is not a barrier to migration of these animals. Of the 31 species of Penaeidae described from Sri Lankan waters, 22 have been found east of the Malacca straits. Regarding the overall distribution of marine species of Penaeidae, identical species have been found in India, Sri Lanka, Malaya, Philippines, Indonesia, Formosa, Japan and Australia (Bruin 1965).

Because of these inconsistancies, Johnson (1960) advocated a dynamic concept

Table 6. Distribution of atyid shrimps and palaemonid prawns from Madagascar to Malaya through Indian Ocean Islands (from Costa 1980).

	Madagascar	Seychelles	Mauritius	Comores	Reunion	Sri Lanka	Andaman & Nicobar Is.	Malayan Archipelago
Caridina micropthalma	×							
C. bouvieri	×							
C. petiti	×							
C. gladiifera	×							
C. angulata	×							
C. nilotica	×							
C. hova	×							
C. longirostris	×							
C. xiphias	×							
C. lamiana	×							
C. edulis	×							
C. norvestica	×							
C. troglopila	×							
C. japonica	×							
C. spathulirostris	×		×					
C. isaloensis	×							
C. calmani	×							
C. weberi				×				×
C. typus	×	×	×	×	×	×	×	×
C. mauritii			×					
C. richtersi			×					
C. serratirostris	×	×	×					
C. brevirostris	×	×						
C. gracilirostris	×					×	×	×
C. singhalensis						×		
C. pristis						×		
C. fernandoi						×		
C. zeylanica						×		
C. costai						×		
C. simoni						×	×	×
C. brachydactyla		×	×	×	×		×	×
C. prasadi							×	
C. excaratoides								×
C. propinqua						×		×
C. tonkinensis								×
C. babulti								×
C. thambapillai								×
C. wycki								×
Atya spinipes						×	×	×

(Table 6 continued)

	Madagascar	Seychelles	Mauritius	Comores	Reunion	Sri Lanka	Andaman & Nicobar Is.	Malayan Archipelago
A. pilipes	X	X	X	X	X		X	
Macrobrachium lar	X	X	X	X	X		X	X
M. australe	X		X		X	X		
M. idae	X	X	X		X			X
M. hirtimanus			X		X			X
M. rosenbergii						X		X
M. malcolmsi						X		
M. idella	X					X		
M. equidens	X					X		X
M. rude	X					X		
M. latimanus						X		
M. scrabriculum	X					X		
M. latidactylus						X	X	X
M. kistnensis						X		
M. srilankense						X		

of zoogeographical regions, instead of a static one, especially when describing the distribution of freshwater fish and crustaceans.

Considering Sri Lanka, the crustacean fauna is a complex one in which one could recognise cosmopolitan elements, Indo-Pacific elements, Indian Ocean elements (Table 6) and Indian elements. Accordingly, the shrimps, prawns and crabs from freshwaters can be considered under several headings from a distributional viewpoint. These are listed as follows:

A. Species which are endemic to Sri Lanka

Genus *Caridina-C. singhalensis*
C. pristis
C. fernandoi
C. zeylanica
C. costai
Genus *Macrobrachium - M. srilankanse*
Genus *Ceylonthelphusa-C. rugosa*
C. soror
C. inflatissima
Genus Oziothelphusa-O. minneriyensis

B. Species which are common to the Indian subcontinent and Sri Lanka

Genus *Macrobrachium-M. kistnensis*
M. malcolmsi
Genus *Oziothelphusa*
O. senex
Genus *Spiralothelphusa-S. hydrodroma*

C. Species which have and Indo-Pacific distribution

Genus *Caridina-C. typus*
C. simoni
C. gracilirostris
Genus *Atya-A. spinipes*
Genus *Macrobrachium-M. rosenbergii*
M. latimanus
M. idae
M. idella
M. equidens
M. latidactylus
M. australe
M. scabriculum
M. rude
Genus *Perbrinckia-P. enodis*

References

Arudpragasam, K.D. & H.H. Costa. 1962. Atyidae of Ceylon 1. Crustaceana 4:7-24.

Bott, R. 1970. Die susswasserkraben von Ceylon (Crustacea, Decapoda). Ark. Zool. Stockholm 22:627-640.

Bruin, G.H.P. de. 1965. Penaeid prawns of Ceylon (Crustacea, Decapoda, Penaeidae). Zool. Meded. 41:74-104.

Bruin, G.H.P. de. 1971. Fluctuations in species composition of penaeid prawns in estuaries. Bull. Fish. Res. Stn Ceylon 22:47-51.

Costa, H.H. 1974. Limnology and fishery biology of the streams at Horton Plains, Sri Lanka (Ceylon). Bull. Fish. Res. Stn Sri Lanka (Ceylon) 25:15-26.

Costa, H.H. 1978. The freshwater Hirudinea of Sri Lanka. Spol. Zeylan. 32:7-17.

Costa, H.H. 1979. The Palaemonidae of the inland waters of Sri Lanka. Ceylon J. Sci. (Bio. Sci.) 13:39-64.

Costa, H.H. 1980a. Results of the Austrian Hydrobiological Mission 1974 Seychelles, Comores, and Mascarene Archipelagos Part III. The ecology and distribution of Decapoda Caridea in the Indian Ocean islands of Seychelles, Mauritius, Comores and Reunion. Ann. Naturhist. Mus. Wien 83:673-700.

Costa, H.H. 1980b. Preliminary studies on the breeding of the giant prawn *Macrobrachium rosenbergii* using locally available diets. Proc. Giant Prawn 1980 Symposium, Bangkok, Thailand.

Costa, H.H. & S.S. de Silva. 1978. Hydrobiology of Colombo (Beira) Lake IV. Seasonal fluctuations in aquatic fauna living on water plants. Spol. Zeylan. 32:55-70.

Dissanayake, A.S. & C.H. Fernando. 1960. *Parathelphusa ceylonensis* C.H. Fern., second intermediate host of *Pleurogenoides sitapuri* (Srivastava). J. Parasit. 46:889-890.

Fernando, C.H. 1960. The Ceylonese freshwater crabs (Potamonidae). Ceylon J. Sci. (Bio. Sci.) 5:191-22.

Fernando, W. 1952. Studies on the Temnocephalidae of Ceylon I. *Caridinicola platei* sp. nov. II *Monodiscus macbridei* sp. nov. Ceylon J. Sci. (B). 15:19-27.

Johnson, D.S. 1960. Some aspects of the distribution of fresh water organisms in the Indo Pacific area and their relevance to the validity of the concept of an Oriental region in zoogeography. pp. 258-267 In: Proc. Cent. Bicent. Cong. R.D. Purchon (Ed.). Singapore.

Junk, W.J. 1977. The invertebrate fauna of the floating vegetation of Bung Borapet, a reservoir in Central Thailand. Hydrobiologia 53:229-238.

Pinto, L. & S. Wignarajah. 1980. Some ecological aspects of edible oyster *Crassotrea cucullata* Born occurring in association with mangroves in Negombo lagoon, Sri Lanka. Hydrobiologia 69:11-19.

Poinar, G.O. & D.W.W. Kannamgara. 1972. *Rhabdochona praecox* sp. nov. and *Proleptus* sp. (Spiruroidea, Nematoda). Ann. Parasit. 47:121-129.

Pretzmann, G. 1973. The Austrian Ceylonese hydrobiological Mission 1970 XIV. The freshwater crabs of the Ceylonese mountain rivers. Bull. Fish Res. Stn Sri Lanka (Ceylon) 24:129-133.

Selvarajah, N. & H.H. Costa. 1979. The distribution of Anostraca and Conchostraca (Crustacea) in the Jaffna Peninsula. Bull. Fish. Res. Stn Sri Lanka 29:79-88.

Weerakkody, J.S. & H.H. Costa. in press. Post embryonic growth and development of *Caridina simoni Bouvier (Decapoda, Atyidae)* reared in the laboratory. Bull. Fish. Res. Stn Sri Lanka.

11. Mountain stream fauna, with special reference to Mollusca

F. Starmühlner

Introduction

Investigations of the aquatic fauna of the running waters in the central mountains of Sri Lanka initially were conducted by the Fisheries Research Station of the Department of Fisheries in Colombo, with the most intensive effort on the freshwater fishes. Mendis & Fernando (1962) have published a guide to the freshwater fauna, with supplements (Fernando, 1963, 1964, 1969a, 1974). A survey of the freshwater invertebrate fauna was given by Fernando (1980). Numerous references are given in these monographs, including some to stream fauna in the mountains. Silas (1952) published a study on the speciation of freshwater fishes. Fernando & Indrasena (1969) surveyed the freshwater fisheries and mentioned the trout fishery in the uplands. Fernando (1969b) listed some freshwater molluscs and gave notes on their distribution and biology and Costa (1974) gave an account on the limnology and fisheries biology of the Horton Plains streams. Costa & Fernando (1967) dealt with the fauna of small mountain streams and Geisler (1967) made some limnological and ichthyological studies in the streams of the Southwest uplands. Descriptions of the mountain stream stations and their fauna are found in the papers by Brinck et al. (1971), Bott (1970), DeLeve (1973a), Enckell (1970) and Kuiper (1979). In 1970, a hydrobiological mission carried out by the Institute of Zoology, University of Vienna, in cooperation with the Department of Zoology, University of Sri Lanka, Vidyalankara Campus, investigated the fauna of the mountain streams of the Southwest region. The results of this mission were published mainly in the Bulletin of the Fisheries Research Station (Bertrand 1973, Costa 1972, Costa & Starmühlner 1972, DeBeauchamp 1973, Deleve, 1973b, Elzen 1972, Hadl 1974, Kaltenbach 1973, Kuiper 1979, Malicky 1973, Polhemus 1979, Pretzmann 1973, Radda 1972, 1973, Reichholf 1973, Starmühlne 1974, St Quentin 1972, 1973, Viets 1972, Weninger 1972, Wewalka 1973, Zwick 1980, Müller Liebenau 1982a, b, c, Jaech 1982a, b). Some aquatic groups are still in need of study, such as some genera of Ephemeroptera, Hydrophilidae (partly studied by Jaech 1982a, b), Blepharoceridae, Chironomidae and the algae.

The interior of Sri Lanka consists of the Central Highlands, rising from elevations of 1,400-1,800 m in the middle and 2,524 m at the highest point at Pidurutala-gala near Nuwara Eliya. The highland, divided into three portions (Knuckles Group in the North, Central Massif and Sabaragamuwa Ridges in the Southwest) is surrounded by four peneplains (Sawicky 1925). Geologically, the Island is a detabled part of the continental Deccan Plateau of ancient crystalline rocks (Cooray 1967, Sievers 1964, see also Chapter 1 this volume). The coast and the outflow of the large streams and rivers cover alluvial soils, containing mainly clay, sand and silt. In the Dry zone of the Northwest, the dominating soils consist of reddish-brown earths, derived from basic or neutral rock material. In the South-west area non-calcic soils developed a deficiency of ferromagnesium minerals in the acid gneisses (Brinck *et al.* 1971). Moorman & Panabokke (1961) stated that in the Wet zone podzolic soils predominate. In the high altitude grasslands and fernlands dark humic horizons are found in the soil. Laterization occurs only in the Western part of the Wet zone.

Climate, Vegetation and Chemical and Physical Conditions of Streams.

Sri Lanka is a tropical island in the Indian Ocean situated within the equatorial belt of calms. The intensity and narrow amplitude of insolation are important factors controlling the climate. In the Wet zone, only slight seasonal variation in tempera-ture, humidity and day length occur. The Southwest and Northeast monsoon rains on the Central massifs of the mountain area result in differential rainfall patterns. In the Wet zone, rainfall is high, ranging from 2-7 m (7 m yearly average). The annual mean air temperature and the mean relative humidity vary only slightly (Brinck *et al.* 1971). Mean annual climatic measurements are given in Table 1.

The Wet zone of the south and southwest lowlands (500-800 m elevation) is covered with evergreen tropical rainforest. About 250 species of trees are found in this region, the most important are *Dipterocarpus* sp., *Doona* sp., *Antocarpus* sp., *Zinnamonnum* sp. and *Ficus* spp. In the valleys of the larger streams, the forests have been replaced by rice, rubber, pepper and fruit plantations. In the uplands (500/800-1,500 m elevation) the wet, warm mountain area is covered by tropical mountain rainforest. Here there is a decrease in the luxuriance of plant growth;

Table 1. Annual mean air temperature, amplitude and humidity.

Area	Air temperature	Amplitude	Humidity	
			Day	Night
Lowlands (Colombo)	27.2 °C	1.7-1.8 °C	68-79%	87-93%
Uplands (Kandy)	24.2 °C	2.7 °C	60-70%	87-94%
Highlands (Nuwara Eliya)	15.4 °C	2.4 °C	60-84%	88-93%

Kina sp. is the dominant plant. With the exception of the Sinharaja Forest, this type of rainforest has been almost totally replaced by rubber plantations and tea estates. The wet, cool highlands (1,500-2,500 m elevation) have high rainfall and humidity. Here there is an even further decrease in the luxuriance of plant growth and at altitudes over 2,000 m, many trees occur in cripple-growth. The *Kina* tree is typical of this region, along with *Cypressus* sp., *Pinus* sp., *Rhododendron* spp. and *Azalea* spp.

Stream gradients, currents and bottom formation

Most running waters in the south and southwest mountain areas drain regions of crystalline rocks (Fig. 1). In the uplands of the southwest, some rivers traverse crystalline chalk layers. Because of their short courses compared with streams on the continents, most running waters in Sri Lanka have steep gradients (Figs. 2,3). The average gradient is 15-20% over a 2,000 m drop in elevation over a distance of 100-130 km from headwater to mouth. In the upper courses, the gradient rises to 50-100% and in the slopes of the highland peneplains, cascades often occur. The most famous waterfall is the Diyaluma Falls (Fig. 3) in the southeast. In the

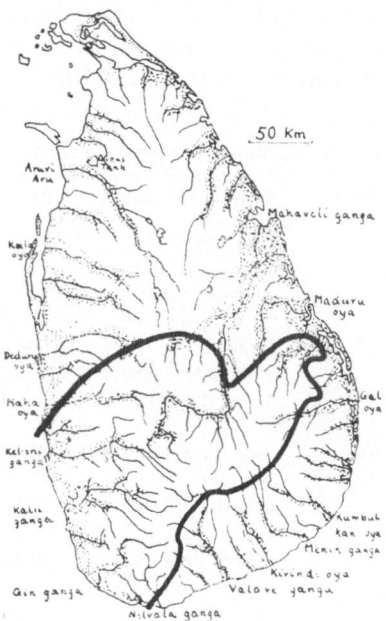

Fig. 1. Map of Sri Lanka showing the main river systems. The dark line indicates the Wet zone of the highlands, uplands and lowlands in the southwest. The dotted areas are flooded frequently during the monsoon periods.

uplands and lowlands the gradient decreases to 5-10 %. The falls cause high current velocities in the headwaters and upper courses, and to some extent in the middle courses. The average flow velocity on the surface reaches 75-100 cm sec^{-1} (Starmühlner 1972). In the cascades and waterfalls, the current becomes turbulent and soars to more than 2 m sec^{-1}. In the pools between the cascades, the surface velocity ranges from 0-20 or 30 cm sec^{-1}. The pools often are deep (50-100 cm). These 'lentic' regions are flooded after heavy showers and they are thus included in the regions of strong current from time to time.

The bottom formation is a result of the flow velocity and the structure of the bottom is important to its inhabitants (Hora 1928, 1930, 1936). The types of bottom formation are given in Table 2. The depth of the bottom of most headwaters with low water levels and widths of 1-10 m ranges from 10-30 cm. In the pools, the bottom depth extends to 50 cm or more. In the middle courses, with widths of 15-40 m, the bottom may reach a depth of one or more metres. Drainage is a reflection of the monsoon rainfalls. Downstream transport of boulders, pebbles, gravel and sand may occur during the Southwest monsoon. An appearance of alluvial sand in the coastal waters is noted when this occurs. During the monsoon, the streams are reddish-brown floods of floating laterite, consisting of clay and iron oxide. The pebbles carried downstream may consist of splinters of precious and semiprecious stones (Fig. 4).

From the headwaters to the mouths of rivers (100-150 km distance, 2,000 m elevation drop) water temperature increases on average, from 15-28 °C. In the headwaters, the average temperature is 16.8 °C, rising to 21.7, 25.4 and 26.6 °C

Table 2. Type of current and bottom formations of mountain streams.

Current speed	Running water type	Bottom formation	Bottom depth	Figure
1.5-3.0 m sec^{-1}	waterfall	standing, smooth granite pyroxen rocks	0.5-5.0 cm	2, 3
1.0-1.5 m sec^{-1}	cascades	large, granite boulders, 1-2 m diameter	5-10 cm	
75-100 cm sec^{-1}	small cascades interrupted by pools	medium granite boulders, gravel	10-30 cm	4, 5, 6
50-75 cm sec^{-1}	lotic area	gravel, pebbles	10-50 cm	
0-30 cm sec^{-1}	lotic areas	sand, stones	20-50 cm	
0-20 cm sec^{-1}	lentic areas, pools, bays	fine sand, mud, silt, debris	50 cm	7

through the upper, middle and lower courses respectively, and reaching temperatures of 28 °C at the mouth. The average increase in water temperature per 100 m drop in elevation is 0.75-1.5 °C. The difference between day and night water temperatures becomes less pronounced from the mountain regions towards the hilly and coastal regions. At an altitude of 1,300-1,500 m, the difference may be 4-5 °C, while at an altitude of 60 m, a difference of only 2-3 °C is noted.

The physiography, geology, geochemistry and climate have great influence on the hydrochemistry of Sri Lanka (Geisler 1967, Weninger 1972). Geologically, Precambrian crystalline series rocks are found in the Wet zone. In the North, some Jurassic and Miocene limestones are found. In this chapter the running waters of the mountains in the Wet zone in the Southwest and Central Highland are considered.

Nearly all of the running waters of the Wet zone in the crystalline series exhibit low concentrations of dissolved minerals and are slightly acidic. Sodium (Na) predominates over calcium (Ca), while Ca predominates over magnesium (Mg). With increasing altitude, the concentration of dissolved minerals decreases. The ranges and average values of several important chemical parameters are given in Table 3. The increase in conductivity, hardness, Ca, Mg and alkalinity is almost linear with decreasing altitude.

Table 3. Chemical parameters of mountain streams (after Weninger 1972).

Parameter	Headwater 1500-2000 m		Upper course 500-1500 m		Middle course 50-500 m		Lower course 5-50 m		Mouth 0-5 m
	Extr.	Av.	Extr.	Av.	Extr.	Av.	Extr.	Av.	
EC_{20}	8.85-26.3	18.27	11-45.6	26.65	18.2-89.2	41.56	27-46.2	44.4	288
pH field	5.4-6.1	5.8	5.5-6.7	5.9	5.9-7.0	6.3	6.0-	6.0	—
electr.	5.6-6.9	6.1	5.5-7.7	6.5	6.6-7.2	6.8	6.4-6.5	6.47	6.4
Hardness (°dH7)	0.08-0.65	0.322	0.5-4.44	1.81	0.21-2.35	0.98	0.36-1.40	0.765	1.38
Alk. (HCO^3) (mval)	0.05-0.38	0.08	0.08-0.53	0.30	0.08-0.97	0.40	0.08-0.36	0.22	0.44
Free CO_2	44.0-55.0	52.25	11.4-44.0	30.1	66-	50.6	—	—	—

Remarks on Fauna

The mean densities of the mesofauna in various reaches of mountain streams and at different current speeds are given in Table 4. The fauna, grouped according to current speed, for the headwater (1,500-2,000 m elevation), upper (500-1,500 m), middle (50-500 m) and lower (0-20 m) courses of the mountain streams are summarized in Tables 5-8, respectively. These animals occupy a wide range of stream microhabitats, as illustrated in Fig. 8, and some species represent only single records in Sri Lanka. Some of these groups are discussed in greater detail in other chapters of this volume (Crustacea, Chapter 10; Ephemeroptera, Chapter 12; Simuliidae, Chapter 13).

Molluscs

Some remarks on biology and distribution

The exploration of the non-marine molluscs of Sri Lanka commenced long ago. Until the 20th century in most cases only their shells were described by conchologists. However, the shells of freshwater molluscs are very variable, depending on the ecological features of the local aquatic environment. The consequence of this was the description of numerous species inhabiting Sri Lanka based only on the shape of the shells and sometimes only based on the examination of a single shell. With the study of the shells of large populations and the anatomy of the soft body parts, including the radula, it was necessary to synonomize many 'species' (Starmühlner, 1974). Descriptions and figures of shells from Sri Lanka have been given by Reeve (1843-1878), Deshayes (1854), Dohrn (1857, 1858), Hanley & Theobald (1876), Godwin-Austin (1882-1920) and in several volumes of the 'Conchylien-Cabinet' edited by Martini & Chemnitz (Kuster 1852, 1862, Brot 1874, 1880, Clessin 1886, Kobelt 1915). Other contributions have been made by Colett (1897, 1898), Satyamurthi (1960), Sivalingam (1949) and Fernando (1969 b). Hubendick (1951) gave a revision of the species of *Lymnaea* of Sri Lanka in his monograph on the

Table 4. Mean densities of the mesofauna in various reaches of mountain streams (Individuals × $10^2/m^2$).

Current speed	Altitude		
	1500-2000 m	500-1500 m	30-500 m
0- 30 cm/sec	3.20- 4.80	4.80	1.60
30- 76 cm/sec	8.0 - 32.0	8.0- 32.0	3.2- 11.2
75-300 cm/sec	320.0 -480.0	320.0-480.0	80.0-120.0

Fig. 2. Headwaters and upper courses of the Gartmore Estate, Dola near Maskeliya. The headwaters arise in primary forest, the upper reach flows through tea plantations.

Lymnaeidae. Some anatomical comments on *Indoplanorbis exustus* have been given by Hubendick (1955) in his monograph on Planorbidae. Starmühlner (1974) published a monograph on the freshwater gastropods on the basis of the material collected by the Austrian-Sri Lankan Hydrobiological Mission (1970), the Lund University-Sri Lankan Expedition (1962) and collections made by the Department of Zoology of the University of Sri Lanka, Vidyalankara Campus. Starmühlner (1974) lists the literature concerning the freshwater gastropods found in Sri Lanka and gives the systematics of the different species. The material on freshwater mussels from the missions listed above has been described by Hadl (1974), and Kuiper (1975) has given a short account on the Sphaeridae.

Starmühlner (1974, 1977, 1978) describes 31 species of fresh (and brackish) water gastropods from Sri Lanka. Twenty-eight species belong to the Prosobranchia (Streptoneura) and 3 species belong to the Pulmonata-Basommatophora (Euthyneura). Their biology and distributions are as follows.

Archaegastropoda-Neritidae

Neritina (Neripteron) auriculata This species settles in the lower reaches of streams (Fig. 9) below stones near the borders in current between 20-50 cm sec^{-1}. The snail tolerates slightly brackish water.

Table 5. Members of the Headwater community (1,500-2,000 m elevation).

Group / Order	0-30 cm/sec	30-75 cm/sec	75-300 cm/sec
Annelida Oligochaeta	*Tubifex tubifex* *Limnodrilus hoffmeisteri* *Lumbricus variegatus*		
Platyhelminthes Turbellaria	*Dugesia nannophallus*	*Dugesia nannophallus*	
Crustacea	*Paratelphusa (Ceylontelphusa) rugosa*		
Insecta Odonata	*Lestes* sp. *Anisoptera* gen. sp.	*Euphaea splendens*	
Coleoptera	*Lacconectes simoni* *Hydaticus luzonicus* *Pelthydrus* sp. *Aulonogyrus obliquus* *Orectochilus* sp.	*Ilamelis foveicollis* *Podelmis quadriplagiata* *Podelmis ovalis* *Podelmis humeralis* *Podelmis cruzei* *Podelmis ater* *Ohyia carinata* *Zailzeviaria zeylanica* *Eubrianax ceylonicus* *Lampyridae* cf. *luciolala*	
Heteroptera	*Sigara nilgirica* *Enithares simplex* *Metrocoris stali* *Strongylovelia formosa* *Pseudovelia gnoma*		
Ephemeroptera	*Indocloeon primum*	*Indocloeon primum* *Indobaetis costai* *Leptophlebiidae* gen. spp.	*Indobaetis costai* *Pseudocloeon orientale* *Pseudocloeon klapaleki*
Diptera	*Chironomus* sp. (cf. *Plumosus*)		*Simulium* sp. **B.** *S.* sp. g. *Gomphostilba (S.)* sp. I

Table 5 (continued)

Group / Order	0-30 cm/sec	30-75 cm/sec	75-300 cm/sec
Plecoptera		*Neoperla angulata* *Phanoperla sp.*	Blephariceridae gen. spp.
Trichoptera		*Apsilochroma diffinis* *Synagapetus sp.* Hydroptilidae gen. sp. Polycentropidae gen. sp. *Economus sp.* Psychomyidae gen. sp. *Marilia* cf. *ceylanica* *Molanna taprobane* *Oecetis sumansara* *Goerodes sp.* *Helicopsyche amarawathi* *H. ruparawathi* *H. sp. E* *H. sp. F* Helicopsychidae	*Chimarra sp.* *Pseudoleptonema sp.* Hydropsychidae gen. spp. *Hydropsyche katugahakanda* *H. annulata* *H. sp. A* *Diplectona* sp. **B** *Diplectonella taprobanes* *Ceylanopsyche* sp. **A** *C.* sp. **B**
Mollusca Gastropoda	*Paludomus (Philopotamis) nigrescens*	*P. (P.) nigrescens*	
Vertebrata Amphibia	*Rhacophorus crucigereques* *Rana temporalis*		

223

Table 6. Members of the Upper Course communities (500-1,5000 m elevation).

Group / Order	0-30 cm/sec	30-75 cm/sec	75-300 cm/sec
Turbellaria	*Dugesia nannophallus*	*Dugesia nannophallus*	
Crustacea	*Atya spinipes* *Caridina fernandoi* *C. nilotica simoni* *C. pristis* *Paratelphusa (C.) rugosa*	*Macrobrachium australe* *M. latimanus* *M. scabriculum*	
Acari		*Arrenurus (Micruacarus) maderaszi* *Torrenticola (Monatractides) pusta* *T. (M.) oxyostoma hamata* *T. (M.) ceylonensis* *Atractides schwoerbeli*	
Insecta Ephemeroptera	*Leptophlebiidae* gen. spp. *Ephemera* sp. *Indocloeon primum* *Pseudocloeon difficilum*	*Leptophlebiidae* gen. spp. *Ecdyonuridae* gen. spp. *Indocloeon primum* *Pseudocloeon difficilum* *Indobaetis costai* *Prosopistoma* sp.	*Pseudocloeon orientale* *Pseudocloeon difficilum* *Indobaetis costai*
Plecoptera		*Neoperla angulata* *N. triangulata* *Phanoperla* sp.	
Odonata	*Libellago greeni* *L. finalis* *Vestalis apicalis nigrescens* *Neurobasis chinensis chinensis* *Lestes elata* *Drepanosticta* sp. *Elattoneuris tenax* *Prodasineura sita*	*Euphaerea splendens* *Zygonyx iris ceylanica*	

225

Table 6 (continued)

Group / Order	0-30 cm/sec	30-75 cm/sec	75-300 cm/sec
	Heliogomphus sp.	*Apsilochroma affinis*	*Chimarra* sp.
	Megalogomphus ceylonicaus	*Synagapetus hanumata*	*Gunungiella madakumbura*
	Paragomphus henryi	*S.* cf. *rudis*	*Oestropsyche vitrina*
	Anax immaculifrons	*Hydroptila kurukepitiya*	*Macronema* sp. A
	Macromia zeylanica	*H.* sp.	*M.* sp. B
	Trihemis festiva	*Oxyethira* sp.	*Pseudolepionema* sp. B
		Hydroptilidae gen. sp.	*Hydropsyche katugahakanda*
Othoptera	*Rhabdoblatta* sp.	*Pseudoneuroclepsis starmuehlneri*	*H. flinti*
	Paranemobius pictus	*Polycentropidae* gen. sp.	*H.* sp. A
	Euscelimena gavialis	*Paduniella mahanawana*	*H.* sp. B
		P. subhakara	*H.* sp. D
Trichoptera		*Psychomyidae* gen. sp.	*Synaptopsyche nikalandugola*
		Marilia ceylanica	*Diplectrona* sp. A
		Trichosetodes meghawanabaya	*Diplectronella taprobanes*
		Adicella sp.	*Hydropsychidae* gen. sp.
		Oecetis belihuloya	*Ceylanopsyche asaka*
		O. hamata	
		O. malighawa	
		O. sumanasara	
		O. sp.	

Table 6 (continued)

Group / Order	0-30 cm/sec	30-75 cm/sec	75-300 cm/sec
		Setodellina punctatissima	
		Leptoceridae gen. sp.	
		Goera katugalkanda	
		G. paragoda	
		Goerodes fuscata	
		G. puncta	
		G. sp.	
		Helicopsyche amarawathi	
		H. rupawathi	
Coleoptera	*Lacconectes simoni*	*Helichus naviculus*	
	Hydaticus luczonicus	*Ceradryops punctatus*	
	Neptosternus sp.	*Potamophilinus impressicollis*	
	Pelthydrus sp.	*Ordobrevia fletcheri*	
	Aulonogyrus obliquus	*Ilamelmis brunnescens*	
	Canthydrus luctuosus	*I. fovecollis*	
	Laccophilus chinensis innefficiens	*I. crassa*	
	Berosus indicus	*Caphalolimnuis ater*	
	Hydrophilidae gen. sp.	*Podelmis quadriplagiata*	
		P. aenea	
		Aruelmis starmuehlneri	
		Zaitzeviaia bicolor	
		Hydrocyphon atratus	
		H. striatus	
		Eubrianax ceylonicus	
		Lampyridae (cf. *Pyrophanes*)	
		Aulonogyrus obliquus	
		Orectochilus sp.	

Table 6 (continued)

Group / Order	0-30 cm/sec	30-75 cm/sec	75-300 cm/sec
Heteroptera	*Microneta memonidesi*		
	M. wroblewskii		
	M. quadristrigata		
	Aphelocheirus clivicolus		
	Heleocoris breviceps		
	Cercometus strangulatus		
	Enithares simplex		
	Ochterus marginatus		
	Limnogonus nitidus		
	Limnometra anadyomene		
	Metrocoris stali		
	Ptilomera cingalensis		
	Rheumatogonus custodiensis		
	Ventidius henryi		
	Timasius splendens		
	Hydrometra greeni		
	Microvelia (Kirkaldya) longicornis		
	M. pererai		
	Rhagovelia ceylanica		
	R. karunaratnei		
	Tetraripis asymmetricus		
Diptera	*Aedes (Finlaya) lineatus*	Orthocladiinae gen. spp.	*Simulium cinquestriatum*
	Dixidae gen. sp.	*Rheotanytarsus* sp.	*S.* sp. b
	Odontomyia sp.		*S.* sp. d
	Psilopinae n. gen., n. sp.		*S.* sp. g
	Limonia sp.		*Gomphostilbia (S.)* sp. I
			G. (S.) sp. III
			Blephariceridae gen. spp.
			Orthocladiinae gen. spp.

Table 6 (continued)

Group / Order	0-30 cm/sec	30-75 cm/sec	75-300 cm/sec
Lepidoptera			*Aulacodes*
Mollusca Gastropoda	*Paludomus (P.) tanschauricus nasutus* *P. (Philopotomis) bicinctus* *P. (Tanalia) loricatus* *P. (T.) neritoides* *Melanoides tuberculata* *Thiara scabra*	*P. (T.) tanschauricus nasutus* *P. (P.) bicinctus* *P. (T.) loricatus* *P. (T.) neritoides*	*P. (T.) loricatus* *P. (T.) neritoides*
Bivalvia	*Pisidium (Odhnerpisidium) annandalei* *P. (O.) prasongi* *P. (Afropisidium) javanum*		
Vertebrata Reptilia	*Natrix piscator asperrimus* *Boiga ceylonensis*		
Amphibia	*Rhacophorus cruciger cruciger* *Rana temporalis* *R. cyanophlictis* *R. limnocharis*		
Pisces			*Danio malabaricus* *Rasbora vaterifloris* *R. daniconius* *Chela laubuca* *Barbus (Puntius) bimaculatus* *B. (P.) dorsalis* *B. (P.) filamentosus* *B. (P.) nigrofasciatus* *B. (P.) sarana* *Cyprinus carpio* *Garra lamta* *Noemachilus notostigma*

Table 6 (continued)

Group / Order	0-30 cm/sec	30-75 cm/sec	75-300 cm/sec
			Lepidocephalus thermalis
			Xenotodon cancila
			Anguilla bicolor
			A. nebulosa
			Aplocheilus dayi
			Poecilia reticulata
			Xiphophorus helleri
			Channa orientalis
			Ophiocephalus gachua
			Glossogobius giuris
			Belontia signata
			Macrognathus aculeatus

Table 7. Members of the Middle Course Community (50-500 m elevation).

Group / Order	0-30 cm/sec	30-75 cm/sec	75-300 cm/sec
Turbellaria	*Dugesia nannophallus*	*Dugesia nannophallus*	
Crustacea	*Atya spinipes* *Caridina fernandoi* *C. nilotica simoni* *C. n. zeylanica* *Paratelphusa (C.) rugosa* *P. (O.) senex*	*Macrobrachium australe* *M. idella* *M. latimanus*	
Acari		Hydrachnellae gen. sp.	
Insecta Ephemeroptera	Leptophlebiidae gen. spp. *Pseudocloeon difficilum*	Leptophlebiidae gen. spp. *Pseudocloeon difficilum* *Indobaetis starmuelhneri* Ecdyonuridae gen. spp.	*Pseudocloeon ambiguum*
Plecoptera		*Neoperla angulata* *N. triangulata* *Phanoperla* sp.	
Odonata	*Libellago greeni* *Vestalis apicalis nigrescens* *Neurobasis chinensis chinensis* *Copera marginipes* *Drepanosticta* sp. *Elattoneura centralis* *Microgomphus wijaya* *Heliogomphus* sp. *Megalogomphus ceylonicus* *Macromia zeylanica* *Trithemis aurora* *T. festiva*	*Euphaea splendens* *Zygonyx iris ceylanica*	

Table 7 (continued)

Group / Order	0-30 cm/sec	30-75 cm/sec	75-300 cm/sec
Orthoptera	*Rhabdoblatta* sp.		
Coleoptera	*Guignotus flammulatus*	*Helichus naviculus*	
	Neotosternus taprobanicus	*Potamophilinus impressicollis*	
	N. starmuehlneri	*P. costatus*	
	Pelthydrus sp.	*Stenelmis brincki*	
	Aulonogyrus obliquus	*S. andersoni*	
	Orectochilus sp.	*Ordobrevia fletcheri flaveolineata*	
		Leptotelmis ceylonica	
		Graphelmis ceylonica	
		Ilamelmis foveicollis	
		Podelmis aenea	
		P. viridianea	
		P. similis	
		Taprobanelmis carinata	
		Zaitzeviaria bicolor	
		Z. zeylanica	
		Unguiseta rubrica	
		Hydrocyphon atratus	
		Eubrianax lioneli	
		Lampyridae gen. spp. (cf. *Pyrophanes*)	
		Aulonogyrus obliquus	
		Orectochilus sp.	
Heteroptera	*Diplonychus rusticus*		
	Cylindrostethus productus		
	Onychotrechus sakuntala		
	Ptilomera cingalensis		
	Rheumatogonus custodiensis		
	Ventidius henryi		
	Hydrometra greeni		

Table 7 (continued)

Group / Order	0-30 cm/sec	30-75 cm/sec	75-300 cm/sec
	Microvelia douglasi		
	M. (Kirkaldyia) longicornis		
	Rhagovelia ceylonica		
	Strongovelia formosa		
	Tetraripis asymmetricus		
	Xiphovelia iota		
	Micronecta anatolica		
	M. quadristrigata		
	M. sanctae-catharinae		
	M. deseriana		
	M. alterna		
	M. punctinotum		
	Hydrotrephes kirkaldyi		
	Triphotrephas indicus		
	Aphelocheirus clivicolus		
	Heleocoris breviceps		
	Cercotmetus strangulatus		
Diptera	*Chlorichaeta tuberculata*	Orthocladiinae gen. spp.	*Simulium quinquestriatum*
		Tanytarsus sp.	*S.* sp. b
		Rheotanytarsus sp.	*S.* sp. d
		cf. *Lithotanytarsus* sp.	*Gomphostilba (S.)* sp. I
			G. (S.) sp. III
			Blephariceridae gen. spp.
Trichoptera		*Synagapetus* cf. *hanumata*	*Chimarra* sp.
		S. sp.	*Oestropsyche vitrina*
		Plethus cursitans	*Macronema* sp. A
		Plethus sp.	cf. *Pseudoleptonema* sp. A
		Oxyethira sp.	cf. *P.* sp. B
		Nyctiphylax sp.	*Hydropsyche katugahakanda*

Table 7 (continued)

Group / Order	0-30 cm/sec	30-75 cm/sec	75-300 cm/sec
		Pseudoneureclipsis thuparama	*H. flinti*
		Polycentropidae gen. spp.	*H.* sp. **A**
		Ecnomus sp.	*H.* sp. **B**
		Psychomyidae gen. spp.	*H.* sp. **C**
		Anisocentropus sp.	*Diplectrona* sp. **A**
		Marilia cf. *ceylanica*	Hydropsychidae gen. spp.
		Trichosetodes argentolineata	
		cf. *T. meghawanabaya*	
		Adicella biramosa	
		Oecetis belihuloya	
		Leptocerida gen. spp.	
		Goerodes sp.	
		Helicopsyche amarawathi	
Lepidoptera		*Pyralidae* gen. sp.	*Aulacodes*
Mollusca Gastropoda	*Paludomus (P.) tanschauricus nasutus*	*Paludomus (P.) bicinctus*	*Paludomus (P.) loricatus*
	P. (P.) chilinoides	*P. (P.) decussatus*	*P. (T.) neritoides*
	P. (Tanalia) solidus	*P. (T.) neritoides*	*P. (T.) solidus*
	P. (T.) loricatus	*P. (T.) solidus*	
	P. (T.) neritoides		
	Melanoides tuberculata		
	Thiara scabra		
	Bulimus (= Bithynia) inconspicua		
	B. (B.) stenothyroides		
	Indoplanorbis exustus		
	Lamellidens lamellatus		
Vertebrata Amphibia	*Rana cyanophilictis*		
	R. temporalis		
	R. tigrina crassa		

Table 7 (continued)

Group / Order	0-30 cm/sec	30-75 cm/sec	75-300 cm/sec
Pisces			*Danio malabaricus*
			Rasbora daniconius
			Esomus danrica
			Chela laubuca
			Barbus (Puntius) bimaculatus
			B. (P.) cummingi
			B. (P.) dorsalis
			B. (P.) filamentosus
			B. (P.) nigrofasciatus
			B. (P.) sarana
			B. (P.) titteya
			B. (P.) vittatus
			Tor khudree longispinus
			Garra lamta
			Noemacheilus botia
			N. notostigma
			Lepidocephalus themalis
			Macrones (=Mystus) keletius
			Clarias dussumieri
			C. teysmanii brachysoma
			Anguilla nebulosa
			Aplocheilus dayi
			Channa orientalis
			Ophiocephalus gachua
			Glossobius giuris
			Sicydium halei
			Anabas testudineus
			Belonti signata
			Mastocembalus armatus
			Macrognathus aculeatus

Table 8. Members of the Lower Course Community (20 m elevation).

Group / Order	0-30 cm/sec	30-75 cm/sec
Crustacea	*Caridina nilotica simoni*	*Macrobrachium australe* *M. idella*
Insecta Coleoptera		*Ilamelmis* sp.
Vertebrata Pisces		*Rasbora danicornius* *Barbus (=Puntius) chola*

236

Fig. 3. Diyaluma Falls, coming from Horton Plains at World's End near Belihuloya. The height of the falls is about 120 m (upper course of the Kirindi-Ganga).

Distribution: Coastal areas of the Indopacific coasts between Madagascar and the Pacific Islands.

Septaria lineata inhabits the lower reaches of streams (Fig. 9), mostly not far from their mouths, settling on the surface of stones, rocks and floating vegetation in moderate currents (50-100 cm sec^{-1}). Distribution: Southern India, Sri Lanka, Malay Archipelago, Philippines to Papua New Guinea, Western Australia and the Pacific Islands.

Fig. 4. Hal-Oya near Ginigathhena (region of Belihuloya), a tributary of the Kelani Ganga, flowing through secondary forests and plantations.

Mesogastropoda-Viviparidae

Bellamya dissimilis var. *ceylanica* inhabits sandy, muddy bottoms of stagnant waters and is also found at the margins of slow running streams in the lowlands. The females are viviparous and bear young the whole year round.

Distribution: India, Burma (var. *ceylanica* : lowlands of Sri Lanka).

Ampullariidae

Pila (Pila) globosa is found as a typical inhabitant of lowland and upland stagnant waters, irrigated paddy fields and sometimes also the muddy margins of slow

Fig. 5. Granite boulders in the centre of the Campden Hill-Dola near Deniyaya. Small cascades alternate with pools, typical of the upper courses of mountain streams.

Fig. 6. Middle course of the Kalu-Ganga near Ratnapura, flowing through secondary forests and plantations.

Fig. 7. Lower course of the Kelani-Ganga near Hewala, with sandy banks, and flowing through coconut plantations.

running streams. They prefer waters with dense growths of water plants such as *Azolla*. It is also found in slightly brackish waters, such as Mundel Lake. If the water disappears during the dry season, the snails dig into the mud or move on land to other aquatic habitats.

Distribution: India, Sri Lanka.

Hydrobiidae

Tricula montana is recorded only in Madgoda in a small mountain stream at 800 m elevation on a forest clad mountain side.

Distribution: India (Bhim Valley), Jhiri Valley (N. Cachar), Assam and central Sri Lanka.

Stenothyridae

Gangetia burmanica occurs in slightly brackish waters near the coast of northern Sri Lanka (flooded rivers, crossing salt meadows, Fig. 9).

Distribution: Burma (coast of the Arakan Mountains), Northern Sri Lanka.

240

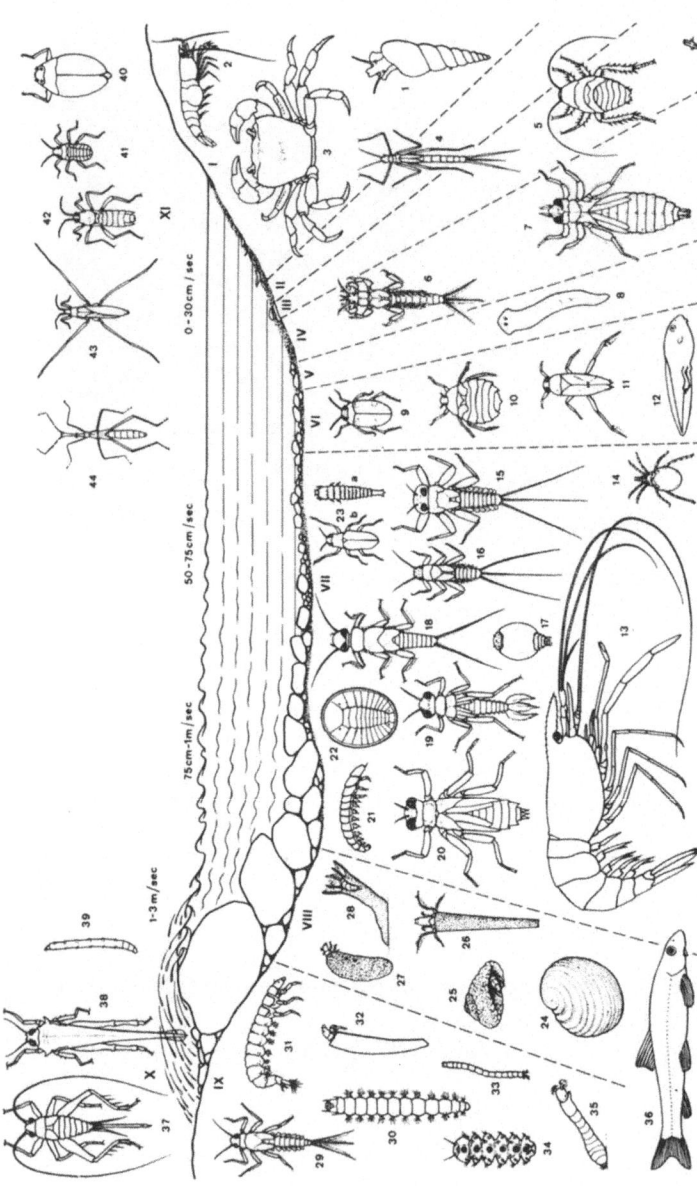

Fig. 8. Schematic cross-sectional diagram of the upper course of a typical mountain stream, showing the microdistribution of the fauna in various habitats. I. Mud and sand surfaces, banks and pools between cascades: 1. *Melanoides tuberculata.* 2. *Caridina* (Atyidae). 3. *Paratelphusa rugosa*; II. Sand surfaces, banks and pools between cascades: 4. Damselflies; *Neurobasis chinensis chinensis*; III. Under stones near banks: 5. *Rhabdoblatta*; IV. In sand, near banks and in pools: 6. *Ephemera*, 7. *Gomphinae*; V. Stone surfaces, near banks and in pools and in medium current: 8. *Dugesia nannophallus*; VI. Swimmers in lentic areas of low current: 9. *Lacconectes simoni*, 10. *Aphelocheirus clivicolus*, 11. *Enithares simplex*, 12. *Rana temporalis*; VII. In gravel in medium current: 13. *Macrobrachium* spp., 14.*Hydrachnellae*, 15. Ecdyonuridae, 16. *Leptophlebiidae*, 17. *Prosopistoma* sp., 18. *Neoperla* spp., 19. *Euphaea splendens*, 20. *Zygonyx iris ceylanica*, 21. *Luciola* sp. (?), 22. *Eubrianax ceylonicus*, 23. *Elminthidae* (12 genera and 31 spp.); VIII. Among boulders and gravel in strong current: 24. *Paludomus (Tanalia) neritoides*, 25. *Helicopsyche* spp., 26. *Leptoceridae*, 27. *Hydroptilidae*, 28. *Rheotanytarsus*; IX. On boulders and rocks in rapids and cascades: 29. *Baetidae*, 30. *Aulacodes* sp., 31. *Hydropsychidae*, 32. *Sericostomatidae*, 33. *Orthocladiinae*, 34. *Blephariceridae*, 35. *Simuliidae*, 36. *Garra lamta*; X. On sprayed rocks: 37. *Paranemobius pictus*, 38. *Euscelimena gavialis*, 39. *Limonia* sp.; XI. Surface dwellers: 40. *Aulonogyrus* and *Orectochilus*, 41. *Microvelia*, 42. Water cricket (different genera and species), 43. Pond skaters (different genera and species), 44. *Hydrometra greeni*.

Buliminidae (=Bithyniidae)

Bulimus (=Bithynia) inconspicua inhabits swamps, pools, ponds, reservoirs (tanks) and the margins of slow running lowland and upland streams with muddy bottoms, which are rich in debris. It is sometimes also found in brackish lagoons with exposed grassy shores (e.g. Mundel Lake).

Distribution: Endemic to Sri Lanka (coast,lowlands and uplands).

B. (=B.) stenothyroides is found in habitats similar to those of *B.* (=B.) inconspicua, but more rarely.

Distribution: Southern India, Sri Lanka.

Mysorella costigera is a rare species, found in stagnant waters of the lowlands with muddy bottoms and dense vegetation.

Distribution: India (Bengal, Bangalore), Sri Lanka.

Synceridae

Syncera (=Assiminea) cf. *hidalgoi* was recorded in brackish water of a flooded river, crossing salt meadows near the coast (Fig. 9) of northern Sri Lanka.

Distribution: *S.* (=A.) hidalgoi is reported from the brackish shores of the Indopacific between Mauritius and New Caledonia.

S. (=A.) cf. *woodmasoniana* is associated with *S.* (=A.) cf. *hidalgoi* and *Gangetia burmanica* in a flooded river, crossing salt meadows near the coast of northern Sri Lanka.

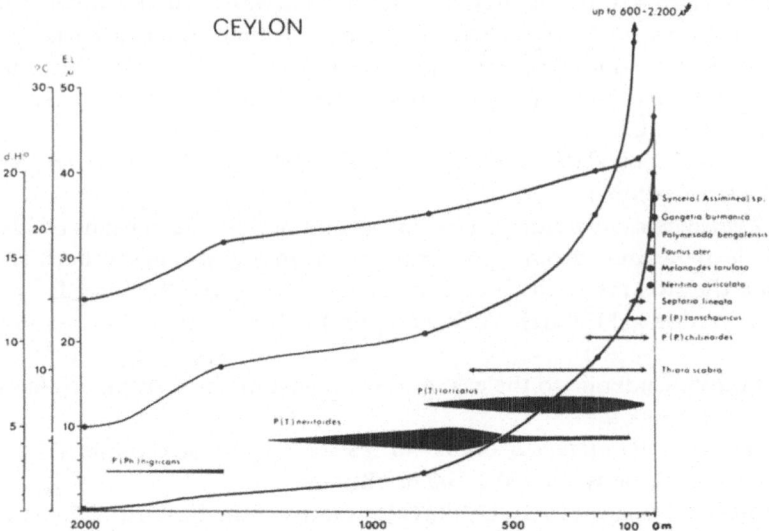

Fig. 9. Diagram showing water temperature (° C), hardness (° dH) and conductivity (EC[20]) of waters, correlated with the longitudinal distribution of gastropods and bivalves (after Starmühlner 1979).

Distribution: Coasts of Lower Bengal, Andaman Islands, Malayan Archipelago; probably the first record from Sri Lanka.

Thiaridae (=Melaniidae)

Faunus ater is found in the coastal areas (Fig. 9) in fresh or slightly brackish water, living on mud flats which dry during low tide.

Distribution: Coasts of the Indopacific between Mauritius and Sri Lanka to New Guinea and New Ireland.

Paludomus (Paludomus) chilinoides (Fig. 10) is very frequent in slow to fairly fast running streams with gravel and sandy bottoms in the lowlands and uplands (up to 600 m) (Fig. 9). The snail prefers streams outside of the crystalline series with an average conductivity of 300-$600 S\mu$ cm^{-1} and a total hardness of 9-$13°$dH (CaO: mean $= 50$ ppm). Their average density is 8-12 individuals m^{-2}, and up to 32-48 ind. m^{-2}. The snails prefer stones and rocks emerging from the water surface and usually extend the anterior parts of their bodies out of the water. On sandy bottom the animals usually come towards the waters edge and crawl up out of the water.

Distribution: Endemic to the lowlands and uplands of Sri Lanka.

P. (P.) inflatus is found in shallow rivers with sandy bottoms.

Distribution: Southern India, Eastern Sri Lanka.

P. (P.) palustris is found in stagnant waters, such as the grassy margins of tanks.

Distribution: Endemic to Northern Central Province: Anuradhapura.

P. (P.) tanschauricus tanschauricus inhabits slow flowing streams and canals, from the lowlands to the uplands (500 m) (Fig. 9) in northern Sri Lanka. The snails usually come towards the waters edge and extend their anterior parts out of the water. They can live out of water for quite some time and even crawl outside water. This behaviour gives the snails the option of temporarily leaving dried streams to burrow in the mud.

Distribution: Central India, Bombay, Madras, Hoogly, Pondycherry, Northern Sri Lanka (first record).

P. (P.) tanschauricus nasutus. This subspecies prefers the margins of streams in the uplands of the central and eastern parts of the mountains in southern Sri Lanka (especially the streams in the limestone zone of the Uva Province: EC$_{20}$: 360μS cm^{-2}; total hardness: $11.6°$dH, CaO: 59.4 ppm). The density of this species is 16-80 ind. m^{-2}.

Distribution: Endemic to the east and southeast of the Central Highlands and uplands of Sri Lanka.

P. (Philopotamis) bicinctus occurs in fast to fairly fast running streams with cascades in forests between 250-1,100 m altitude.

Distribution: Endemic to Sri Lanka (mountains of the Eastern, Central and Uva Provinces).

P. (Ph.) decussatus is found as a rare inhabitant of fast running streams.

Distribution: Endemic to Sri Lanka in the uplands of the Uva Province. *P. (P.)*

nigricans is found exclusively in cool brooks and torrents of the highlands (Fig. 9) at altitudes of 800-2,000 m (water temperature 15 ° C) in slightly acidic and very soft waters (EC$_{20}$: 26μS cm^{-2}; total hardness: 0.65$_4$dH; CaO: 2.6ppm).

Distribution: Endemic to Sri Lanka (crystalline highlands of the Central Province).

P. (Ph.) regalis is found in fast running, stony streams of the uplands. It is a rare species.

Distribution: Endemic to Sri Lanka (uplands of the South and Western Province). *(Ph.) sulcatus* inhabits moderate to fast flowing streams with gravel, boulders and sand. Its density may range from 1-2 ind. m^{-2} to 4-12 ind. m^{-2} in currents between 30-75 cm sec^{-1} (EC$_{20}$: 29-35μS cm^{-2}; total hardness: 0.6-1.0° dH; CaO: 2 ppm).

Distribution: Endemic to Sri Lanka (uplands of the Southwest).

P. (Tanalia) loricatus (Fig. 12) occurs with many variations (Fig. 13) as a typical inhabitant of fast running mountain streams in the uplands of the crystalline mountains of the southwest (Fig. 9). This species prefers stones and rocks and currents of 30- 100 cm sec^{-1}, where it moves near the water surface. Sometimes the animal crawls outside the water on rocks emerging from the cascades. It is found in streams with average temperatures of 18-27.3 ° C, slightly acidic to neutral, very soft with low mineral content (EC$_{20}$: 23-46μS cm^{-2}; total hardness: 0.5-1.2° dH; CaO: 2.00-6.38 ppm). This species prefers the main tributaries of the mountain stream systems, over the smaller streams, where *P. (T.) neritoides*, the second most frequent species of this subgenus, dominates. The distribution of these two species does overlap; *P. (T.) loricatus* was found also in slightly polluted streams surrounded by cultivated areas and settlements, where the current and turbidity were suitable. Their density varied from 16-32 ind. m^{-2} to 48-64 ind. m^{-2}. Juvenile specimens were found in some localities near the margins with densities of 320-480 ind. m^{-2}.

Distribution: Endemic to Sri Lanka (crystalline uplands of the southwest).

P. (T.) neritoides (Fig. 14) occurs frequently, with many variations (Fig. 13) in the smaller streams, torrents and brooks running in the mountainous uplands and highlands of the Southwest (Fig. 9). Like the other species of *Paludomus* they crawl mostly just under the surface of the water and sometimes move out of the water around waterfalls. Their density ranges from 48-160 ind. m^{-2} (current: 30-100 cm sec^{-1}). Juveniles are found mostly near the margins in currents between 10 cm sec^{-1} and 30 cm sec^{-1}, with densities between 480-960 ind. m^{-2}. *P. (T.) neritoides* has been recorded at altitudes of 50 m (valleys of mountain streams) and 1,500 m. The temperature in these streams ranges from 18.8-27.2 ° C (average 20-25 ° C) and the water is soft and slightly acidic (EC$_{20}$14.6-89.0μS cm^{-2}; total hardness: 0.25-2.35° dH; CaO: 1.12-15.60 ppm).

Distribution: Endemic to Sri Lanka (crystalline uplands and highlands of the Southwest).

P. (T.) solidus is the only representative of this subgenus found in calcareous

244

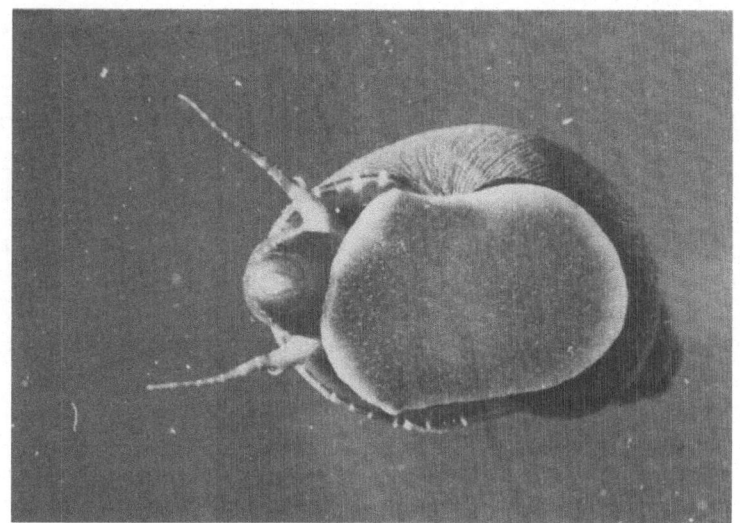

Fig. 10.   Paludomus (Paludomus) chilinoides, central view.

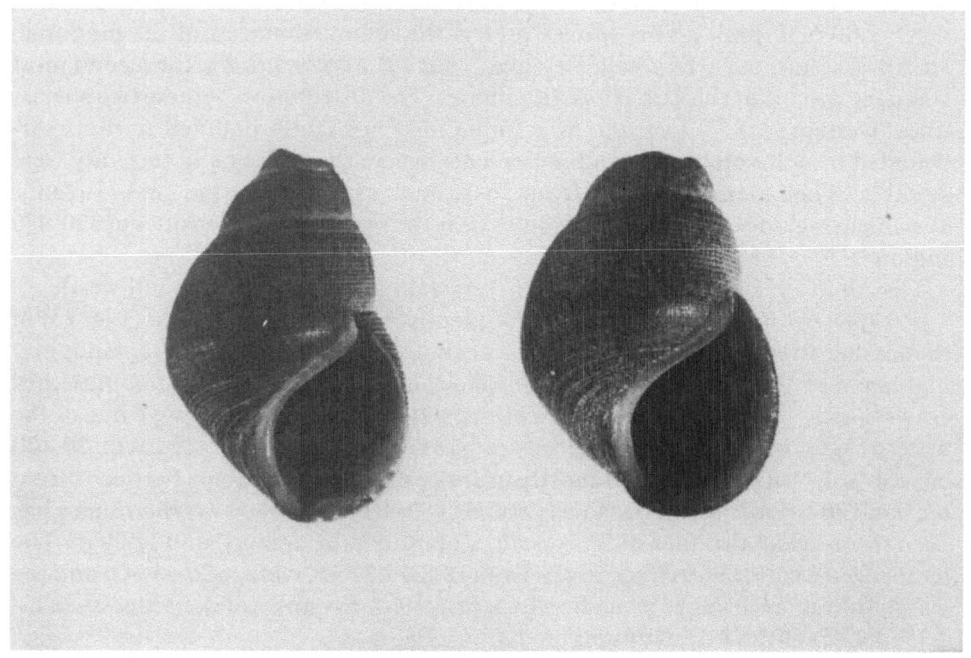

*Fig. 11.   Shells of *Paludomus (Philopotamis) nigricans*.*

Fig. 12. Shell of *Paludomus (Tanailia) loricatus* var. *funiculatus.*

streams of the limestone areas of the hilly regions of the Uva Province (near Buttala and Wellawaya). The stream chemistry shows alkaline, moderately hard water (EC_{20}: $360\mu S$ cm^{-2}; total hardness: 11.6° dH; CaO: 59.4ppm; SiO_2: 54.8 ppm). The shells of this species are different from the shells of species occurring in the acidic waters of the crystalline series; they are not decollated and eroded. The density of this species is 80-160 ind. m^{-2} in currents of 30-100 cm sec^{-1}, with 640-960 juveniles m^{-2} near the margins in currents of 0-30 cm sec^{-1}.

Distribution: Endemic to Sri Lanka (southeast, Uva Province).

Fig. 13. Variations, with transitions, in the shells of *Paludomus (T.) neritoides* (upper row) and *P. (T.) loricatus* (lower row) (after Starmühlner 1974).

Fig. 14. Shell of *Paludomus (T.) neritoides* var. *gardneri.*

Thiara (Plotia) scabra occurs in slow running streams near their margins or in pools of fast running streams. They are found also in stagnant waters in the lowlands and uplands between 60-700 m in altitude (Fig. 9). It was recorded in waters with an average temperature of 21-27.2 °C (EC$_{20}$: 35-295μS cm^{-2}; total hardness: 1.1-9.2°dH; CaO: 2-52 ppm). Their density was 80-96 ind. m^{-2} on sand and stones, coated with mud and debris in currents of 10-30 cm sec^{-1}.

Distribution: Widely distributed on the coasts and islands of the Indopacific and South and Southeast Asia between East Africa in the west and the South Pacific Islands in the east.

Melanoides (Melanoides) tuberculata, like *Thiara scabra* is an eurytopic tropical species found in Sri Lanka from the coast to the lowlands and uplands. It occurs mostly in stagnant water, but also in the margins of slow and moderately fast flowing streams, in currents of 10-30 cm sec^{-1}. This species is very resistant to high water temperatures and organic pollution and has also been recorded from slightly brackish waters. It occurs mostly in waters with an average temperature of 21-26 °C (EC$_{20}$: 35-89μS cm^{-2}; total hardness: 1.2-2.3°dH; CaO: 4.0-15.6 ppm). The snails do not occur in acidic water with low CaO concentrations. Their density is sometimes as high as 800-960 ind. m^{-2}.

Distribution: North Africa, Asia Minor, Iran, India, Sri Lanka, Indian Ocean Islands, Malay Peninsula and Archipelago, South China, Northern Australia, Papua New Guinea to the Pacific Islands. It must be assumed that this parthenogenetic species, which is a common inhabitant of irrigated rice fields has been passively and unintentionally spread the by inter-local, or even inter-continental transport of young rice plants).

M. (Stenomelania) torulosa inhabits the mouths (Fig. 9) of streams near the coast. This species has free-living, veliger larvae.

Distribution: Coasts of India, Sri Lanka, Andaman Islands, Malay Archipelago, Philippines to the Solomon Islands.

Basommatophora-Lymnaeidae

Radix (Cerasina) luteola var. *pinguis* occurs in stagnant waters, such as pools, irrigated paddy fields, swamps, ponds and reservoirs (tanks) as well as in slow running streams with muddy bottoms and dense vegetation and debris. This species was found from the coastal regions to the lowlands and uplands (1,200 m).

Distribution: India, Burma, Thailand, Andaman Islands; only the variety *pinguis* occurs in Sri Lanka.

Bulinidae

Indoplanorbis exustus occurs from the coast (also in slightly brackish water such as lagoons and marshes) to the lowlands and uplands (1,200 m). This snail prefers stagnant waters, such as irrigated paddy fields, marshes, pools, ponds and reservoirs (tanks) with dense aquatic vegetation. The snail is often very abundant in cushion-like clumps of filamentous algae on the margins of streams and in pools filled with flood waters (temperature 25.6°; EC_{20}: 89μS cm^{-2}; total hardness: 2.35°dH, CaO: 15.6ppm).

Distribution: Lower Mesopotamia, Iran, India, Sri Lanka, Burma, Malay Peninsula and Archipelago, Indo-China, Thailand.

Planorbidae

Gyraulus (Gyraulus) conveciusculus. All species of *Gyraulus* found and described from Sri Lanka are synonyms of this variable species, which is widely distributed in the Oriental region (Starmühlner, 1974). This species occurs from the coast to the lowlands in pools, irrigated paddy fields, ponds, reservoirs (tanks) and sometimes at the margins of slow running streams with sandy, muddy bottoms and dense aquatic vegetation. One locality where the species was recorded was slightly brackish.

Distribution: Lower Mesopotamia, Eastern Iran, Pakistan, India, Sri Lanka, Burma, Vietnam, Japan, Thailand, Malay Peninsula and Archipelago, Philippines and Papua New Guinea.

Pelycypoda

Hadl(1974) has described four species of freshwater mussels from the collection of the Swedish Lund University Expedition (1962) and from the Austrian-Sri Lanka Mission (1970). Three of these belong to the family Unionidae, and one to the Family Corbiculidae.

Unionidae

Lamellidens lamellatus occurs in slow flowing streams with sandy, muddy bottoms. This species is known from two localities, one in the north and the other in the southeast, crossing the limestone areas of the Island. The chemistry of these streams is known: temperature: 25.5 °C; EC_{20} 295-605μS cm^{-2}; total hardness: 9.2-12.9°dH; CaO: 50.9-52.0 ppm.
    Distribution: India (Bengal), Sri Lanka.
    L. testudinarius occurs in slow flowing streams with sandy, muddy bottoms, and has been recorded thus far only near the coast.
    Distribution: India, Burma, Sri Lanka.
    Parreysia corrugata is found in slow flowing streams with sandy bottoms.
    Distribution: India (Bengal, Deccan), Sri Lanka.

Corbiculidae

Polymesoda ceylonica (=bengalensis) is found in backwaters near the coast, and tolerates slightly brackish water.
    Distribution: This extremely variable species seems to be distributed from Sri Lanka to the coasts of the Indo-Pacific. Hadl(1974) states that the other Indo-Pacific 'species' or 'subspecies' possibly fall within the range of variation of this species and large numbers of specimens would be necessary to clarify the relationship of this variable species.

Sphaeridae

Kuiper (1979) reported on *Pisidium* (Sphaeriidae), the pea-shell cockle, of Sri Lanka:
    Pisidium (Afropisidium) javanum is known only from a small mountain stream near Udawla, east of Kandy, in the Central Province.
    Distribution: Burma, Laos, Thailand, Malay Archipelago (Java, Sumba).
    Pisidium (Odhnerpisidium) annandalei is known from only one locality, a stream near the Kandy Reservoir.
    Distribution: Sicily, Greece, Asia Minor, Israel, Iran to Japan, Philippines, Papua New Guinea and the Bismarck Archipelago. Three specimens have been collected from Sri Lanka (Kuiper 1979).

Pisidium (O.) prasongi has been collected from the same locality as *P. (A.) javanum*, a small stream near Teldeniya, Udawela, east of Kandy at 450 m altitude.
Distribution: Thailand, Sri Lanka.

Vertical zonation of the freshwater Molluscs

A) Crystalline Central Highlands (1,500-2,000 m), Central Province.

A-1) Brooks and torrents: *Paludomus (Philopotamis) nigricans*

B) The precipice from the lower regions of the crystalline Central Highlands to the uplands (200-800/1,200 m), Sabaragamuwa, Central, Uva, South and Western Provinces.

B-1) Brooks, torrents, smaller streams:
Strong current (50- 100 cmsec⁻¹): *Paludomus (Tanalia) neritoides* (var. *typica, dilatatus*), *P. (T.) loricatus* (var. *funiculatus, pictus*), rare species: *P. (Ph.) sulcatus, P. (Ph.) regalis, Tricula montana.*
At the margins of creeks and pools between cascades (0-30 cm sec⁻¹): Juveniles of *P. (T.) neritoides* and *loricatus, Thiara (Plotia) scabra, Melanoides (Melanoides) tuberculata, Bulimus* (=Bithynia) *stenothyroides, Indoplanorbis exustus.*

B-2) Larger streams with moderate to strong currents:
Strong current (50-100 cm sec⁻¹): *P. (T.) loricatus* (var. *typica, erinaceus*), *P. (T.) neritoides* (var. *tennanti, gardneri*)
At margins of creeks and pools between cascades (0-30 cm sec⁻¹): juveniles of *P. (T.) neritoides* and *loricatus, T. scabra, M. (M.) tuberculata, B. stenothyroides, I. exustus, Pisidium (Afropisidium) javanum, P. (O.) annandalei, P. (O.) prasongi.*

B-3) Brooks and smaller streams in the limestone formation of the Uva Province:
Strong current (50->100 cm sec⁻¹): *P. (T.) solidus*
At the margins of creeks and pools (0-30 cm sec⁻¹): Juveniles of *P. (T.) solidus, P. (Paludomus) tanschauricus nastus, P. (P.) chilinoides, B. inconspicua, Lamellidens lamellatus.*

B-4) Stagnant waters (marshes, irrigated paddy fields, pools, ponds in the uplands of the Sabaragamuwa, Central, Uva, South and Western Provinces): *Bellamya dissimilis* var. *ceylanica, Pila globosa* (all var.), *B. inconspicua, M. (M.) tuberculata, T. scabra, Radix (Cerasina) luteola* var. *pinguis, I. exustus, L. lamellatus, Parreysia corrugata.*

C) Streams and stagnant waters in the transition region between uplands and lowlands and in the lowlands (20/50-100/200 m); Eastern, Uva, Southern and Western Provinces.

C-1) Streams: Moderate current (30-50 cm sec^{-1}): *P. (P.) chilinoides, P. (P.) bicinctus, P. (P.) decussatus, P. (P.) inflatus, P. (P.) tanschauricus nasutus.*

Low current on the margins (0-30 cm sec^{-1}): Juveniles of P. (P.) species and *B. inconspicua, M. (M.) tuberculata, T. scabra, R. luteola* var. *pinguis, I. exustus, Gyraulus (G.) convexiusculus* (all var.), *L. lamellatus, L. testudinarius, P. corrugata.*

C-2) Stagnant waters: *B. dissimilis* var. *ceylanica, P. globosa* (all var.), *B. inconspicua, M. (M.) tuberculata, T. scabra, R. luteloa* var. *pinguis, I. exustus, G. (G.) convexiusculus* (all var.).

D) Streams and Stagnant waters from the lowlands to the coast (2/5-20 m); Southern and Western Provinces.

D-1) Streams:

Moderate current (20-50 cm sec^{-1}): *Neritina (Neripteron) auriculata, Septaria lineata, M. (Stenomelania) torulosa, Polymesoda ceylonica.*

Low current on the margins (0-20 cm sec^{-1}): The same species listed for the low current at the margins in section C-1 and *P. ceylonica.*

D-2) Stagnant waters: The same species listed in C-2 and *Mysorella costigera.*

E) Mouths of streams on the coast (0-2/5 m), during high tides these regions are sometimes slightly brackish; Southern and Western Province: *Faunus ater, P. ceylonica.*

F) Slow running streams in the lowlands (5-100 m); of the north of the Central, Northwestern and Northern Provinces:

F-1) Streams:

Moderate current (30 cm sec^{-1}): *P. (P.) chilinoides, P. (P.) tanschauricus tanschauricus.*

Low current at the margins (0-30 cm sec^{-1}): Juveniles of *P. (P.)* species listed above and *P. globosa* (all var.), *T. scabra, M. (M.) tuberculata, I. exustus, L. lamellatus.*

F-2) Stagnant waters (reservoirs, tanks, pools, irrigated paddy fields): *B. dissimilis* var. *ceylanica, P. globosa* (all var.), *B. inconspicua, M. costigera, M. (M.) tuberculata, T. scabra, R. luteola* var. *pinguis, I. exustus, G. convexiusculus* var. *compressus.*

G) Mouths of streams on the coast (0-2/5 m), sometimes slightly brackish during high tides; Northwestern Province: *F. ater, P. ceylonica*

H) Brackish water lagoon on the Northwest coast (Mundel Lake): *P. globosa* var. *carinata, B. inconspicua, M. (M.) tuberculata, I. planorbis exustus.*

I) Streams running through salt meadows; Northern Province: *Gangetia burmanica, Syncera* cf. *hidalgoi, S.* cf. *woodmasoni, G. convexiusculus* var. *compressa.*

Zoogeography

Of the 31 species, including the subspecies of *Paludomus (P.) tanschauricus nasutus,* of aquatic gastropods found in Sri Lanka, 37% are endemic and 33% of these are species or subspecies of *Paludomus.* These include: *Bulimus inconspicua, P. (P.) chilinoides, P. (P.) palustris, P. (P.) tanschauricus nasutus, P. (P.) bicinctus, P. (P.) decussatus, P. (Philopotamis) nigricans, P. (Ph.) regalis, P. (Ph.) sulcatus, P. (Tanalia) loricatus, P. (T.) loricatus, P. (T.) neritoides* and *P. (T.) solidus.*

Thirty-two percent of the aquatic gastropods have an Indian (Southern Indian)-Sri Lankan distribution. These are: *Bellamya dissimilis* (with var. *ceylanica* in Sri Lanka), *Pila globosa* (all var. in Sri Lanka), *Tricula montana, Gangetia burmanica, Bulimus stenothyroides, Mysorella costigera, P. (P.) inflatus, P. (P.) tanschauricus tanschauricus* and *Radix luteola* (var. *pinguis* in Sri Lanka and Southern India).

Twenty-five percent of the known species are widely distributed on the coasts of the Indo-Pacific: *Neritina (Neripteron) auriculata, Septaria lineata, Syncera* cf. *hidalgoi, Faunus ater, Thiara scabra, Melanoides (M.) tuberculata* (occurs also in North Africa, Asia Minor, Iran) and the *Gyraulus (G.) convexiusculus* group (occurs from Mesopotamia and Iran to China, Japan and New Guinea).

About 6% of the known species have an Indo-Malayan distribution: *Syncera* cf. *woodmasoniana* and *Indoplanorbis exustus* (in the West to Mesopotamia).

The highest number of endemic species is shown in *Paludomus.* Its members in the subgenus *Paludomus* are distributed in the Indo-Malayan region extending to the Philippines. The subgenus *Philopotamis* is restricted to Sri Lanka (and possibly to Sumatra). The subgenus *Tanalia* is endemic to the mountainous areas in the South and Southwest of Sri Lanka. The two species with many variations in shell size and shape, *P. (T.) loricatus* and *P. (T.) neritoides,* have evolved during the long period of separation and isolation in the mountain streams (Fig. 9). A third species, *P. (T.) solidus,* with a shell the size of a young *loricatus,* is restricted to the calcareous streams in the southeast of the Central mountains. This suggests that these three species have developed from the ancestors of the Tribe Paludomeae as a result of isolation in the 'island-like' mountains in the south. They have adapted to the conditions of the fast flowing mountain streams. The occurrence of the Tribe Paludomeae with the genera *Paludomus* and the anatomically truly related *Cleopatra* in Central and East Africa, Seychelles, Comores, Madagascar, Sri Lanka, South India, Further India (Malay Peninsula) and Malay Archipelago suggests

252

that this ancient tribe is a relict of the 'Gondwanian' stock.

In summary, it can be stated that the freshwater gastropods of Sri Lanka show very close affinities with the Indian, especially the South Indian fauna. There also are affinities and transitions, especially in the genera *Paludomus, Pila* and *Bellamya*, with the Ethiopian fauna, especially with that of the Seychelles and Madagascar. Most of the species living near the coast in the lower reaches and mouths of streams, such as the Neritidae, *Syncera, Faunus ater*, and *Stenomelania* are widely represented on the coasts of the Indo-Pacific. Also widely distributed in these areas are the parthenogenetic species of the Thiaridae (=Melaniidae).

Of the freshwater mussels, the Unionidae and Corbiculidae, no endemic species are known. The three species of the Unionidae recorded from Sri Lanka have an Indian-Sri Lankan distribution. *Lamellidens testudinarius* extends to Burma. *Polymesoda ceylanica* (Corbiculidae) seems to be widely distributed in many variations on the coasts of the Indo-Pacific.

Three species of pea cockles, *Pisidium* are known from Sri Lanka. Two species belong to the subgenus *Afropisidium*, which is widely distributed in the tropical and subtropical zones. *P. (A.) javanum* is known, as reported above, from the Malay Archipelago, Thailand, Laos, Burma and Sri Lanka. The subgenus *Odhnerpisidium* is distributed mainly in Asia and south, as well as north of the Himalayas. *P. (O.) annandalei* is reported from Sicily and Greece in the west to Asia Minor, Israel, Iran, further in the east to Japan, the Philippines, Papua New Guinea and the Bismarck Archipelago, but it has not been found in Australia, New Zealand and Africa. The third species, *P. (O.) prasongi* is known only from some rivers in Thailand and also from Sri Lanka. Further work must be done in order to clarify the distribution of these small mussels.

References

Beauchamp, P. de. 1973. Results of the Austrian-Ceylonese Hydrobiological Mission 1970. Part X. Freshwater triclads (Turbellaria, Tricladida) from Ceylon. Bull. Fish. Res. Stn Sri Lanka (Ceylon) 24:89-93.

Bertrand, H.P.I. 1973. Results of the Austrian-Ceylonese Hydrobiological Mission 1970. Part XI. Larvae and pupae of water beetles collected in the Island of Ceylon. Bull. Fish. Res. Stn Sri Lanka (Ceylon) 24:95-112.

Bott, R. 1970. Die Süsswasser-Kraben von Ceylon (Crustacea, Decapoda). Ark. Zool. Stockholm 22:627-640.

Brinck, P., H. Andersson & L. Cederhom. 1971. Report No. 1 from the Lund University Ceylon Expedition in 1962: Introduction. Ent. Scand. Suppl. 1:1-36.

Brot, A. 1874. In: Martini & Chemnitz, Syst. Conch. Cab. 1(24): Die Melaniaceen (Melanidae). Bauer & Raspe, Nurnberg. 480 pp.

Brot, A. 1880. In: Martini & Chemnitz, Syst. Conch. Cab. 1(25): Die Gattung *Paludomus auct.* Bauer & Raspe, Nurnberg. 52 pp.

Clessin, S., H.C. Kuster & W. Dunker. 1886. In: Martini & Chemnitz, Syst. Conch. Cab. 1(17): Die Familie der Limnaeiden. Bauer & Raspe, Nurnberg. 430 pp.

Colett, O. 1897. Contribution to Ceylon Malacology. J. Roy. Asiat. Soc. (Ceylon Branch) 15:12-22.

Colett, O. 1898. Contribution to Ceylon Malacology. II. J. Roy. Asiat. Soc. (Ceylon Branch) 16:153-154.

Cooray, P.G. 1967. An introduction to the geology of Ceylon. Spol. Zeylan. 31. Colombo. 324 pp.

Costa, H.H. 1972. Results of the Austrian-Ceylonese Hydrobiological Mission 1970. Part V. Decapoda, Caridea. Bull. Fish. Res. Stn Sri Lanka (Ceylon) 23:127-135.

Costa, H.H. 1974. Limnology and fishery biology of the streams at Horton Plains, Sri Lanka (Ceylon). Bull. Fish. Res. Stn Sri Lanka (Ceylon) 25:15-26.

Costa, H.H. & E.C.M. Fernando. 1967. The food and feeding relationships of the common meso- and macrofauna in the Maha Oya, a small mountainous stream near Peradeniya (Ceylon). Ceylon J. Sci. Biol. Sci. 7:74-90.

Costa, H.H. & F. Starmühlner. 1972. Results of the Austrian-Ceylonese Hydrobiological Mission 1970. Part I. Preliminary Report: Introduction and description of the stations. Bull. Fish. Res. Stn Sri Lanka (Ceylon) 23:43-76.

Deleve, J. 1973a. Report No. 25 from the Lund University Ceylon Expedition in 1962: Coleoptera; Dryopidae et Elminthidae. Ent. Scand. Suppl. 4:5-23.

Deleve, J. 1973b. Results of the Austrian-Ceylonese Hydrobiological Mission 1970. Part VII. Dryopidae and Elminthidae of Ceylon. Bull. Fish. Res. Stn Sri Lanka (Ceylon) 24:69-74.

Deshayes, P.G. 1854. Description of new species of shells from the Collection of H. Cuming. Proc. Zool. Soc. Lond. 22:13-23.

Dohrn, H. 1857. Description of 13 new species of Paludinacea from Ceylon in the collection of H. Cuming. Proc. zool. Soc. Lond. 25:123-125.

Dohrn, H. 1858. Descriptions of new species of land and freshwater shells from the collection of H. Cuming. Proc. Zool. Soc. Lond. 26:133-135.

Elzen, P.M.M., van. 1972. Results of the Austrian-Ceylonese Hydrobiological Mission 1970. Part IV. List of Reptilia and Amphibia collected in the mountains of Southwest Ceylon, with notes on finding localities. Bull. Fish. Res. Stn Sri Lanka (Ceylon) 23:113-125.

Enckell, P.H. 1970. Parastenocarididae (Copepoda, Harpacticoida) from Ceylon. Ark. Zool. Stockholm 22:545-556.

Fernando, C.H. 1963. A guide to the freshwater fauna of Ceylon. Suppl. 1. Bull. Fish. Res. Stn Ceylon 16:29-38.

Fernando, C.H. 1964. A guide to the freshwater fauna of Ceylon. Suppl. 2. Bull. Fish. Res. Stn Ceylon 17:177-211.

Fernando, C.H. 1969a. A guide to the freshwater fauna of Ceylon. Suppl. 3. Bull. Fish. Res. Stn Ceylon 20:18-25.

Fernando, C.H. 1969b. Some freshwater molluscs from Ceylon with notes on their distribution and biology. Bull. Fish. Res. Stn Ceylon 20:135-140.

Fernando. C.H. 1974. A guide to the freshwater fauna of Ceylon. Suppl. 4. Bull. Fish. Res. Stn Sri Lanka (Ceylon) 25:27-81.

Fernando, C.H. 1980. The freshwater invertebrate fauna of Sri Lanka. Centenary Volume, Colombo Museum, Sri Lanka 35:15-42.

Fernando, C.H. & H.H.A. Indrasena. 1969. The freshwater fisheries of Ceylon. Bull. Fish. Res. Stn Ceylon 20:101-134.

Fernando, L.J.D. 1948. The geology and mineral resources of Ceylon. Bull. Imp. Inst. Lond. 46:303-325.

Geisler, R. 1967. Limnologisch-ichthyologische Beobachtungen in Southwest Ceylon. Int. Revue. ges. Hydrobiol. 2:559-572.

Godwin-Austen, H.H. 1882-1920. Land and freshwater Mollusca of India. Vol. 1 (1882-1888); Vol. 2 (1897-1914); Vol. 3 (1920), London. Hadl, G. 1974. Results of the Austrian-Ceylonese Hydrobiological Mission 1970. Part XVIII. Freshwater mussels (Bivalvia). Bull. Fish. Res. Stn Sri Lanka (Ceylon) 25:183-189.

Hanley, S. & W. Theobald. 1876. Conchologica Indica. L. Reeve & Co., London. 65 pp.

Hora, S.L. 1928. Animal life in torrential streams. J. Bombay nat. Hist. Soc. 32:111-126.

Hora, S.L. 1930. Ecology, bionomics and evolution of the torrential fauna with special reference to the organs of attachment. Phil. Trans. Roy. Soc. B. 218:171-282.

Hora, S.L. 1936. Nature of the substratum as an important factor in the ecology of the torrential fauna. Proc. nat. Inst. Sci. India 2:45-47.

Hubendick, B. 1951. Recent Lymnaeidae. Kungl. Svenska Vetenskapsakad. Handl. Fj. Ser. 3, 233 pp.

Hubendick, B. 1955. Phylogenie in Planorbidae. Trans. Zool. Soc. Lond. 28:453-542.

Jaech, M. 1982a. Neue Dryopidae und Hydraenidae aus Ceylon, Nepal, Neu Guinea und der Turkei (Col.) Koleopterische Rundschau 56:83-114.

Jaech, M. 1982b. Die Koleopterenfauna der Berbüche von Südwest-Ceylon. (Unter besonder Berücksichtigung der Taxonomie Ökologie sowie des allgemein geographischen Verbreitung der Madagascar taxa. Diss. Inst. f. Zool., Universität Wien.

Kaltenbach, A. 1973. Results of the Austrian-Ceylonese Hydrobiological Mission 1970. Part XIII. Some remarkable ripicol insects of the Ceylonese Fauna. Bull. Fish. Res. Stn Sri Lanka (Ceylon) 24:125-128.

Kobelt, W. 1915. In: Martini & Chemnitz, Syst. Conch. Cab. 1: Die gattung *Ampullaria* N.F. Bauer & Raspe, Nurnberg. 236 pp.

Kuiper, J.G.J. 1979. Report No. 47 from the Lund University Ceylon Expedition in 1962. Mollusca, Lamellibranchiata, Sphaeriidae: *Pisidium* from Sri Lanka (Ceylon).

Kuster, H.C. 1852. In: Martini & Chemnitz, Syst. Conch. Cab. 1: Die gattung *Paludina*. Bauer & Raspe, Nurnberg. 96 pp.

Kuster, H.C. 1862. In: Martini & Chemnitz, Syst. Conch. Cab. 1. Bauer & Raspe, Nurnberg. 77 pp.

Malicky, H. 1973. Results of the Austrian-Ceylonese Hydrobiological Mission 1970. Part XVI. The Ceylonese Trichoptera. Bull. Fish. Res. Stn Sri Lanka (Ceylon) 24:153-177.

Mendis, A.S. & C.H. Fernando. 1962. A guide to the freshwater fauna of Ceylon. Bull. Fish. Res. Stn Ceylon 12:1-160.

Moorman, F.R. & C.R. Panabokke. 1961. Soils of Ceylon. Trop. Agriculturist 117:4-65.

Müller-Liebenau, I. 1982a *Indobaetis*: A new genus of Baetidae from Sri Lanka (Insecta, Ephemeroptera) with two new species. Gewäss. Abwäss. 68/69:26-34.

Müller-Liebenau, I. 1982b. Five new species of *Pseudocloeon* Klápalek, 1905 (Fam. Baetidae) from the Oriental Region (Insecta, Ephemeroptera) with some general remarks on *Pseudocloeon*. Arch Hydrobiol. 95(1/4):283-298.

Müller-Liebenau, I. 1982c. A new genus and species of Baetidae from Sri Lanka (Ceylon): *Indocloeon primum* gen. n., sp. n. (Insecta Ephemeroptera). Aquatic Insects 4(3):125-129.

Polhemus, J.J. 1971. Aquatic and semiaquatic Hemiptera of Sri Lanka from the Austrian Indo-Pacific Expedition 1970/71. Bull. Fish. Res. Stn Sri Lanka (Ceylon) 29:89-113.

Pretzmann, G. 1973. Results of the Austrian-Ceylonese Hydrobiological Mission 1970. Part. XIV. The freshwater crabs of the Ceylonese mountain rivers. Bull. Fish. Res. Stn., Sri Lanka (Ceylon) 24:129-133.

Quentin, D. St. 1972. Results of the Austrian-Ceylonese Hydrobiological Mission 1970. Part VI. A new Drepanosticta from Ceylon (Odonata: Insecta). Bull. Fish. Res. Stn Sri Lanka (Ceylon) 23:137-139.

Quentin, D. St. 1973. Results of the Austrian-Ceylonese Hydrobiological Mission 1970. Part XII. Contributions to the ecology of the larvae of some Odonata from Ceylon. Bull. Fish. Res. Stn Sri Lanka (Ceylon) 24:113-124.

Radda, A.C. 1972. Ceylonfahrt-der Fische wegen. Vivarium 2:6-10.

Radda, A.C. 1973. Results of the Austrian-Ceylonese Hydrobiological Mission 1970. Part XV. Collection on Fishes (Osteichthyes). Bull. Fish. Res. Stn Sri Lanka (Ceylon) 24:135-151.

Reeve, L.A. 1847-1878. Conchologica Iconographica: *Paludomus*, 1847; *Neritina*, 1856; *Ampullaria*, 1856; *Navicella*, 1856; *Melania*, 1861; *Limnaea*, 1872; *Planorbis*, 1878. London.

Reichholf, J. 1973. Results of the Austrian-Ceylonese Hydrobiological Mission 1970. Part VIII. Larval stages of water moths (Lepidoptera, Pyralidae, Nymphulinae) from torrents of Ceylon and some South Pacific Islands. Bull. Fish. Res. Stn Sri Lanka (Ceylon) 24:75-81.

Sarasin, F. 1910. Über die Geschichte der Tierwelt von Ceylon. Zool. Jb., Suppl. 12:1-160.

Satyamurthi, S.T. 1960. Land and freshwater Mollusca in the Collection of the Madras Museum. Bull. Madras Govt. Mus. New Ser., Nat Hist. Sect. 6:1-174.

Sawicky, L. 1925. On the geomorphology of Central Ceylon. Trav. de l'Institut Geogr. de l'Univ. de Cracovie, Krakau.

Sievers, A. 1964. Ceylon. Gesellschaft und Lebensraum in den orientalischen Tropen. Bibliothek Geogr. Handbücher. Franz Steiner Verlags-Gmbh. Wiesbaden. 398 pp.

Silvas, E.G. 1952. Further studies regarding Hora's Satpura hypothesis. (2) Taxonomic assessment and levels of evolutionary divergence of fishes with the so-called Malayan affinities in Peninsular India. Proc. nat. Inst. Sci. India 18:432-448.

Sivalingam, V. 1949. Some Ceylonese freshwater snails and human schistosomiasis. Ceylon J. Sci (D) 6:184-185.

Starmühlner, F. 1974. Results of the Austrian-Ceylonese Hydrobiological Mission 1970. Part XVII. The freshwater gastropods of Ceylon. Bull. Fish. Res. Stn Sri Lanka (Ceylon) 25:97-181.

Starmühlner, F. 1977. The genus *Paludomus* in Ceylon. Proc. 5th Malac. Congr. Malacologia (A) 16:261-264.

Starmühlner, F. 1979. Distribution of freshwater molluscs in mountain streams of tropical Indo-Pacific Islands (Madagascar, Ceylon, New Caledonia). Proc. 6th Malac. Congr. Malacologia (A) 18:245-255.

Viets, K.O. 1972. Results of the Austrian-Ceylonese Hydrobiological Mission 1970. Part III. Über einige Wassermilben aus Ceylon (Hydrachnellae, Acari). Bull. Fish. Res. Stn Sri Lanka (Ceylon) 23:101-111.

Weninger, G. 1972. Results of the Austrian-Ceylonese Hydrobiological Mission 1970. Part II. Hydrochemical studies on mountain rivers in Ceylon. Bull. Fish. Res. Stn Sri Lanka (Ceylon) 23:77-100.

Wewalka, G. 1973. Results of the Austrian-Ceylonese Hydrobiological Mission 1970. Part IX. Dytiscidae (Coleoptera). Bull. Fish. Res. Stn Sri Lanka (Ceylon) 24:83-87.

Zwick, P. 1980. The genus *Neoperla* (Plecoptera: Perlidae) from Sri Lanka. Oriental Insects 14:263-269.

12. Ephemeroptera of Sri Lanka: an introduction to their ecology and biogeography

M. D. Hubbard & W. L. Peters

Introduction

The Ephemeroptera fauna of Sri Lanka is not well known. Treatment of the mayflies of Sri Lanka has been sparse and sporadic since Walker with *Caenis perpusilla* and *Baetis taprobanes* in 1853 was the first to describe any mayflies from the island. Hagen (1858) then described 8 new Sri Lankan species, upon which he elaborated in additional works in 1859 and 1873. Eaton (1871, 1883-1888) included some Sri Lankan species in his work, although they comprised only a minor portion in his cosmopolitan taxonomic treatments. A few other authors (e.g. Banks, Chopra, Kimmins, Navás, Lestage, Ulmer; see Hubbard & Peters 1978, for references) dealt briefly or indirectly with mayflies of Sri Lanka.

It was not until the 1960's and 1970's that there emerged even a basic picture of the mayfly fauna of Sri Lanka. Peters (1967) and Peters & Edmunds (1970) published on the taxonomy of the Leptophlebiidae and Prosopistomatidae of Sri Lanka. Fernando (1965) compiled a list of all reported Sri Lankan mayflies; this was followed by inclusion of Sri Lanka in the more comprehensive catalog of the Ephemeroptera of the Indian Subregion by Hubbard & Peters (1978), which gives the most complete account to date of the extent of the Sri Lanka mayfly fauna. Both of these last publications suffered from dependence on inadequate studies of the Sri Lankan fauna.

The taxonomic understanding of the Sri Lankan mayflies still is sketchy, as most of the published papers were based on only a few specimens encountered in collections or supplied by general collectors. The ecological and biogeographical understanding of the Sri Lankan mayflies obviously lags far behind the taxonomic knowledge.

However, several major efforts concentrating on the aquatic fauna, beginning with the Lund University Expedition in 1962 and the Austrian-Ceylonese Hydrobiological Mission of the University of Vienna and the Vidyalankara University in 1970, and continuing with several investigators (including the senior author) associated with the Smithsonian Institution's 'Biosystematic Studies of the Insects of Ceylon' project, have now yielded a large number of specimens which allows us to start toward an understanding of the Ephemeroptera fauna of Sri Lanka.

We are still in the process of studying these specimens and it will be some time before the work is completed. It must therefore be borne in mind that many of the data reported here are preliminary in nature and will change somewhat as our research progresses and the results are published. Several of these papers should be forthcoming shortly, but others will not appear for some time. It is quite clear to us that there is a need for much more collecting and rearing of mayflies in Sri Lanka if we are to understand fully the taxonomic situation there. Ecological research on Sri Lankan Ephemeroptera has been almost non-existent, and much could be learned from such urgently needed research in the future.

Ecology

The island of Sri Lanka is roughly ovate, with an area of approximately 65,600 square kilometres. The south central portion is occupied by the hill country (Fig. 1-4), reaching from about 300 m to an altitude of 2,524 m. Surrounding the hill country below about 300 m altitude, and especially extensive to the north, are the lowlands (Fig. 5-8). The lowland area to the east and north of the mountains of the hill country, known as the dry zone, receives much less rainfall than the hill country

Fig. 1. Hill country stream south of Kandy at 750 m altitude.

and the southern and western lowlands, which together are known as the 'wet zone'.

The principal river systems radiate out from the hill country like the spokes of a wheel. The streams usually are quite rapid in their upper reaches, due to the steepness of the drop in elevation as they leave the hill country. Upon reaching the low country, the rivers slow and broaden.

Water temperatures in the hill country streams range from about 14 °C to over 25 °C. Fig. 8 illustrates representative water temperatures in relation to altitude (taken from September through April). Temperatures at lower elevations can vary greatly depending on the steepness of the drop from the higher elevations; those streams with a rapid descent are cooler than those with a slower rate of descent. Water temperatures near the river mouths can reach almost 30 °C.

The hill country contains myriads of streams and rivers. These streams usually have a substrate of rock, or stones and gravel. The species richness of Ephemeroptera is much greater in this region than in the lowlands. Except for *Chromarcys*, *Povilla*, and to some extent *Ephoron*, the Sri Lankan mayflies are found throughout the streams of the hill country (Fig. 9), with no obvious vertical zonation. Much of the hill country aquatic fauna also extends deep into the lowlands in the streams at the bottom of their steep drop from the hill country highlands, especially in the southern and western wet zone lowlands. The hill country can be considered to be much more extensive for the mayfly fauna than for terrestrial organisms.

Fig. 2. Belihul Oya at about 650 m altitude after its steep fall from the Horton Plains.

Fig. 3. Stream in the Sinharaja rain forest.

Fig. 4. Waterfall in a hill country stream at about 600 m altitude.

Fig. 5. Tributary of the Kelani Ganga, a typical *Prosopistoma* habitat.

The northern and eastern dry zones have been little collected for mayflies, although the few localities which have been collected point to a reduced fauna in most areas in comparison with the upland wet zone fauna. *Povilla* and *Chromarcys* are typical low country genera. *Choroterpes* is known from marshes and streams in the northern dry zone, as are the Caenidae.

There are almost no natural lakes in Sri Lanka, but there are a tremendous number of tanks and reservoirs, both large and small. Most of these standing waters have only a moderate mayfly fauna, mainly Baetidae and Caenidae. The mainfly fauna of paddy fields appears somewhat limited, probably because of the fluctuating water regime, although Fernando (1980) found Ephemeroptera to be not uncommon in paddy field samples.

Many of Sri Lanka's aquatic habitats remain in a semi-natural state; others show vivid evidence of the influence of man. These differences often are reflected in the mayfly fauna. It was readily apparent, for instance, after collecting a number of localities in the hill country, that those streams which periodically carried a silt load from rainfall runoff were distinguished by a dearth of mayflies. This was especially common in many of the streams which ran through tea plantations. There is no ground cover among the tea bushes and the red soil readily runs off into nearby streams after heavy rains. Similar streams close by which did not receive the silty runoff had a healthy mayfly fauna. This same phenomenon also was evident in

Fig. 6. A slow-moving lowland river.

Fig. 7. The Kelani Ganga at about 60 m altitude.

263

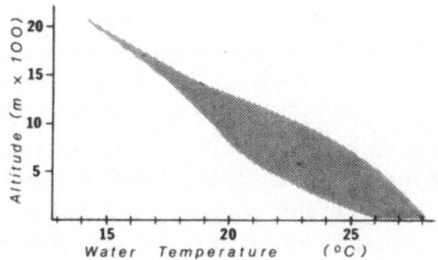

Fig. 8. Representative water temperatures in Sri Lankan streams (data from Costa & Stamühlner 1972 and Hubbard unpublished).

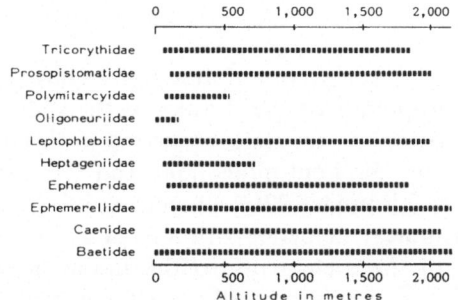

Fig. 9. Altitudinal distribution of the families of Ephemeroptera in Sri Lanka.

streams at lower elevations. In several localities, where buffalo were watered in the stream, periodic heavy silt loads reduced the mayfly population greatly compared to nearby areas.

Edmunds & Edmunds (1980) indicate that the most common pattern for tropical mayflies is emergence just after dark and transformation to imagos before dawn. Swarming then takes place from before dawn through the early morning hours. Light trapping and observations by the senior author in Sri Lanka indicate that most of the mayfly subimagos emerge in the first hours of darkness. The only emergence of subimagos observed in the field took place at or shortly after dusk. Subimagos placed in rearing cages generally had transformed to imagos by the next morning. Early morning light trapping was generally quite unproductive; most of the mayflies have stopped coming to the light by about three hours after sunset, regardless of the altitude. Swarming (other than at light traps) was never observed in two months in the field. This may be because of remote swarming (either in height or distance) or because swarming took place only in darkness. Little is known of emergence patterns of the mayflies of the low country, although light trapping suggests that the mayflies of the wet lowlands follow the same pattern as those of the hill country.

There is no obvious pattern of seasonal development yet evident in the mayflies of Sri Lanka. Although most of the specimens we have examined were collected from November through April, this time period coincides with the major aquatic collecting effort in Sri Lanka. Examination of specimens collected at various times throughout the year reveal no obvious pattern of development of mayfly nymphs. We also were unable to discern any altitudinal, seasonal or temporal differences in development, although the sporadic collecting that has been done seems unlikely to reveal such patterns as lunar periodicity if they are present. It appears from what data we do have that the Ephemeroptera in Sri Lanka emerge, as a general rule, throughout the year in an aseasonal pattern.

Biogeography

As is apparent in this chapter, few taxonomic studies have been completed on any group of the Ephemeroptera from Sri Lanka or the Indian Peninsula. Detailed phylogenetic studies of Ephemeroptera occurring in this area of the world have been completed only for the Leptophlebiidae and these studies are only at the generic level (Peters & Edmunds 1970). Therefore, it is impossible at this time to discuss the biogeographical elements within Sri Lanka and the probable more recent relationships of these elements to the fauna of the Indian Peninsula. However, we can discuss in some detail the origins of the mayfly fauna of Sri Lanka at the generic level.

The fauna of any area, especially such a small area as Sri Lanka, can only be examined in the context of being part of a larger geographic distribution. The presence or absence of a genus or species in Sri Lanka, or how a species arrived there, depends on its place in a much larger pattern. One must examine the phylogeny of a group to determine the history of its dispersal. Consequently the biogeography of the mayflies of Sri Lanka can be no better than their phylogeny.

Mani (1974) discussed in detail the biogeography of the Indian Peninsula. He concluded that the biogeographical elements of the present day fauna of the Peninsula could be divided into two groups. One group is composed of the derivatives of older faunas which differentiated in a southern landmass (Gondwanaland) and the other group is composed of the derivatives of a relatively younger fauna, which differentiated mainly in Asia and comprises essentially the Tertiary mountain faunas. The mayfly fauna of Sri Lanka and the Indian Peninsula indicates similar biogeographical patterns.

Few references exist on the biogeography of mayfly genera occurring in Sri Lanka. The phylogeny and biogeography of various genera of the Leptophlebiidae are discussed by Peters & Edmunds (1970), Sivaramakrishnan & Peters (in press a, in press b) and Sivaramakrishnan (in press). Edmunds (1979) discussed the biogeographical relationships of the Oriental and Ethiopian mayflies. He concluded that faunal interchange between the two areas occurred by the northern movement of the Indian Peninsula at an early time, and by dispersal through the Middle East. Additional comments on biogeographical relationships of various families and genera are given by Edmunds (1972, 1975).

The following comments summarize our knowledge of the biogeography of the Sri Lankan mayflies.

A. Leptophlebiidae

1. *Kimminsula-Kimminsula* and several related, but undescribed, genera from Sri Lanka and southern India are true Gondwanian derivatives (Peters & Edmunds 1970). Phylogenetic studies indicate a common ancestor between the *Kimminsula* group and undescribed genera in Madagascar. The ancestor of these modern genera must have been present during the time that Madagascar and the Indian Peninsula were connected. At present, these genera are the only confirmed Gondwanian derivatives in Sri Lanka.
2. *Choroterpes-Choroterpes* appears to be northern in origin and *Choroterpes s.s.* gave rise to *C. (Euthraulus)* which is now widespread throughout the Oriental and Ethiopian Regions. Sri Lankan species of *C. (Euthraulus)* probably are closely related to those of the Indian Peninsula.
3. *Megaglena-Megaglena* is a member of the *Thraulus* lineage and presently is endemic to Sri Lanka.
4. *Isca-Isca* is a highly specialized member of the Atalophlebiinae and is known throughout the mainland of the Oriental Region and probably evolved within the region. The subgenus *I. (Tanycola)* is presently endemic to Sri Lanka.

B. Baetidae

1. *Baetis* -The Baetidae probably had their principal evolutionary development in Gondwanaland with South America as the prime centre of evolution, and Africa as a secondary evolutionary centre (Edmunds 1979). The exact dispersals of the baetid groupings between the Ethiopian and Oriental Regions are not known. The species of *Baetis* in Sri Lanka probably are closely related to other species in the Indian Peninsula.
2. *Cloeon* -The Sri Lankan species probably are related to other species in the Indian Peninsula.
3. *Indobaetis* -At present Indobaetis is endemic to Sri Lanka and appears to be most closely related to the *Baetis muticus* group in Europe (Müller-Liebenau & Morihara 1982). This is the first recorded influence of the Palearctic Region on the mayfly fauna in the Indian Peninsula and Sri Lanka.
4. *Indocloeon* -This genus is endemic to Sri Lanka and is thought to be most closely related to *Procloeon* and *Centroptilum*

5. *Procloeon*-The Sri Lankan species of *Procloeon* probably are closely related to other species in the Indian Peninsula.
6. *Pseudocloeon*-Several distinct phylogenetic lineages from various areas of the world have been assigned to *Pseudocloeon* (Müller-Liebenau 1981). Four species of true *Pseudocloeon* have now been described from Sri Lanka. This phyletic lineage is known only from the Oriental Region.

C. *Caenidae*

1. *Caenis*-The Caenidae probably have spread from the Oriental Region to much of the rest of the world (Edmunds 1979). *Caenis* is widespread in the Indian Peninsula and Sri Lanka, as in many areas of the world, and the Sri Lankan species of *Caenis* probably are closely related to those in the Peninsula.
2. *Clypeocaenis*-*Clypeocaenis* presently is thought to be endemic to the Indian Peninsula and Sri Lanka.

D. *Ephemerellidae*

1. *Teloganodes*-*Teloganodes* occurs throughout the Oriental Region and apparently evolved within the region. Its closest relatives occur in Africa (Edmunds 1972, 1975) and its ancestor probably spread to Asia via the Asia Minor land bridge.

E. *Ephemeridae*

1. *Ephemera*-*Ephemera s.s.* probably dispersed from the Oriental to the Ethiopian Region (Edmunds 1979). At least four species of *Ephemera s.s.* occur in and apparently are endemic to Sri Lanka (Hubbard in press a). These species probably are closely related to those living in the Indian Peninsula.

F. *Heptageniidae*

1. *Compsoneuriella*-*Compsoneuriella* is widespread in Africa, some areas of the Malagasy Republic, and southeast Asia. It is possible that *Compsoneuriella* is present in Sri Lanka and southeast Asia due to transport by the drift of the Indian Peninsula (Edmunds 1979).

G. *Oligoneuriidae*

1. *Chromarcys*-*Chromarcys*, known from Sumatra, China and Thailand, in addition to Sri Lanka, is closely related to the African and Madagascan *Elassoneuria*. They probably evolved in Africa from an *Isonychia*-like ancestor (Edmunds 1975, 1979). *Chromarcys* was carried to Asia with drifting India and then spread east from the Indian Peninsula on its union with Asia.

H. Polymilarcyidae

1. *Povilla-Povilla* evolved in Africa as a sister group of *Asthenopus* when South America separated from Africa. It then spread through the Middle East land bridge to Sri Lanka and the rest of Asia where the genus radiated into the present species (Hubbard unpublished).
2. *Ephoron-Ephoron* probably dispersed from the Oriental Region to the Ethiopian Region (Edmunds 1979). The species of *Ephoron* occurring in Sri Lanka is either *E. indicus*, which is widespread throughout the Oriental Region or, if it is not *E. indicus*, then this is indicative of a closely evolved species group.

I. Prosopistomatidae

1. *Prosopistoma-Prosopistoma* is fairly diverse in Africa and the Malagasy Republic, and occurs also from Sri Lanka to the Philippines and Australia with one species in Europe. The Asian species appear to be closely related and may have arisen from the speciation of the original ancestor that arrived in Asia from Gondwanaland via the Indian Peninsula (Edmunds 1975, 1979).

J. Tricorythidae

1. *Neurocaenis* -The Tricorythidae appear to have evolved primarily in Gondwanaland (Edmunds 1979). The ancestor of *Neurocaenis* probably spread northward via the Middle East (Edmunds 1972, 1975). *Neurocaenis* is widespread throughout the Oriental Region.

Systematic accounts

Baetidae

Six genera of Baetidae, *Baetis, Cloeon, Indobaetis, Indocloeon, Procloeon* and *Pseudocloeon* have been reported from Sri Lanka, and twelve Sri Lankan baetid species have been described up to this time. Hagen (1858) described *Cloe consuetus* and *Cloe solidus*, both now placed in *Baetis*. These two species, known only from the adults, are endemic to Sri Lanka. A third endemic species of *Baetis, B. feminalis*, also known only from the adults, was described by Eaton in 1885. *Cloeon marginale* was described from the female by Hagen in 1858 (as *Cloe marginalis*). This species, whose type locality is in Sri Lanka, has since been reported from as far east as Taiwan and the Philippines. *Procloeon bimaculatum*, described in *Cloeon* by Eaton from Sri Lankan adults in 1885, has since also been reported from as far east as Taiwan and the Philippines. Müller-Liebenau & Morihara (1982) described two species of the endemic genus *Indobaetis*, viz., *I.*

costata and *I. starmuehlneri* from nymphs. The monotypic endemic *Indocloeon primum* was described by Müller-Liebenau (1982a). Müller-Liebenau (1982b) described four species of true *Pseudocloeon* which are endemic to Sri Lanka (*P. ambiguum, P. difficilum, P. klapaleki* and *P. orientale*).

Examination of the specimens of Sri Lankan Baetidae available to us indicates the presence of at least a few more species in the genera *Baetis, Cloeon* and *Procloeon*, and of two distinct '*Pseudocloeon*-type' genera.

The Baetidae are common in every sized stream throughout Sri Lanka. They have been collected at altitudes of over 2,000 m in the Horton Plains, to almost sea level, from a known temperature range from about 14 °C to >27 °C. They occur in tanks and reservoirs as well as the lotic habitats of streams and rivers. They probably are the most ubiquitous family of Ephemeroptera on the island, also the most diverse, occurring in the hill country, the wet lowlands, and throughout the dry zone lowlands.

Caenidae

Caenis perpusilla was described from Sri Lanka by Walker in 1853. This species has since been reported from Bangladesh and India. In addition to this species, we have specimens of an undescribed species in our collection. The genus *Clypeocaenis* also is present in our collections from Sri Lanka.

The Caenidae have been collected from streams and rivers from over 2,000 m in the hill country down to about 60 m in the wet lowlands. They also have been found in streams and tanks in the dry lowlands almost to sea level.

Nymphs of the Caenidae have been collected in Sri Lanka in February, March, April, June, October and December, and adults are known from January, February, March, April, June and November.

Ephemerellidae

Only the subfamily Teloganodinae of the Ephemerellidae is known from Sri Lanka. Two species of the genus *Teloganodes, T. tristis* (originally described as *Cloe*) and *T. major*, have been described from the island. Both species were based only on females. *Teloganodes tristis* has since been reported from Java, Sumatra and the Philippines.

A luminous male imago, identified by the Rev. A.E. Eaton as *Teloganodes* was captured at Kitulgalle, at 1,500 ft. This specimen was said to be luminous, 'sufficiently to serve for its capture on a very dark night' (Ent. Soc. Lond. 1882). The senior author was unable to find any luminous mayflies in Sri Lanka, even though a special search was made.

We have seen specimens of *Teloganodes* from about 2,100 m in the Horton

Plains down to about 90 m. The water temperatures ranged from about 14 °C to about 26 °C. Always occurring on rock in swiftly flowing waters from large rivers to small streams, the nymphs of *Teloganodes* have been collected by the senior author even on vertical rock faces in waterfalls.

Nymphs of *Teloganodes* have been collected in Sri Lanka in February, March, April, September and December, and adults have been collected in March.

Ephemeridae

The genus *Ephemera* is the only genus of Ephemeridae found in Sri Lanka. There are at least four endemic species of *Ephemera* on the island, all apparently belonging to the subgenus *Ephemera s.s.* (Hubbard in press a). Two of these species, *E. supposita* and *E. postica* (originally described as a *Rhoenanthus*), previously were known. We also have specimens of at least two undescribed species. Reports of *E. supposita* from Taiwan were based on misidentifications.

Ephemera is found in flowing waters where sandy substrates predominate and stable, moderately cool water temperatures occur (16 °C to 25 °C). It has been collected from about 1,800 m in the hill country to about 60 m in the wet lowlands. Adults have been collected in February, March, April and October, and nymphs have been collected in February, March, April and November.

Heptageniidae

We have specimens of adults and nymphs of an undescribed species of *Compsoneuriella* from Sri Lanka. The Heptageniidae previously have been unreported from the island. *Compsoneuriella* has been collected in Sri Lanka from habitats ranging from medium sized streams to large rivers. The localities ranged from about 700 m altitude in the hill country to about 50 m in the wet zone and the northern edge of the hill country, temperature range about 20 °C to 25 °C. Both nymphs and adults have been collected in March.

Leptophlebiidae

The Leptophlebiidae probably are the most diverse family of mayflies in Sri Lanka, rivalled only by the Baetidae. The genera *Choroterpes, Isca* and *Kimminsula*, along with several related but undescribed genera, are known from Sri Lanka. All the known Sri Lankan species are endemic and belong to the subfamily Atalophlebiinae.

There appear to be several species of *Choroterpes* in Sri Lanka. All belong to the subgenus *Euthraulus*. Only one species, *Choroterpes (Euthraulus) signata* (originally described in *Cloe*), has yet been described.

Choroterpes has been collected from streams at altitudes over 1,800 m in the hill country to about 60 m in the wet lowlands, temperature range of about 14 °C to 26 °C. *Choroterpes* also has been taken from streams in the dry lowlands of the Eastern Province, and from a marsh in the Northwestern Province at an altitude of 2-5 m.

The adults of *Choroterpes* are known from January, February, March, April, June and September, and the nymphs have been collected in February, March, April, June, September, October, November and December.

Megaglena brincki has been described from Sri Lanka. At least one additional species probably is present on the island. Nymphs are known from February, March, November and December, and adults have been collected in January, February, March and April. *Megaglena* is known from stream habitats with a

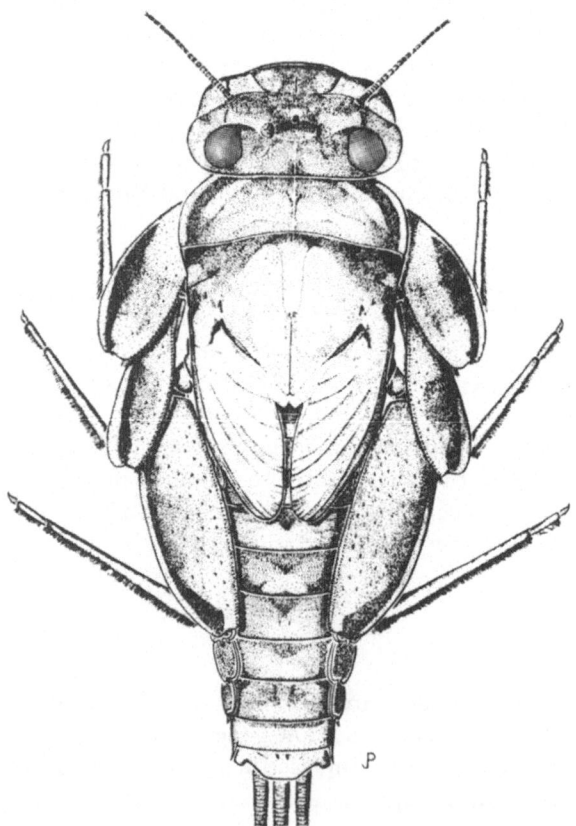

Fig. 10. Nymph of *Kimminsula* sp., a representative of the only confirmed Gondwanian derivatives (after Peters & Edmunds 1970).

temperature range of about 14 °C to about 26 °C, from about 2,000 m in the hill country, down to about 60 m in the wet lowlands.

Isca (Tanycola) serendiba has been described from Sri Lanka. Further study may reveal at least one more species of *Isca* on the island. *Isca* has been collected in streams from about 2,000 m altitude in hill country to 60 m in the wet lowlands, temperature range of about 14 °C to 26 °C. Adults have been collected also at a paddy field, although it is probable that they emerged from streams nearby and not from the paddy field itself.

The adults have been collected in January, February and March, and the nymphs are known from January, February, March, April, November and December.

Kimminsula (Fig. 10), and several related but undescribed genera, contain four described and several undescribed species from Sri Lanka. The species described originally as *Potamanthus femoralis, Potamanthus fasciatus, Potamanthus annulatus* and *Baetis taprobanes* were last placed in either *Atalophlebia* or *Kimminsula*, but definitely do not belong to the former genus, which does not occur in southern Asia. Some generic rearrangement in regard to the species currently placed in *Kimminsula* will also be forthcoming as our research progresses.

Species of the '*Kimminsula*-group' have been collected from small to medium-sized streams from an altitude of about 1,800 m in the hill country to about 60 m in the wet lowlands, temperature range about 16 °C to 26 °C. Adults are known from February, March, April, September, October and December, and nymphs have been collected in January, February, March, April, August, September, October, November and December.

Oligoneuriidae

The subfamily Chromarcyinae of the Oligoneuriidae is represented in Sri Lanka by an undescribed species of the genus *Chromarcys*. *Chromarcys* has been collected in both the wet lowlands and the dry lowlands in moderately sized streams below about 150 m in Sri Lanka, although Peters (unpublished) found it only in torrential mountain streams in Thailand. Adults have been collected in March, October, November and December, and nymphs have been collected in April.

Polymitarcyidae

The Polymitarcyidae are represented in Sri Lanka by one species of the asthenopodine genus *Povilla* and one species of the polymitarcyine genus *Ephoron*. *Povilla* are burrowers in mud, vegetable matter, and even wood. Nymphs of this undescribed endemic species of *Povilla* have been collected burrowing into teak lock gates and at 120 m altitude in the wet country from a medium sized stream

(Hubbard in press b). The nymphs have been collected in April and October.

The species of *Ephoron*, originally described as *Anagenesia greeni* by Banks (1914), has since been thought to be synonymous with *Ephoron indicus*. Further study is needed to confirm this synonymy. Adults have been collected from reservoirs, and large rivers to medium sized streams in March, April and October from 500 m altitude to 60 m in the wet zone, and from about 60 m in the northern dry zone.

Potamanthidae

Banks (1914) described *Rhoenanthus posticus* from Sri Lanka. However, examination of the types reveals that *R. posticus* is in fact and *Ephemera*, and that the Potamanthidae are not known to occur in Sri Lanka.

Prosopistomatidae

One species of *Prosopistoma*, the endemic *P. lieftincki*, has been described from Sri Lanka. The somewhat cryptic nymph of *Prosopistoma* is found on flat rock surfaces in small swiftly flowing streams in the hill country and wet lowlands. We have seen specimens collected from 2,000 m to 90 m, at temperatures of 14 °C to 26 °C. The nymphs have been collected in February through April and August through October.

Tricorythidae

The tricorythid species *Neurocaenis jacobsoni* (originally described in *Tricory-thus*) has been reported from Sri Lanka (Ulmer 1924). Further study of speciments may reveal two species of *Neurocaenis* in Sri Lanka. The assignment of Sri Lankan specimens to *N. jacobsoni*, whose type locality is Java, needs re-evaluation.

Neurocaenis has been collected in the hill country at about 1,800 m to 50 m in the wet lowlands. It has been collected from large rivers to moderately sized streams, in a temperature range of about 16-25 °C.

The adults often fly in great numbers at dusk, usually the earliest of the crepuscular flyers to arrive at light traps. Adults of *Neurocaenis* have been collected in Sri Lanka in January, February, March, April and November, and nymphs have been collected in September, October and December.

Acknowledgements

Many people have contributed to the research programs from which this chapter evolved. In particular we would like to thank Dr T. Gunawardane and Mr P.B. Karunaratne of the National Museum in Colombo, and Mr T. Wijesinhe and Miss I. Kotelawala who were of great help to the senior author in his field work in Sri Lanka. Prof. F. Starmühlner, Vienna, Prof. P. Brinck, Lund, and Dr G.F. Edmunds Jr., University of Utah, were instrumental in making available many specimens from Sri Lanka for our research. Field work by the senior author in Sri Lanka was in cooperation with and supported by the Smithsonian Institution's 'Biosystematic Studies of the Insects of Ceylon' project, K.V. Krombein, principal investigator. The writing of this chapter was supported by a research program (FLAX 79009) of CSRS, U.S. Department of Agriculture. Figure 10 reproduced by permission of Pacific Insects.

References

Banks, N. 1914. New neuropteroid insects, native and exotic. Proc. Acad. Nat. Sci. Phil. 66:608-618.

Costa, H.H. & F. Starmühlner. 1972. Results of the Austrian-Ceylonese Hydrobiological Mission 1970 of the 1st Zoological Institute of the University of Vienna (Austria) and the Department of Zoology of the Vidyalankara University of Ceylon, Kelaniya. Part I. Preliminary Report: Introduction and description of the stations. Bull. Fish. Res. Stn Sri Lanka (Ceylon) 23:43-76.

Eaton, A.E. 1871. A monograph on the Ephemeridae. Trans. Ent. Soc. Lond. 1871:1-164.

Eaton, A.E. 1883-1888. A revisional monograph of Recent Ephemeridae or mayflies. Trans. Linn. Soc. Lond. 2nd Ser. Zool. 3:1-352.

Edmunds, G.F. Jr. 1972. Biogeography and evolution of Ephemeroptera. Ann. Rev. Ent. 17:21-42.

Edmunds, G.F. Jr. 1975. Phylogenetic biogeography of mayflies. Ann. Miss. Bot. Gardens 62:251-263.

Edmunds, G.F. Jr. 1979. Biogeographical relationships of the Oriental and Ethiopian mayflies. pp. 11-14 In: K. Pasternak & R. Sowa (Eds.) Proc. 2nd. Int. Conf. Ephem. Państowowe Wydawnictwo Naukowe, Warzawa-Kraców.

Edmunds, G.F. Jr. & C.H. Edmunds. 1980. Predation, climate, and emergence and mating of mayflies. pp. 277-285 In. J.F. Flannagan & K.E. Marshall (Eds.) Advances in Ephemeroptera Biology. Plenum Press, N.Y.

Entomological Society of London. 1882. [Report of meeting]. Trans. Ent. Soc. Lond. 1882: xii.

Fernando, C.H. 1965. A guide to the freshwater fauna of Ceylon. Suppl. 2. Bull. Fish. Res. Stn Ceylon 17:177-211.

Fernando, C.H. 1980. The freshwater zooplankton of Sri Lanka, with a discussion of tropical freshwater zooplankton composition. Int. Rev. ges. Hydrobiol. 65:85-125.

Hagen, H.A. 1858. Synopsis der Neuroptera Ceylons. Verh. Zool. Bot. Ges. Wien 8:471-488.

Hagen, H.A. 1859. Synopsis der Neuroptera Ceylons.II. Verh. Zool. Bot. Gess. Wien 9:199-212.

Hagen, H.A. 1873. Notes on the Ephemeridae, by Dr H.A. Hagen; compiled (with remarks) by the Rev. A.E. Eaton, M.A. Trans. Ent. Soc. Lond. 1873:381-406.

Hubbard, M.D. in press a. Ephemeroptera of Sri Lanka:Ephemeridae. Syst. Entomol.

Hubbard, M.D. in press b. A revision of the genus *Povilla* (Ephemeroptera:Polymitarcyidae). Aq. Insects.

Hubbard, M.D. & Peters, W.L. 1978. A catalogue of the Ephemeroptera of the Indian Subregion. Orient. Insects, Suppl. 9:1-43.Mani, M.S. 1974. Biogeography of the Peninsula. pp. 614-647 In: M.S. Mali (Ed.) Ecology and biogeography in India. Monogr. Biol. 23. Dr W. Junk. The Hague.

Müller-Liebenau, I. 1981. Review of the original material of the baetid genera *Baetis* and *Pseudocloeon* from the Sunda Islands and the Philippines described by G. Ulmer, with some general remarks (Insecta: Ephemeroptera). Mitt. Hamb. Zool. Mus. 78:197-208.

Müller-Liebenau, I. 1982a. A new genus and species of Baetidae from Sri Lanka (Ceylon): *Indocloeon primum* gen.n., sp.n. (Insecta,Ephemeroptera). Aq. Insects 4:125-129.

Müller-Liebenau, I. 1982b. Five new species of *Pseudocloeon* Klapalek, 1905, (Fam. Baetidae) from the Oriental Region (Insecta, Ephemeroptera) with some general remarks on *Pseudocloeon*. Arch. Hydrobiol. 95:283-298.

Müller-Liebenau, I. & D.K. Morihara. 1982. *Indobaetis*: a new genus of Baetidae from Sri Lanka (Insecta:Ephemeroptera) with two new species. Gewäss. Abwäss. 68/69:26-34.

Peters, W.L. 1967. New species of *Prosopistoma* from the Oriental Region (Prosopistomatoidea:-Ephemeroptera). Tijdschr. Entomol. 110:207-222.

Peters, W.L. & G.F. Edmunds Jr. 1970. Revision of the generic classification of the Eastern Hemisphere Leptophlebiidae (Ephemeroptera). Pac. Insects 12:157-240.

Pictet, F.J. 1843-1845. Histoire naturelle générale et particuliere des Insectes Névroptères. Famille des Ephémérines. J. Kessman & Ab. Cherbuliez, Geneva. 300 pp.

Sivaramakrishnan, K.G. in press. A new genus and two new species of Leptophlebiidae from southern India (Ephemeroptera).

Sivaramakrishnan, K.G. & W.L. Peters. in press a. *Notophlebia*: one new species and reassignment of the nymph of *Nathanella* (Ephemeroptera:Leptophlebiidae).

Sivaramakrishnan, K.G. & W.L. Peters. in press b. A new genus and species of Atalophlebiinae (Ephemeroptera: Leptophlebiidae) from southern India.

Ulmer, G. 1913. Note V. Ephemeriden aus Java, gesammelt von Edw. Jacobson. Notes Leyden Mus. 35:102-120.

Ulmer, G. 1924. Ephemeropteren von den Sunda-Inseln und den Philippinen. Treubia 6:28-91.

Walker, F. 1853. Ephermeridae. pp. 533-585 In: List of the specimens of neuropterous insects in the collection of the British Museum. Part III. (Termitidae-Ephemeridae).

13. Notes on Simuliidae (Diptera) in Sri Lanka

D. M. Davies

Introduction

The Simuliidae or black flies of Sri Lanka have received little attention, although there are records in the British Museum (Natural History) (BMNH) and in the literature (Brunetti 1912) of adult simuliids of both sexes being collected at Peradeniya in 1911. A few additional adults were collected from 1915-1929, some of which (14) are in the National Museum of Sri Lanka (Colombo), others in the BMNH, and two in the Berlin Museum (Enderlein 1921).

In 1962, immature simuliids were taken from streams and adults from light traps by the Swedish Lund University Ceylon Expedition (Brinck *et al.* 1970), and in 1970 the University of Vienna (Austria) and the Vidyalankara University of Ceylon gathered immature simuliids during a hydrological study in the country (Costa and Starmühlner 1972). The last two collections have been kindly loaned to me. Costa and Fernando (1967) found simuliids in a stream on the Peradeniya Campus of the University, but none in the stomachs of other aquatic insects or fish.

In 1975, I spent three months (July-October) in Sri Lanka, for one month helped by Professor Hilary Crusz, Peradeniya Campus of the University and later under the auspices of the Smithsonian Institute, as part of the Ceylon Insect Survey. During this time, while travelling over 4,800 km by minibus and jeep in the southern two-thirds of the country, I collected immature black flies at 199 stream sites. From some of these sites (Fig. 1) a total of 1,500 adults were reared. In addition, 16 collections of adults were made, two from humans and most others by a black-light (UV) trap at night. A few other simuliids collected by others during the Ceylon Insect Survey, as well as specimens from the BMNH, also were made available to me.

In scattered locations several streams that appeared to be good simuliid breeding sites were devoid of larvae, pupae and exuviae. No simuliids were found in streams and rivers sampled in September around Anuradhapura, around Kala Oya (near Wilpattu Nature Reserve) and south from there to near Kurunegala, and around Moratuwa (27 km south of Colombo) where it is flat and streams flow slowly.

276

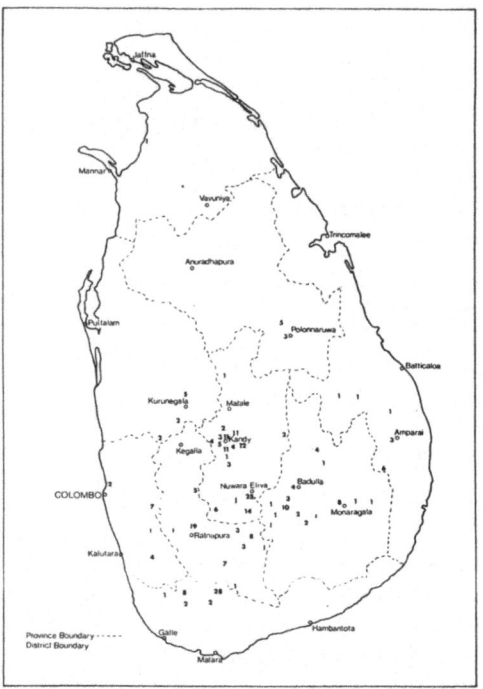

Fig. 1.   Map of Sri Lanka showing sites where Simuliidae of any stage were collected. The number at each point refers to the number of collections made around that area. Additional stream sites were sampled but yielded no simuliids (see text), and these have not been designated on the map.

Several streams may lack aquatic insects because they flow intermittently from irrigation dams, or because of marked irregularities in precipitation, especially in the north.

In many parts of the world female black flies severely attack humans, domestic and feral mammals and birds for their blood (Crosskey 1973). Bloodsucking simuliids of several species are known to transmit *Onchocerca volvulus,* which causes serious human disease in Africa, North Yemen, Saudi Arabia, and Central and South America (Crosskey 1973, Raybould & White 1979). This disease is not found in Sri Lanka (Chou *et al.* 1954). Lewis (1972, 1973, 1974) discusses black-fly attacks on humans in Pakistan, India and Nepal. He found these largely due to *Simulium indicum.* He refers to this species as one of the few in the area that bite man. It is a serious pest, but there has been no report of human onchocerciasis in these countries. He mentions other simuliid species attacking humans and domestic animals in India and Pakistan, but none of these species has been reported in Sri Lanka.

Sri Lankan Simuliidae

My preliminary study of the Sri Lankan Simuliidae indicates the presence of the following:

Simulium (Eusimulium) aureohirtum
Simulium (Gomphostilbia) ceylonicum
Simulium (Gomphostilbia) pattoni
Simulium (Gomphostilbia) sp. I
Simulium (Gomphostilbia) sp. II
Simulium (Gomphostilbia) sp. III
Simulium (Morops) sp.
Simulium (Simulium) consimile
Simulium (Simulium) near *palmatum*
Simulium (Simulium) sp. a
Simulium (Simulium) sp. b
Simulium (Simulium) sp. c
Simulium (Byssodon) sp.

Chou *et al.* (1954) and Senadhira (1969) refer to the specimens collected in Sri Lanka as belonging to the following five species:

1. *S. atratum* is probably *S. (G.) pattoni*
2. *S. ceylonica (=* S. (G.) ceylonicum)
3. *S. grisescens* and
4. *S. striatum.* Specimens of these two species may be any one of *S. (S.) consimile, S. (S.)* nr. *palmatum or S. (S.) striatum*
5. *S. nilgiricum* Puri, 1932. None of our specimens appear to be this species, although specimens so labelled may prove to be either our *S. (S.)* sp. b or *S. (S.)* sp. c.

A paper on the taxonomy of Sri Lankan Simuliidae is in preparation.

Notes on species

S. (E.) aureohirtum occurs at altitudes of 160-1875 m mainly in the central and east-central parts of the country, from south and east of Kandy, to Belihul-Oya and around Ella, and from Gal Oya National Park to Amparai. However two streams, overflows from a causeway connected to Parakrama Samudra (tank) at Polonnaruwa, yielded moderate populations on 14 September 1975. Breeding occurs in small, often slow-flowing streams, 15-90 cm wide, with much emergent and trailing vegetation on which the larvae and pupae often are attached. This species usually was collected near outlets of a pool or tank, but sometimes from streams running through open, relatively flat areas. Such aquatic habitats may be subjected to high temperature, and immatures of this species occurred at water temperatures of 23-33 °C. Adults have been taken in relatively low numbers throughout the year.

The females may develop their first batch of eggs without bloodfeeding (autogeny), as has been shown elsewhere (Takaoka & Noda 1979). They probably feed on avian blood for further egg batches as judged by their toothed tarsal claws, thought to aid in crawling through feathers (Shewell 1955). This species occurs from Pakistan in the west to Java in the south, and to the Philippines and southern Japan in the east (Lewis 1973, Crosskey 1974, Takaoka & Noda 1979).

S. (G.) ceylonicum was described by Enderlein (1921) from two females collected by S. Nietner in Ceylon. No specific locality is mentioned by Enderlein nor on the labels of two female syntypes in the Berlin Museum, one of which was established as a lectotype by Crosskey (1967). From my study, this species frequents streams more often at higher altitudes than S. (G.) pattoni, being found on Horton Plains (2,000 m) and around Nuwara Eliya, Rangala, Ella and Deniyaya, as well as at Labugama, east of Colombo (150 m).

S. (G.) pattoni is the most common and widely distributed species in Sri Lanka. It extends east from Agalawatta (Kalutara District) to Morawaka (Matara District) and to Wellawaya (Monaragala District) to Amparai. It is found on the east side to Maha-Oya west to Rangala (near Knuckles), north to Polonnaruwa, and northwest across to 33 km northeast of Mannar and then south to Kurunegala and to Labugama. It is less common in the hill country but is found near Nuwara Eliya (1,800 m). In his original description of the species and on labels in the collection of BMNH, Senior-White (1922) reports five females taken from a bull's belly and scrotum at dusk at Coonoor, south India. The specimens from Sri Lanka previously were considered as S. (G.) atratum, although Edwards (1934, p. 122) suggested that they might be S. (G.) pattoni, differing from his earlier idea that they might be S. (G.) ceylonicum (Edwards 1927).

There are two other less common new species of subgenus Gomphostilbia, one of which, S. (G.) sp. I, occurs in the south from near Ratnapura and Deniyaya, east to Ella and to Monaragala. S. (G.) sp. II is found in the same areas, but only occasionally in the same streams, and in addition occurs east to Agalawatta and north to Peradeniya in the Mahaweli Ganga. S. (G.) sp. III was infrequent at Agalawatta and Deniyaya and a little north of Ratnapura. Another species, possibly in the subgenus Morops or close to it, is even less frequent, being mainly in the east from Mahiyangana to Maha-Oya, and from just south of Bibile to 32 km east of Monaragala at Siyambalanduwa. However, a female was taken from a black light trap at Ratnapura and another netted 1.5 km south of Hiniduma (Galle District).

S. (G.) consimile, a species close to S. (S.) striatum, was found more often in rivers of 5-16 m width although it occurred in an overflow (20-25 cm wide) from an irrigation canal near Mahiyangana. Its range extends from Ratnapura to Deniyaya and to Badulla and Perideniya at altitudes from 60 to 600 m. It was described originally from India. A few specimens that may be S. (-S.) striatum also were found in Sri Lanka but further study is needed to establish this with certainty.

S. (S.) cf. palmatum breeds more usually in rivers or larger streams of 2-15 m in width and of moderate flow, but it occurs also in small streams of 10-60 cm in width. It is found from Ratnapura east to Wellawaya and Badulla, to Nuwara Eliya and Peradeniya at altitudes from 120 to 1,320 m. A female bit me 0.5 km south of Peradeniya University at 5 p.m. on 26 August 1975. The true *palmatum* is found in south India.

Simulium (Simulium) sp. a is a large fly that breeds usually in moderate to swift rocky streams, often in cascades down rock faces in mountainous terrain at altitudes from 250 to 1,700 m. Because of its somewhat specialised habitat, it is never abundant. It is in streams of width from 0.1 to 10 m often with only *ca.* 1 cm depth but sometimes in deeper waters with a little submerged or trailing vegetation. On the west it occurs as far north as near Kurunegala, to Labugama (near Colombo), south to near Haycock (Galle District) and Deniyaya, north and east to near Ella and Bibile and in the region of Kandy near Peradeniya and around Knuckles. A female also was collected biting P. Brinck (unpublished notes) 5 km northeast of Belihul-Oya (Ratnapura District) at 900 m on March 1 1962, and from the author 0.5 km south of Peradeniya University at 5 p.m. on 26 August 1975.

S. (S.) sp. b is the second most abundant simuliid species and is widely distributed, being found from Kurunegala, to Peradeniya and Knuckles, to Nuwara Eliya and Hakgala, to Badulla and Ella and to Deniyaya-Rakwana and north to Ratnapura. It occurs at altitudes from 400-1,800 m in rapid trickles on rock faces and in streams from 10 cm to 10 m in width with rocky bottoms or with much trailing vegetation and fallen leaves. Several specimens of both sexes were taken in black light traps.

S. (S.) sp. c is a blackish species (possibly a complex) that breeds in streams of various currents, usually fast to moderate, and width from 0.3 to 6 m with some submerged and trailing vegetation. It occurs more often at 1,000 to over 2,000 m, but sometimes at lower altitudes in streams draining higher plateaus. It is most frequent in the region of Nuwara Eliya, east towards Badulla, south to Horton Plains and west to Hatton, but does occur around Knuckles (east of Kandy) and in the south from north of Balangoda (Maratenna) to Rakwana and Deniyaya (Enselwatta). One female fed on a human at 10 a.m. on 24 September 1975 on the higher ground near a cliff on the west side of Nuwara Eliya. The vegetable farmers there said they are often bitten by these flies. Two other females gorged with blood were taken in a black light trap at Nuwara Eliya. This may be the species that attacked Sandrasagara (1951) on Horton Plains in June, although Chou *et al.* (1954) said that I. M. Puri had identified the specimens as *S. nilgiricum.*

A rare species, probably in the subgenus *Byssodon* Enderlein, was found in a stream at Maha-Oya and in one 16 km east of there (Amparai District). This seems to be in the *griseicolle* Becker species-group that is found in Africa and the Middle East, as far as Iraq (Crosskey 1969).

Although black flies of three species have been reported biting humans in Sri Lanka, this is infrequent and mainly in the hill country. At times immature simuliids are abundant in streams and doubtless are important in the ecological food web.

280

Acknowledgements

Professor Hilary Crusz, University of Sri Lanka (Peradeniya) and Dr Thelma Gunawardhane, National Museum of Sri Lanka gave me invaluable advice and made arrangements for most of the collecting trips. Dr Crusz provided laboratory space for me at the University and transportation during the first month. Dr Karl V. Krombein, U.S. National Museum, Washington, made financial and other arrangements for me to participate in the Ceylon Insect Survey for two months with provision of a vehicle, driver and two collectors. Mrs Helen Gyorkos was of great assistance in the identification of specimens. Dale O'Quinn, Margaret Gadsby, Lorna MacKinnon and Catherine O'Neill were of technical help. The research was supported by a grant from the Natural Sciences and Engineering Research Council of Canada.

References

Brinck, P., H. Anderson & L. Cederholm. 1970. Introduction: Report I (pp. IV-XXXVI). Reports from the Lund University Ceylon Expedition in 1962. Vol. 1. Entomol. Scand. Suppl. pp. I-XXXVI, 1-292.

Brunetti, E. 1912. A new species of blood-sucking fly (Simulium) from Ceylon. Spol. Zeylan. 8:90-91.

Chou, C.Y., E.S. Thevasagayam & K. Tharu-marajah. 1954. Insects of public health importance in Ceylon. Rev. Equat. Entomol. Parasitol. 2(1/2):105-150.

Costa, H.H. & E.C.M. Fernando. 1967. The food and relationships of the common meso and macrofauna in the Maha Oya, a small mountainous stream at Peradeniya, Ceylon. Ceylon J. Sci. (Biol. Sci.) 7:74-90.

Costa, H.H. & F. Starmühlner. 1972. Part I. Preliminary report. Introduction and description of the stations. In: Results of the Austrian-Ceylonese Hydrobiological Mission 1970 of the 1st Zoological Institute of Vienna (Austria) and the Department of Zoology of the Vidyalankara University of Ceylon, Kelaniya. Bull. Fish. Res. Stn Sri Lanka (Ceylon) 23:43-76.

Crosskey, R.W. 1967. The classification of *Simulium* Latrielle (Diptera:Simuliidae) from Australia, New Guinea and the Western Pacific. J. nat. Hist. 1:23-51.

Crosskey, R.W. 1969. A re-classification of the Simuliidae (Diptera) of Africa and its islands. Bull. Brit. Mus. (Nat. Hist.). Entomol. Suppl. 14: 195 pp.

Crosskey, R.W. 1973. 3. Diptera-Nematocera. a. Simuliidae (Black-flies). pp. 109-153 In: Insects and other Arthropods of Medical Importance. K.G.V. Smith (Ed.). Brit. Mus. (Nat. Hist.), London.

Crosskey, R.W. 1974. Family Simuliidae. pp. 423-430 In: A Catalog of Diptera of the Oriental Region. Vol. 1, Suborder Nematocera. M.D. Delfinado & D.E. Hardy (Eds.).

Edwards, F.W. 1927. Some nematocerous Diptera from Ceylon. Spol. Zeylan. 14: 117-128. (Simuliidae pp. 120-121).

Edwards, F.W. 1934. Deutsche limnologische Sunda-Expedition. The Simuliidae (Diptera) of Java and Sumatra. Arch. Hydrobiol. Suppl.13:92-138.

Enderlein, G. 1921. Neue aussereuropaische Simuliiden. Sber. Ges. naturf. Freunde Berl., 4-5:77-81.

Lewis, D.J. 1972. *Simulium indicum* in Nepal (Diptera:Simuliidae). Senckenbergiana Biol., 53:387-390.

Lewis, D.J. 1973. The Simuliidae (Diptera) of Pakistan. Bull. entomol. Res., 62:453-470.

Lewis, D.J. 1974. Man-biting Simuliidae (Diptera) of northern India. Israel J. Entomol., 9:23-53.

Raybould, J.N. & G.B. White. 1979. The distribution, bionomics and control of onchocerciasis vectors (Diptera:Simuliidae) in eastern Africa and the Yemen. Tropenmed. Parasitol., 30:505-547.

Sandrasagara, T.R. 1951. A note on the blood-sucking *Simulium* Ceylon. J. Bombay nat. Hist. Soc., 50:421-422.

Senadhira, M.A.P. 1969. The parasites of Ceylon. V. Arthropoda. A host check list. Ceylon-vet. J., 17:3-25.

Senior-White, R. 1922. Notes on Indian Diptera. Mem. Dept. Agric. India (Entomol. Soc.) 7(9):83-169. (Simuliidae pp. 88, 126-131).

Shewell, G.E. 1955. Identity of the black fly that attacks ducklings and goslings in Canada (Diptera:-Simuliidae). Can. Entomol., 87:345-349.

Takaoka, H. & Noda, S. 1979. Autogeny of the black fly *Simulium (Eusimulium) aureohirtum* (Diptera:Simuliidae). J. Med. Entomol., 15:183-184.

14. Ecology of rocky shores and estuaries of Sri Lanka

K. D. Arudpragasam

Introduction

The island of Sri Lanka has a coastline of about 1,760 km, which is mostly low lying and geologically polygenetic. Beyond the coastline the continental shelf extends for a distance which varies from 9–45 km at an average depth of 36 fathoms. The shelf is narrow around the southern part of the island but widens out to the north and merges with the Indian Continental Shelf (Cooray 1967).

The coastline along the southwestern and northeastern parts of the island is characterised by a succession of sandy bays, each protected on either side by rocky headlands. These are transverse coastlines running against the strike of the rocks (e.g. Bentota, Beruwela, Hikkaduwa). The southwest coast also has complex systems of 'lakes' and lagoons which are drowned valley systems (e.g. Ratgama, Bolgoda, Koggala).

The northwestern and southeastern coastlines, by contrast, are emerging coastlines. Barrier beaches and sand spits enclose a succession of lagoons, 'lakes', swamps and tidal flats, or divert river mouths northwards or southwards. Sand dunes are common along much of the south and east coast and on the west coast north of of Chilaw. Dunes reach their highest development south of Point Pedro, in the north. Dunes are oriented strongly in a southwest-northeast direction, in the direction of monsoonal winds.

Except on the northern coastline, the mouths of many rivers indent the coastline and form estuaries.

Tides are of the mixed variety (Fig. 1). Semidiurnal tides are dominant during and around the spring tides but a gradual transition takes place to diurnal tides during and around neap tides. The highest tides of the year are around 0.88 m and the range is about 0.70 m at its maximum. The heights of the two daily tides differ from each other. In considering the ecology of rocky shores, it is important to have information about times of low tides in relation to daytime temperatures. The highest daytime temperatures are usually experienced around 2.00 pm. In Colombo and Galle harbours the lower of the low waters occurs each day in the mornings

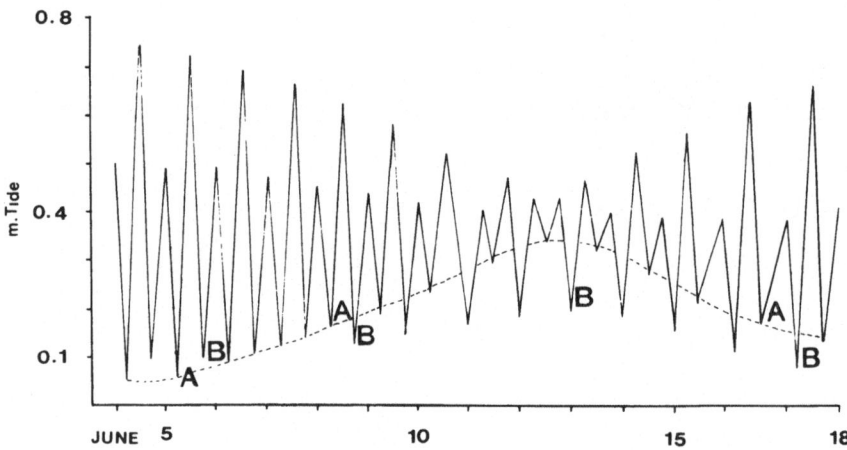

Fig. 1. Tidal movements predicted for Colombo Harbour for the period June 4-18, 1967.

between 7.00 and 11.00 am from January to September and between 7.30 and 9.00 pm from October to December. Therefore, at no period of the year are shores on the west coast fully exposed during the hottest period of the year. At Trincomalee Harbour on the East coast, the low water occurs after midnight from February to August and after midday from September to January. There is a probability of east coast shores being exposed to extreme daytime temperatures during the latter part of the year.

The phase of the tide shows marked differences between east and west coasts. It is roughly high water along the west coast between Manaar Island and Galle when it is low water on the east coast between Trincomalee and Point Pedro. Between Galle and Trincomalee the time of high water lags behind as one moves around the south and east coasts towards Trincomalee.

The major ocean currents south of Sri Lanka show a predominantly westerly trend during the northeast monsoon season and a predominantly easterly trend during the southwest monsoon season.

In the offshore areas, currents are influenced more by wind and other seasonal effects than by tide induced effects. The general pattern of currents on the west coast is southerly during the northeast monsoon and northerly during the southwest monsoon. Tidal currents become significant only in and around lagoons and estuaries, filling and draining these water bodies. Around the mouths of rivers, river currents influence the coastal current systems.

Rocky shores

Distribution

From the description of the Sri Lanka coastline given above, it is evident that a majority of Sri Lankan shores are sandy. The distribution of rocks around the island is shown in Fig. 2. Rocky shores are made up of granitic gneiss, sandstone or coral debris or a combination of these.

From Colombo southwards to Ambalangoda, small patches of granitic rocks are seen separated by stretches of sand. Between Ambalangoda and Tangalle, rocky patches are more numerous, while between Tangalle and Hambantota, stretches of sand predominate. At Hambantota, rocks show a black and white banding of gneiss with acid bands of quartz and microcline with orthoclase (Coates, 1935). Deposits of coral occur between Ambalangoda and Matara. A fringing reef exists a short distance from the shore over most of this section of coast. This coastline is exposed to the full force of the Southwest monsoon and is subjected to erosion.

Proceeding northwards from Colombo, a prominent sandstone reef fringes the coastline to the headland off Negombo, where it runs out to sea. Sandstone appears again south of Chilaw and along the inner edge of the spit flanking the Chilaw Lagoon. These sandstone reefs are built up of quartz and shell fragments

Fig. 2. Distribution of littoral rocks in Sri Lanka. (Important locations mentioned in the text are marked.)

compressed and bound together. On their seaward faces, massed tubes of sabella-riid worms, themselves made up of sand and shell fragments cemented together, are often seen. North of Puttalam there is a coastal band of limestone which is exposed at intervals (e.g. Kalpitiya). This band increases in width northwards, where it forms the whole of the Jaffna Peninsula.

The northern coastline of the island borders the relatively calm waters of the Palk Strait. The limestone bed of the peninsula is exposed along the shoreline, forming a low cliff at Keerimalai. The limestone itself is a Miocene formation and it is possible to collect fossil molluscs and crabs along the shore around Keerimalai. A fringing reef runs a short distance from the shore over most of this coastline.

On the eastern coast, except for the region around Trincomalee and occasional patches of rock at Kalkudah and Kirinde, sand extends from Point Pedro to Hambantota. The only prominent cliffs along the island's coastline are those at Trincomalee, where a configuration of bay, headland and cliffs exist.

Zonation

Zonation on rocky shores in Sri Lanka has been described by Atapattu (1968) and Arudpragasam (1970). Both authors have based the recognition of zones on indicator organisms rather than on tidal levels. The use of indicator organisms in this manner has been much discussed (Stephenson & Stephenson 1952, Guiler 1953, Lewis 1962, 1964, Southward 1958); the consensus seems to favour the use of indicator organisms. The case is clearly stated by Lewis (1962), who points out that, while tidal and other changes in the sea level are primarily responsible for the appearance of zones, the complete physical environment is a balance of many additional factors. Since it is not possible to measure the entire physical environment, the definition of zones according to one factor, the tides, is inappropriate. He emphasises that there is no alternative but to use the dominant zone forming organisms.

Atapattu (1968), in the course of a study of the molluscan fauna of littoral rocks around the island, observed the patterns of zonation on 25 shores along the west, southwest and south coasts, 7 shores on the east coast, and 10 shores on the north coast. She has divided the shores of Sri Lanka into three groups in terms of zonation, indicator organisms and molluscan fauna. These are:
Group I
  shores of the west, southwest and south and exposed shores of the east;
Group II
  sheltered shores of the east coast and
Group III
  shores of the north coast.

Table 1 summarises the zones and indicator organisms with respect to these three groups. The original scheme proposed by Atapattu (1968) is based on the universal zones as proposed by Stephenson & Stephenson (1949). The scheme shown in Table 1 is modified from that of Atapattu (1968) to fit the scheme of zonation as proposed by Lewis (1964).

The zones recognised are the littoral fringe, the eulittoral and the sublittoral. On Group I shores the eulittoral tends to be demarcated into three subzones; an upper, a mid and a lower eulittoral zone with a more or less clearly defined belt of oysters. The littoral fringe of Group II shores is characterised by the littorine *Littorina novaezealandiae* and by neritids. Group III shores also show neritids in the littoral fringe and a scarcity of oysters and barnacles in the eulittoral.

Group I shores (Fig. 3)

Three species of littorines are found in the littoral fringe: *Littorina undulata, Nodilittorina pyramidalis* and *N. granularis. L. undulata*-occupies the upper part of the zone and indicates its upper limit. *N. pyramidalis* lives lower down while *N. granularis* extends well into the eulittoral. Where the rock is granitic, the background colour of the rock is black due to growths of Myxophyceae.

The eulittoral shows subzonation. The lower eulittoral is a zone recognisable by the presence of discrete patches of weed growths, including-species of *Laurencia, Gracilaria, Caulerpa, Chaetomorpha, Gelidium, Ulva,* and *Jania.* The fauna includes the serpulid *Potatoleios crosslandi, Cellana radiata* and other molluscs

Table 1. Shore types, zones and indicator organisms.

	Group I	Group II	Group III
Littoral	*Littorina undulata* Littorines	*Littorina novaezealandia* Littorines & neritids	*Littorina undulata* Neritids
Eulittoral			
Upper	Barnacles **Barnacles**		*Planaxis & Cerithium*
Mid	Oysters **Oyster Belt**		Planaxids & Cerithiids
Lower	Patchy weeds **Limpets**		
Sublittoral	Dense weeds	No indicators	Zooanthid and Weed beds

288

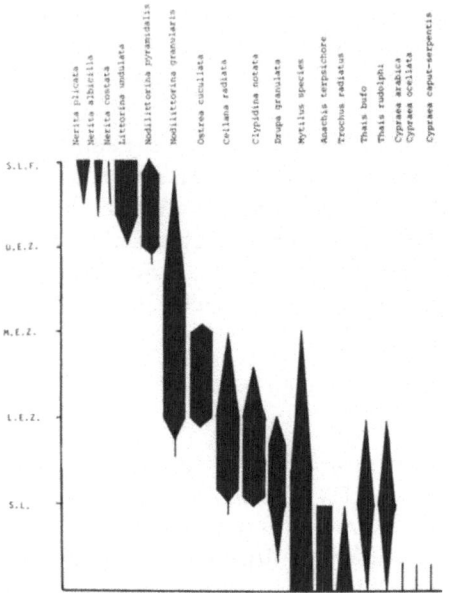

Fig. 3. Zonation of molluscs on a Group I rocky shore.

including *Anachis terpsichore, Trochus radiatus, Thais alveolata* and *Drupa granulata. N. granularis* begins to appear in the upper part of this subzone.

The mid-eulittoral zone is usually characterised by a belt of oysters, *Crassostrea cucullata.* The upper limit of the oysters marks the upper edge of this subzone. Associated with the oysters are *Cellana radiata* and *N. granularis.* Barnacles are also seen here. Below the oyster belt limpets and the fissurellid *Clypidina notata* are seen and *N. granularis* reaches its greatest density on the shore. On some shores patches of *Perna* sp., and *Brachidontes variabilis* are found in this region.

The upper eulittoral is a subzone of barnacles and littorines, mainly *N. granularis.* A few specimens of *N. pyramidalis* also may be encountered here. Its upper limit is the limit of barnacles occurring in quantity.

The upper limit of the sublittoral is marked by the upper limit of dense weed growths, covering more or less the entire rock surface. The composition of the weeds is variable according to location, but includes *Sargassum, Gracilaria, Padina, Laurencia, Gelidium, Jania,* and *Turbinaria.* Molluscs include *Trochus radiatus, Anachis terpsichore,* and *Thais alveolata.* Other animals seen include crabs, amphipods, isopods and polychaetes. These are also present in the lower eulittoral but in much smaller numbers.

It should be understood that the above is the broad pattern recognisable on Group I shores. Obviously there will be considerable variation on individual shores in terms of detail. Arudpragasam (1970), in comparing two shores, one of sandstone and the other of granitic rock, with differences in slope and exposure, has pointed out major differences. Sandstone appears not to favour the settlement of many barnacles and encourages littorines. Therefore, a clear belt of barnacles above the oysters may not develop on such shores. Moreover, on the more or less horizontal sandstone platform, the oysters tend to scatter to a greater degree than on a sloping rock surface. Further, the lower eulittoral and especially the upper sublittoral on sandstone tend to be populated by sea urchins which form individual burrows on the surface.

Group II shores

These include very sheltered shores of the east coast, of which two were examined in Trincomalee bay by Atapattu (1968). These shores are exposed to very little wave action.

One feature of these shores is the scarcity of growths of the larger weed types, even below the water line. Rock surfaces are covered with growths of *Stigonera* and *Lyngbya*. In the littoral fringe, the creamy white littorine, *L. novaezelandiae* is prominent and indicates the limits of the zone. Also found in the zone are *L. undulata* and the neritids *Nerita chameleon, N. abicilla, N. undata, N. polita* and *N. plicata*. Small numbers of *Nodilittorina pyramidalis* also are found.

The eulittoral shows no subzonation. Oysters, *Crassostrea cucullata* are present but do not form a belt. Other molluscan species seen here include *N. polita, N. undata, N. albicilla, Arca plicata* and *Planaxis sulcatus*.

Below the water line the main molluscan species are *Thais rudolphi, T. bituber-culata, Anachis terpsichore, Trochus radiatus, Cypraea arabica* and *C. moneta*.

Group III shores

These are shores of limestone. At most locations they form a narrow fringe along the shoreline and are not extensive enough for the full development of zones. At a few locations, the rocks along the shoreline extend almost to the reef flat of the reef that lies a short distance from the shore along the whole of this coast. The pattern of zonation seen along the shore also applies in large measure to this reef flat and in fact the zones are more completely laid down here.

The upper limit of the littoral fringe is marked by *L. undulata*. Small numbers of *N. pyramidalis* also are seen. However the dominant molluscs of the fringe are the neritids. There are four species: *Nerita chameleon, N. plicata, N. albicilla* and *N. undata*. The lower limit of the littoral fringe is marked by the upper limit of the planaxids and cerithiids of the eulittoral.

The eulittoral is characterised by the presence of large numbers of *Planaxis sulcatus, Cerithium morus* and *C. clypeomorus. Nerita undata* also is present in fair numbers. Oysters, *Crassostrea cucullata* are present but are always scattered and barnacles are scarce. Lower down in the eulittoral, *Drupa margariticola, D. granulata, Thais tissoti, T. bituberculatus* may be found. On some shores large populations of echiuroids are seen sheltering among the many crevices in the limestone. These crevices also provide shelter for numerous annelids, crustaceans and flatworms.

On the few shores where the shoreward rocks continue below the waterline at low tide characteristic beds of zoanthids are seen with small growths of *Codium* and *Padina*. On the reef flat the general pattern is as described above but there is usually no littoral fringe type zone and the beds of zoanthids and *Codium, Padina* and *Sargassam* are prominent.

Estuaries

The island has a network of rivers that rise in the central highlands. Many of the larger rivers flow down to the west, southwest and southern coasts and estuarine conditions exist in their lower reaches. The Kelani ganga, with its mouth imme- diately north of Colombo Harbour, and the Kalu ganga a few miles south of Colombo are two of the larger rivers draining on the west coast.

Though the Department of Irrigation of the Government of Sri Lanka has conducted studies on many of the outfall areas on waves, tides, water levels, littoral drift and sand bar formation, there is very little published information. Jayasinghe (1979), Arudpragasam & Jayasinghe (1980) and Jayasinghe & Arudpragasam (1980a, b) have studied some aspects of the hydrobiology of the Moratuwa Pana- dura estuary. This is a small estuarine system located a few miles south of Colombo. It runs parallel to the coastline from its mouth, northwards, for about 8 km. It then turns inland to join the north lake of the Bolgoda lake system. The North Lake is connected to the south lake by means of the Bolgoda river. The principal streams draining into the system are the Maha Oya into the north lake and the Panape ela into the south lake. The estuary and lake system extends about 30 km from the mouth.

Although the Panadura Moratuwa estuary is a very small system, in the absence of other published information, the broad features of this system can be used to get an idea of estuarine conditions on the west coast of the island.

Salinity structure and variation

Arudpragasam & Jayasinghe (1980) have shown that in the Moratuwa Panadura estuary a clear pattern of distribution and seasonal change exists which is deter- mined mainly by rainfall and wind patterns. The following sequence of changes has been described considering two stations, one at the mouth and the other 8 km upstream.

Fig. 4 shows changes in surface and bottom salinity recorded over a period of one year. Rainfall and wind patterns also are indicated.

During the period January to early April, which includes the end of the northeast monsoon and the succeeding intermonsoonal period, salinities at the mouth fluctuate around 30% in the water column. During the same period salinity levels increase at the surface and bottom upstream. The highest yearly surface salinities are recorded in April. In April heavy showers and fresh water input reduce the surface salinity near the mouth to around 15% and bottom salinities to around 20%. At the upstream station the change is more drastic and the water column has a salinity around 5%, with the bottom water being slightly more saline. During the period of May to September, which includes the Southwest monsoon period and the succeeding inter monsoonal period, surface salinities are low both at the mouth (around 10%) and upstream (around 5%). However, at both the mouth and upstream, this is a period of increasing bottom salinity. The highest levels of bottom salinity are reached during this period, by late September. Surface salinities are also rising at this time. This pattern is brought about by the fresh water input of the monsoon which keeps the surface salinities depressed, while the predominantly westerly winds push sea water up the estuary in the bottom layers. The pattern changes drastically once more in late October and November, when heavy fresh water input from the intermonsoonal rains, combined with changing wind patterns result in the entire water column becoming practically fresh, both near the mouth as well as upstream. Depressed values persist through December and then begin to rise.

The appearance of this sequence of changes is obviously dependent on the success and timing of monsoonal and intermonsoonal rains and winds. However, as these changes observed in the Moratuwa Panadura estuary are related to factors that have the same broad influences on the west and southwest coasts of the island, it is reasonable to assume that similar patterns of change will occur in other estuaries along these coasts. Of course, in the larger estuaries where the volume of fresh water is much larger, substantial differences are expected. Patterns similar to the one reported are known to occur in the Negombo lagoon, north of Colombo.

Studies on salinity distribution over the entire Bolgoda system in early Bolgoda system in early March, with salinity at the mouth around 30% showed surface salinity around 10% 10 km upstream and around 5% 28 km upstream in the south lake. In the southwest monsoon months salinities at the bottom could be higher, as explained above.

Studies during a tidal cycle have shown that differences between high water and low water values 8 km upstream in the Moratuwa Panadura estuary are in the range of 1-2%, while near the mouth, they are in the range of 10-15%.

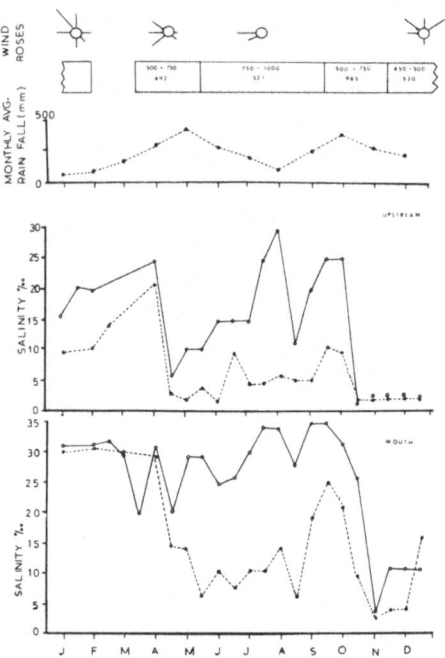

Fig. 4. Salinity, rainfall and wind direction in the Moratuwa Panadura estuary over a twelve month period. (Broken lines ± surface, solid lines ± bottom.)

Zooplankton of the Moratuwa Panadura estuary

Jayasinghe & Arudpragasam (1980) have described the zooplankton of this estuary. Fig. 5 shows values for the average salinity, total numbers of zooplankton and displacement volume over a 12 month period. The highest density of zooplankton was recorded in October and the lowest density in November and December. A sharp increase in density was recorded in the post southwest monsoon period, while densities were only moderate in the post northeast monsoon period. Calanoid and cyclopoid copepods were present in the zooplankton throughout the year, as were nauplii. Hydroid medusae and chaetognaths were seen in July and August, months of high bottom salinity. Rotifers appeared during the low salinity months of October to December and in April. Cladocera were present from September through May and ostracods were present from October to February. Calanoid copepods dominated the plankton except during the months of August, September, November and March, during which times they were displaced by nauplii.

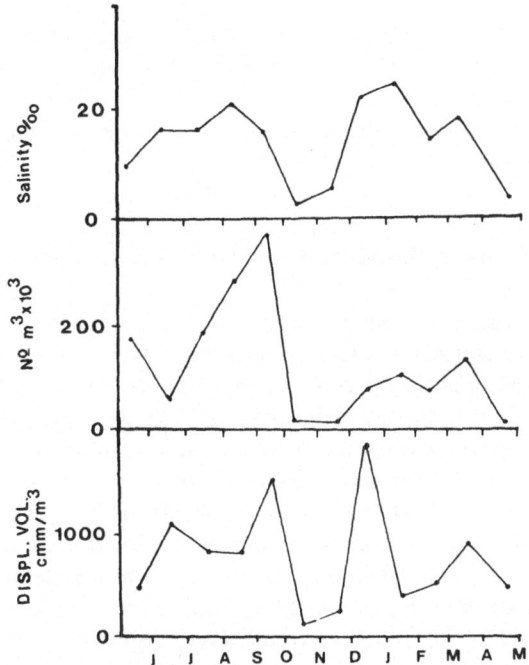

Fig. 5. Average salinity, total zooplankton numbers and zooplankton volume in the Moratuwa Panadura estuary over a twelve month period.

Bottom sediments and benthos in the Moratuwa Panadura estuary

The bottom sediments in the 8 km upstream from the mouth were determined according to the triangular plot of Lane and Hill (1975). In the first kilometre upstream, the dominant textural type was silty, clay-like sand. Over the next three kilometres, clay-like sand predominated and sandy silt-like clay over the succeeding two kilometres. In the six to eight kilometre region the sediment was mainly sandy, clay-like silt.

Thirty-two species of macrobenthos were collected from 4 locations over the lower eight kilometres of the estuary. The largest number of species were found near the mouth (25 spp.) and the smallest number furthest upstream (16 spp.). The largest numbers of benthic animals were found near the mouth and the smallest number furthest upstream. At the upstream station numbers were highest during the low salinity months of October, November, April and May. At the station nearest to the mouth, the largest numbers of animals were collected during the period of the southwest monsoon and the period immediately after, as well as during the early months of the year, these being high salinity months.

The benthos belonged mainly to three major taxa. These were the Mollusca, Crustacea and Polychaeta. At the station furthest upstream polychaetes formed the dominant taxon followed by Crustacea. At the mouth Mollusca formed the dominant group followed by Crustacea. At intermediate stations the Crustacea were the most numerous.

Prawn and fish fauna of the Moratuwa Panadura estuary

Bruin (1971) has studies the penaeid prawns of this estuary using samples from the stake nets called **Ja-kotu**. The species found were *Penaeus indicus, P. monodon, P. latisulcatus, Metapenaeus dobsoni, M. monoceros, M. affinis* and *M. elegans*. In terms of overall abundance the order was *M. dobsoni* followed by *M. elegans* and *P. indicus. P. monodon* was found in very small numbers. With regard to distribution in the estuary, *M. elegans* was collected only from the station furthest upstream. *P. latisulcatus, M. monocercos* and *M. affinis* were collected only near the mouth. *P. indicus* and *M. dobsoni* occurred -from the mouth to beyond 8 km upstream but they were most numerous about 6 km upstream.

The fish fauna of the estuary was also studied from **Ja-kotu** samples by Arudpragasma & De Silva (unpublished data). Over a period of one year, 90 species of fish were recorded at four stations in the lower 8 km of the estuary. Many of the species were found only in small numbers. By far the most numerous species was *Ambassis dayi*. Other species that occurred in fair numbers were *Siganus javus, Harengula ovalis, Anchoviella commersoni* and *Leiognathus equulus*. Species of fish that were present in the estuary throughout the year were *A. dayi*, But is but is, *A. commersoni, Pertica filamentosa, Glossogobius giuris, L. equulus, Secutor insidiator, S. ruconius, Monodactylus argenteus, S. javus, Chelenodon fluviatilis* and *C. patoca*. Certain species of fish were collected only near the mouth. These were *Archamia lineolatus, Tylosurus strongylurus, Pseudorhombus javanicus, Thrissocles malabarica, Suggmundus* sp., *Johnius diacanthus* and *Arothron reticularis*. All these species occurred in very small numbers. Two species were collected only at the station that was furthest upstream. These were *Apogon endekataemia* and *Oplopomus oplopomus*.

Fishing activity in estuaries

The type of fishing activity in the Moratuwa Panadura estuary described by Jayasinghe (1979) is typical of other shallow estuaries. Stake nets known locally as **Ja-kotu** are common. These are made up of systems of barriers and traps. The barriers are made of tats and supported by stakes driven into the river bed. They are placed across the estuary extending from the banks with an opening in the deepest part for the passage of boats. Traps which can be raised or lowered fit into openings in the barrier. Fish are diverted by the barriers into the traps which are lowered into position at night. The traps are often provided with oil lamps suspended near the water surface in order to attract fish. In the Moratuwa Panadura estuary there are numerous **Ja-kotus** extending far upstream. The prawn catch is valuable and the fish catch is sold as bait for sea fisheries or consumed.

Another common type of fishing activity in estuaries and lagoons is the brush pile fishery. Bunches of dead twigs and branches are located in shallow parts of the estuary and left undisturbed for a period of weeks. During this period the pile attracts populations of prawns, crabs and small fish. For actual fishing the pile is surrounded by an appropriate mesh net and the brush pile removed.

Associated with the **Ja-kotus** is another fishing device; the **Atoli dela** or barrier net. Two corners of the net are fastened to two stakes and the free corners are lowered into the water to a depth of about one metre. Once fish move into the submerged portion of the net, the net is raised. Other fishing methods in use in estuaries include cast nets, drag nets and hand lines.

Sand bar formation

Rivers bring down large quantities of sand which is deposited at or near the river mouth. This sand is transported by littoral drift northward during the southwest monsoon season and southward during the northeast monsoon period. When drainage from the river is low sand accumulates at the mouth and sand bars are formed, to be washed away again at times of high river outflow.

The Moratuwa Panadura estuary was subject to blockage by sand bars and a groyne was constructed to remedy this. However, the mouth is still partially blocked at certain times of the year. Studies on this sand bar have been carried out by the Department of Irrigation, although the results remain unpublished. The sand bar reaches its greatest development in February, which is also the driest month of the year on the west coast. It is at a minimum in October and November when river flow is at its maximum. The size of the sand bar declines through the southwest monsoon period.

The sand carried down to the sea by the major rivers and transported by littoral drift is an important factor in the building up of shores. It has been estimated that the Kelani river emptying north of Colombo Harbour brings down about 600,000 m^3 annually. This sand is transported to beaches northward beyond Negombo. Similarly, the sand brought down by the Kalu ganga and the Moratuwa Panadura oya nourish the beaches of Panadura, Moratuwa and Mt. Lavinia. However, this sand is a convenient source for building activity, a fact that has encouraged sand mining from near river mouths. Of the estimated total bedload of about 600,000 m^3 transported from the Kelani river about 400,000 m^3 is removed annually by sand mining. As a result, the coastal area northward from Handala to

Negombo is subject to erosion, affecting beaches with tourist potential. Similarly, sand mining in the Panadura area has led to erosion affecting the Moratuwa area.

Pollution

There is little published information on the impact of human and industrial activity on estuaries in Sri Lanka. A considerable part of the sewage from the city of Colombo is emptied into the Kelani estuary, but its impact is not known. Studies have been made on the polluting effect of a paper mill located on the east coast at Valaichenai by the Ceylon Institute for Scientific and Industrial Research and by the Ministry of Fisheries. In this case the effluent of about 3 million gallons day^{-1} is discharged into the Pothuveli Aru, a wet season stream. The effluent reaches the Valaichenai Lagoon and brings with it lignin and lignins sulphates and floating clumps of pulp. Studies have shown high BOD and COD levels and depressed oxygen concentration in the effluent. These is an accumulation of sludge on the bottom of the lagoon in adjacent areas, which is detrimental to the bottom fauna.

References

Arudpragasam, K.D. 1970. Zonation on two shores on the west coast of Ceylon. J. mar. biol. Assn. India. 12:1-14.

Arudpragasam, K.D. & Jayasinghe, J.M.P.K. 1980. Salinity distribution in the Moratuwa-Panadura estuary and the Bolgoda system. Proc. Sri Lanka Assn. Adv. Sci. 36 Part I, 41.

Atapattu, D.H. 1968. Studies on molluscan fauna of rocks of Ceylon. Ph.D. thesis, Univ. of Ceylon, Colombo.

Coates, J.S. 1935. The geology of Ceylon. Ceylon J. Sci. (B) 19:101-187.

Cooray, P.G. 1967. An introduction to the geology of Ceylon. Nat. Mus. Ceylon. Publ. 324 pp.

De Bruin, G.H.P. 1971. Fluctuation in species composition of penaeid prawns in estuaries. Bull. Fish. Res. Stn Ceylon 22:47-51.

Guiler, E.R. 1953. Intertidal classification in Tasmania. J. Ecol. 41:381-384.

Jayasinghe, J.M.P.K. & K.D. Arudpragasam. 1980a. A preliminary study of zooplankton of the Moratuwa-Panadura estuary. Proc. Sri Lanka Assn. Adv. Sci. 36 Part I, 41.

Jayasinghe, J.M.P.K. & K.D. Arudpragasam. 1980b. Observations on the bottom sedi-ments and benthos of the Moratuwa-Panadura estuary. Proc. Sri Lanka Assn. Adv. Sci. 36 Part I, 42.

Lane, E.D. & E.M. Hills. 1975. The marine resources of Anaheina Bay Calif. Dept. Fish. Game. Bull. 165:17-175.

Lewis, J.R. 1962. The littoral zone on rocky shores - a. Biological or physical entity? Oikos 12:280-301.

Lewis, J.R. 1964. The ecology of rocky shores. The English Univ. Press, London. Southward, A.J. 1958. The zonation of plants and animals of rocky sea-shores. Biol. Rev. 33:137-177.

Stephenson, T.A. & A. Stephenson. 1949. The universal features of zonation between tide marks on rocky coasts. J. Ecol. 37:289-305.

Stephenson, T.A. & A. Stephenson. 1952. Life between tide marks in North America. II. Northern Florida and Carolinas. J. Ecol. 40:1-49.

15. Coastal lagoons

S.S. De Silva & C.D. De Silva

Introduction

Lagoons, estuaries and backwaters of the tropics are among the most productive ecosystms (Qasim 1969, Odum 1970), possibly second only to coral reefs. Barnes (1974) suggested that estuarine food webs are characterized by a comparatively smaller number of energy pathways and therefore a higher energy flow per pathway. Also, there is a large energy input in the form of detritus. These are the most important contributory factors to the high productivity of these ecosystems. In tropical lagoons a major proportion of the animal species useful to man; shellfish and finfish species, obtain their food energy through direct grazing or from detritus. As such, lagoons and estuaries provide an opportunity for manipulating the yields through aquacultural practices. These yields could easily supersede those from other aquatic ecosystems. Apart from their high productivity, lagoons and estuaries also provide a habitat for the juvenile stages of many shellfish and finfish. They act as nursery grounds and even as a 'fattening ground' for some important catadromous fish species, such as the milkfish (*Chanos chanos*) and the grey mullet (*Mugil cephalus*).

Being an island, the importance of lagoons and estuaries as a natural resource in Sri Lanka, biologically, economically and culturally, is not second to other aquatic environments. It has a coastline of approximately 1,760 km and a shelf area of about 31,000 km^2 and 121,460 ha (surface area) of lagoons, estuaries and backwaters of which one quarter are lagoons. In this paper, a lagoon is defined as an area of relatively shallow water situated in a coastal environment and having access to the sea, but separated from the open sea by a barrier. Lagoons are formed by a succession of stages, ranging from the formation of barrier beaches through barrier splits, these actions tending to enclose lagoons and tidal flats.

The major lagoons of the island are those of Jaffna, Puttalam, Negombo, Batticaloa and Trincomalee and are named accordingly. Apart from these, a number of smaller lagoons are found on the eastern and western coasts. Each lagoon is characterized by its own topography and they vary in area, hydrography

298

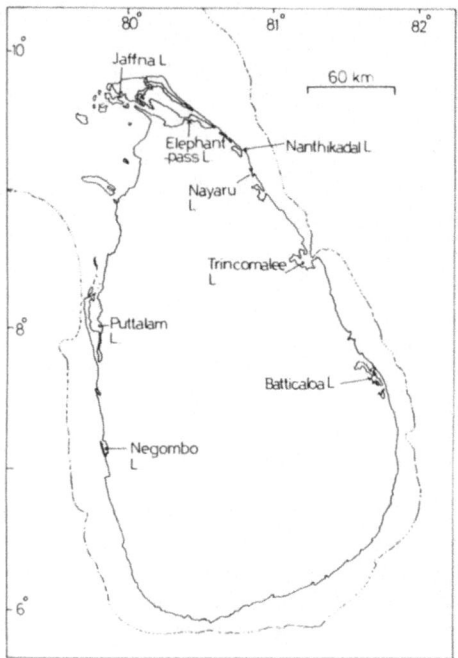

Fig. 1. The location of major lagoons of Sri Lanka, and the position of the continental slope around the Island.

and hydrobiology. The location of major lagoons in Sri Lanka is shown in Fig. 1. The Jaffna lagoon differs from all other major lagoons in that it receives no river discharge, whilst Trincomalee lagoon is very deep in contrast to the others, The topography of the Jaffna and Negombo lagoons have been described by Sachithananthan & Perera (1970) and Silva & De Silva (1981) respectively. In addition, Perera & Sachithananthan (1977) studied the topography of Nathikadal and Nayuru lagoons, two small lagoons on the east coast.

All lagoons in Sri Lanka support profitable shellfish, and at least subsistence finfish fisheries. Human settlements have developed around all the lagoons and these fisheries provide part of their livelihood. Apart from direct usage, lagoons also are used for transportation and anchorage of boats, and, unfortunately, for sewage disposal.

Hydrography

The earliest hydrographical study was on Tamblegam 'lake', an inshore bay of the Trincomalee lagoon, on the east coast of the island (Pearson 1910-1911). Since then hydrographical studies, primarily confined to seasonal physical and chemical variation of the surface waters, have been carried out in the Jaffna (Sachithanan-than 1969, Arudpragasam 1974), the Puttalam (Durairatnam 1963) and the Negombo (Silva & De Silva 1982) lagoons.

The tidal fluctuation around the coast is small, about 1 m during the spring tides. It is apparent that the hydrography of the lagoons is determined primarily by:
(a) The effects of climatic changes brought about by two monsoons; (1) on the west coast, the southwest monsoon and (2) on the east coast, the northeast monsoon, together with their associated intermonsoonal changes. Sewell (1929) has described temporal and spatial climatic changes prevalent in the Indian Ocean.
(b) The nature and extent of the connection with the sea. This is variable and is exemplified in Fig. 2, where details of the three major lagoons are shown.

Salinity and temperature

Salinity variation in time and space in lagoons is the single most important factor determining the diversity of the flora and fauna. As a result of the low tidal amplitude around the island, the salinity fluctuations in the lagoons tend to be

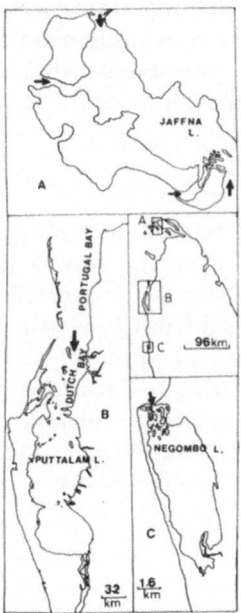

Fig. 2.   Detailed maps of the Jaffna (A), Puttalam (B) and Negombo (C) lagoons and their locations relative to the rest of the Island.

primarily influenced by the rains. Temperature, though important in determining the floral and faunal diversity in sub-tropical lagoons (Allen & Horn 1975) is of little importance in the tropics.

Jaffna lagoon

Sachithananthan (1969) found that maximum salinity and temperature of the surface water occurs from August to July. Peak temperatures were recorded in April. The lowest salinities of 28% were recorded in the November to January period, whilst the maximum of 37% was recorded in June. Monsoon (northeast) rainfall, exchange of water between the lagoons and the currents in Palk Bay were thought to influence the salinity pattern of surface waters in Jaffna lagoon (Sachithananthan 1969).

Using the overall salinity distribution, Arudpragasam (1974) recognized two sections in the lagoon; the northwestern part with a higher salinity and the southeastern part with a lower salinity and little direct movement between the two parts except during the northeast monsoon. The pattern of water flow in and out of the lagoon was thought to be influenced by tides and monsoon winds and shows a close relationship with changing wind systems. The relationship of rainfall, salinity and temperature at one point in the lagoon is shown in Fig. 3.

A considerable diurnal variation in salinity also was observed in the Jaffna lagoon (Arudpragasam 1974). Salinity differences within the lagoon and between the lagoon and the sea, and the prevailing water flow pattern in and out of the lagoon were thought to be responsible for diurnal variations. In general, such variations were smallest near the sea mouth when the flow was unidirectional.

Puttalam lagoon

The mean monthly surface temperature of the lagoon varied from 27.6 to 30.8 °C, with peaks occurring in April, May and June whilst the lowest temperature was recorded in November, December and January (Durairatnam 1963). Durairatnam (1963) found that salinity ranged from 29.9‰ in May to 36.4‰ in October and from 20‰ in November to 22.8‰ in April, i.e. higher salinity during the southwest monsoon and lower salinity during the northeast monsoon (Fig. 3). Apart from high evaporation during the southwest monsoon, waters with higher salinity are brought into the lagoon from the central Indian Ocean and the southern part of the Arabian Sea during this period (Sewell 1937).

Negombo lagoon

Peak surface and bottom temperatures were recorded in March and April (Samarkoon & Raphael 1972, Pinto & Wignaraja 1980, Silva & De Silva 1982). Silva & De Silva (1982) did not find thermal stratification but reported higher bottom

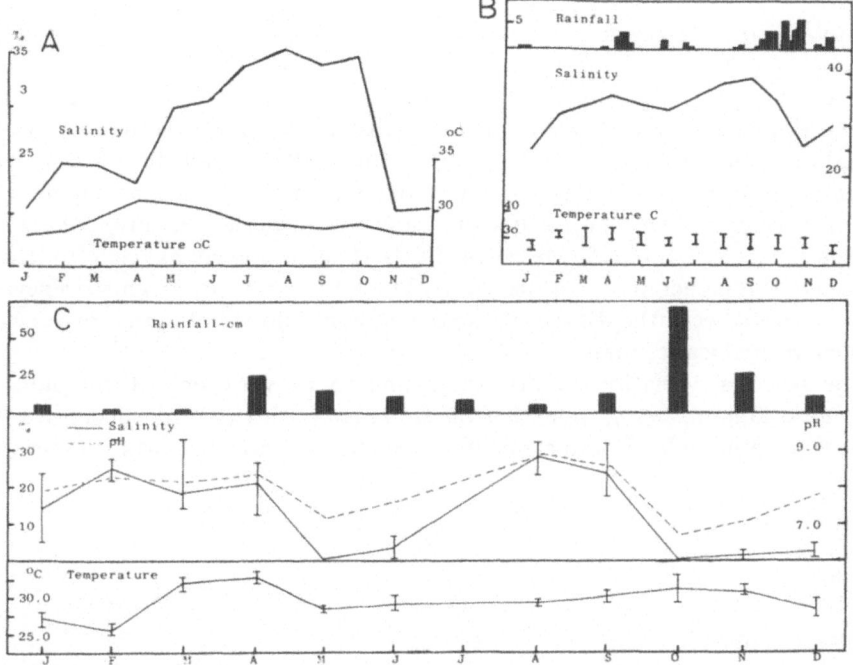

Fig. 3.   Seasonal variation in the surface temperature, salinity and pH and rainfall (where applicable) in three lagoons (vertical lines indicate the range: A ± Puttalam lagoon; B ± Jaffna lagoon; C ± Negombo lagoon, adapted from Durairatnum (1963), Arudpragasm (1974) and Silva & De Silva (1982) respectively).

temperatures for most months of the year in most places. The surface temperatures decreased from May to September and again from November to January (Fig. 3) and this was correlated with frequent winds (Pinto & Wignaraja 1980). Diurnal variations in temperature of up to 5 °C have been recorded.

A typical annual salinity curve, with two peaks separated by troughs during the rainy season, is recognizable (Fig. 3). Also, the lagoon is characterized by the fact that most of it turns into a freshwater 'lake' for a short period of time (De Silva & Silva 1980, Silva & De Silva 1982). It was reported also that the mean salinity is higher during the southwest monsoon than during the northeast monsoon period. This is because the highly saline waters are brought into the lagoon from the Indian Ocean and the southern part of the Arabian Sea by the predominant north-south currents (Sewell 1929).

Diurnal salinity changes are less pronounced with increasing distance from the lagoon mouth (Silva and De Silva 1982).

pH and dissolved oxygen

Apart from the seasonal and diurnal variation in salinity and temperature, physico-chemical characteristics bearing on the hydrography of the lagoon ecosystems have been poorly documented. Silva & De Silva (1981) studied the seasonal and diurnal variation of pH and dissolved oxygen in the Negombo lagoon. The dissolved oxygen in surface and bottom water was variable, both in space and time except in March (Silva & De Silva 1982). The dissolved oxygen content, as expected, was higher in the surface layers. These authors also recognized a negative correlation between the dissolved oxygen content and salinity, the former being influenced by local rainfall.

The seasonal variation in pH was found to be very small throughout the Negombo lagoon, being highest near the sea mouth (9.3). pH decreased with increasing input of fresh water. Farthest from the sea mouth, in May, it was about 6.5.

Flora

Phytoplankton

Seventy-five species were recorded from the Puttalam lagoon by Durairatnam (1963). He noted that more than one phytoplankton maximum per year occurred during the southwest monsoon period; the species dominance varied from month to month. For example, blooms of *Thalassiosira subtilis*, *Rhizosolenia alata* and *R. imbricata* occurred in November, February and March-May respectively. Diatom peaks were observed in June and October, each peak consisting of several species.

De Silva & Wijeyaratne (1977) reported 18 genera of diatoms, 8 genera of green algae, 6 genera of blue-greens and 1 genus of Xanthophyceae in the gut contents of young grey mullet, *Mugil cephalus* from the Negombo lagoon. Even though the food habits of this species were followed throughout the year, the authors did not estimate relative seasonal abundance of the phytoplankton and they provide only a seasonal record of the genera.

Seasonal variation in the phytoplankton abundance in the Negombo lagoon was investigated by Rajapaksa (unpublished data). She found that the biomass ranged from 0.35 to 23.0 g m^{-3} wet weight, or 0.228 to 14.95 g m^{-3} dry weight (Tamiya 1975), suggesting a high productivity. There also was an apparent correlation of the cyclic changes in biomass with salinity; biomass being higher at lower salinities (Fig. 4). Edwards (1978) found also that production was curtailed during the dry season in Mexican lagoons, when the salinities were high.

Those genera whose contribution to the phytoplankton biomass exceeded 5% of the total in any one month are given in Table 1. A total of 18 genera are listed. The

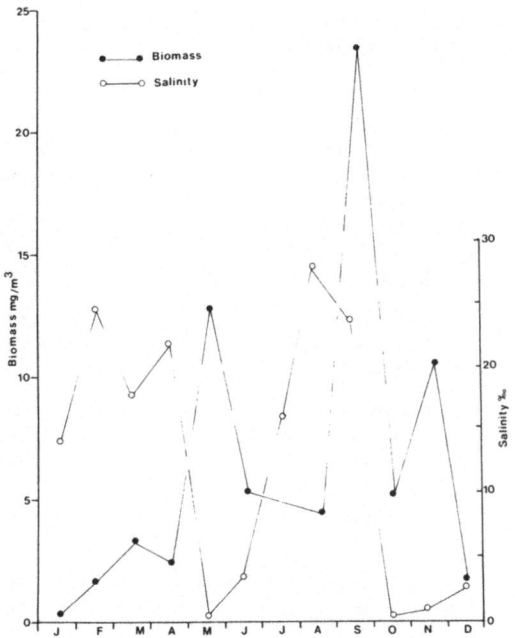

Fig. 4. Seasonal changes in the phytoplankton biomass and salinity in the Negombo lagoon.

seasonal percentage contribution to the total by different taxa (Fig. 5) indicates that in most months green algae are dominant, whilst diatoms become relatively important during high salinity periods. Dinoflagellates and blue-green algae are less important contributors to the total biomass and do not show an apparent correlation with salinity changes.

Macroflora

Gracillaria lichenoides has been reported in abundance in calm, undisturbed areas protected from the open sea in Puttalam lagoon (Sivapalan 1975). Pinto & Wignaraja (1980), in the course of their study on the edible oyster in the mangrove inlets of the Negombo lagoon, observed that the vegetation was dominated by *Avicennia marina* whilst *Rhizophora mucronata* and *R. apiculata* formed a continuous border on the islets. Apart from the above, *Halophile ovata* and *H. beccarri*, small leaved grasses and the taller sea grasses *Halodule wrightii* and *H. uninervis* have been described from the habitats of the indigenous cichlid fishes (*Etroplus* spp.) in the Negombo lagoon (Ward & Wyman 1977). Sedentary vegeta-

Table 1.   Seasonal abundance of the most common species that contributed 5% of more to the total phytoplankton biomass in Negombo lagoon.

Month	Phytoplanktonic species	Mean % of total
December '76	Pediastrum	32.18
	Cyclotella	30.90
	Pinnularia	7.20
	Spirogyra	7.18
	Cymalopleura	5.30
	Stauroneis	5.30
January	Cyclotella	53.14
	Gomphonema	24.57
	Ceratium	17.43
March	Cyclotella	24.00
	Tabellaria	20.70
	Spirogyra	19.53
	Stauroneis	13.12
	Navicula	10.58
	Cymbella	8.00
April	Cyclotella	66.15
	Ankistrodesmus	8.53
May	Microspora	50.84
	Cyclotella	9.73
	Micrasterias	5.32
	Mougetia	5.19
June	Microspora	35.29
	Campylodiscus	25.25
	Micrasterias	9.13
	Cyclotella	5.74
July		
August	Spirotaenia	41.30
	Campylodiscus	35.90
	Merismopedia	10.66
September	Spirotaenia	41.30
	Microspora	8.16
	Campylodiscus	5.00
October	Pediastrum	37.19
	Ulothrix	17.56
	Pandorina	12.64
	Ankistrodesmus	12.00
	Micrasterias	5.47
November	Pediastrum	80.53
	Navicula	7.53

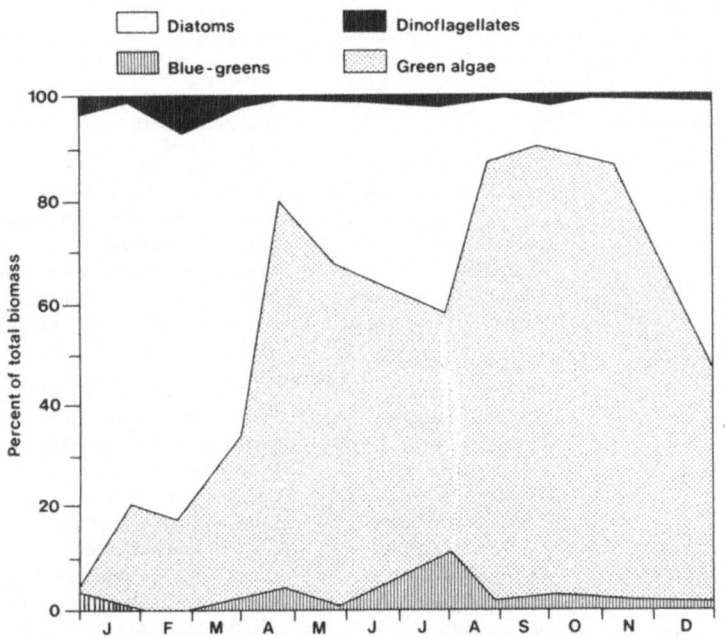

Fig. 5. Seasonal changes in the percentage contribution of the major floral taxa to the total phyto-planktonic biomass in the Negombo lagoon.

tion of Nanthikadal and Nayaru lagoons, two small lagoons on the east coast with surface areas of 60.45 km^{-2} and 6.3 km^{-2} respectively, were mapped by Perera & Sachithananthan (1977). Weeds of the genera *Naias, Ruppia* and *Chara* were recorded from Nathikadal lagoon whilst the mangrove shrub *Ceriops* sp. was recorded from the area above mean sea level in Nayura lagoon. The mangrove flora associated with lagoons and brackish waters of the Island has been dealt with by Arulchelvam (1968).

Fauna

In spite of the wide fluctuations of environmental parameters, faunistically speaking, lagoons provide a source or a bank, continually importing and exporting to both the ocean and the inflowing rivers. Barnes (1974) believed that the above factors were responsible for lagoons and estuaries retaining their low order of maturity unlike other environments, which pass from a low to a high order of maturity with time. Not surprisingly, faunistic studies of lagoons and estuaries all over the world have been biased towards a vertebrate component (see Dahlbert & Odum 1970, Haedrich & Haedrich 1974, Oviatt & Nixon 1973, Allen & Horn 1975).

Table 2. Invertebrate fauna recorded from lagoons in Sri Lanka.

Group	Species	Habitat
Porifera/Monoaxonid	not known	Associated with oyster populations[8]
Annelida/Polychaeta	*Hydroides inornata*	On oyster shells[6]
	Neopomatus uschakovi	Attached to shells, stones and upper surfaces of petiole bases of fallen branches[6 & 8]
	N. similis	Mangrove swamps, stones, bricks etc.[6]
	Neanthes manatensis	Associated with oyster populations[8]
	Trypanosyllis zebra	Associated with oyster populations[8]
Mollusca/Bivalvia	*Anadara clathrate*	—
	Larkinia rhombea	Soft substrata: shallow and deep water[5]
	Barbatia obliquata	Firmly attached to rocks[5]
	Pinctata radiata	Firmly attached to rocks[5]
	Pinna bicolor	Soft mud[1 & 5]
	Placuna placenta	Soft mud[1 & 5]
	Crassostra belcheri	Inter and sub-tidal zones[5]
	Gaffrarium tumidum	Littoral and sub-littoral[5]
	Dosinia cretacea	Burrowing, shallow water[5]
	Marcia opima	Burrowing, undisturbed shallow water[5]
	M. hiantina	Muddy bottom[5]
	Gari variegata	Burrowing vertically in sandy mud[5]
	Brachydontes variabilis	Associated with oyster populations[8]
	Spondylus descalis	Associated with oyster populations[8]
	Martesia striata	Associated with oyster populations[8]
Mollusca/Polyplacophora	*Squamopleura imitator*	Associated with oyster populations[8]
Mollusca/Gastropoda	*Euchelus asper*	Associated with oyster populations[8]
	Cellana radiata	Associated with oyster populations[8]
	Nerita polita	Associated with oyster populations[8]
	Littorina scabra	Associated with oyster populations[8]
	Cliothon onaleniensis	Soft mud[4]
	Cirithidea cingulata	Soft mud[4]
	Corbula solidula	Soft mud[4]
	Stenothyra sp.	Soft mud[4]
	Cuspiolaria sp.	Soft mud[4]
	Modiolus sp.	Soft mud[4]
Arthropoda/Crustacea	*Scylla serrata*	High saline lagoons[7]
	Neptunus pelagicusi	High saline lagoons[7]
	Penaeus canaliculatus	High saline lagoons[7]
	P. monodon	High saline lagoons[7]
	P. semisulcatus	High saline lagoons[7]
	P. indicus	High saline lagoons[7]
	Metapenaeus dobsoni	High saline lagoons[7 & 3]
	M. elegans	High saline lagoons[3]
	M. burkenroadi	High saline lagoons[3]
	M. mutatus	High saline lagoons[3]
	M. monoceros	High saline lagoons[3]
	Macrobrachium rosenbergi	High saline lagoons[2]
	M. equidens	High saline lagoons[2]
	M. srilankense	High saline lagoons[2]

Table 2 (continued)

Group	Species	Habitat
Arthropoda/ Brachyura	*Baruna socialis*	Associated with oyster populations[8]
	Pyseidognathus deianira	Associated with oyster populations[8]
	Metapograpsus messor	Associated with oyster populations[8]
Arthropoda/ Macrura	*Alpheus edwardsii*	Associated with oyster populations[8]
Arthropoda/ Cirripedia	*Balanus amphitrite*	Associated with oyster populations[8]
Arthropoda/ Isopoda	*Ligia exotica*	Littoral[8]
	Cirolana willeyi	Associated with oyster populations[8]
Arthropoda/ Amphipoda	*Credocus* sp.	Associated with oyster populations[8]
	Lembos sp.	Associated with oyster populations[8]

(1. Anonymous 1958, 2. Costa 1980, 3. De Bruin 1965, 4. de Silva, S. S. pers. obs., 5. Fernando 1977, 6. Pillai 1960, 7. Pillai 1965, 8. Pinto & Wignaraja 1980).

Invertebrates

Invertebrate species recorded from the lagoons of Sri Lanka are listed in Table 2. The list is almost totally confined to annelids, arthropods and molluscs, and more specifically to polychaetes, crustaceans, gastropods and bivalves. This brings to the forefront the severe dearth of ecological studies of an important ecosystem in the island.

No account of the invertebrate fauna of coastal and brackish waters of Sri Lanka could afford to neglect the citation of the pearl oyster fishery, even though the pearl oyster does not typically inhabit lagoons. The pearl oyster fishery of the Gulf of Manaar was last revived in 1958 and has been reviewed by Sivalingam (1961). In contrast with the pearl oyster, the window-pane oyster, *Placuna placenta* is found in some of the lagoons (e.g. Tamblegam lake) and it yields a fairly regular fishery for which divers are employed (Anon. 1958). These bivalves yield smaller pearls than the pearl oyster and the abrupt changes in salinity brought about by precipitation seem to act as a stimulus for the breeding of the species. However, prevalence of low salinity in the lagoon for a prolonged period harms the population (Anon. 1958).

The study on the edible oyster population, *Crassostrea cucullata* in the Negombo lagoon (Pinto & Wignaraja 1980) remains one of the few detailed studies of the molluscan population in any of the lagoons. These workers found two peaks, related to the intermonsoonal periods. Breeding was favoured at higher salinities ($27°5\%_{00}$) and temperature ($32.1°1°C$). Distribution of oysters in the islets in the lagoon was thought to be related to the current speed which was responsible for accumulating a suitable substrate. Pinto & Wignaraja (1980) also attempted to categorize the fauna associated with the edible oyster populations (see also Table 2) into sessile, boring and free moving and evaluated the distribution of four species

308

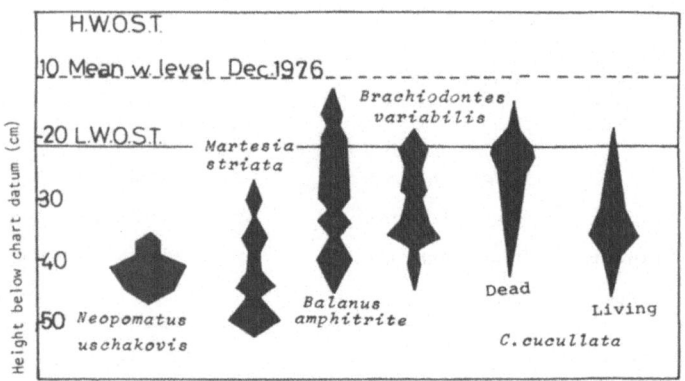

Fig. 6. Distribution of five sessile invertebrate species in relation to the water level in the Negombo lagoon (from Pinto & Wignaraja 1980).

of this fauna in relation to the water level (Fig. 6). Fernando (1977) described 32 species of bivalves belonging to 12 families from estuarine and coastal areas, Of these, 14 species belonging to 7 families were recorded from lagoons (Table 2).

The association of polychaetes, crustaceans and molluscs on the edible oyster, *Ostrea virginica* from the Batticaloa lagoon was documented by Perera & Arudpragasam (1966). They observed that chironomid larvae, isopods and two annelid worms inhabit the algal mat that covers the exposed parts of the shells of the oyster, whilst the barnacle, *Balanus amphitrite*, the serpulid *Neopomatus uschakovi* and a bivalve (*Mytilus* sp.) were found growing attached to the shell. As well, two species of the worm *Poldora*, a species of nereid and a clionid sponge were found to bore into the shell valves. The nature of the burrows, distribution of burrows and a quantitative estimate of the degree of infection also were dealt with (Perera & Arudpragasam 1966). Pillai (1960), in his study of marine and brackish water serpulids, recorded 6 species from brackish waters, including a new genus, *Neopomatus*. Of these, only four species were obtained from the lagoons (Table 2) and these included two new species of the new genus *Neopomatus*. The rest were recorded from coastal areas, mainly from the pearl banks in the Gulf of Manaar.

It is surprising that there have been only a few detailed studies on the crustacean population of any of the lagoons, even though they constitute the most important group commercially. De Bruin (1965) reported striking differences in the relative abundance of penaeid species, particularly those of the genus *Metapenaeus*. *M. dobsoni* and *M. elegans* were numerically most abundant in lagoons in the southern, southwestern and northeastern part of Sri Lanka (De Bruin 1965). *M. burkenroadi*, *M. ensis* and *M. mutatus* were scarce in these lagoons, whilst in the northern and northeastern lagoons these species were the most abundant. He also pointed out that the key to the discontinuous distribution of the genus *Metapenaeus* lies in the salinity tolerance of the individual species. Samarkoon & Raphael (1972) surveyed the 'seed' availability in the Negombo lagoon and found that the postlarvae and juveniles of *Penaeus indicus, P. semisulcatus, M. dobsoni* and *M. elegans* occur in sufficiently large numbers during the period of September to November (Table 3), whilst *P. monodon* and *P. latisulcatus* were recorded in smaller numbers for a short period. These authors were unable to find a direct correlation between the availability of shrimp seed and salinity, whilst the size and number of the *Penaeus* species occurring in different areas were correlated broadly with the nature of the bottom sediment and the vegetation.

Table 3. Fin fish species recorded from lagoons and estuaries of Sri Lanka and their typical habitat (adapted from Pillai 1965, and de Silva & Silva 1979b; M-marine; F-freshwater; E-truly estuarine; C-catadromous; m-migratory; u-uncertain).

Family	Species	Habitat
Acanthuroidae	*Acanthurus mata*	m
	A. strigosus	m
Albulidae	*Albula vulpes*	m
Ambassidae	*Ambassis commersoni*	E
	A. gymnocephalus	E
	A. urotaenia	E
Anguillidae	*Anguilla bicolor*	C
	A. nebulosa	C
Antherinidae	*Allanetta forskali*	m
	Pranesus duodecimalis	m
Bagridae	*Macrones gulio*	E
Belonidae	*Tylosurus strongylurus*	m
Bothidae	*Pseudorhombus arsius*	m
Carangidae	*Alectis ciliaris*	M
	Carangoides gymnostethoides	M
	C. malabaricus	M
	Caranax ignobilis	M
	C. sansun	M
	Chorinemus lysan	M
	C. tala	M
	Decapterus russeli	M
	Megalaspis cordyla	M
	Trachinotus blochi	M
	T. russeli	M
Chanidae	*Chanos chanos*	C
Cichlidae	*Etroplus maculatus*	F
	E. suratensis	E
	Sarotherodon mossambicus	F

Table 3 (continued)

Family	Species	Habitat
Clupeidae	*Harengula ovalis*	M
	Kowala coval	M
	Macrura kelee	M
	Opisthopterus tardoore	M
	Pellona ditchela	M
Cynoglossidae	*Cynoglossus lingua*	m
	C. macrolepidotus	M
Cyprinodontidae	*Panchax panchax blochii*	m
	P. melastigma	
Dorosomidae	*Nematalosa nasus*	E
Dussumieridae	*Ehirava fluviatilis*	E
Eleotridae	*Butis butis*	E
	Eleotris fusca	m
	Eleotrides muralis	m
	E. sexguttatus	m
	Ophiocara porocephala	E
Elopidae	*Elops machnata*	u
Engraulidae	*Thrissina baelama*	M
Gerridae	*Pertica filamentosa*	M
Gobiidae	*Acentrogobius griseus*	M
	Callogobius hasseltii	M
	Glossogobius giuris	m
	Mugilogobius valigouva	E
	Stigmatogobius sadanundio	m
Hemiramphidae	*Hemirhamphus marginatus*	m
	Hyporhamphus gaimardi	m
	Zenarchopterus dispar	M
Kuhlidae	*Kuhlia taeniuris*	m
Labridae	*Thalassoma amblycephala*	M
Lagocephalidae	*Amblyrhynchotes hypselogenion*	M
Latidae	*Lates calcarifer*	m
	Psammoperca waigeinsis	m
Leognathidae	*Leiognathus equulus*	m
	L. fasciatus	M
	L. splendens	M
Lobotidae	*Lobotes surinamensis*	m
Lutianidae	*Lutianus argentimaculatus*	m
	L. johni	M
	L. russeli	M

Table 3 (continued)

Family	Species	Habitat
Megalopidae	*Megalops cyprinoides*	E
Menidae	*Mene maculata*	M
Mugilidae	*Mugil cascasia*	C
	M. cephalus	C
	M. dussumieri	C
	M. kelaartii	C
	M. macrolepis	C
	M. oligolepis	C
	M. prasia	C
	M. strongycephalus	C
	M. tade	C
	M. waigiensis	C
	Valamugil buchanani	C
Mullidae	*Upeneus vittatus*	M
Muraenesocidae	*Muraenesox cinereus*	E
Muraenidae	*Gymnothorax polyuranodon*	M
	Pseudechidna brummeri	E
	Thyrsoidea macrura	m
Ophicthydae	*Caecula orientalis*	u
	Callechelys longipinnis	E
	Ophicthus rhytododermatoides	u
	Pisodonophis canonivorous	u
Periophthalmidae	*Periophthalamus koelreuteri*	E
Plectorhynchidae	*Plectorhynchus grieseus*	m
Plotosidae	*Plotosus canius*	u
Polynemidae	*Eleutheronema tetradactylum*	m
	Polynemus indicus	m
	P. heptadactylus	m
	P. plebeius	m
Scatophagidae	*Scatophagus argus*	E
Sciaenidae	*Johnius dicanthus*	M
Serranidae	*Epinephelus fario*	M
	E. rasciatus	M
	E. merra	M
	E. tauvina	M
Siganidae	*Siganus javus*	m
	S. oramin	m
	S. vermiculatus	m
Sillaginidae	*Silago maculata*	m

312

Table 3 (continued)

Family	Species	Habitat
Soleidae	*Brachirus orientalis*	M
Spraidae	*Sparus berda*	m
Sphyraenidae	*Sphyraena jello*	M
	S. obtusata	M
Symbranchidae	*Symbranchus bengalensis*	m
Syngnathidae	*Dorichthys cunculus*	E
	Michrophis brachyurus	m
	Syngnathus specifer djarong	E
Tachysuridae	*Aroides dussumieri*	E
	Hexanematichthys sona	E
	Netuma serratus	u
	Osteogeneiosus militaris	m
	O. sthenocephalus	m
	Pseudarius jatius	E
	P. jella	m
	Tachysurus caelatus	m
	T. maculatus	E
	T. subrostratus	m
Tetraodontidae	*Chelonodon fluviatilis*	E
	C. patoca	m
	Monotretus cutcutia	F
Theraponidae	*Autisthes puta*	m
	Eutherapon theraps	E
	Pelates quadrilineatus	E
	Therapon jarbua	m
Toxotidae	*Toxotes chatareus*	E
Tricanthidae	*Tricanthus biaculeatus*	M
	T. brevirostris	M
Trichuridae	*Trichiurus haumela*	M
	T. savala	M

Eight species of palaemonids have been recorded from brackish waters, although only *Macrobrachium equidens* and *M. rosenbergii* have been collected from a lagoon (Negombo) by Costa (1980). Seven species of lagoon prawns (Table 2) are commercially valuable and of these *Metapenaeus monoceros* (Anon. 1958) and *M. elegans* (De Bruin 1965) are capable of completing their life cycles within the lagoon. In all other species the larval and adult life is spent in the sea.

Crabs also constitute a commercially important fishery in the lagoons. *Scylla serrata* ranks foremost. *S. serrata* seed occurs throughout the year (Raphael 1972). She recorded a survival of 35% in her culture experiments for 8 months. A simple bait trap is used for catching crabs. The device consists of a 2 foot diameter cane circle over which a wide meshed net is stretched and fixed. The trap ends are suspended from a float (usually coconut husks) with a long string, which ends in three or four short lengths tied to the cane hoop to keep the trap horizontal, without tilting when it is drawn up. Several of these traps are laid from a boat and are periodically lifted to catch crabs feeding on the bait.

Vertebrates

The most important vertebrates in the lagoons in terms of biomass and commercial value are fish. Estuaries, lagoons and near shore areas have a major nursery function; young fishes aggregate in these areas and benefit from the availability of food and protection from predators (Gunter 1938, Lobell 1939). Negombo lagoon remains the only lagoon which has been studied in some detail with respect to its fish fauna and the biology and ecophysiology of some of its constituent species.

Pillai (1965) recorded 125 species of fish from brackish waters, of which 80 species were immigrant marine species. In a detailed study of the fish fauna of the Negombo lagoon, De Silva & Silva (1979a) recorded 62 species, belonging to 36 families, of which 33 typically were marine and 2 were freshwater. The total number of species occurring in the lagoon was salinity dependent, tending to increase with increasing salinity, whilst the distribution of species within the lagoon at any one time probably was dependent on the food availability or other ecological factors. Fish species recorded from brackish waters of Sri Lanka are listed in Table 3. Allen & Horn (1975) reported that the abundance and diversity of the fish fauna in a Colorado lagoon is correlated with temperature, whilst Whitfield (1980) found that in the Mhbanga Estuary, South Africa, the changes in the standing crop of the food had an effect on the diversity of the fish fauna.

Most studies have revealed that tropical lagoon fish faunas are dominated by a small number of species, these are primary consumers or detritivores (Allen & Horn 1975), Warburton 1978, Edwards 1978, McErlean *et al.* 1978). The dominant species in Sri Lanka are *Mugil cephalus* (a detritivore) and *Chanos chanos*, an algal feeder.

Methods of capture

Fishing methods and devices adopted in the lagoons of Sri Lanka are very diverse. They vary from simple push nets, gill nets, and cast nets to very complicated and elaborate *kraals* which are almost permanently fixed structures. Most types of fishing gear and the associated methods employed, such as the fish kraal are very efficient, but at the same time destructive to the stock due to lack of size and species selectivity. More recently Sachithananthan & Thevathason (1970) described the stake nets operated in the Jaffna lagoon whilst Ward & Wyman (1975) described the brush pile fishery in the Negombo lagoon.

Biological studies

Biological studies on finfish species have been confined mostly to the main commercial species. Ramanathan (1969) concluded from surveys conducted in the brackish water areas of Manaar, Puttalam and Negombo that two peak seasons of fry production exist for the milkfish, *Chanos chanos* (April to June and October to November). Catches are considerably higher from April to June. It also was observed that fry catches increased on full moon days and were lowest during the first quarter. Ramanathan & Jayamaha (1972) reported that *Chanos* fry most frequently occurred in tidal pools 40-50 cm deep, along with other fish such as *Megalops cyprinoides, Elops machnata, Sarotherodon mossambicus* and mugilids. They noted also that *Chanos* fry were restricted to waters with a salinity of 36-38‰

Investigations on the biology of young *Mugil cephalus* populations of the Negombo lagoon showed that fry abundance varied from month to month with peaks in December to January, May and September to October and coincided with the rainy seasons (De Silva & Silva 1978b). These investigations indicated also that the fry migrate into the lagoon at 10-20 mm length and when they have reached a length of 30-35 mm, they tend to move into deeper waters. Also, there is a diurnal migration of fry into shallower water probably to escape fish predators. In addition, peak seasons of occurrence of other finfish fry in the Negombo lagoon have been documented (Table 4).

The observations of De Silva & Silva (1979b) appear to support the experimental evidence that *M. cephalus* fry grow best in salinities of about 20‰ (De Silva & Perera 1976). These together with aspects of the biology of this species have been reviewed by De Silva (1980).

The biology of adult *M. cephalus* populations in the Negombo lagoon were investigated by Silva & De Silva (1981). They showed that spawning migrations were related to salinity and that the seasonal changes in the observed sex ratio could be caused by the differential timing of the pre- and post spawning migrations of the sexes. Also, the distribution of adults in the lagoon is a function of size to depth and that the fecundity of this species varies from 0.45-4.2 million in fish ranging in length from 32-56 cm.

Ecological and behavioural interactions of *Etroplus suratensis* and *E. maculatus*, cichlid species indigenous to Asia (southern India and Sri Lanka) were studied by Ward & Wyman (1977). *E. suratensis* is cleaned by *E. maculatus* and the latter in turn benefitted through the eggs and eleuthero embryo of its egg-young predatory habit. These authors recorded altruistic multiple parent care for the first time in fishes, where several adults care for a single brood, presumably spawned by only two adults. In addition to ethological observations, the authors also commented on the habitat of each species in the Negombo lagoon and postulated that *E. suratensis* could be a seasonal breeder as compared with *E. maculatus*. Ward & Samarakoon (1981), however, showed that both of these species are seasonal

Table 4. Time of peak availability of shellfish and finfish seed of some commercially important species in the lagoons and coastal waters of Sri Lanka (The commonly used name is given in parentheses; + = caught through the year).

Species	Month(s)	Area	Authority
Metapenaeus dobsoni	*Aug./Sept. (post-larvae)* *Oct./Nov. (juveniles)*	Negombo L.	Samarakoon & Raphael (1972)
M. elegans	*Aug./Sept. (post-larvae)* *Oct./Nov. (juveniles)*	Negombo L.	Samarakoon & Raphael (1972)
Penaeus semisulcatus	*April/May/Oct. (seed)*	Negombo L.	Samarakoon & Raphael (1972)
P. indicus	*July/Aug./Sept. (post-larvae)*	Negombo L.	Samarakoon & Raphael (1972)
Mugil cephalus (grey mullet)	*May/Dec.*	Negombo L.	De Silva & Silva (1979a)
M. buchanani	*Nov./Dec.*	Negombo L.	De Silva & Silva (1979a)
Ambassis gymnocephalus (Nakedhead glassy perchlet)	*Jan./Feb.*	Negombo L.	De Silva & Silva (1979a)
Periophthalamus koelreuteri	+	Negombo L.	De Silva & Silva (1979a)
Glossogobius giuris (bar-eyed goby)	+	Negombo L.	De Silva & Silva (1979a)
Chanos chanos (milk fish)	April/June; Oct./Nov.	Gulf of Manaav	Ramanathan & Jayamaha (1972)

breeders, reproducing when conditions are favourable for nest construction and maintaining visual contact with offspring in the drier pre-monsoonal and monsoonal seasons when turbidity is low and salinity is high. They found also that *E. suratensis* does not forage during nesting. Nest site selection was determined mainly by factors favouring offspring survival.

Except for the study by De Silva & Wijeyaratne (1977) on the feeding habits of young *M. cephalus*, studies on the use of food resources by lagoon fish communities are lacking. Whitfield (1980) reported that benthic floc (detritus and associated microorganisms) accounted for 83% of the food resource of the Mhlanga Estuary in South Africa, which in turn supported 93% of the fish biomass. Bottom substrates and microorganisms were the main food of 94% of the fish of the Sakumo estuarine lagoon in West Africa (Pauly 1975).

Community studies

The importance of community studies lies in understanding the dynamics of the ecosystem and in planning and exocution of management strategies of aquatic environments. It is surprising, however, that so few community studies of this nature have been carried out in tropical lagoons all over the world (Boltt 1975, Darnell 1976, Day & Morgan 1956, Edwards 1978, Whitfield 1980). Nearly all these studies have indicated the high relative importance of the benthic floc as an energy source in comparison with other aquatic systems. The source of the benthic floc varies according to the river inlets, land drainage, the littoral mangrove flora and the sub-littoral vegetation, in each lagoon. There are no data available on lagoon bioenergetics in Sri Lanka, although it is resonable to assume that the benthic floc plays an important role as one of the main energy sources. It should also be borne in mind that the energetics of the lagoon may vary depending also on the extent of the connection with the sea. For example, Edwards (1978) found the semi-enclosed lagoon system in Mexico to be more productive, tending to retain high nutrient levels, in comparison with those which are more open and/or closed for most of the year.

Studies on trophic relationships with reference to the ichthyofauna in particular also have contributed to the understanding of the dynamics in lagoons. Most such studies have been carried out in South African lagoons with reference to the mugilids, which constitute the major biomass (Masson & Marais 1975, Whitfield & Blaber 1978, Blaber 1977, Blaber & Whitfield 1977, Whitfield 1980). These studies also have revealed the varying mechanisms that are adopted by different species and intra-specifically at different stages of life in avoiding competition for food.

Similar studies are needed urgently in Sri Lanka in order to make an initial evaluation of their use for aquacultural practice as well as an aid to understanding of the trophic dynamics.

Although it is difficult to generalize, observations made by the authors on the fisheries of the Negombo lagoon suggest that predatory fish constitute a significant proportion of the ichthyomass. Apart from fish, crabs also act as dominant predators on the fishes in the lagoons, of which very little is known (Verstein 1977).

Aquaculture

There has been much debate in the past as to the suitability of Sri Lankan lagoons for intensive aquaculture (Raphael 1977, Arudpragasam 1975). The small and insufficient tidal amplitude around the coast has been recognized as one of the main constraints for the development of aquaculture. It is suggested, however, that although this physical constraint does exist, the primary reasons have been the lack of governmental support and incentives as well as technical expertise.

The high productivity in the lagoons (De Silva & Silva 1979c) would undoubtedly make milkfish, mullet and shellfish culture profitable. The insufficient tidal amplitude would not make either cage or pen culture of desirable species unprofitable. Observations of Ramanathan (1969) and De Silva & Silva (1979a) indicate the availability of fairly significant resources of mullet and milkfish fry in the coastal shallows and lagoons, respectively. A similar evaluation of the abundance of shrimp and fry of other culturable finfish species such as *Siganus* sp., is needed urgently to evaluate the approximate area of cage and pen culture enterprises that are operable on naturally available fry resources. The sociological attitudes of the indigenous population to commercial culture operations may have to be overcome by way of incentives provided by the Government.

Preliminary culture trials of the grey mullet (De Silva & Perera 1979), milkfish (Samarakoon 1972) and the crab, *S. serrata* (Raphael 1972) in lagoons have indicated the viability of such operations. Samarakoon (*loc. cit.*) compared the yield from selective harvesting and selective harvesting with replenishment of stock and recorded yields of 799 and 1,160 kg ha^{-1} yr^{-1} respectively.

Sri Lanka has a number of small lagoons which remain closed from the sea for much of the year. Use of these for extensive culture operations, supported by artificial stocking of desirable species also needs to be explored. Most of these lagoons are rather small and as such the authors hold the view that these present the greatest prospect for extensive culture, thus bringing about an upsurge in the fish and shellfish production of the Island.

Acknowledgements

We wish to thank Mrs R. Rajapaksa for permitting the use of unpublished data and Mr E.I.L. Silva for help in the compilation of Table III, and for reading an early draft of the manuscript.

References

Allen, L. G. & M.H. Horn. 1975. Abundance, diversity and seasonality of fishes in Colorado lagoon, Alamitos Bay, California. Estuarine Coastal Mar. Sci. 3:3171-380.

Anon. 1958. Brackishwater fisheries. Bull. Fish. Res. Stn Ceylon 8:29-31.

Arudpragasam, K.D. 1974. Seasonal and diurnal variation in salinity and water movement in Jaffna lagoon. Estuarine Coastal Mar. Sci. 2:251-259.

Arudpragasam, K.D. K.D. 1975. The lagoons and coastal waters of Sri Lanka. Proc. Sri Lanka Assoc. Adv. Sci. Part II 13:65-78.

Arulchelvam, K. 1968. Mangroves. Ceylon Forester 8:59-92.

Barnes, R.S.K. 1974. Estuarine Biology. Edward Arnold, London.

Blaber, S.J.M. 1977. The feeding ecology and relative abundance of mullet (Mugilidae) in Natal and Pondoland estuaries. J. Linn. Soc. Lond. 9:259-275

318

Blaber, S.J.M. & A.K. Whitfield. 1977. The feeding ecology of juvenile mullet (Mugilidae) in south east African estuaries. J. Linn. Soc. Lond. 9:277-284.

Boltt, R.E. 1975. The benthos of some African lakes. Part 5. The recovery of the benthic fauna of St. Lucia Lake following a period of excessively high salinity. Trans. R. Soc. S. Africa 41:295-323.

Costa, H.H. 1980. The Palaemonidae of the inland waters of Sri Lanka. Ceylon J. Sci. (Bio. Sci.) 13:39-64.

Dahlberg, M.D. & E.P. Odum. 1970. Annual cycle of species occurrence, abundance and diversity in Georgia estuarine fish populations. Am. Midl. Nat. 83:382-392.

Darnell, R.M. 1961. Trophic spectrum of an estuarine community based on studies of Lake Pontchartrain, Louisiana. Ecology 42:553-568.

Day, J.H. & J.F.C. Morgan. 1956. The ecology of South African estuaries part 7. The biology of Durban Bay. Ann. Natal Mus. 13:259-312.

De Bruin, G.H.P. 1965. Penaeid prawns of Ceylon (Crustacea, Decapoda, Penaeidae). Zool. Meded. 41:73-104.

De Silva, S.S. 1980. Biology of young grey mullet: a short review. Aquaculture 19:21-37.

De Silva, S.S. & P.A.B. Perera. 1976. Studies on the young grey mullet, *Mugil cephalus* L. I. Effects of salinity on food intake, growth and food conversion. Aquaculture 7:327-328.

De Silva, S.S. & E.I.L. Silva 1979a. Fish fauna of Negombo lagoon. Distribution and seasonal variation. Bull. Fish. Res. Stn Sri Lanka 29:1-9.

De Silva, S.S. & E.I.L. Silva. 1979b. Biology of young grey mullet, *Mugil cephalus* L. populations in a coastal lagoon in Sri Lanka. J. Fish. Biol. 15:9-20.

De Silva, S.S. & E.I.L. Silva. 1979c. Studies on the fish ponds at Pitipana, Negombo. I. Seasonal and diurnal variation of some hydrobiological factors. Bull. Fish. Res. Stn Sri Lanka 29:63-78.

De Silva, S.S. & Wijeyaratne, M.J.S. 1977. Studies on the young grey mullet, *Mugil cephalus* L. II. Food and feeding. Aquaculture 12:157-168.

Durairatnam, M. 1963. Studies on the seasonal cycle of sea surface temperatures, salinities and phytoplankton in Puttalam Lagoon, Dutch Bay and Portugal Bay along the west coast of Ceylon. Bull. Fish. Res. Stn Ceylon 16:9-24.

Edwards, R.R.C. 1978. Ecology of a coastal lagoon complex in Mexico. Estuarine Coastal Mar. Sci. 6:75-92.

Fernando, D.H. 1977. Lamellibranchiate fauna of the estuarine and coastal areas in Sri Lanka. Bull. Fish. Res. Stn Sri Lanka (Ceylon) 27:29-54.

Gunter, G. 1938. Seasonal variations in abundance of certain estuarine and marine fishes in Louisiana with particular reference to life histories. Ecol. Monogr. 8:313-346.

Haedrich, R.L. & S.O. Haedrich. 1974. A seasonal survey of the fishes in the Mystic River, a polluted estuary in downtown Boston, Massachusetts. Estuarine Coastal Mar. Sci. 2:59-73.

Lobell, J.J. 1939. A biological survey of the salt waters of Long Island 1938. Report on certain fishes. 28th Ann. Rep. N.Y. State Cons. Dept. Suppl. I:62-96.

Masson, H. & J.F.K. Marais. 1975. Stomach content analysis of mullet from Swartkops Estuary. Zool. Afr. 10:193-208.

McErlean, A.J., S.G. O'Connor, J.A. Mihursky & C.I. Gibson. 1973. Abundance, diversity and seasonal patterns of estuarine fish populations. Estuarine Coastal Mar. Sci. 1:1936.

Odum, W.E. 1970. Utilization of the direct grazing and plant detritus food chains by the striped mullet (*Mugil cephalus*). pp. 222-240 In: J.J. Steel (Ed.) Marine Food Chains. Oliver & Boyd, Edinburgh.

Oviatt, C.A. & S.W. Nixon. 1973. The demersal fish of Narragansett Bay: an analysis of community structure, distribution and abundance. Estuarine Coastal Mar. Sci. 1:361-378.

Pauly, D. 1975. On the ecology of a small West African lagoon. Bericht Deutsch. wiss. Komm. Meeres. 24:46-62.

Pearson, J. 1910-1911. Biological survey of Trincomalee Harbour. Admin. Rept. Ceylon Part 4:3-4.

Pearson, J. 1925. Fish appliances of Ceylon. Bull. Ceylon Fisheries 1:65-134.

Perera, M.M. & Arudpragasam, K.D. 1966. Animals living in association with *Ostrea virginica*, the edible oyster at Batticaloa. Ceylon J. Sci. (Bio. Sci.) 6:20-25.

319

Perera, W.K.T. & K. Sachithananthan. 1977. Topography of Nanthikadal and Nayaru lagoons. Bull. Fish. Res. Stn Sri Lanka (Ceylon) 27:9-15.

Pillai, T.G. 1960. Some marine and brackish water serpulid Polychaeta from Ceylon, including new genera and species. Ceylon J. Sci. (Bio. Sci.) 3:1-2.

Pillai, T.G. 1965. Brackishwater fishery resources. Bull. Fish. Res. Stn Ceylon 18:75-85.

Pinto, L. & Wignaraja, S. 1980. Some ecological aspects of the edible oyster *Crassostrea cucullata* (Borno) occurring in association with mangroves in Negombo lagoon, Sri Lanka. Hydrobiologia 69:11-19.

Qasim, S.Z. 1969. Organic production in a tropical estuary. Proc. Indian Acad. Sci. 59:51-94.

Ramanathan, S. 1969. A preliminary report on *Chanos* fry surveys carried out in the brackish water areas of Manaar, Puttalam and Negombo. Bull. Fish. Res. Stn Ceylon 20:79-85.

Ramanathan, S. & D.E.S. Jayamaha. 1972. On the collection, transport and acclimatization of the fry of *Chanos chanos* for brackishwater pond culture in Ceylon. pp. 241-250 In: T.V.R. Pillay (Ed.) Coastal Aquaculture of the Indo-Pacific Region. Fishing News Books, Surrey.

Raphael, Y.I. 1972. A preliminary report on the brackishwater pond culture of *Scylla serrata* (Forskal) in Ceylon. p. 395 In: T.V.R. Pillay (Ed.) Coastal Aquaculture of the Indo-Pacific Region. Fishing News Books, Surrey.

Raphael, Y.I. 1977. Brackishwater aquaculture in Sri Lanka. Proc. Indo-Pacific Fish. Conn. 17th Sess. Colombo, Sri Lanka Sect. 3:127-130.

Sachithananthan, K. 1969. Salinity and temperature of the surface waters in the Jaffna lagoon. Bull. Fish. Res. Stn Ceylon 20:87-89.

Sachithananthan, K. & W.K.T. Perera. 1970. Topography and substratum of the Jaffna lagoon. Bull. Fish. Res. Stn Ceylon 21:75-85.

Sachithananthan, K. & A. Thevathason. 1970. Sirahu-Valoi - A passive fishing gear in Ceylon. Bull. Fish. Res. Stn Ceylon 21:87-95.

Samarakoon, J.I. & Y.I. Raphael. 1972. On the availability of seed of culturable shrimps in the Negombo lagoon. pp. 251-259 In: T.V.R. Pillay (Ed.) Coastal Aquaculture in the Indo-Pacific Region. Fishing News Books, Surrey.

Samarakoon, J.I. 1972. On the experimental culture of milkfish *Chanos chanos* (Forskal) as tuna bait in Ceylon. p.454 In: Coastal Aquaculture in the Indo-Pacific Region. T.V.R. Pillay (Ed.). Fishing News Books, Surrey

Sewell, R.B.S. 1929. Geographic and oceanographic research in Indian waters. V. Temperature and salinity of the surface waters of the Bay of Bengal and Andaman sea, with reference to Laccadive Sea. Mem. Asiat. Soc. Bengal. 9:207-356.

Sewell, R.B.S 1937. Oceans India. pp. 17-41 In: An outline of the Field Sciences of India. India Sci. Assoc.

Silva, E.I.L. & S.S. De Silva. 1981. Aspects of the biology of adult grey mullet, *Mugil cephalus* L. populations in a coastal lagoon of Sri Lanka. J. Fish. Biol. 19.

Silva, E.I.L. & De Silva, S.S. 1982. Studies on the coastal lagoons of Sri Lanka. I. Hydrography of Negombo and Koggala lagoons. Bull. Fish. Res. Stn Sri Lanka.

Sivalingam, S. 1961. The 1958 pearl oyster fishery, Gulf of Manaar. Bull. Fish. Res. Stn Sri Lanka 11:7-28.

Sivapalan, A. 1975. Cultivation of *Gracilaria lichenoides* in Puttalam Lagoon. Bull. Fish. Res. Stn Sri Lanka 26:1-4.

Tamiya, H. 1975. Green micro algae. pp. 35-39 In: Food Protein Sources. N.Y. Pirie (Ed.). Cambridge Univ. Press.

Vernstein, R.W. 1977. The importance of predation by crabs and fishes on the benthic infauna in Chesapeake Bay. Ecology 58:1199-1217.

Warburton, K. 1978. Community structure, abundance and diversity of fish in a Mexican coastal lagoon system. Estuarine Coastal Mar. Sci. 7:497-519.

Ward, J.A. & J.I. Samarakoon. 1981. Reproductive tactics of the Asian cichlids of the genus *Etroplus* in Sri Lanka. Env. Biol. Fish. 6:5-103.

320

Ward, J.A. & R.L. Wyman. 1975. The cichlids of the resplendent isle. Oceans 8:42-47.
Ward, J.A. & Wyman, R.L. 1977. Ethology and ecology of cichlid fishes of the genus *Etroplus* in Sri Lanka; preliminary findings. Env. Biol. Fish. 2:137-145.
Whitfield, A.K. 1980. A quantitative study of the trophic relationships within the fish community of the Mhalanga Estuary, South Africa. Estuarine Coastal Mar. Sci. 10:417-436.
Whitfield, A.K. & S.J.M. Blaber. 1978. Food and feeding ecology of piscivorous fishes at Lake St. Lucia, Zululand. J. Fish. Biol. 12:61-70.

16. Parasites of endemic and relict vertebrates: a biogeographical review

H. Crusz

Introduction

'Among parasites, two of the most obvious and easily examined variables in the biology of related species are their geographical distributions and their hosts... A defensible evaluation of causal agents in parasite evolution may be formed if knowledge is available on the evolutionary history of the areas in which related parasites occur, as well as the mobility, geneology, and ecology of the hosts'.

Daniel R. Brooks, 1979

A consideration of the parasites of the fauna of islands such as the Galapagos or Madagascar (Chabaud & Brygoo 1964, Vassiliades 1970) clearly shows that their isolation has had its effects in the evolution of particular parasite species peculiar to the peculiar hosts found on them. The Galapagos are volcanic in origin, while Madagascar is an old continental island. Sri Lanka, on the other hand, is a more recent continental island, despite which it harbours some interesting endemic and relict species of hosts. Furthermore, Madagascar was, according to the drift theory, connected before the Cretaceous with Africa, Sri Lanka, India and Malaysia. In fact, the connection between Sri Lanka and Madagascar would seem to have lasted longer than that which either of them had with Malaysia and Indonesia (Melville 1970). This is reflected in some ways in the host and parasite faunas of these regions.

Ihering was the first to speak of 'phylogenetic specificity' (1891) and later (1903) to correlate data from hosts and their parasites with problems of dispersal and zoogeography, two ideas which were reviewed by Stunkard (1967, 1970). Before Stunkard, however, Baylis (1938) had questioned the 'von Ihering method' when it came at least to the helminth fauna, where according to him, the behaviour and environment of the hosts play a more important role than their phylogenetic relationships. The truth, in fact, would be more complex, and would seem to lie, as always, somewhere between these two aspects (see Brooks 1979).

In any case, a consideration of the parasites of Sri Lanka's endemics and relicts, as such, was long overdue. Early work, howsoever valuable and necessary, tended to confine itself mostly to recording the occurrence of parasites in a somewhat haphazard fashion, or in relation to other parameters such as host-phyla and so on. The time has come to focus attention on, and try to correlate parasitology with, concepts such as endemicity and relictness of hosts. The series of papers that have appeared recently (Crusz *et al.* 1970-1978) represent a mere scratch on the surface in this direction, mere pointers to further work, the final evaluation of which would necessarily be far off. The attempt, however, was ineluctable.

The general trend of this chapter follows the work of Crusz & Nugaliyadde (1978), and opportunity is taken here of reviewing and adding to the previously recorded material, presenting it in a more immediately recognizable and vivid form for the drawing of conclusions (e.g. Table 1) and indicating some lines of research for the future.

Distribution of fauna and flora

The best survey of the distribution of Sri Lanka's fauna and flora would be one that takes into account most if not all of the relevant parameters. This would be achieved if it is based on a recognition of 7 vegetational zones, following Gaussen *et al.* (1964), Müller-Dombois & Sirisena (1967), Fernando (1968) and Müller-Dombois (1968). For this reason the most plausible analysis of Sri Lanka's mammalian distribution is the one that utilizes this zonal classification (Eisenberg & McKay 1970). It is argued that climate, and to some extent soil, interact to determine vegetation form, which in turn influences mammalian distribution in a given area. This principle would apply to other life forms as well, both plant and animal. The 7 vegetational zones are, as indicated by the symbols on the map (Fig. 1):

A_1 - Monsoon Scrub Jungle - extreme north and north west.
A_2 - Monsoon Scrub Jungle - extreme south east
B - Monsoon Forest and Grassland
C - Intermonsoon Forest
D_1 - Rain Forest and Grassland - below 914 m (5,000 ft)
D_2 - Rain Forest and Grassland - 914 to 1,524 m
D_3 - Rain Forest and Grassland - above 1,524 m (5,000 ft)

Endemics and relicts

While the concept of endemicity poses hardly any problems, that of relictness has to be viewed more carefully. Some biologists seem even not to pay much heed, apparently for no sound reason, to the existence of relict forms, betraying thereby a certain want of courage in considering degrees of endemism and the finer aspects of the isolation of plant and animal species.

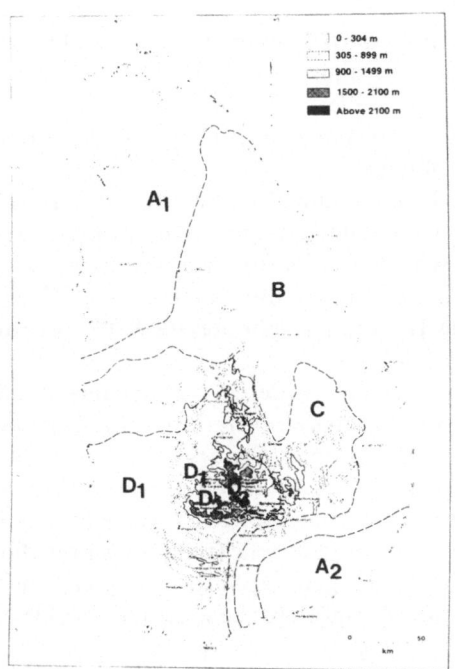

Fig. 1. Sketch-map of Sri-Lanka (Ceylon) showing vegetational zones (see page 322).

As far as Sri Lanka is concerned, the existence on it of relict forms has been accepted for a considerable time. Sarasin (1910) referred to them as 'die Reste', and of 22 genera of Sri Lanka and South Indian amphibians and reptiles he wrote: 'Diese 22 Gattungen dürfen als ein Rest der alten, vortrappzeitlichen Tierwelt angesehen werden'. He recognized 8 of these genera as being Sri Lankan relicts. This tradition was continued chiefly in the work of Taylor (1950) on snakes, Kirtisinghe (1957) on amphibians, and Ripley (1949, 1980) on birds. Ripley (1980) prefaced a treatment of Sri Lanka's avian relicts with the remark that 'relicts among birds are a somewhat "soft" form of characterization in scientificterms, in that most authors interested in avian speciation or evolution feel constrained to use the word with caution'. Nevertheless, he boldly proceeds to explore the concept and justify it.

Delamare-Debouteville & Botoséanu (1970), Sedon (1971) and Prentice (1976) contributed to a refinement of these concepts. An endemic species is one that is confined to a particular region and is found nowhere else, chiefly owing to geographic isolation. Prentice, however, has warned that any attempt to further categorize endemics are bound to overlap with concepts of relictness, where, according to the respective criteria, there could be two categories of the latter, namely relics (reliques) and reclicts (relictes) (Delamare-Debouteville & Botoséanu 1970).

Relics are living remnants of a main stock (or of basal lines of evolutionary branches) which has differentiated elsewhere. They are living fossils in the paleontological sense. Relicts, however, are living representatives of environments that have disappeared for various reasons. They are therefore living fossils in the biogeographical or ecological sense.

Ekman (1915) and Thienemann (1950) contributed to a further refinement of these concepts. A species is relict if its presence cannot be explained other than by the fact that the form which gave rise to it has disappeared from the locality under environmental conditions that are now adverse to it. This definition applies to geographical relicts, as Delamare-Debouteville & Botoséanu (1970) already have indicated.

When the time factor also is considered, we have recent relicts, which became so owing to recent activities such as deforestation, urbanization, industrial pollution, etc., and ancient relicts whose past history is much older, going back even to geological times. There also are glacial and pluvial relicts, marine relicts, tertiary relicts, etc., categorized according to criteria revealed in the names. One fact must be emphasized. While the notion of relic (relique) is morphological-phylogenetic, and that of relict (relicte) is biogeographical-ecological, the two are not mutually exclusive. They often can be applied to the same species (Delamare-Debouteville & Botoséanu 1970).

Although the terms 'relic' and 'relict' are concise, it would seem preferable to follow some other authors (e.g. Seddon 1971) and call them by the more explicit terms 'evolutionary relict' and 'geographical relict', respectively. Evolutionary relicts are recognized only on their phylogenetic relationships and the fossil evidence of the antiquity of the groups to which they belong. Unlike geographical relicts, their geographical distribution is secondary.

Derivation of biota

Sri Lanka's plants and animals mostly are derived from peninsular India, but there are many endemics among them, of which some could be characterized as geographical relicts. Biogeographically, India and Sri Lanka have even been regarded as lands of 'vanishing relicts' (Ripley 1980, after Mani 1974) where 'in combination with the geological maturity, the senile topography, faunal stability, and the relative evolutionary stagnation, the effects of the present day conditions tend greatly to accelerate and intensify' this phenomenon.

Fauna

There are 557 known species of inland vertebrates on Sri Lanka (59 fish, 37 amphibian, 139 reptile, 237 resident birds and 85 mammal species) of which 137

(25%) are endemic. The vertebrate fauna of the dry zones (Fig. 1: A_1, A_2, B and C) is related closely to that of India, while that of the wet hill zones (D_1, D_2 and D_3) is more peculiar to the island, some of it showing, however, affinity with the fauna of the Malabar tract of India, and with that of Malaysia and Indonesia, as indicated for instance by the fish *Belontia*, the mountain lizard *Cophotis*, the water lizard *Varanus monitor*, and the snake *Cylindrophis*, and with the fauna of Africa and Madagascar, as indicated by the limbless skinks of the subfamily Acontianinae, and a species of chameleon which also is represented in South India. The endemic fauna inhabits chiefly the rain forest zones (D_1, D_2 and D_3). About 25% of this fauna is confined to the upper montane zone (D_3) and represents the most distinctive and conservative element of it, which has been least influenced by any recent invasions from South India (see Table 1, where this is reflected in the locality column).

Flora

About 28.5% of Sri Lanka's flora is endemic, 65% is of Indian and Himalayan affinity, and the remaining 6.5% is of Malaysian, African and Australian affinity (Abeywickrama 1956, 1959, 1980). Half the endemic species are confined to the wet zones (D_1, D_2 and D_3), of which more exist at the lower elevations than on the hills, making the distribution somewhat different from that of the highly endemic hill fauna.

Comparitive endemism and Parasitism

A comparison of the degrees of endemism in Sri Lanka and an old contintental island such as Madagascar leads to interesting conclusions. 86% of Malagasy plant species are endemic (Nicholson 1970, after Bathie 1936). Among animals 95% of the reptile species and subspecies (Blanc 1970) and 65% of the bird species (Nicholson 1970) are endemic, with several endemic families, genera and species among the 70 Malagasy mammals (Nicholson 1970). The contrasting figures reflect the age and degree of isolation of the two islands. They could help in providing an 'index or coefficient of insularity' (Crusz 1973). A preliminary bold step in this direction was taken when habitat values and endemicity were worked out numerically for the fauna of the rain forests of Sri Lanka (Senanayake, Soulé & Senner 1977, Senanayake 1977).

After considerations such as these one could examine the role of comparative parasitology (especially the study of the parasites of endemic and relict hosts vis-à-vis those of other hosts in investigating animal relationships, and the antiquity of faunas on islands relative to that of the faunas of their neighbouring land masses. The work of parasitologists such as Ihering (1891, 1903), Metcalf (1923,

1929), Dogiel (1947), Cameron (1952), Vanzolini & Guimares (1955), Rausch (1957), Chaubaud & Brygoo (1964), Inglis (1967), Stunkard (1967, 1970), Vassili-ades (1970), Durette-Desset (1971), Crusz & Nugaliyadde (1978) and Brooks (1979) provide guidelines for the assessment of such a role, although we could hope for definitive and more convincing conclusions about it probably only after the analysis of the data takes on a strict mathematical form: an event which is, from all accounts, a considerable way in the future.

In this context, the appeal made to parasitologists by Boucot (1976) is especially relevant. It would be interesting to check the validity of his speculations in the Sri Lankan context, i.e.:
1) 'hosts with a large distribution area have a much larger number of parasitic species than is the case for endemic species',
2) if the evolution of parasites lagged behind that of host species, there would be 'even more of a disparity in number of parasitic species known from related cosmopolitan as opposed to endemic species of host', and
3) if, as he predicts, the parasites of highly endemic taxa evolve at a far greater phyletic rate than is the case for cosmopolitans, the 'the parasites on endemic hosts should tend to be much more "advanced" than is the case for related types occurring on cosmopolitan hosts'.

Discussion and conclusions

The host-parasite checklist (Table 1) and the list of numbers of inland vertebrates of Sri Lanka (Table 2) have been compiled from the most recent and authoritative sources available, and serve chiefly as guides to further research on this subject. The sequence of host names in Table 1 follows that given by Munro (1955) for fishes, Kirtisinghe (1957) for amphibians, Deraniyagala (1953, 1955) for reptiles, Ripley (1961) for birds and Eisenberg & McKay (1970) for mammals. These groups are discussed briefly below, and assessed, as far as present data allow, from the point of view of taxa distribution, endemicity/relictness and parasites (proto-zoans and helminths).

Pisces

There are 62 species and subspecies of inland fishes (excluding introduced forms) in Sri Lanka, of which 28 (45%) are endemic. These comprise 3 endemic genera, 16 endemic species and 12 endemic subspecies, occuring mostly in the rainforest zone (D₁) below 914 m. Of them, the cyprinid *Horadandia atukorali*, the anabantid *Malpulutta kretseri*, and the dussumieriid *Ehirava fluviatilis* (a riverine and estua-rine species) have the greatest claim to being relicts; all three genera are monotypic, apparently with no close relatives outside Sri Lanka. There are few records of parasites from these endemic and relict species.

Among aquatic animals, it is the inland species that are relevant to this study. Sri Lanka abounds in streams, (estuaries and lagoons) and hot-water springs, which are its natural inland-water components. There are no natural lakes. The man-made lakes (wewas or reservoirs), channels, canals and paddy fields are its artificial components. The forest water-holes (villus), especially those in the dry-zone area, are of uncertain origin and appear to be of a secondary nature.

Therefore, one has now to look chiefly to rivers, streams and hot-water springs, for relict forms whose most likely habitats, however, would have disappeared long ago owing to man's interference through cultivation, damming of rivers, village expansion and urbanization. Even on the highest hills, where such forms must have existed, the destruction of wilderness for planting coffee and tea, and the introduction of exotic forms such as trout for sport, would have led to their extinction. Hot-water springs and remote mountain streams may well be the only present-day habitats of truly relict fishes. Senanayake (in press) has described an isolated species of *Barbus* which may be one of these.

Amphibia

Of the 37 Sri Lankan amphibian taxa, 19 (51%) are endemic, almost entirely confined to the rainforest zones, especially D_2 and D_3, above 914 m and 1,524 m respectively. They are represented by one endemic genus (*Nannophrys*) and 19 endemic species. There are 3 species of caecilians, all endemic, of the genus *Ichthyophis*, other species of which occur in India and south-east Asia. As in the case of the relict birds, discussed later, these limbless amphibians have affinities with the Indo-Malaysian caecilians rather than with those of peninsula India (Nussbaum & Gans 1980). It is noteworthy, however, that the trematode *Gorgoderina carli*, and the neamatode *Rhabdias escheri*, originally described from a South Indian hill caecilian, occurs also in Sri Lanka caecilians.

While ranids of the genus *Nannophrys* could be counted as relict species, the high endemicity and limited distribution of other forms as well indicate a degree of isolation amountin to relictness. This is particularly true for a species like *Rhacophorus microtympanum*, with its direct development on land, without free swimming aquatic larvae (Kirtisinghe 1946). Endemic *Rhacophorus* spp. also are interesting for their acanthocephalans, which show affinities with those in montane frogs in Madagascar, and for a monogenean trematode closely related to an eastern species.

Reptilia

The reptiles are the most interesting of the fauna from the point of view of relict forms. There are 159 taxa on the island, of which 98 (62%) are endemic. The

Table 1. Endemic/Relict Host-Parasite Checklist.

(Key to notations: Host column: two asterisks = relict; one asterisk = endemic.
Locality column: letters refer to vegetational zones shown in Fig. 1.
Parasite column: asterisks denotes that the parasite has so far been recorded only from Sri Lanka.)

Host	Locality	Parasite
PISCES		
Family: Dussumieriidae		
**1. *Ehirava *fluviatilis*	D₁	
Family: Cyprinidae		
2. *Chela laubuca *lankensis*	A₁ A₂ B C D₁	
3. *Esomus danrica *brevibarbatus*	D₁	
4. *Garra *ceylonensis *ceylonensis*	A₁ B C D₁ D₂ D₃	
5. *Garra *ceylonensis *phillipsi*	D₁ D₂	
**6. *Horadandia *atukorali*	D₁	
7. *Labeo *fisheri*	D₁	
8. *Labeo percellus *lankae*	A₁ B	
9. *Barbus *bimaculatus*	A₁ B C D₁ D₂ D₃	*Dactylogyroides bimaculati*
10. *Barbus *cumingi*	D₁	
11. *Barbus melanampux *sinhala*	D₁	
12. *Barbus *nigrofasciatus*	D₁	
13. *Barbus *pleurotaenia*	D₁	
14. *Barbus ticto *melanomaculatus*	A₁	
15. *Barbus *titteya*	D₁	
16. *Rasbora *vaterifloris*	D₁	
17. *Tor khudree *longispinis*	D₁ D₂ D₃	
Family: Cobitiidae		
18. *Lepidocephalus *jonklaasi*	D₁	
19. *Noemacheilus *notostigma*	D₁	

Table 1 (continued)

Host	A_1	A_2	B	C	D_1	D_2	D_3	Parasite
Family: Clariidae								
20. *Clarias *nebulosus*	–	–	–	–	D_1	–	–	*Procamallanus planorbis*
21. *Clarias teysmanni *brachysoma*	–	–	–	–	D_1	–	–	*Procamallanus spiculogubernaculus*
								Zeylanema sweeti
Family: Cyprinodontidae								
22. *Panchax lineatus *dayi*	–	–	–	–	D_1	–	–	
Family: Ophicephalidae								
23. *Channa *orientalis*	–	–	B	–	D_1	–	–	
24. *Ophicephalus marulius *ara*	–	–	–	–	D_1	–	–	
25. *Ophicephalus gachua *kelaarti*	A_1	–	–	–	D_1	–	–	*Spinitectus corti*
Family: Anabantidae								
26. *Anabas testudineus *kavaiya*	A_1	–	B	C	D_1	–	–	*Zeylanema anabantis*
								Zeylanema kulasirii
								Zeylanema sweeti
27. *Belontia *signata*	A_1	–	B	C	D_1	–	–	
**28. **Malpulatta *kretseri*	–	–	–	C	D_1	–	–	
AMPHIBIA								
Family: Bufonidae								
1. *Bufo *atuloralei*	–	A_2	B	–	D_1	–	–	
2. *Bufo *kelaarti*	–	–	–	–	–	–	D_3	
Family: Ranidae								
3. *Rana *corrugata*	–	–	–	–	D_1	D_2	–	
4. *Rana *gracilis*	–	–	–	–	D_1	D_2	D_3	
5. *Rana *greeni*	–	–	–	–	–	D_2	D_3	
**6. **Nannophrys *ceylonensis*	–	–	–	–	D_1	–	–	
**7. **Nannophrys *marmorata*	–	–	–	–	–	D_2	–	
**8. **Nannophrys *guentheri*	–	–	–	–	D_1	–	–	

Table 1 (continued)

Host	Locality				Parasite
		C	D_1	D_2	D_3	
Family: Rhacophoridae					
9. Rhacophorus *cruciger	C	D_1	–	–	Opalina virgula
	–				Nyctotherus sp.
	–				Nyctotherus papillatus
10. Rhacophorus *eques	–	–	D_2	D_3	*Mesocoelium monas
					*Polystoma hakgalense
	–	D_1	D_2	–	*Acanthocephalus srilankensis
11. Rhacophorus *nasutus	–	D_1	D_2	–	*Cosmocercoides rickae
12. Rhacophorus *microtympanum	–	D_1	D_2	D_3	*Acanthocephalus srilankensis
13. Philautus *schmardanus	–	–	D_2	D_3	*Acanthocephalus srilankensis
Family: Microhylidae					
14. Ramanella *palmata	–	–	–	D_3	
15. Ramanella *obscura	–	–	D_2	–	
16. Microhyla *zeylanica	–	–	–	D_3	
Family: Caeciliidae					
17. Ichthyophis *glutinosus	C	D_1	D_2	–	Mesocoelium monas
					Gorgoderina carli
					Rhabdias escheri
					Raillietnema loveridgei
					Raillietnema multipapillata
					Pelodera chabaudi
					*Meteterakis sinharajensis
18. Ichthyophis *pseudoangularis	–	D_1	D_2	–	Gorgoderina carli
	–	D_1	D_2	D_3	Rhabdias escheri
19. Ichthyophis *orthoplicatus	C	–	–	–	Pelodera chabaudi
	–	–	D_2	D_3	Meteterakis sinharajensis

Table 1 (continued)

Host	Locality							Parasite
	A_1	A_2	B	C	D_1	D_2		
REPTILIA								
Family: Aniliidae								
1. Cylindrophis *maculatus	A_1	A_2	B	C	D_1	–	–	
Family: Boidae								
2. Python molurus *pimbura	A_1	A_2	B	C	D_1	D_2	–	*Bothridium pithonis* *Ophidascaris filaria* *Polydelphis anoura*
3. Eryx conicus *brevis	A_1	–	B	–	–	–	–	
Family: Uropeltidae								
4. Uropeltis *melanogaster	–	–	–	–	D_1	–	–	*Oswaldocruzia gansi* Cosmocercella uropeltidarum
5. Uropeltis *phillipsi	–	–	–	C	D_1	D_2	–	Cosmocercella uropeltidarum *Oswaldocruzia gansi*
6. Uropeltis *ruhunae	–	–	–	–	D_1	–	–	
**7. *Pseudotyphlops *philippinus	–	–	B	C	D_1	–	–	*Meteterakis sinharajensis*
8. Rhinophis *blythi	–	–	–	–	D_1	D_2	D_3	Cosmocercella uropeltidarum
9. Rhinophis *dorsimaculatus	A_1	–	–	–	–	D_2	–	
10. Rhinophis drummondhayi	–	–	C	–	D_1	D_2	–	Cosmocercella uropeltidarum *Oswaldocruzia gansi*
11. Rhinophis *oxyrhynchus	A_1	–	B	–	–	–	–	
12. Rhinophis *philippinus	–	–	–	–	D_1	–	–	Cosmocercella uropeltidarum
13. Rhinophis *punctatus	A_1	–	B	–	D_1	–	–	
14. Rhinophis *trevelyanus	–	–	–	–	D_1	–	–	
15. Rhinophis *tricoloratus	–	–	–	–	D_1	–	–	
16. Platyplectrurus madurensis *ruhunae	–	–	–	–	D_1	–	–	*Oswaldocruzia gansi*

Table I (continued)

Host	A₁	A₂	B	C	D₁	D₂	D₃	Parasite
Family: Typhlopidae								
17. Typhlops *ceylonicus	–	–	–	–	D₁	–	–	
18. Typhlops *lankaensis	–	–	B	–	–	–	–	
19. Typhlops *leucomelas	–	–	B	–	D₁	–	–	
20. Typhlops *malcolmi	–	–	B	–	–	–	–	
21. Typhlops *mirus	–	–	B	–	D₁	–	–	
22. Typhlops *tenebrarum	–	–	B	–	–	–	–	
23. Typhlops *veddae	–	–	B	–	–	–	–	
24. Typhlops *violaceus	–	–	B	–	–	–	–	
Family: Colubridae								
25. Sibynophis subpunctatus *ceylonicus	A₁	–	–	–	D₁	–	–	
26. Liopeltis calamaria *calamaria	A₁	–	B	C	D₁	D₂	–	
27. Ptyas mucosus *maximum	A₁	A₂	B	C	D₁	D₂	D₃	Haemogregarina sp.
28. Oligodon *calamarius	–	–	–	–	D₁	D₂	–	
29. Oligodon *sublineatus	–	–	–	–	D₁	D₂	–	
30. Lycodon striatus *sinhaleyus	–	–	–	C	D₁	D₂	–	
**31. *Cercaspis *carinatus	A₁	–	–	–	D₁	–	–	
**32. *Aspidura *brachyorrhos	–	–	–	–	D₁	D₂	D₃	
**33. *Aspidura *copei	–	–	–	–	–	D₂	–	
**34. *Aspidura *drummondhayi	–	–	–	–	D₁	D₂	–	
**35. *Aspidura *guentheri	–	–	–	–	D₁	D₂	–	
**36. *Aspidura *trachyprocta	–	–	–	C	D₁	D₂	D₃	Kalicephalus brachycephalus
**37. *Haplocercus *ceylonensis	–	–	–	C	D₁	D₂	–	
38. Dendrelaphis *caudolineolatus	–	–	–	–	D₁	D₂	–	
39. Dendrelaphis *oliveri	–	–	B	–	–	–	–	
40. Chrysopelia ornata *sinhaleya	A₁	–	–	–	D₁	–	–	Haemogregarina sp. *Oochoristica cryptobothria
41. Chrysopelea *taprobanica	A₁	–	B	C	–	–	–	
42. Bioga *barnesi	–	–	–	–	D₁	D₂	D₃	
43. Bioga ceylonensis *ceylonensis	–	–	B	C	D₁	D₂	–	Haemogregarina sp.

Table 1 (continued)

Host	Locality							Parasite
	A₁	A₂	B	C	D₁	D₂	D₃	
44. Ahaetulla nasuta *nasuta	A₁	A₂	B	C	D₁	D₂	–	Haemogregarina mirabilis
45. Ahatulla pulverulenta *pulverulenta	A₁	A₂	B	C	D₁	D₂	–	Polydelphis brachycheilus
46. Xenochrophis *asperrimus	A₁	A₂	B	C	D₁	D₂	–	
Family: Elapidae								
**47. *Balanophis *ceylonensis	–	–	–	C	D₁	D₂	–	
48. Macropisthodon plumbicolor *palabariya	–	–	–	C	D₁	D₂	D₃	
49. Bungarus *ceylonicus *ceylonicus	A₁	A₂	B	C	D₁	D₂	–	Haemogregarina sp.
50. Bungarus *ceylonicus *karavala	–	–	–	C	D₁	D₂	–	Ophidascaris naiae
51. Calliophis melanurus *sinhaleyus	A₁	A₂	B	C	D₁	–	–	
52. Naja naja *naja	A₁	A₂	B	C	D₁	D₂	–	
Family: Viperidae								
53. Vipera russelli *pulchella	A₁	A₂	B	C	D₁	–	–	Kalicephalus willeyi
54. Echis carinata *sinhaleya	A₁	A₂	B	–	–	–	–	
55. Agkistrodon *nepa	A₁	–	–	–	D₁	D₂	D₃	
56. Trimeresurus *trigonocephalus	–	–	–	C	D₁	D₂	D₃	*Ophiotaenia phillipsi
								Kalicephalus viperae chungkingensis
								Polydelphis sewelli
Family: Crocodylidae								
57. Crocodylus palustris *kimbula	A₁	A₂	B	C	–	–	–	
Family: Gekkonidae								
58. Cnemaspis jerdoni *scalpensis	–	–	–	–	D₁	–	–	
59. Cnemaspis kandianus *tropidogaster	–	–	–	–	D₁	–	–	
60. Cnemaspis *podihuna	–	–	B	–	–	–	–	
61. Gymnodactylus *frenatus	–	–	B	–	D₁	–	–	
62. Geckoella *triedrus	–	–	–	–	D₁	–	–	
63. Geckoella *yakhuna *yakhuna	A₁	–	–	–	–	–	–	
64. Geckoella *yakhuna *zonatus	–	–	B	–	–	–	–	

Table 1 (continued)

Host	Parasite	A_1	A_2	B	C	D_1	D_2	D_3
65. Hemidactylus brooki *parvimaculatus	Paradistomum sp.	–	–	–	–	D_1	–	–
66. Hemidactylus triedrus *lankae	Haemogregarina treidri	A_1	A_2	B	C	D_1	–	–
	Trypanosoma pertenue	–	–	–	–	–	D_2	–
67. Hemidactylus *depressus		A_1	–	B	C	D_1	D_2	–
68. Calodactylodes *illingworthi		–	–	B	–	–	–	–
Family: Agamidae								
69. Calotes *ceylonensis	Paradistomoides ceratophorae	A_1	A_2	B	C	D_1	–	–
70. Calotes *liocephalus	*Cosmodercoides rickae	–	–	–	–	D_1	D_2	D_3
71. Calotes *liolepis	*Entomelas cruszi	–	–	–	–	D_1	D_2	–
72. Calotes *nigrilabris	*Strongyluris calotis	–	–	–	–	D_1	D_2	D_3
	*Entomelas cruszi							
73. Otocryptis *wiegmanni	Paradistomoides ceratophorae	A_1	A_2	B	C	D_1	D_2	–
	Meteterakis longispiculata							
	Strongyluris chamaeleonis							
74. Cophotis *ceylanica	Paradistomoides ceratophorae	–	–	–	–	–	–	D_3
	*Cosmocercoides rickae							
	*Meteterakis baylisi							
	Strongyluris chamaeleonis							
	*Acanthocephalus serendibensis							
**75. *Ceratophora *stoddarti	Paradistomoides ceratophorae	–	–	–	–	–	–	D_3
	*Cosmocercoides rickae							
	*Meteterakis baylisi							
**76. *Ceratophora *tennenti	Paradistomoides ceratophorae	–	–	–	–	–	D_2	D_3
	*Cosmocercoides rickae							
	*Meteterakis baylisi							
**77. *Ceratophora *aspera	*Zeylanurotrema lyriocephali	–	–	–	–	D_1	D_2	–
**78. *Lyriocephalus *scutatus	*Cometerakis lyriocephali	–	–	–	–	D_1	D_2	–
	Meteterakis longispiculata							
	*Meteterakis sinharajensis							

Table 1 (continued)

Host	Locality							Parasite
	A_1	A_2	B	C	D_1	D_2	D_3	
Family: Scincidae								
79. Mabuya carinata *lankae	A_1	–	–	–	–	–	–	Trichomastix mabuiae, Trichomonas mabuiae, Filaria tuberosa
80. Dasia *haliana	A_1	–	B	C	D_1	–	–	
81. Sphenomorphus *megalops	A_1	–	B	–	–	–	–	
82. Sphenomorphus *striatopunctatus	–	–	–	–	D_1	D_2	D_3	
83. Sphenomorphus *taprobanensis	–	–	–	–	D_1	D_2	D_3	
84. Sphenomorphus *deignani	–	–	–	–	D_1	–	–	
85. Sphenomorphus *dorsicatenatus	–	–	–	–	D_1	–	–	
86. Sphenomorphus *fallax	A_1	A_2	B	C	D_1	–	–	
87. Riopa *singha	–	–	–	–	D_1	–	–	
**88. *Chalcidoseps *thwaitesi	–	–	–	–	D_1	D_2	–	Meteterakis sp.
**89. *Nessia *burtoni	–	–	–	–	D_1	–	–	
**90. *Nessia *didactyla	–	–	–	–	D_1	–	–	
**91. *Evesia *monodactyla	–	–	–	–	D_1	–	–	
**92. *Bipedos *sarasinorum	–	–	B	C	D_1	–	–	
**93. *Bipedos *smithi	–	–	–	–	–	D_2	–	Thelandros sp.
**94. *Anguinicephalus *deraniyagalai	–	–	B	–	–	–	–	
**95. *Anguinicephalus *layardi	–	–	–	–	D_1	–	–	
**96. *Anguinicephalus *hickanala	A_1	–	B	–	–	–	–	
Family: Lacertidae								
97. Cabrita leschenaulti *lankae	A_1	–	B	–	–	–	–	
Family: Varanidae								
98. Varanus monitor *kabaragoya	A_1	A_2	B	C	D_1	–	–	Acanthotaenia shipleyi, Gnathostoma sp., Tanqua tiara

Table 1 (continued)

Host	A_1	A_2	B	C	D_1	D_2	D_3	Parasite
AVES								
Family: Accipitridae								
1. Accipiter trivirgatus *layardi	A_1	A_2	B	C	D_1	D_2	D_3	Contracaecum haliaeti
2. Ichthyophaga ichthyaetus *plumbeiceps	A_1	–	B	C	–	–	–	
3. Spilornis cheela *spilogaster	A_1	A_2	B	C	D_1	D_2	D_3	
Family: Phasianidae								
4. Francolinus pictus *watsoni	–	–	B	C	–	–	–	*Eimeria lafayetei
5. Gallus *lafayetti	A_1	A_2	B	C	D_1	D_2	D_3	*Eimeria indentata
								*Eimeria symmetrica
								*Eimeria dissanaikei
								*Tyzzeria galli
								Plasmodium juxtanucleare
								*Leucochloridium ceylonicum
								*Morishitium sinhaladvipa
								Raillietina tetragona
								Heterakis pusilla
6. Gallopersix *bicalcarata	A_1	A_2	B	C	D_1	D_2	D_3	
7. Perdicula asiatica *ceylonensis	A_1	A_2	B	C	D_1	–	–	
Family: Turnicidae								
8. Turnix suscitator *leggei	A_1	A_2	B	C	D_1	D_2	–	
Family: Charadriidae								
9. Charadrius alexandrinus *seebohmi	A_1	A_2	B	C	–	–	–	*Panuwa lobivanelli
10. Vanellus indicus *lankae	A_1	A_2	B	C	D_1	–	–	
Family: Recurvirostridae								
11. Himantopus himantopus *ceylonensis	A_1	A_2	B	C	–	–	–	Infula burhini
								*Malika himantopodis

Table 1 (continued)

Host	Locality							Parasite
Family: Columbidae								
12. Chalcophaps indica *robinsoni	A_1	A_2	B	C	D_1	D_2	D_3	
13. Columba *torringtoni	–	–	–	–	D_1	D_2	D_3	
14. Streptopelia chinensis *ceylonensis	A_1	A_2	B	C	D_1	–	–	
15. Treron bicincta *leggei	A_1	A_2	B	C	D_1	–	–	
16. Treron phoenicoptera *phillipsi	A_1	A_2	B	C	–	–	–	
17. Treron pompadora *pompadora	A_1	A_2	B	C	D_1	D_2	–	
Family: Psittacidae								
18. Loriculus *beryllinus	–	–	B	C	D_1	D_2	D_3	Plasmodium sp.
								Proteosoma sp.
								*Paronia biuterina
								*Paronia coryllidis
								*Biporouterina psittaculae
19. Psittacula *calthorpae	–	–	B	C	D_1	D_2	D_3	
Family: Cuculidae								
20. Cacomantis sonnerati *waiti	A_1	A_2	B	C	D_1	–	–	
21. Centropus *chlororhynchus	–	–	–	–	D_1	–	–	
22. Cuculus varius *ciceliae	–	–	–	–	D_1	D_2	D_3	
23. Surniculus lugubris *stewarti	A_1	A_2	B	C	D_1	D_2	–	
Family: Strigidae								
24. Bubo nipalensis *blighi	A_1	A_2	B	C	D_1	D_2	D_3	
25. Bubo zeylonensis *zeylonensis	A_1	A_2	B	C	D_1	D_2	D_3	*Haplorchis pearsoni
26. Glaucidium radiatum *castanonotum	–	–	–	–	D_1	D_2	D_3	
27. Otus scops *leggei	A_1	A_2	B	C	D_1	D_2	–	
28. Phodilus badius *assimilis	–	–	–	–	D_1	D_2	–	
29. Strix leptogrammica *ochrogenys	A_1	A_2	B	C	D_1	D_2	D_3	
Family: Caprimulgidae								
30. Caprimulgus asiaticus *eidos	A_1	A_2	B	C	D_1	–	–	
31. Caprimulgus indicus *kelaarti	–	–	–	C	D_1	D_2	D_3	
32. Caprimulgus macrurus *aequabilis	–	–	B	C	D_1	–	–	

Table 1 (continued)

Host	Locality							Parasite
	A₁	A₂	B	C	D₁	D₂	D₃	
Family: Apodidae								
33. *Apus affinis *singalensis*	A₁	A₂	B	C	D₁	–	–	
34. *Apus melba *bakeri*	A₁	A₂	B	C	D₁	D₂	D₃	
Family: Trogonidae								
35. *Harpactes fasciatus *fasciatus*	–	–	B	C	D₁	D₂	D₃	*Hymenolepis uragahaensis*
Family: Alcedinidae								
36. *Alcedo meninting *phillipsi*	–	–	B	C	D₁	–	–	
Family: Coraciidae								
37. *Eurystomus orientalis *iris*	–	–	B	C	–	–	–	
Family: Bucerotidae								
38. *Tockus griseus *gingalensis*	A₁	A₂	B	C	D₁	–	–	
Family: Capitonidae								
39. *Megalaima *falvifrons*	–	–	B	C	D₁	D₂	D₃	*Haemoproteus* sp. *Proteosoma* sp.
40. *Megalaima rubricapilla *rubricapilla*	A₁	A₂	B	C	D₁	–	–	
Family: Picidae								
41. *Chrysocolaptes festivus *tentus*	A₁	–	B	–	–	–	–	
42. *Chrysocolaptes lucides *stricklandii*	A₁	A₂	B	C	D₁	D₂	D₃	*Krimi chrysocolaptis*
43. *Dendrocopos nanus *gymnopthalmus*	A₁	A₂	B	C	D₁	–	–	
44. *Dinopium benghalense *jaffnense*	A₁	–	B	–	–	–	–	
45. *Dinopium benghalense *psarodes*	–	–	B	C	D₁	D₂	–	
46. *Picus chlorolophus *wellsi*	–	–	B	C	D₁	D₂	D₃	
Family: Hirundinidae								
47. *Hirundo daurica *hyperythra*	A₁	A₂	B	C	D₁	–	–	
Family: Oriolidae								
48. *Oriolus xanthornus *ceylonensis*	A₁	A₂	B	C	D₁	–	–	

Table 1 (continued)

Host	Locality							Parasite
	A₁	A₂	B	C	D₁	D₂	D₃	
Family: Dicruridae								
49. Dicrurus adsimilis *minor	A₁	–	–	–	–	–	–	
50. Dicrurus caerulescens *insularis	–	A₂	B	C	D₁	–	–	
51. Dicrurus caerulescens *leucopygialis	–	–	–	–	D₁	–	–	
52. Dicrurus paradiseus *ceylonensis	A₁	A₂	B	C	D₁	–	–	
53. Dicrurus paradiseus *lophorhinus	–	–	–	–	–	–	–	
Family: Sturnidae								
54. Acridotheres tristis *melanosternus	A₁	A₂	B	C	D₁	D₂	D₃	Haemoproteus sp.
								*Lankesterella lainsoni
								Plasmodium circumflexum
								Plasmodium vaughani
								Trypanosoma sp.
								Hymenolepis ellisoni
55. Gracula *ptilogenys	–	–	–	–	D₁	D₂	D₃	
56. Sturnus *senex	–	–	–	–	D₁	D₂	D₃	
Family: Corvidae								
**57. Kitta *ornata	–	–	–	–	D₁	D₂	D₃	
Family: Campephagidae								
58. Coracina novaehollandiae *layardi	A₁	A₂	B	C	D₁	D₂	–	
59. Hemipus picatus *leggei	–	–	–	–	D₁	D₂	D₃	
60. Tephrodornis pondicerianus *affinis	–	–	–	–	D₁	D₂	–	
Family: Pycnonotidae								
61. Hypsipetes indicus *guglielmi	–	–	–	–	D₁	D₂	–	
62. Hypsipetis madagascariensis *humii	–	–	B	C	D₁	D₂	–	
63. Pycnonotus caffer *haemorrhousus	A₁	A₂	B	C	D₁	D₂	D₃	
64. Pycnonotus luteolus *insulae	A₁	A₂	B	C	D₁	–	–	
65. Pycnonotus melanicterus *melanicterus	A₁	A₂	B	C	D₁	–	–	
**66. Pycnonotus *penicillatus	–	–	–	–	–	D₂	D₃	

Table 1 (continued)

Host	Locality							Parasite
	A_1	A_2	B	C	D_1	D_2	D_3	
Family: Muscicapidae								
67. Acrocephalus stentoreus *meridionalis	A_1	–	B	C	D_1	–	–	
**68. Bradypterus *palliseri	–	A_2	B	C	D_1	D_2	D_3	
69. Chrysomma sinensis *nasalis	A_1	A_2	B	C	D_1	D_2	D_3	
70. Cisticola juncidis *omalura	–	–	–	–	D_1	D_2	D_3	Spiloptera sp.
71. Copsychus malabarica *leggei	A_1	A_2	B	C	D_1	D_2	D_3	
72. Dumetia hyperythra *phillipsi	A_1	A_2	B	C	D_1	D_2	D_3	
**73. Garrulax *cinereifrons	–	–	–	–	D_1	D_2	D_3	
74. Muscicapa *sordida	–	–	–	–	D_1	–	–	
75. Muscicapa tickelliae *jerdoni	–	–	–	C	D_1	D_2	–	
**76. Myiophoneus *blighi	–	–	–	–	D_1	–	–	
77. Orthotomus sutorius *fernandonis	–	–	–	–	–	D_2	D_3	
78. Orthotomus sutorius *sutorius	A_1	A_2	B	C	D_1	D_2	–	
**79. Pellorneum *fuscocapillum *babaulti	A_1	A_2	B	–	D_1	–	D_3	
**80. Pellorneum *fuscocapillum *fuscocapillum	–	–	–	–	D_1	D_2	–	
**81. Pellorneum *fuscocapillum *scortillum	–	–	–	–	D_1	–	–	
82. Pomatorhinus schisticeps *holdsworthi	–	–	–	–	D_1	–	–	
83. Pomatorhinus schisticeps *melanurus	–	–	–	–	D_1	–	–	
84. Prinia hodgsonii *pectoralis	–	–	–	–	D_1	–	–	
85. Prinia socialis *brevicaudata	–	–	–	–	D_1	–	–	
86. Prinia subflava *insularis	–	–	B	C	D_1	–	–	
87. Prinia sylvatica *valida	–	–	B	C	D_1	–	–	
88. Rhopocichla atriceps *nigrifrons	–	–	–	C	D_1	D_2	D_3	
89. Rhopocichla atriceps *siccatus	A_1	A_2	B	C	D_1	D_2	D_3	
90. Saxicola caprata *atrata	–	–	–	–	D_1	–	–	
91. Saxicoloides fulicata *leucoptera	–	–	–	–	D_1	–	–	
92. Terpsiphon paradisi *ceylonensis	–	–	–	–	D_1	–	–	
93. Turdus merula *kinnisii	–	–	–	–	D_1	D_2	D_3	
94. Turdoides affinis *taprobanus	A_1	A_2	B	C	D_1	D_2	D_3	
95. Turdoides *rufescens	–	–	–	–	D_1	D_2	D_3	

Table 1 (continued)

Host	A₁	A₂	B	C	D₁	D₂	D₃	Parasite
96. Zoothera dauma *imbricata	–	–	–	–	–	D₂	D₃	
97. Zoothera *spiloptera	A₁	A₂	B	C	D₁	D₂	D₃	
Family: Dicaeidae								
98. Dicaeum agile *zeylonicum	A₁	–	B	–	–	–	–	
99. Dicaeum erithrorhynchus	A₁	A₂	B	C	D₁	D₂	D₃	
**100. Dicaeum *vincens	–	–	–	–	D₁	–	–	
Family: Nectariniidae								
101. Nectarinia lotenia *lotenia	A₁	A₂	B	C	D₁	D₂	D₃	
102. Nectarinia zeylonica *zeylonica	–	–	–	–	D₁	–	–	
Family: Zosteropidae								
103. Zosterops *ceylonensis	–	–	–	–	D₁	D₂	D₃	
Family: Ploceidae								
104. Lonchura kelaarti *kelaarti	–	–	–	–	D₁	D₂	D₃	
MAMMALIA								
Family: Soricidae								
1. Suncus etruscus *fellowes-gordoni	A₁	A₂	B	C	D₁	D₂	–	
2. Suncus murinus *kandianus	A₁	A₂	B	C	D₁	D₂	–	Pseudophysaloptera formosana formosana
3. Suncus murinus *montanus	A₁	A₂	B	C	D₁	D₂	D₃	*Hymenolepis calva *Pseudhymenolepis eisenbergi *Vampirolepis montana
4. Suncus murinus *zeylanicus	–	A₂	B	C	D₁	D₂	–	
5. Crocidura *miya	–	–	–	–	–	D₂	D₃	
**6. *Feroculus *feroculus	–	–	–	–	–	–	D₃	
**7. *Solisorex *pearsoni	–	–	–	–	–	D₂	D₃	*Vampirolepis solisoricis
Family: Pteropodidae								
8. Cynopterus brachyotis *ceylonensis	A₁	A₂	B	C	D₁	D₂	D₃	

Table 1 (continued)

Host	Locality							Parasite
	A₁	A₂	B	C	D₁	D₂	D₃	
Family: Megadermatidae								
9. Megaderma spasma *ceylonense	A₁	A₂	B	–	D₁	D₂	–	
Family: Rhinolophidae								
10. Rhinolophus luctus *sobrinus	A₁	A₂	B	–	D₁	–	–	
Family: Vespertilionidae								
11. Pipistrellus ceylonicus *ceylonicus	–	–	–	C	D₁	D₂	D₃	
12. Murina cyclotis *eileenae	–	–	–	–	D₁	D₂	–	
13. Kirivoula hardwickei *malpasi	–	–	–	–	–	D₂	–	
Family: Molossidae								
14. Tadarida plicata *insularis	–	–	–	–	–	D₂	–	
Family: Lorisidae								
15. Loris tardigradus *tardigradus	–	–	–	–	D₁	–	–	*Phaneropsolus lakdivensis
								Subulura perarmata
16. Loris tardigradus *grandis	A₁	–	–	–	D₁	D₂	–	Subulura perarmata
17. Loris tardigradus *nordicus	–	A₂	B	–	–	–	–	
18. Loris tardigradus *nycticeboides	–	–	–	–	–	–	D₃	
Family: Cercopithecidae								
19. Presbytis *senex *senex	A₁	–	B	–	–	–	–	Plasmodium kochi
20. Presbytis *senex *vetulus	–	–	–	–	D₁	D₂	–	
21. Presbytis *senex *nestor	–	–	–	–	D₁	–	–	
22. Presbytis *senex *monticola	–	–	–	–	–	D₂	D₃	
23. Presbytis *senex *harti	A₁	–	–	–	–	–	–	Plasmodium shortti
24. Macaca *sinica *sinica	A₁	A₂	B	C	–	–	–	*Plasmodium cynomolgi ceylonensis
								Plasmodium fragile
								*Plasmodium simiovale
								Streptopharagus pigmentatus
								Theileria celli

Table 1 (continued)

No.	Host	A_1	A_2	B	C	D_1	D_2	D_3	Parasite
25.	Macaca *sinica *aurifrons	—	—	—	—	D_1	D_2	—	
26.	Macaca *sinica *opisthomelas	—	—	—	—	—	—	D_3	
	Family: Leporidae								
27.	Lepus nigricollis *singhala	A_1	A_2	B	C	D_1	D_2	D_3	Mosgovoyia pectinata Multiceps serialis Brugia (Brugiella) buckleyi Micipsella sp. Trichostrongylus pigmentatus Capillaria hepatica
	Family: Sciuridae								
28.	Petaurista petaurista *lanka	—	—	—	—	D_1	D_2	D_3	
29.	Petinomys fuscocapillus *layardi	—	—	B	C	D_1	D_2	—	
30.	Funambulus palmarum *kelaarti	A_1	A_2	B	—	—	—	—	
31.	Funambulus palmarum *brodiei	A_1	A_2	—	—	—	—	—	Capillaria hepatica
32.	Funambulus palmarum *matugamensis	—	—	B	C	D_1	—	—	
33.	Funambulus layardi *layardi	—	—	B	C	D_1	D_2	—	
34.	Funambulus sublineatus *obscurus	—	—	—	—	D_1	D_2	D_3	
35.	Ratufa macroura *macroura	—	—	—	—	—	D_2	D_3	
36.	Ratufa macroura *melanochra	—	—	—	—	D_1	D_2	D_3	
37.	Ratufa macroura *dandolena	A_1	A_2	B	C	D_1	—	—	Hepatocystis vassali var. ratufae
	Family: Muridae								
38.	Vandeleuria oleracea *nolthenii	—	—	—	—	—	D_2	D_3	
39.	Rattus rattus *kandianus	A_1	A_2	B	C	D_1	D_2	D_3	
40.	Rattus rattus *kelaarti	—	—	—	—	—	D_2	D_3	
41.	Rattus *montanus	—	—	—	—	—	—	D_3	
42.	Rattus *ohiensis	—	—	—	—	—	D_2	D_3	
43.	Mus cervicolos *fulvidiventris	A_1	A_2	B	C	D_1	D_2	D_3	
44.	Mus *fernandoni	A_1	A_2	B	C	—	D_2	—	
45.	Mus *mayori *pococki	—	—	—	—	D_1	D_2	—	

Table 1 (continued)

Host	Locality							Parasite
46. Mus *mayori *mayori	—	—	—	—	—	D₂	D₃	Brachylaemus advena Hymenolepis sp.
47. Golunda ellioti *ellioti	—	A₂	B	C	D₁	D₂	—	
48. Golunda ellioti *nuwara	—	—	—	—	—	D₂	D₃	
49. Bandicota bengalensis *gracilis	A₁	A₂	B	C	D₁	D₂	—	
50. Tatera indica *ceylonica	A₁	A₂	B	C	D₁	—	—	
Family: Canidae								
51. Canis aureus *lanka	A₁	A₂	B	C	D₁	D₂	—	Babesia gibsoni Echinochasmus liliputanus Echinococcus granulosus
Family: Mustelidae								
52. Lutra lutra *nair	A₁	A₂	B	C	D₁	D₂	D₃	
Family: Viverridae								
53. Viverricula indica *mayori	A₁	A₂	B	C	D₁	—	—	Brachylaemus fuscatus Paragonimus compactus Pharyngostomum cordatum Testifrondosa cristata Ancylostoma braziliense
54. Paradoxurus *zeylonensis	A₁	A₂	B	—	D₁	D₂	—	Paragonimus siamensis Trogloterma sp.
55. Herpestes edwardsi *lanka	A₁	A₂	B	C	—	—	—	
56. Herpestes fuscus *flavidens	—	—	—	—	—	D₂	D₃	
57. Herpestes fuscus *maccarthiae	A₁	—	B	—	—	—	—	
58. Herpestes fuscus *ceylanicus	A₁	—	B	—	—	—	—	
59. Herpestes smithi *zeylanius	A₁	A₂	B	C	D₁	D₂	—	Euparadistomum buckleyi

344

345

Table 1 (continued)

Host	Locality							Parasite
	A_1	A_2	B	C	D_1	D_2	D_3	
Family: Elaphantidae								
60. *Elephas maximus *maximus*	A_1							*Fasciola jacksoni*
								Decrusia additictia
								Equinurbia sipunculiformis
								Grammocephalus hybridatus
								Murshidia falcifera
								Murshida longicaudata
								Murshida murshida
								Parabronema smithi
								Quilonia renniei
Family: Cervidae								
61. *Axis axis *ceylonensis*	A_1				D_1	–	–	*Babesia pattoni*
								Paramphistomum bathycotyle
								Taenia hydatigena
								Artionema sp.
								Setaria axis

greatest number of them are confined to the rainforest zone D_1, below 914 m, but most of the relict forms occur in the montane zones D_2 and D_3.

The Uropeltidae are a relict family of snakes confined to Sri Lanka and the hill country of southwestern India (Gans & Baic 1977). They are 'adapted primarily to the shaded, low-temperature soils of the montane tropical rain forests' (Gans 1976). In Sri Lanka they are represented by 12 endemic species and one endemic subspecies belonging to four genera, including the endemic *Pseudotyphlops*, all living mostly in the rainforest zones, especially D_1, below 914 m (Gans 1966). They have been shown (Crusz & Ching 1975, Baker & Crusz 1980) to have two characteristic species of nematodes, *Oswaldocruzia gansi* and *Cosmocercella uropeltidarum*, which generally coexist in the same host but in different habitats, i.e. the small intestine and rectum respectively (see also Gans 1976).

The colubrid snake genera, *Cercaspis, Aspidura, Haplocercus,* and *Balanophis* all are endemic. Of these, *Aspidura* has 5 species, all restricted to the rainforest zones, mostly D_1 and D_2. Although *Aspidura trachyprocta* has been reported from the Maldives (Laidlaw 1902), it is doubtful whether it could be considered as anything other than a very recent introduction there (Taylor 1950). Hence all the *Aspidura* species would belong to the category of Sri Lankan relicts.

The group that is perhaps most markedly relict is the Lyriocephalinae, a subfamily of agamid lizards represented in Sri Lanka by the endemic genera *Lyriocephalus* (monotypic) and *Ceratophora* (with 3 endemic species), and by the endemic species *Cophotis ceylanica*, whose only other congener is *Cophotis sumatrana* of Sumatra and Java. They would have had a wider distribution within the island, but are now confined mostly to rainforest zones (D_2 and D_3) above 914 m, with *Cophotis ceylanica* and *Ceratophora stoddarti* occurring in D_3, above 1,524 m. The parsite complex of the Lyriocephalinae appears to be marked by a microevolution of nematodes of the subfamily Meteterakinae, and by the trematode genus and species *Zeylanurotrema lyriocephali*.

Limbless skinks of the subfamily Acontianinae are represented in Sri Lanka by 9 endemic species belonging to 5 endemic genera, all relict forms. They inhabit the rainforest region D_1. Two species are found in D_2, and 3 outside the rainforest area, in zones A_1. Of these latter, *Anguinicephalus hickanala* of zone A_1 is perhaps the only relict animal to be found anywhere in the dry zone. The Acontianinae in Sri Lanka show a microevolution from 4-limbed forms, found mostly at higher altitudes, to limbless forms found at the lowest altitudes (Deraniyagala 1965). The other species of this subfamily, belonging to quite different genera, occur only in Africa and Madagascar. The parasites of the Acontianinae of Sri Lanka have just begun to be investigated. A *Meteterakis* sp has been found in *Nessia didactyla* in Dewatura (860 m) near Namuna Kula, and what appears to be a *Thelandros* sp. has been recovered from *Bipedos smithi* in the Gammaduwa area (720 m) (Crusz & Santiapillai in prep.).

Aves

There are 248 species and subspecies of birds in Sri Lanka, excluding visitors, stragglers, vagrants and species of doubtful status. 104 (42%) of these are endemic, with no endemic genera, 20 endemic species and 84 endemic subspecies. Eight of these endemics are relict species showing affinities with eastern Himalayan or Oriental forms, many of which are derivatives of Malaysian or southern Chinese species (Ripley 1980).

Only one relict species, *Zoothera spiloptera*, has an islandwide distribution. One of the three subspecies of the relict *Pellorneum fuscocapillum* is confined to the entire dry zone (A_1, A_2 and B). The other relicts are variously distributed in the rainforest zones D_1, D_2 and D_3, with *Pycnonotus penicillatus* occurring only in D_2 and D_3, above 1,524 m.

Special mention must be made of *Phaenicophaeus pyrrhocephalus*, which was long regarded, even by Ripley (1949) as an endemic and relict species is Sri Lanka. It had, however, been obtained by Stewart, and recorded by Legge, in the Cochin-Travancore area, and Stuart-Baker (1927), Wait (1931) and Biddulph (1956) have accepted this. Biddulph, referring in convincing detail to his close sighting of the bird in the Madurai district of the Madras Presidency, thus confirmed these earlier reports of its occurrence in South India. Ripley (1961, 1980) has since accepted this position.

Although parasites have from time to time been recorded from Sri Lankan birds, including some of the endemic birds, there has apparently been no parasitological study of the relict species. There is now in progress. Preliminary observations, however, seem to support a key concept regarding the paucity of parasites in relict hosts. This may account for the fact that Burt (1936-1977) has reported none from relict birds, although he may well have examined at least some of them for parasites.

Table 2. Numbers of Inland Vertebrates of Sri Lanka.

	Genera	Endemic genera	Species	Endemic species	Species & subspecies	Endemic species & subspecies
Pisces (Excluding introduced forms)	32	3 (9%)	59	16 (27%)	62	28 (45%)
Amphibia	10	1 (10%)	37	19 (51%)	37	19 (51%)
Reptilia	68	12 (19%)	139	72 (51%)	159	98 (62%)
Aves (Resident only)	165	0	237	20 (8%)	248	104 (42%)
Mammalia	51	2 (4%)	85	10 (12%)	114	61 (54%)
	326	18 (6%)	557	137 (25%)	620	310 (50%)

348

Mammalia

The mammals of Sri Lanka comprise 114 taxa, of which 61 (54%) are endemic. The endemics include two monotypic genera (of soricid shrews, *Feroculus* and *Solisorex*), 10 species and 51 subspecies. Most of them inhabit the rainforest montane zones D_1, D_2 and D_3.

Two of the 3 endemic species of shrews are in zone D_2 and D_3, while the third, *Feroculus feroculus*, is confined to the highest zone D_3. This, together with the fact that *F. feroculus* and *Solisorex pearsoni* are so isolated as to be extremely rare, qualify them as relict forms. The other shrews, represented by 4 endemic subspecies, have an islandwide distribution.

Among the other mammals, the subspecies of the endemic carcopithecid monkeys *Presbytis senex* and *Macaca sinica* show an interesting geographical distribution in Sri Lanka, at least one subspecies of each being restricted to the dry zones (A and B), one of the lower montane zones (D_1 and D_2), and one to the highest zone (D_3). A similar pattern of distribution is shown by the endemic species of *Rattus* and *Mus*.

Most of the named parasites of the endemic mammals occur also in hosts outside Sri Lanka, but those of the relict shrews tend to be distinctive. However, as in the case of the endemic and relict species of the other vertebrate groups, much more work on their parasites has to be done, before firmer conclusions can be drawn.

Acknowledgements

The valuable assistance rendered in the preparation of the manuscript of this chapter by Mrs Anoma Santiapillai, Research Assistant and Mrs Priscilla Pereira, Secretary of the Department of Zoology, and Mr S.M.B. Amunugama, Draughtsman of the Department of Zoology, University of Peradeniya, is gratefully acknowledged. Also valuable was the kind cooperation of Vice-Chancellor Prof. B.L. Panditharatne and Dean of Science Prof. H.W. Dias in supporting the awards of University Research Grants for the many field study trips and other facilities this work required.

References

Abeywickrama, B.A. 1956. The origin and affinities of the flora of Ceylon. Proc. 11th Ann. Sess. Ceylon Assoc. Adv. Sci. Part 2:99-121.

Abeywickrama, B.A. 1959. The evolution of the flora of Ceylon. Proc. 14th Ann. Sess. Ceylon Assoc. Adv. Sci. Part 2:217-219

Abeywickrama, B.A. 1980. The flora of Sri Lanka (Ceylon). Spol. Zeylan. 35 (Colombo Museum Centenary Volume):1-8.

Baker, M.R. & H. Crusz. 1980. *Cosmocercella uropeltidarum* (Crusz & Ching, 1975) n. comb. (Nematoda:Cosmocercidae) parasite des serpents (Uropeltidae) d'Asie. Bull. Mus. natn. Hist. nat. Paris. 4 sér. Sect. A(3), 2:719-722.

Bathie, H.P. de la. 1936. Biogéographie des plantes de Madagascar. Soc. édit. Géo. mar. et Col. édit. Paris. 156 pp.

Baylis, H.A. 1938. Helminths and evolution. pp. 249-270 In: Evolution. Essays on Aspects of Evolutionary Biology. G.R. de Beer (Ed.). Oxford Univ. Press.

Biddulph, C.H. 1956. Occurrence of the redfaced malkoha. *Phoenicophaus pyrrhocephalus* (Pennant) in Madura district, Madras Presidency. J. Bombay nat. Hist. Soc. 53:697-698.

Blanc, C.P. 1970. Intérèt des Reptiles endémiques Malgaches. (Paper read at the) Conférence Internationale sur l'Utilisation Rationelle et la Conservation de la Nature, Tananarive, October 1970: 8 pp (mimeographed).

Boucot, A.J. 1976. Request for information on parasites. J. Parasit. 62:38.

Brooks, D.R. 1979. Testing hypotheses of evolutionary relationships among parasites: the digeneans of crocodiles. Am. Zool. 19:1225-1238.

Burt, D.R.R. (1936-1977). Ceylon J. Sci. (B) Zoology 19, 21 and 22 and Zool. J. Linn. Soc. 53, 58 and 61.

Cameron, T.W.M. 1952. Parasitism, evolution and phylogeny. Endeavour 11:193-199.

Chabaud, A.G. & Brygoo, E.R. 1964. L'endémisme chez les helminthes de Madagascar. C.R. Soc. Biogéogr. 356:3-13.

Cooray, P.G. 1967. An introduction to the geology of Ceylon. Nat. Mus. Ceylon Publ. Colombo. 324 pp.

Crusz, H. 1973. Nature conservation in Sri Lanka (Ceylon). Biol. Conserv. 5:199-208.

Crusz, H. & C.C. Ching. 1975. Parasites of the relict fauna of Ceylon V. New species of nematodes from uropeltid snakes. Ann. Parasit. hum. comp. 50:339-349.

Crusz, H. & C.C. Ching 1976. Parasites of the relict fauna of Ceylon VI. More new helminths from amphibians and reptiles, with a new host-record and redescription of *Acanthocephalus serendibensis* Crusz & Mills, 1970. Ann Parasit. hum. comp. 50:531-558.

Crusz, H. & E.V. Mills. 1970. Parasites of the relict fauna of Ceylon I. *Acanthocephalus serendibensis* sp. nov. from the Ceylon horn-nosed lizard, *Ceratophora stoddarti* Gray. Ann. Parasit hum. comp. 45:13-19.

Crusz, H. & L. Nugaliyadde. 1978. Parasites of the relict fauna of Ceylon VII. General considerations and first host-parasite checklist. C.R. Soc. Biogéogr. 477:85-106. Addenda and corrigenda. *Ibid.* (1980) 489-492:87-88.

Crusz, H. & V. Sanmugasunderam. 1972. Parasites of the relict fauna of Ceylon II. New species of cyclophyllidean cestodes from small hill-vertebrates. Ann. Parasit. hum. comp. 46:575-588.

Crusz, H. & V. Sanmugasunderam. 1974a. Parasites fo the relict fauna of Ceylon III. Nematodes from a rhacophorid frog and reptiles of the hill country. Ann. Parasit. hum. comp. 48:767-795.

Crusz, H. & V. Sanmugasunderam. 1974b. Parasites of the relict fauna of Ceylon IV. *Zeylanurotrema lyriocephali* gen. et sp. nov. and other trematodes from mountain lizards and a rodent. Ann. Parasit. hum. comp. Ann. Parasit. hum. comp. 48:797-810.

Crusz, H. & A. Santiapillai. 1982. Parasites of the relict fauna of Ceylon VIII. Helminths from *Ichthyophis* spp. (Amphibia:Gymnophiona). Ann. Parasit. hum. comp. 57:317-327.

Delamare-Deboutteville, C. & L. Botosanéanu. 1970. Formes primitives vivantes: Musée de l'Evolution. Hermann, Paris. 232 pp.

Deraniyagala, P.E.P. 1953. A coloured atlas of some vertebrates from Ceylon 2. Tetrapod Reptilia. Nat. Mus. Ceylon. Publ. Colombo. 101 pp.

Deraniyagala, P.E.P. 1955. A coloured atlas of some vertebrates from Ceylon 3. Serpentoid Reptilia. Nat. Mus. Ceylon Publ. Colombo. 121 pp.

Deraniyagala, P.E.P. 1965. Some aspects of the fauna of Ceylon. J.R. Asiatic Soc. (Ceylon Br.) N.S. 9:165-219.

350

Dogiel, V.A. 1947. The importance of the parasitological evidence for the solution of the zoogeographical problems. Zool. Zh. 26:481-492.

Dubost, G. & C.M. Hladik. 1971. Les gros mammifères de la réserve de Wilpattu à Ceylan. 2ᵉᵐᵉ Partie, Sci. Nat. 105:3-11.

Durette-Desset, M.C. 1971. Essai de classification des Nématodes Héligmosomes. Corrélations avec la paléobiogéographie des hôtes. Mem. Mus. Nat. Hist. Natur. Ser. A. 69:1-126.

Eisenberg, J.F. & G.M. McKay. 1970. An annotated checklist of the recent mammals of Ceylon with keys to the species. Ceylon. J. Sci. (Biol. Sci.) 8:69-99.

Ekman, S. 1915. Vorschläge und Erörterungen zur Reliktenfrage in der Hydrobiologie. Ark. Zool. 9:17.

Fernando, S.N.U. 1968. The natural vegetation of Ceylon. Lake House Bookshop, Colombo. 85 pp.

Gans, C. 1966. Liste der rezenten Amphibien und Reptilien:Uropeltidae. Das Tierreich 84:1-29.

Gans, C. 1976. Aspects of the biology of uropeltid snakes. pp. 191-204 In: Morphology and biology of reptiles. A. d'A. Bellairs & C.B. Cox (Eds.). Linn. Soc. Lond.

Gans, C. & D. Baic. 1977. Regional specialization of reptilian scale surfaces: relation of texture and biologic role. Science 195:1358-1350.

Gaussen, H., P. Legois, M. Viart & L. Labroue. 1964. International Map of Vegetation, Ceylon. Ceylon Surv. Dept Publ. Colombo.

Ihering, H. von. 1891. On the ancient relations between New Zealand and South America. Trans. Proc. N.Z. Inst. 24:431-455.

Ihering, H. von. 1903. Die Helminthen als Hilfsmittel der zoogeograpischen Forschung. Zool. Anz. 26:42-51.

Inglis, W. 1967. The geographical and evolutionary relationships of Australian trichostrongyloid parasites and their hosts. J. Linn. Soc. (Zool.) 47:327-347.

Kirtisinghe, P. 1946. The presence in Ceylon of a frog with direct development on land. Ceylon J. Sci. (B) 23:109-112.

Kirtisinghe, P. 1957. The amphibia of Ceylon. Publ by the author. Colombo. 112 pp.

Laidlaw, F.F. 1903. Amphibia and Reptilia. pp. 119-122 In: The Fauna and Geography of the Maldive and Laccadive Archipelagos. J.S. Gardiner (Ed.). Cambridge Univ. Press.

Mani, M.S. 1974. Ecology and Biogeography in India. Dr W. Junk, The Hague.

Melville, R. 1970. Significance of the Madagascan flora among floras of the world. (Paper read at the) Conf. Int. l'Util. Rat. Cons. Nat., Tananarive, October 1970. 17 pp. (mimeographed).

Metcalf, M.M. 1923. The Opalinid Ciliate Infusorians. Smithsonian Inst. U.S. Nat. Mus. Bull. 120:484 pp. Washington D.C.

Metcalf, M.M. 1929. parasites and the aid they give in problems of taxonomy, geographical dsitribution, and paleogeography. Smithsonian Mus. Coll. 81(8):36 pp.

Müller-Dombois, D. 1968. Ecogeographic analysis of a climate map of Ceylon with particular reference to vegetation. Ceylon Forester 8:39-58.

Müller-Dombois, D. & V.A. Sirisena. 1967. Climate map of Ceylon. Ceylon Surv. Dept Publ. Colombo.

Munro, I.S.R. 1955. The marine and fresh water fishes of Ceylon. Dept. Ext. Aff. Publ. Canberra. 351 pp.

Nicholson, E.M. 1970. Role Possible de Madagascar au Sein d'un Réseau Mondial de Centres Biologiques. (Paper read at the) Conf. Int. l'Util. Rat. Cons. Nat. Tananarive, October 1970: 11 pp. (mimeographed).

Nussbaum, R.A. & C. Gans. 1980. On the *Ichthyophis* (Amphibia:Gymnophiona) of Sri Lanka. Spol. Zeylan. 35 (Colombo Mus. Cent. Comm. Vol.):137-154.

Prentice, H.C. 1976. A study in endemism: *Silene diclinis*. Biol. Cons. 10:15-30.

Rausch, R.L. 1957. Distribution and specificity of helminths in microtine rodents: evolutionary implications. Evolution 11:361-368.

Ripley, S.D. 1949. Avian relicts and double invasions in peninsular India and Ceylon. Evolution 3:150-159.

Ripley, S.D. 1961. A synopsis of the birds of India and Pakistan together with those of Nepal, Sikkim, Bhuttan and Ceylon. Bomb. Nat. Hist. Soc. Madras. 703 pp.

Ripley, S.D. 1980. Avian relicts of Sri Lanka. Spol. Zeylan. 35 (Colombo Mus. Cent. Comm. Vol.):197-202.

Sarasin, F. 1910. Über die Geschichte der Tierwelt von Ceylon. Zool. Jahrb. Suppl. 12(1):1-160.

Seddon, B. 1971. Introduction to Biogeography. Duckworth, London. 220 pp.

Senanayake, F.R. 1977. Habitat values and endemicity in the vanishing rain forests of Sri Lanka. Nature, Lond. 268:568.

Senanayake, F.R. in press. A new species of cyprinid fish from Sri Lanka. Ceylon J. Sci. (Biol.Sci.) 15.

Senanayake, F.R., M. Soulé, & J.W. Senner. 1977. Habitat values and endemicity in the vanishing rain forests of Sri Lanka. Nature Lond. 265:351-354.

Stuart-Baker, E.C. 1927. The fauna of British India including Ceylon and Burma. Birds IV. Taylor & Francis, London. 472 pp.

Stunkard, H.W. 1967. Platyhelminthic parasites of invertebrates. J. Parasit. 53:673-682.

Stunkard, H.W 1970. Trematode parasites of insular and relict vertebrates. J. Parasit. 56:1041-1054.

Taylor, E.H. 1950. The snakes of Ceylon. Univ. Kansas Sci. Bull. 33:519-603.

Thienemann, A. 1950. Verbreitungsgeschichte der Süsswassertierwelt Europas. Die Binnengewässer. Stuttgart.

Vanzolini, P.E. & L.R. Guimarães. 1955. South American land mammals and their lice. Evolution 9:345-347.

Vassiliades, G. 1970. Nématodes parasites d'oiseaux malgaches. Ann. Parasit. hum. comp. 45:47-88.

Wait, W.E. 1931. Manual of the birds of Ceylon. 2nd edit. Nat. Mus. Colombo. 494 pp.

17. Ecological aspects of some parasitic diseases in Sri Lanka

A.S. Dissanaike

Introduction

As the term 'ecology' refers to the relationships between organisms and their environment, it cannot be used for a disease, and so the phrase 'ecology of disease' is incorrect, as pointed out by Audy (1958). However, many have used it in that context in the past. This chapter therefore discusses some ecological aspects of two important vector borne diseases of man in Sri Lanka, namely malaria and filariasis, and briefly refers to a few others which are of ecological interest because animal reservoirs are involved, making them important (or potential) zoonoses.

Ecological stimuli for disease

For a disease or infection to establish itself in man or for that matter in any vertebrate host, several stimuli are necessary. The response of the host to these stimuli shows up as disease or not, depending on its resistance or genetic makeup. The environment is complex and has three components; physical biological and socio-economic (Fig. 1). 'As such, a disease entity is the result of no simple contact between a host and an agent of disease. A great variety of environmental factors have a part in its origin, and a balance of affairs determines the issue' (Gordon 1958).

The environmental stimuli include climatic ones (such as temperature, humidty, winds, rainfall) and other physical factors (such as topography, rivers, soil factors, and the nature of the land, whether low-lying plains or mountainous). Biological stimuli, in addition to the causative organism itself, would include reservoir hosts of the parasites, the vectors which transmit them from one vertebrate host to another, and the vegetation; in other words, all the living organisms that make up the macroenvironment, as it is sometimes referred to. The microenvironment also is important and consists of the genetic and biochemical make-up of the individual host and its innate or acquired resistance or susceptibility to the invading parasite.

To social and economic factors also are of importance in that movement of

354

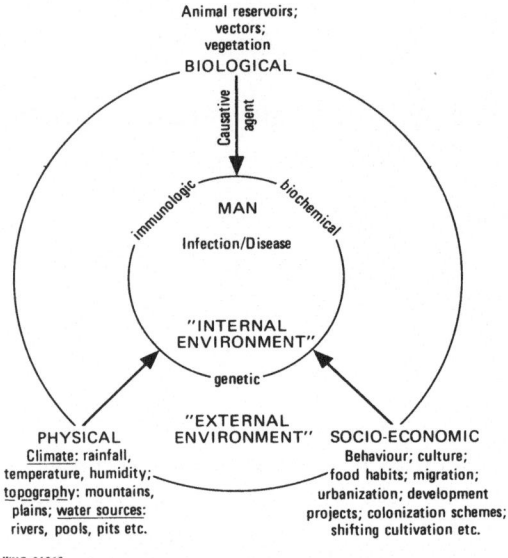

WHO 81968

Fig. 1. Diagrammatic representation of the internal and external environmental stimuli and factors responsible for disease.

people, urbanization, development schemes and projects, and behavioural and cultural practices including food habits, all tend to enhance or diminish the prevalance, maintenance and spread of a disease.

Some parasitic diseases in Sri Lanka

Let us now consider some human parastitic diseases in Sri Lanka from the ecological point of view. Malaria provides the best example of a disease in which many ecological factors have played a role in the epidemiology in the past and continue to do so, as recently pointed out by Schweinfurth (1977).

Malaria

Parasites and vector

Three protozoan parasites are responsible for malaria in Sri Lanka: *Plasmodium vivax, P. falciparum* and *P. malariae*, causing, respectively, benign tertian, malignant tertian and quartan fevers. They are intraerythrocytic parasites which have a phase of development in the parenchyma cells of the liver before they enter the erythrocytes to start the clinical manifestations of the disease, which are too well known to be repeated here.

These parasites are transmitted from an infected person to a susceptible one by female anopheline mosquitos, in which vectors the sexual cycle takes place. In Sri Lanka, the only such vector identified with certainty is *Anopheles culicifacies* (Carter & Jacocks 1929, Carter 1930), although several other anophalines which are known vectors in enighbouring countries are present, such as *A. maculatus, A. varuna* and *A. subpictus.*

As Carter (1929, 1930) showed, *A. culicifacies* is essentially a dry zone species with an extensive distribution in the jungle-covered plains of the north, east and southwest parts of the country, and rarely is found above an elevation of 2,500 ft (760 m). It feeds towards evening, especially at dusk until about 9 pm, and again briefly at dawn and has the habit of remaining in dwellings after biting, often for long periods in dark corners and crevices, a fact that is made use of in residual spraying with insecticides.

Its breeding places include slow-moving streams and rivers, but particularly sand and rock pools in the beds of rivers and irrigation channels, wells, temporary rain pools, and water in borrow-pits, quarries and gem pits. It prefers clear fresh water containing little aquatic vegetation and is a sun-loving species.

Physiography and climatic zones

As with many other diseases, there are for malaria both endemic areas and epidemic regions where the disease manifests itself very differently, and a knowledge of the topography, climate and drainage of these regions is essential to understand the differences, especially in relation to the vector ecology.

The south-central part of the country is mountainous, sloping gradually to sea level from over 7,000 ft (2,100 m). Areas over 1,000, 3,000 and 5,000 ft (300, 900 and 1,500 m) are indicated in Fig. 2. The rest of the country is more or less flat or undulating, sloping from the foothills of the mountains to the sea on all sides. The mountain slopes are scored in all directions by ravines cut by streams joining to form the main rivers of the island, most of which originate around 2,000-3,000 ft (600-900 m) above sea level. The mountain mass determines the chief rainfall variations in the whole island.

From the point of view of malaria, Sri Lanka has in the past been divided into three main zones, depending on the rainfall during the south-west monsoon from May to September (Rustomjee, 1944), as shown in Fig. 2. The rainfall during this monsoon is over 40 inches in the wet zone, between 20 and 40 inches in the intermediate zone and below 20 inches in the dry zone. The climatic zones, which show wide variation in altitude in addition to rainfall (Fig. 2) must also be considered in relation to the rivers running through them, particularly in the wet and intermediate zones. The most important rivers in relation to malaria transmission in the past have been the Mahaweli Ganga, the Deduru Oya, the Maha Oya, the Kelani Ganga in the north, and the Nilwala Ganga in the south (see Fig. 2).

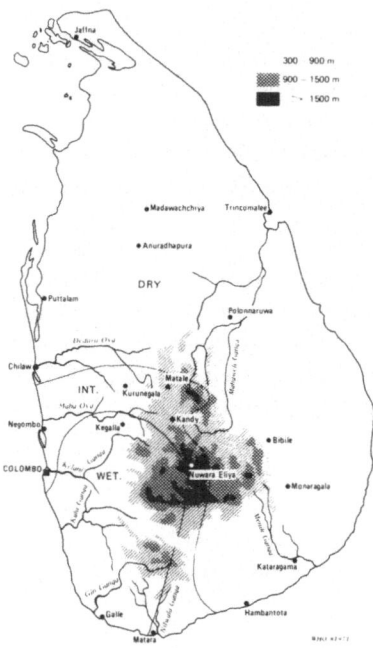

Fig. 2. Map of Sri Lanka showing the main rivers, contours over 300 metres and climatic zones (dry, wet and intermediate) in relation to malaria.

The wet zone covers an area of about 4,000 sq. miles (10,350 km²) in the south-west quadrant of the island; the dry zone covers most of the remaining part (19,000 sq. miles or 49,200 km²), while the intermediate zone (2,250 sq. miles or 5,820 km²) occupies a narrow strip in between. The dry zone receives its rainfall mainly from the north-east monsoon (October-June), whereas the other two zones are served by both monsoons.

Malaria was severe in the past both in intensity and distribution in the dry zone, except in the north and north-west portions of the peninsula of Jaffna district and in areas where the altitude is over about 3,000 ft (900 m). The dry zone was therefore largely hyperendemic and not subject to violent outbreaks of malaria, whereas the wet and intermediate zones, which were normally either slightly affected or not at all, were liable to epidemics.

As the wet zone has an abundant rainfall throughout the year, its rivers are nearly always flowing and so offer poor conditions for breeding of the vector. It is therefore relatively malaria-free normally. In the dry zone, as the rain falls mainly in the north-east monsoon, long periods of drought occur during the south-west monsoon. Malaria, in addition to being endemic there, has seasonal peaks during the north-east monsoon when the rains cause water collections that are satisfactory for breeding of the vector. On the other hand, when there is an abnormal pattern of rainfall, such as a failure or delay of the south-west monsoon, pooling of water occurs in the river beds in the wet and intermediate zones, and causes *A. culicifacies* to move into these zones and malaria too spreads in them.

Deforestation

Some years ago, Tyssul-Jones (1951) suggested that, as a result of deforestation of the mountain tracts in Sri Lanka for cultivation of tea and rubber, extensive soil erosion resulting in the loss of absorptive capacity of the soil followed by a quick run-off of rain water, led to the marked ebb and flow of rivers. He considered that if the forests had remained in their primeval state, sufficient rain water would have been held in the soil to tide over rainfall deficiencies during critical months and particularly in the Kurunegala, Matale, Kandy and Kegalle districts. These districts are associated with rivers whose catchment areas are situated in the lower hills and which have been most severely deforested.

Epidemiology and epidemics

Historically, epidemics of malaria have been reported once every 4-6 years. This cyclic periodicity reached its peak in the 1934/35 epidemic when there were over one million cases and around 80,000 deaths, mainly due to *P. falciparum* (Gill 1935, Briercliffe 1935). Seasonal increases of malaria transmission have been related to monsoons, usually marked when the pattern of rainfall was abnormal. In addition, periodic exacerbations due to waning immunity had given rise to the cyclical epidemics, and when, as in the 1934/35 epidemic, an abnormal rainfall pattern coincided with the cyclical upsurge, the epidemic was severe in the extreme.

After the introduction of DDT spraying in 1947 by the Anti-malaria Campaign there was a marked decrease in malaria transmission with a fall in maternal and infant mortality rates (Karunaratne 1959), although Meegama (1969) and Frederiksen (1970) do not agree that malaria control was mainly responsible. In 1958, a systematic eradication campaign was carried out over a greater part of the country with spraying on a total coverage basis. Initially more than a million cases were reported annually, but by 1963 only 18 cases were diagnosed for the whole island.

When total-coverage DDT spraying ceased in 1963/64, only 8 indigenous infections were identified and scattered cases appeared in 1964, mainly from *P. falciparum* and *P. malariae*. This was the prelude to the 1967/68 epidemic. In the early months of 1966 a few cases of *P. vivax*-like infections arose in the Kurunegala district, and on epidemiological grounds they were first thought to be cases of simian malaria. Simian malaria parasites had been discovered for the first time from neighbouring localities (Dissanaike 1965, Dissanaike *et al.* 1965a, 1965b) and that possibility had therefore to be considered. However, it was soon evident that the infections were definitely *P. vivax* and this was confirmed by the writer, who in 1966 inoculated several blood samples from suspected cases from Galgamuwa, close to the two enzootic foci of simian malaria, into clean monkeys, with no resulting infection. The epidemic soon got under way, and in successive months the numbers of *P. vivax* infections rose to over 1,000 in December 1967 and 19,000 in July 1968, when it was clear that there was an epidemic on hand. The major centre was in the area drained by the Deduru Oya and Maha Oya, which flow from the central hills to the west, and the Mahaweli Ganga draining to the east. The districts mainly involved were Kurunegala, Matale, Anuradhapura and Polonnaruwa. This had been the major centre of the 1934/35 epidemic. Later a minor centre arose in the Kataragama area related to pilgrims attending traditional festivals.

The 1966/68 epidemic was essentially a *P. vivax* one and seldom had such a major outbreak in a tropical country been dominated by this parasite. the reasons attributed to this were that *P. falciparum* was selectively less prominent because spraying operations had concentrated on the *P. falciparum* areas (this parasite being the first to disappear after spraying), and the effectiveness of the camoquin-primaquine drug regimens.

It is of interest to compare the two major epidemics in Sri Lanka referred to above, because from the ecological point of view they had much in common and provide information on the disease pattern in the country.

Briefly, in both years there was a failure of the south-west monsoon compared to previous years and as a result many of the rivers and streams were reduced to stagnant pools pf clear water, ideal for vector breeding. The vectors therefore invaded the normally malaria-free areas, carrying with them the infection to populations that had not experienced malaria for many years. However, the 1967/68 epidemic was mild owing to the rarity of *P. falciparum* and perhaps also to the efficiency of modern drugs. There was also a difference in the area of spread. Whereas the 1934/35 epidemic had spent its force by the time it reached the periphery where the population living under hyperendemic conditions had a protective immunity, in the 1967/68 epidemic no such immunological barrier existed, although it arose from the same centres of Kurunegala, Metale, Anuradhapura and Polonnaruwa with high transmission potentials.

In addition there was the sudden influx of populations to the gem mining area of Elahera in the Polonnaruwa district. There also were land development camp areas and clusters of 'chenas', where temporary clearing by burning jungle for cultivation provided foci of infection in the human population.

Soon after the 1967/1968 epidemic it was observed that the vector had in many cases acquired resistance to DDT and in 1975 malathion replaced it in the residual spraying programme. Resistance to insecticide was no doubt due to exposure and selection of resistant strains of the vector over the years of spraying.

The internal environment and malaria

The influence of the internal environment on malaria is seen in Glucose 6-phosphate dehydrogenase (G6PD) deficiency. G6PD is an enzyme whose function is necessary for viability of erythrocytes. Persons with a deficiency of this enzyme are specially sensitive to the haemolytic effect of certain drugs like the 8-aminoquino-lines (e.g. primaquine). The distribution of G6PD deficiency closely parallels the distribution of falciparum malaria and evidence for the role of *P. falciparum* as the selective agent comes from its geographical distribution, which correlates with that of the frequency of the gene.

'It is reasonable to suppose that, when a gene that is severly deleterious in the homozygous form reaches a high frequency in a population, it must be selected by conferring some sort of advantage to individuals carrying the gene in the heterozygous combination. Such an advantage might reveal itself either in increased survival of heterozygotes up to the age of reproduction or through increased fertility of marriages involving only one heterozygote' (WHO 1966).

In the Kegalle district in Sri Lanka, Nagaratnam *et al.* (1969) using the methaemoglobin reduction test in selected hospital patients hinted at a possible relationship between G6PD deficiency and malaria. A high frequency of the deficiency was detected in the North Central Province (Medawachchiya/Anuradhapura) by Abeyaratne *et al.* (1976), who found also that in the ancient villages, especially Vihara Bulankulama, it was much higher than in recently colonized areas like Vijeapura and Kandalama. The Sinhalese and Ceylon Moors were found by them to have a significantly higher frequency compared to the Ceylon Tamils. All these findings were attributed to the differences in the history of exposure to malaria over several hundred years.

Although the B type G6PD deficiency appears to be common in most of the cases, more studies are needed and are in fact now being undertaken in other parts of the country such as Hambantota, Bibile and Moneragala, which are hyperendemic areas with a higher incidence of *P. falciparum* since the resurgence of malaria in 1967.

Development projects and malaria

The Mahaweli development project and the accelerated scheme introduced by the Government of Sri Lanka for developing agriculture and electric power involves diversion of the Mahaweli Ganga for irrigation of new areas to augment those already available in the dry zone. This and the Uda Walawwe project are in the dry zone and will irrigate several thousands of acres of land. This involves aggregation of labour where outbreaks of malaria take place. In addition, *A. culicifacies* may spread to man-made irrigation channels and streams especially during drought when pooling of river beds and channels takes place, leading to increased man-vector contact.

All this points to an urgent need for cooperation between the irrigation and health services. Evidence of such cooperation was seen in 1976 (Abeyesundere 1978) when, with the diversion of the Mahaweil Ganga at Polgolla, the river downstream on its original course started pooling as a result of the drought and a localized increase of malaria occurred. The Mahaweli Board flushed out the river, destroying larvae, once a week, and this is now done on a routine basis during drought periods, water supplies permitting. Such measures have been recommended for other areas in the accelerated scheme area by Abeyesundere (1978), who suggested cooperation for spraying, drug prophylaxis and other measures.

Filariasis

Parasites and vectors

This is another mosquito-borne disease in Sri Lanka which illustrates the importance of ecological factors.

The parasites causing lymphatic filariasis are nematodes belonging to the Filarioidea. The adult worms are found in the lymphatic channels and the microfilariae circulate in the blood, exhibiting a periodicity which often is nocturnal and therefore ideally adapted to the biting behaviour of the vectors.

The two parasites that were present in Sri Lanka were *Wuchereria bancrofti* and *Brugia malayi*, the former causing genital lesions in addition to other manifestations including elephantiasis of the extremities caused by the latter. Both are transmitted by culicine mosquitoes; *W. bancrofti* by *Culex quinquefasciatus* and *B. malayi* by *Mansonia* spp. Fig. 3 shows the distribution of the two parasites over the years and illustrates the changing patterns of the two infections which are of ecological interest.

In 1947, the Government started an Anti-filariasis campaign, and investigations

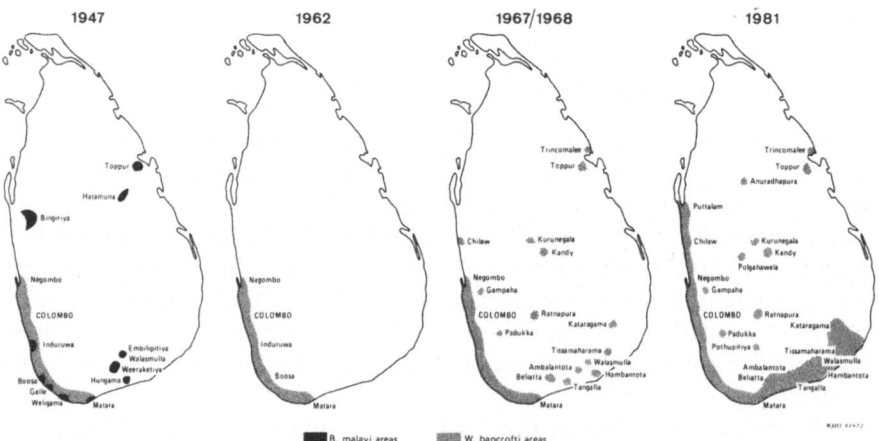

Fig. 3. Sri Lanka, showing the changing pattern of filariasis during 1947, 1962, 1967/68 and 1981; the latter showing the spread of infection outside the south-west coastal belt.

and control measures instituted since then have apparently resulted in the complete disappearance of *B. malayi*. Abdulcader (1967) has stated that this has been due to a combination of factors: treatment of cases with diethylcarbamazine (DEC); larval control of the vectors (Dassanayake & Chow 1954) by weedicide (Phenoxylene 30) spraying of the waterplant *Pista stratioites*, to the roots of which the larvaland pupal stages of the mosquitos prefer to attach; the indirect of residual spraying of DDT in malarious areas, some of which also were *B. malayi* areas (Fig. 3).

Bancroftian filariasis and *C. quinquefasciatus*

As the maps indicate, the areas endemic for *W. bancrofti* have spread from a restricted narrow belt of about 400 sq. miles (1,000 km^2) extending from Negombo, in the north, to Matara in the south of the so-called 'South-west coastal belt'. Along the coast, the spread has now occurred from Puttalam to Hambantota and even Kataragama, in addition to several focal areas outside this belt.

Ecological aspects

It is interesting to consider the ecological and other factors that have affected this change in relation to vectors and their behaviour

Culex quinquefasciatus breeds in catchpits attached to bucket latrines, cesspools, septic tanks, coconut husk-pits, coconut trenches, arecanut pots, discarded receptacles, roof gutters, drains etc., most ofwhich are man-made. In fact it breeds in any foul collection of water rich in decaying organic matter. The adults rest on furniture, clothes and hangings but not generally on walls, so that residual spraying of insecticide for adult control has little value.

It is a predominantly urban mosquito and bites readily indoors as well as outdoors, beginning after sunset and continuing through the night with a peak after midnight.

High rainfall in the south-west part of the island along with other factors has encouraged settlement of people in coastal towns. The dense population and suitable climatic conditions in an unbroken chain of coastal towns, often with poor sanitation, have been ideal for *C. quinquefasciatus* and hence transmission of *W. bancrofti*.

In addition to the 'endemic belt', scattered cases in adjacent areas in a number of towns undoubtedly are due to movements of people. The spread from coastal areas is possible when human population density and the number of introduced cases have reached a critical level, depending on climatic conditions such as temperature and humidity, so that the life-span of the vector is not affected in such a way as to prevent transmission. For instance, Abdulcader *et al.* (1965) showed that *W. bancrofti* microfilariae do not develop beyond the first stage in the vector at Nuwara Eliya where the elevation is around 6,000 ft (1,800 m) although the vector commonly is found there. However, the same strain from Nuwara Eliya would support the development of *W. bancrofti* to the infective stage in Colombo. They found also that *C. quiquefasciatus* is present even at altitudes above 7,000 ft. (2,100 m).

That the natural physical environment is not the only factor of importance in filariasis has been amply borne out by studies on the side-effects of human behavioural activities and underlying sociocultural or psychologic considerations (Dunn 1976). Lambrecht (1974), for instance, has shown that in Sri Lanka, the highest proportion of breeding of the vector, *C. quinquefasciatus* is in peridomestic man-made sites, and that 41.3% of all breeding places for all mosquitos in controlled areas were in discarded receptacles, while 28.4% (the majority) of *C. quinquefasciatus* larval collections were from such receptacles; in uncontrolled areas, 36.4% (again the majority) of *C. quinquefasciatus* collections were from such sites.

Among the health-enhancing behaviour spotlighted by Dunn (1976), are those that are deliberate, such as voluntary and intentional migration to an area of lesser endemicity, while non-deliberate behaviour would include proper house construction, migration and settlement in an area of lesser endemicity, dispersal of infective persons and water and refuse management practices that tend to minimize vector breeding and vector density relative to man. Human activity patterns that maximize contact with biting vectors, both deliberate and non-deliberate are numerous and include sleeping habits.

Urbanization, if properly carried out with due attention to sanitation and disposal of waste, would prevent transmission; but, as shown by Gratz (1973), in many instances inadequate and unsatisfactory measures for disposal of waste water, improperly sealed septic tanks etc., have resulted in man-made breeding places for the vector. Gratz (1973) aptly sums up the situation thus, 'efforts should be made to introduce whatever environmental improvements are possible to the detriment of those species of mosquitoes that have taken advantage of the adverse ecological changes which have resulted from rapid and uncontrolled urbanization in the tropics'.

It is therefore obvious that filariasis-related human behavioural research is necessary for improving the present unsatisfactory state of filariasis control where various indices of infection and disease have reached limit levels as perhaps has occurred in Sri Lanka, and where microfilaria rates in controlled areas, for example, have remained below 1% for several years with transmission still going on.

Filariasis also is a good example of an infection in which the internal environment in man can decide the outcome. As pointed out by Otteson (1980), lymphatic infections provide a spectrum of manifestations from microfilaraemic individuals with no obvious signs or symptoms to those with no microfilaraemia and gross manifestations such as hydrocoele elephantiasis or with occult manifestations such as Tropical Pulmonary Eosinophilia. In highly endemic parts of Sri Lanka too, no satisfactory explanation is available for the fact that with equal exposure to bites by infective mosquitos, some persons show classical clinical manifestations such as elephantiasis while others have occult infections or are not affected at all. HL-A types have been suggested but there is yet no proper study done to confirm this, although it is clear that the immunological responses differ from individual to individual, some promoting disease manifestations and others not (Otteson 1980).

Zoonotic infections

(a) Proven zoonoses

1. Hydatid infection

Echinococcus granulosus, a small tapeworm of dogs and related carnivores, has as its intermediate hosts herbivorous animals such as cattle, goats and buffaloes in which the larval stage (called hydatid cysts) are found. Man can act as an intermediate host when he accidentally swallow the eggs of this tapeworm that are passed

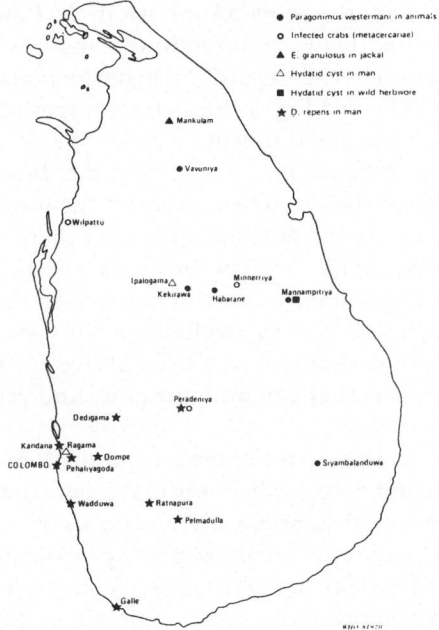

Fig. 4. Map of Sri Lanka showing areas from where *Paragonimus westermani* (in leopards and cats ●); metacercariae (in crabs ○); indigenous human hydatid infection △); hydatid cysts (in wild herbivores ■); *Echinococcus granulosus* (in jackal ▲); and *Dirofilaria repens* (in man *), have been reported.

in stools of infected dogs. The cycle is completed when the fertile hydatid cyst in any of the intermediate hosts is eaten by a dog, in whose small intestines the adult worms develop.

Hydatid cysts in man are important in certain countries including South India and are mostly found in the liver, but may occasionally be found in the lungs, spleen, brain and rarely in other sites. Several cases of hydatid infection have been reported in Sri Lanka, but they have nearly always been in foreigners, mostly from South India (Dissanaike 1962), who have no doubt acquired the infection else-where. Of the seventeen cases reported in the literature (Dissanaike 1968, 1981), only two were identified as indigenous infections; one in a Sinhalese patient from Ipalogama (in the North Central Province) who had a cyst in the orbit (Karuna-ratne & Thamber 1963) and the other in the right lung of a Sinhalese boy in Kandana, near Colombo (Dissanaike 1968).

The possibility of local human infection had been seen both in cattle, buffaloes and goats in the 'domestic environment' and in a monkey *Macaca simica* (Dissa-naike 1958) and a sambhur *Rusa unicolor unicolor* (Dissanaike & Paramananthan 1962) in the jungle areas of the North Central Province (Mannampitiya), while the adult worms had been found in dogs (Seneviratna 1955, Dissanaike 1961) and in a jackal from Mankulam in the Northern Province (Dissanaike & Paramananthan 1960) (Fig. 4). This is therefore a proven zoonosis in Sri Lanka where infection has been naturally transmitted between animals and man, with animals as mainte-nance hosts. The situation with regard to hydatid infection was summarized recently (Dissanaike 1981), and Fig. 5 shows the sources of human infection both in the domestic and sylvatic environments.

The close association between man and dog in the domestic environment and between man and wild carnivores, when he enters the jungle, makes it more than likely that more infections occur. What is important then is that there are ecologi-cal foci of infection in Sri Lanka both in the intermediate domestic as well as the sylvatic environment. Human infection will depend on the closeness of the rela-tionship and the likelihood of man swallowing the tapeworm eggs which are known to be resistant to desiccation and can survive for up to six weeks in the external environment where they contaminate grass and vegetation in areas where infection is present.

Hydatid infection also raises an important question of host-specificity which Baer (1951) states 'does not necessarily imply that a parasite always occurs in or on one and the same species of host, because it may occur in or on hosts that are more or less directly related to one another either phylogenetically or ecologically'. Baer considered that 'host-specificity be regarded as essentially one of the more special-ized aspects of animal ecology, an association in which the host is to the parasite what the biotope is to the free-living organism'. It is clear that for many zoonotic infections of man like the above, provided the human host habitat is physiological-ly suitable for harbouring the parasite (in this case the hydatid cyst), the close ecological surroundings and relationships decide whether and how often an infection takes place. The adult worms of course will only develop in dogs or related carnivores as the parasite is more specific to them than to the intermediate host.

2. Dirofilariasis

A further example of such animal-acquired infections can be cited from Sri Lanka. *Dirofilaria repens* is a common filarial parasite of dogs in the island. The adult worms live in the subcutaneous tissues and the microfilariae circulate in the blood stream. The vector intermediate hosts are several species of mosquitos. Canine infection is common in many urban areas where man lives in close contact with the dog and microfilaria rates in these animals were found to be as high as 60% (Dissanaike 1961). Little wonder than that mosquitoes infected from these dogs could bite man and transmit the infective larvae to him. Man being an unsuitable host in whose internal environment the occasional *D. repens* infective larvae will reach the adult stage, only one or at most two worms occur in a human host. A dozen cases have already been reported from various sites and age groups from several towns (Fig. 4), including Colombo, Dedigama, Dompe, Galle, Nugegoda, Peheliyagoda, Pelmadulla, Peradeniya, Ragama, Ratnapura and Wadduwa (Dissanaike 1971, Dissanaike *et al.* 1972). Several more cases must no doubt be diagnosed though not now reported. The importance of this zoonotic infection has recently been spotlighted with over a hundred cases of subcutaneous dirofilariasis being recorded from many parts of the world (Dissanaike 1979).

A potential zoonosis

One parasitic infection worth mentioning which is a potential human one is that of the lung fluke, *Paragonimus westermani*. This parasite occurs in pairs in cysts of the lungs of its natural definitive hosts, which are members of the cat family. These flukes have two intermediate hosts, a freshwater snail which harbours the stages from sporocysts to cercariae, and a freshwater crustacean, commonly a crab, in which the metacercarial stages encyst. The definitive hosts become infected by eating the second intermediate host uncooked when the metacercariae continue development to reach the final stage as adults in the lungs. In Sri Lanka, leopards, *Panthera pardus fusca,* were found infected in the Vavuniya, Habarana and Kekirawa, in the North and North-Central Provinces (NP, NCP) and also a rusty spotted cat *Felix rubiginosa* in Mannampitiya (NCP) (Dissanaike & Paramananthan 1962). Of six leopards examined, four were infected, some with several hundred flukes. The second intermediate hosts also have been identified (Dissanaike 1968, Kannangara & Karunaratne 1969) as *Oziothelphusa senex, Spiralothelphusa hydrodroma* and *Ceylanthelphusa rugosa* in Wilpattu, Peradeniya, Minneriya and Siyambalanduwa (Fig. 4). These crabs are widely distributed in Sri Lanka, as shown by Fernando (1960), who has made an excellent study of the fresh-water crabs (Potamonidae). Their systematics has been revised by Bott

366

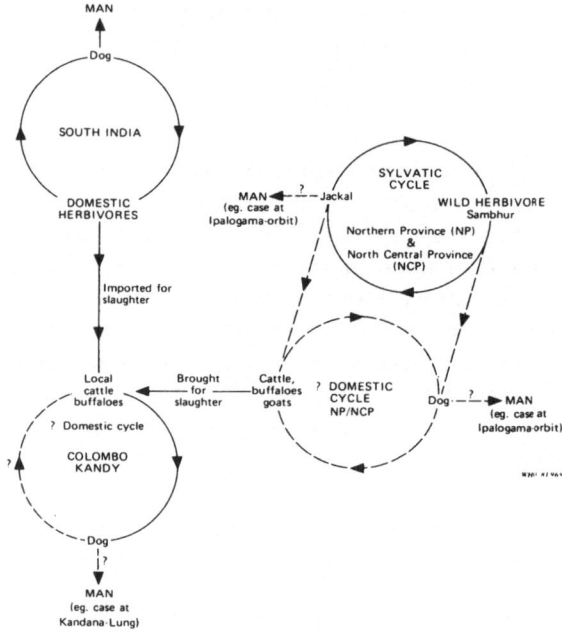

Fig. 5. Diagram showing likely sources of human hydatid infection in Sri Lanka. (Dotted lines represent parts of life-cycle not yet proven.)

(1970). The fact that no human cases have been reported so far can be explained by human behavioural and cultural factors which prevent persons in Sri Lanka eating these crustaceans raw or underdone. Nevertheless, the close ecological relationship and the wide geographical distribution of infected crabs detected points to a potential zoonosis.

The relative high infection rate in leopards and other wild cats may perhaps be difficult to explain when one wonders whether their food habits include eating crabs. However, the recent finding by Miyazaki *et al.* (1978) of natural infection of muscle of wild boar in Japan by immature stages of *P. westermani* would suggest the possibility that these animals act as paratenic hosts from which the leopards become infected. A search for such stages in wild boar in Sri Lanka will be helpful in addition to an investigation of snail intermediate hosts, especially in habitats where infected crabs have been indentified.

Discussion and conclusions

these few illustrative examples of human disease that depend a great deal on ecological factors have demonstrated their importance in the prevalence, incidence, intensity and spread of the infections. In the zoonoses, the necessity to be aware of the potential sources of infection need hardly be stressed. The prevention and control of the diseases or infections depend very much on a knowledge of the physical, biological and socio-economic factors or stimuli, some of which are beyond human control, but many that can be controlled or modified to man's advantage (Fig. 1). One of the problems of control that has not been discussed is that the social environment gets complicated by the fact that administrative borders rarely may coincide with natural areas of the ecosystem.

Eradication of malaria, which at one time seemed to be an attainable goal, must for many years to come be considered a difficult if not impossible task, even in an island like Sri Lanka, unless new tools such as vaccines and better drugs are developed with their proper delivery and acceptance by a suitably educated and motivated population.

As long as development schemes, colonization of newly developed areas. deforestation, shifting cultivation (chemas), transmigration, for pilgrimages, gemming and other purposes, whether temporary or permanent, and other such malaria enhancing activities continue, malaria will remain a problem with the suitable ecological factors necessary for its maintenance.

Although brugian filariasis has been virtually eliminated from the country, the same has not been the case with bancroftian infection. Here again one awaits better tools for diagnosis and treatment and a better understanding of behavioural factors which enhance or diminish the disease combined with community participation, before eradication can be achieved. In addition to human behaviour, vector behaviour and bionomics must be better understood for this purpose.

The zoonotic infections are less serious at the moment and they can be prevented by proper education and a knowledge and awareness of their presence in certain ecological situations, with the avoidance of certain behavioural and food habits, which result in such infection in man.

Sri Lanka provides several good examples of human parasitic diseases that have shown a changing pattern over the years due to socioeconomic and physical influences and that depend on ecological factors for their existence and spread, and these must be taken note of, particularly in future development programmes, if severe outbreaks and spread of existing infections are to be avoided.

References

Abdulcader, M.H.M., P. Rajakone, K. Tharumarajah & R. Mahadeva. 1965. Vectorial capacity of *Culex pipiens fatigans* in Ceylon. J. trop. Med. Hyg. 68:254-256.

Abdulcader, M.H.M. 1967. Present status of *Brugia malayi* infection in Ceylon. J. trop. Med. Hyg. 70:199-200.

Abeyaratne, K.P., S. Premawansa, L. Rajapakse, D.F. Roberts & S.S. Papiha. 1976. A survey of glucose-6-phosphate -dehydrogenase deficiency in the North Central Province of Sri Lanka (formerly Ceylon). Amer. J. Physical Anthropol. 44:135-138.

368

Abeyesundere, A.N.A. 1978. The importance of control of malaria in the Mahaweli Development Project. Bull. nat. Sci. Counc. Sri Lanka 3:14-15.

Audy, J.R. 1958. Medical ecology in relation to geography. Brit. J. clin. Pract. 12:2-10.

Baer, J.G. 1951. Ecology of animal parasites. Univ. Illinois Press, Urbana.

Bott, R. 1970. Die Süsswasserfauna von Ceylon (Crustacea, Decapoda). Ark. Zool. Stockholm 22:627-640.

Briercliffe, R. 1935. The Ceylon Malaria Epidemic, 1934-35. Sess. paper 22, Ceylon Govt Press, Colombo.

Carter, H.F. 1930. Further observations on the transmission of malaria by anopheline mosquitos in Ceylon. Ceylon J. Sci. (D) 2:159-176.

Carter, H.F. & W.P. Jacocks. 1929. Observations on the transmission of malaria by anopheline mosquitos in Ceylon. Ceylon J. Sci. (D) 2:67-85.

Dassanayake, W.L.P. & C.Y. Chow. 1954. The control of *Pistia stratioites* in Ceylon by means of herbicides. Ann. trop. Med. Parasit. 48:129-134.

Dissanaike, A.S. 1957. Some preliminary observations on *Echinococcus* infection in local cattle and dogs. Ceylon med. J. 4:69-75.

Dissanaike, A.S. 1958. On hydatid infection in Ceylon toque monkey, *Macaca sinica*. Ceylon vet. J. 6:33-35.

Dissanaike, A.S. 1961. On some helminths of dogs in Colombo and their bearing on human infections, with a description of a new trematode *Heterophyopsis yehi* sp. nov. (Heterophyidae). Ceylon J. med. Sci. 10:1-12.

Dissanaike, A.S. 1962. Observations on *Echinococcus* infection in Ceylon. pp. 223-240 In: Proc. 1st Int. Symp. on Sci. Knowl. Trop. Paras., Univ. Singapore 5-9 November, 1962.

Dissanaike, A.S. 1965. Simian malaria parasites of Ceylon. Bull. Wld. Hlth. Org. 32:593-597.

Dissanaike, A.S. 1968. The proper study of mankind. Proc. Ceylon Assoc. Adv. Sci. 2:115-142.

Dissanaike, A.S. 1971. Human infections with *Dirofilaria*, a recent filarial parasite of animals in Ceylon, with a brief review of recent cases. Ceylon med. J. 16:91-99.

Dissanaike, A.S. 1979. Zoonotic aspects of filarial infections in man. Bull. Wld. Hlth. Org. 57:349-357.

Dissanaike, A.S. 1981. Parasitic Zoonoses with special reference to Sri Lanka.

Dissanaike, A.S. & D.C. Paramananthan. 1960. On the occurrence of *Echinococcus granulosus* (Batsch, 1786) in a Ceylon jackal. Ceylon vet. J. 8:82-86.

Dissanaike A.S. & D.C. Paramananthan. 1962. On the occurrence and significance of hydatid cysts in the Ceylon sambhur *Rusa unicolor unicolor*. Ceylon J. med. Sci. 11:1-7.

Dissanaike, A.S., P. Nelson & P.C.C. Garnham. 1965a. Two new malaria parasites, *Plasmodium cynomolgi ceylonensis* subsp. nov., and *Plasmodium fragile* sp. nov. from monkeys in Ceylon. Ceylon J. med. Sci. 14:1-9.

Dissanaike, A.S., P. Nelson, & P.C.C. Garnham. 1965b. *Plasmodium simiovale* sp. nov., a new simian malarial parasite from Ceylon. Ceylon J. med. Sci. 14:27-32.

Dissanaike, A.S., V.P. Lykov, I. Sri Skandarajah Sivayoham & M.C.S. Perera. 1972. Four more cases of human infection with *Dirofilaria* (*Nochtiella*). Ceylon J. med. Sci. 17:105-116.

Dunn, F.L. 1976. Human behavioural factors in the epidemiology and control of *Wuchereria* and *Brugia* infections. Bull. publ. Hlth. Assoc. Malaysia 10:34-44.

Fernando, C.H. 1960. The Ceylonese fresh water crabs (Potamonidae). Ceylon J. Sci. (Biol. Sci.) 3:191-22.

Frederiksen, H. 1970. Malaria eradication and the fall in mortality. Pop. Stud. 24:111-113.

Gill, C.A. 1935. Report on the malaria epidemic in Ceylon in 1934-35. Sess. pap. 23, Ceylon Govt Press, Colombo.

Gordon, J.E. 1958. Medical ecology and the public health. Am. J. med. Sci. 235:337-359.

Gratz, N.G. 1973. Mosquito-borne disease problems in the urbanization of tropical countries. pp.445-495 In: CRC Critical reviews in environmental control.

Kannangara, D.W.W. & G.M.S. Karunaratne. 1969. *Paratelphusa ceylonensis* and *Paratelphusa rugosa* as crab hosts of the human lung fluke *Paragonimus westermani* (Kebert, 1878) in Ceylon. Ceylon J. med. Sci. 18:31-32.

Karunaratne, W.A. 1959. The influence of malaria control on vital statistics in Ceylon. J. trop. Med. Hyg. 62:79-85.

Karunaratne, W.A.E. & S.V. Thamber 1963. Hydatid disease with special reference to the orbit. Trans. opthal. Soc. Ceylon 6: 16-20.

Lambrecht, F.L. 1974. Entomological aspects of filariasis control in SriLanka. Bull. Wld. Hlth. org. 51:133-143.

Meegama, S.A. 1969. The decline in maternal and infant mortality and its relation to malaria eradication. Pop. Stud. 23:289-302.

Miyazaki, I., K. Terasaki & K. Iwata. 1978. Natural infection of muscle of wild boar in Japan by immature *Paragonimus westermani* (Kerbert, 1878). J. Parasit. 64:559-560.

Nagaratnam, N., P.K. Leelawathie & W.M.T. Weerasinghe. 1969. Enzyme glucose-6-phosphate dehydrogenase (G6PD) deficiency among Sinhalese in Ceylon as revealed by methaemoglobin reduction test. Indian J. med. Res. 57:569-572.

Ottesen, E.A. 1980. Immunopathology of lymphatic Filariasis in man. Springer Seminars in Immunopathology 2:373-385. Springer-Verlag.

Rustomjee, K.J. 1944. Observations upon the epidemiology of malaria in Ceylon. Sess. pap. 24. Ceylon Govt Press, Colombo.

Schweinfurth, U. 1977. The importance of the geoecological approach in the control of diseases in a tropical country. Yonsie Reports on Trop. Med. 8:85-94.

Seneviratna, P. 1955. A checklist of the helminths in the Department of Veterinary Pathology, Univ. of Ceylon, Peradeniya. Ceylon vet. J. 3:32-37.

Tyssul-Jones, T.W. 1951. Deforestation and epidemic malaria in the wet and intermediate zones of Ceylon. Indian J. Malar. 5:135-161.

World Health Organization 1966. Haemoglobinopathies and allied disorders. Report of a WHO Scientif Group. Wld. Hlth. Org. tech. Rep. Ser. 338.

18. Composition and origin of the ground beetle fauna (Coleoptera, Carabidae)

T.L. Erwin

Introduction

The large and complex ground-beetle fauna of Sri Lanka has never been taxonomically or ecologically analysed. Andrewes (1929, 1935) and Fowler (1912) began the taxonomy, but completed only a small part of the task. Despite the lack of formal taxonomy, it is the purpose of the present paper to provide a preliminary analysis of our present knowledge of the Sri Lankan ground-beetle fauna in terms of biogeography and natural history and contrast it with that of New Guinea and in part, Japan, which lie just outside the eastern end of the Oriental Region. It is now possible and timely to do this because over the last 15 years, three Institutions (named below) devoted much time and resources towards collecting insects in Sri Lanka. I have taken the opportunity to accumulate all the Carabidae (*s. lat.*) from these Institutions as well as others and, although the taxonomy in a formal sense remains to be done, it is now possible to discuss the probable origins and composition of the fauna in a general way. Darlington, in a series of papers (1952-1971), discussed the origin and composition of the ground-beetle fauna of New Guinea, concluding that nearly two-thirds of that country's fauna (where origin of groups could be determined) had its origin in the Oriental region; some of the species range as far as India, but very few (16) to Sri Lanka. The same seems to be the case with part of the Japanese fauna (cf. Habu 1967, 1973, 1978).

The family Carabidae particularly lends itself to comparative faunal analyses for several reasons: 1) it is geographically widespread, 2) it has numerous species, 3) it is especially diverse ecologically, and 4) it is increasingly becoming better known taxonomically (Darlington 1943, Ball 1979, Erwin 1981a). Current estimates place the size of the family at ca. 40,000 described species and many more are known from museum collections and/or are in the process of being described. Latitudinally, the family ranges from well above the Arctic Circle (78°56′N on West Spitsbergen; Bengston *et al.* 1975) to southernmost South America (ca. 55°S at Puerto Williams, Tierra del Fuego; Darlington 1965). Carabid beetles are present on all large islands and all continents except Antarctica (presently!); they are on most small islands including such isolated Pacific ones as those of the Hawaiian archipelago, Bonin Islands, Gilbert Islands, Easter Island and even Rodrigue in the Indian Ocean. In terms of life style, carabids range from blind troglobites to terricoles to arboricoles, from subaquatic to semidesert forms, from vicious carnivorous predators to seed eaters. They range in size from 0.7 mm (Anillina) to 66.0 mm (Scaritini) and are diverse structurally, chromatically and chemically (defense secretions, see Moore & Wallbank 1968, Moore 1979, Anashansley *et al.* 1969 and Eisner *et al.* 1977) (Erwin 1981b). In addition, they are abundant in most habitats and most species are relatively easy to rear, thus making them excellent study organisms.

In Sri Lanka, ground-beetles are found from sea level on sea beaches to the top of the highest mountain at 2,500 m. Much of the known fauna is in some way associated with waterside habitats (Erwin 1979a), and grassland species comprise a significant portion of the fauna. Canopy forms are few, as are forest floor species. Aptery is rare in the carabid beetles, but does occur in a few lowland and highland forms. Several species of apterous blind Anillina recently have been found in bamboo forests in deep litter. At present, 36 tribes of Carabidae are represented in Sri Lanka, only half of the world's representation. Andrewes (1928) and Fowler (1912) recorded 116 genera and 368 species from the Island. Based on new material studied by me (which probably does not include some species mentioned by Andrews and Fowler), the number of genera is at least 140 and the number of species 525. The real fauna of Sri Lanka probably exceeds 600 species.

As is common throughout most of Asia, the Harpalini are dominant in species and in numbers of individuals collected. Pterostichini, Tachyinas and Lebiinii also are common in species with Scaritini, particularly *Clivina* , common in waterside habitats. In Table 4, the taxa of New Guinea and Sri Lanka are contrasted. I find them to be harmonic, that is all groups which are numerous in one place are likewise in the other. Each area has a few small tribes not found in the other, however this is a minor imbalance and easily explained by historical factors. What is peculiar, as indicated by land area comparisons, is that even though New Guinea has 7 times the landmass it has only 25 % more species known than are known from Sri Lanka. Several explanations are possible. First, New Guinea has not been adequately collected through all seasons, particularly the uplands. Secondly, canopy species have not been sampled at all, except at forest ecotones (Erwin 1979a) and by black light. Darlington's (1968) discovery of 56 species of *Demetrida*, in part an arboreal group, indicates that canopy radiation of certain lebiites may parallel that in the Amazonian tropics (Erwin 1982, in press). Darlington (1971) outlines several reasons why the New Guinea fauna is not larger than it is although he also expressed surprise it is so big based on his 'rule of thumb' when judging it against certain of the West Indian island faunas. In the last 10 years, since Darlington's contribution on New Guinea, much has become available in carabid faunal studies (Lindroth 1961-69, Erwin *et al.* 1979, Erwin 1981a) which indicate that carabid faunas thus far studied have been vastly underestimated due mainly to limited collecting up to the time of publishing. Based on my studies of the ground beetles of Central America (Guatemala through Panama), an analysis of over

100,000 specimens, I estimate that fauna to be approximately 2,000 species in a land mass of 522,000 km².

The topography is not as diverse as New Guinea, and Central America is presently connected to continents on both sides; the fauna is a blend of southern and northern elements, but then so is that of New Guinea. Therefore, I would conclude that the fauna of New Guinea should be at least 1,500 species of Carabidae. If the Sri Lanka fauna of 525 is fairly close to the real number of species occurring there (and that is likely because the number is based on 15 or more years of intensive collecting) and the New Guinea fauna is about 1,500, then Darlington's rule of thumb would be substantiated (i.e. a land mass 10 times as great as another should have twice the number of species as the smaller). However, the real unknown element is that the uplands of New Guinea are comprised of many dissected and high ranges and the fauna of them mostly is uncollected. If there is a great deal of species replacement, then there could be many more than 1,500 species on the Island (P.J. Darlington, pers. comm.).

Classification of Caraboidea

(Tribes known to occur in Sri Lanka in italics)

SUPERFAMILY CARABOIDEA

I. Family Trachypachidae
    1. Tribe Trachypachini
II. Family Carabidae
  A. Subfamily Omophroninae
    2. Tribe *Omophronini*
  B. Subfamily Paussinae
    a. Supertribe Metriitae
      3. Tribe Metriinae
    b. Supertribe Brachinitae
      4. Tribe *Crepidogastrini*
      5. Tribe *Brachinini*
    c. Supertribe Paussitae
      6. Tribe Nototylini
      7. Tribe Protopaussini
      8. Tribe *Ozaenini*
      9. Tribe *Paussini*
  C. Subfamily Carabidae
    a. Supertribe Nebriitae
      10. Tribe Cicindisini
      11. Tribe Opisthiini
      12. Tribe Nebriini

13. Tribe Notiokasini
14. Tribe Notiophilini
b. Supertribe Loriceritae
  15. Tribe Loricerini
c. Supertribe Siagonitae
  16. Tribe *Siagonini*
  17. Tribe Enceladini
d. Supertribe Hiletitae
  18. Tribe Hiletini
e. Supertribe Elaphritae
  19. Tribe Migadopini
  20. Tribe Elaphrini
f. Supertribe Carabitae
  21. Tribe *Carabini*
  22. Tribe Ceroglossini
  23. Tribe Pamborini
  24. Tribe Cychrini
g. Supertribe Cicindelitae
  25. Tribe Megacephalini
  26. Tribe Ctenostomatini
  27. Tribe Mantichorini
  28. Tribe Collyrini
  29. Tribe Cicindelini
h. Supertribe Scaratitae
  30. Tribe Promecognathini
  31. Tribe Salcediini
  32. Tribe Scaritini
  33. Tribe Morionini
i. Supertribe Broscitae
  34. Tribe Broscini
j. Supertribe Apotomitae
  35. Tribe *Apotomini*
k. Supertribe Melaenitae
  36. Tribe *Melaenini*
l. Supertribe Psydritae
  37. Tribe Nomiini
  38. Tribe Psydrini
m. Supertribe Rhysoditae
  39. Tribe *Rhysodini*
n. Supertribe Trechitae
  40. Tribe *Trechini*
  41. Tribe Zolini
  42. Tribe *Pogonini*

43. Tribe Gehringiini
44. Tribe *Bembidiini*
o. Supertribe Patrobitae
   45. Tribe Patrobini
p. Supertribe Pseudomorphitae
   46. Tribe Pseudomorphini
q. Supertribe Pterostichitae
   47. Tribe *Pterostichini*
   48. Tribe Zabrini
   49. Tribe Cnemacanthini
r. Supertribe Panagaeitae
   50. Tribe *Panagaeini*
   51. Tribe Agonicini
   52. Tribe *Disphaericini*
   53. Tribe Peleciini
   54. Tribe Bascanini
s. Supertribe Callistitae
   55. Tribe Cuneipectini
   56. Tribe Idiomorphini
   57. Tribe *Callistini*
   58. Tribe *Oodini*
   59. Tribe *Licinini*
t. Supertribe Harpalitae
   60. Tribe *Harpalini*
u. Supertribe Ctenodactylitae
   61. Tribe Ctenodactylini
v. Supertribe Lebiitae
   62. Tribe Amorphomerini
   63. Tribe *Perigonini*
   64. Tribe Catapiesini
   65. Tribe Graphiperini
   66. Tribe *Tetragonoderini*
   67. Tribe *Masoreini*
   68. Tribe *Pentagonicini*
   69. Tribe *Odacanthini*
   70. Tribe *Lebiini*
w. Supertribe Dryptitae
   71. Tribe *Dryptini*
   72. Tribe *Zuphiini*
   73. Tribe Galeritini
x. Supertribe Anthiitae
   74. Tribe *Helluonini*
   75. Tribe Anthiini

76. Tribe *Helluodini*
y. Supertribe Orthogonitae
77. Tribe *Orthogonini*

Synopsis of the Sri Lankan fauna

The information compiled below is taken from various literature sources, mainly Erwin (1979a) and from collection records. References to adults of most tribes annotated below can be found in Darlington (1971) and those to immature stages can be found in Lindroth (1961-69). Species numbers listed in this synopsis, except in a few cases where cited, are from material actually studied rather than catalogue counts, therefore counts must be regarded as a minimum confirmed number of species. Tribes are arranged alphabetically; see above for phylogenetic listing.

Tribe Apotomini - the members of this tribe's single genus, *Apotamus*, are tropically and warm-temperate adapted in the Oriental, Ethiopian and Palearctic Regions. A single species of *Apotamus* recently was discovered in the interior of Brazil (Xingu) (Erwin 1980), thus the distribution may now be described as pantropical. ey live in riparian habitats, in and on the ground in open, grassy country and often are found in flood debris and at light. Larvae are unknown. One species is found commonly in Sri Lanka at low elevations.

Tribe Bembidiini - The numerous members of this tribe's several genera are everywhere in nearly all conceivable habitats. This group, although its members are small beetles, is a dominant member of most faunas. Habitats range from the intertidal zone to tree tops, caves to mountain tops, in the arctic, and around desert springs. Larvae of some genera are described. This group is prone to brachyptery at higher elevations and latitudes. In Sri Lanka, all three subtribes occur: The Bembidiina are represented by two species of *Bembidion*, both coastal lowland forms; the Anillina are represented by several species of blind hypogean forms found in deep litter of bamboo forests (*Argiloborus*~7 species, *Pelocharis*~1 species, from Jeannel 1960); and the Tachyina are represented by numerous species of several genera which occur on most parts of the Island at low and middle altitudes (*Tachyta*~1, *Elaphropus*~27, '*luxatus* group'~1, *Tachys*~3, *Paratachys*~5, *Polyderis*~5, *Lymnastis*~1).

Tribe Brachinini - The members of this tribe's several genera are tropically and warm and cool-temperate adapted. They are found throughout the world although rarely on oceanic islands or in boreal or south moorland conditions. They live on the ground, usually near water, or in mesic forests, and are ectoparasitic in the larval stage. Larvae of two genera are described. Four genera are found in Sri Lanka: *Mastax*~1, *Stenaptinus*~3, and *Styphlomerinus*~3, and *Brachinus*~6.

Tribe Callistini (including Chlaenini - The numerous members of this tribe's several genera are tropically and warm and cool-temperate adapted in all regions. They usually live on the ground near water, but not always; some appear to be true forest dwellers. Larvae of some genera are described. Three genera are recorded from Sri Lanka: *Chlaenius*~29, *Callistomimus*~1, *Hololius*~1.

Tribe Carabini - The numerous members of the few genera in this tribe are warm and cool-temperate adapted in all regions. Many live on the ground; others are adapted to tree-climbing when in pursuit of food. Species are found at both high and low altitudes; those which are monticolous often are flightless, but in the genus *Carabus* many lowland species also are flightless. One species, *Calasoma orientalis* Hope, recently was discovered in Sri Lanka at low elevations. It is likely that this species was introduced recently through the activities of man.

Tribe Cicindelini - The numerous members of this tribe's several genera are broadly adapted to all zones and occur in all regions exclusive of the high Arctic and Antarctic. They live mostly on the ground or in the undercanopy of tropical forest. Many of those living on the ground are near water, fresh or saline. Larvae are described. Two genera are recorded from Sri Lanka: *Prothyma*~2, *Cicindela*~30.

Tribe Collyrini - The members of this tribe's four genera are tropically adapted in the Oriental Region and northern-most Australian Region. They are strictly arboreal in both larval and adult stages. Larvae are described. Four genera occur in Sri Lanka: *Tricondyla*~2, *Derocrania*~7, *Collyris*~1, *Neocollyris*~2.

Tribe Crepidogastrini - The members of this tribe's several genera are adapted to drier tropical uplands and southern Africa south of the tropics, and in southern India and Sri Lanka. It is probable that their life history is similar to Brachinini. Larvae are unknown. All species are flightless. Two species of the genus *Tyronia* are known from Sri Lanka; individuals were collected from beneath stones on grassy areas near ponds.

Tribe Disphaericini - The members of this tribe's two genera are tropical and warm-temperate adapted in the south of Africa and India. They live in dry lowland forested country under stones, at least in part, and in upland wetter situations. Larvae are unknown, but I predict that they are ectoparisitoid and the adults probably eat millipeds (Erwin 1979b). All species are flightless. Two species of *Disphaericus* occur in Sri Lanka.

Tribe Dryptini - The members of this tribe's three genera are tropically adapted in the Neotropical, Ehtiopian, Oriental and Australian Regions. They live among grasses and are somewhat arboreal. Larvae are described. Two genera are found in Sri Lanka: *Drypta*~2, *Desera*~1

Tribe Galeritini - The members of this tribe's few genera are tropically and warm and cool-temperate adapted in all regions except the Australian. They live on the ground usually in forested areas. Larvae are described. One species of *Galerita* is known from Sri Lanka.

Tribe Harpalini (including Amblystomini) - Members of this tribe's numerous genera are adapted to all zones in all regions. They live on the ground for the most part and in a variety of habitats; many species are seed-eaters both as adults and larvae. Larvae of several groups are described. Twenty-three genera are recorded for Sri Lanka, but undoubtedly new ones will be encountered when the taxonomy is completed; in these 23 genera, 92 species have been recorded. In terms of individuals, collecting techniques have turned up more specimens of Harpalini than any other taxon.

Tribe Helluodini (including Physocrotaphini) - The few members of this tribe's three genera are tropically adapted in the Oriental and northern Australian Regions. They live on the ground in leaf litter or under bark of dead trees; one Sri Lankan species lives with termites. Larvae are undescribed. Two genera are recorded from Sri Lanka: *Helluodes*~3, *Pogonglossus*~3.

Tribe Helluonini - The members of this tribe's few genera are tropically and warm and cool-temperate adapted in all regions of the world. They live on the ground and in the undercanopy under bark; some live with army ants. Larvae are described. Four genera are recorded from Sri Lanka: *Macrochilus* ~5, *Cregaris*~1, *Omphra*~, *Colfax* ~1. Members of all three species of the lowland genus *Omphra* are flightless.

Tribe Idiomorphini (including Perochnoristhini) - The members of this tribe's four genera are tropically adapted in the Oriental and Ethiopian Regions. The taxonomy and assignment of genera is at present arbitrary, and in the literature often merged with Orthogonini. *Orthogonius* and related taxa are depressed beetles with truncate elytra (see below), while *Glyptus, Idiomorhus*, and their relatives are convex with rounded elytral apices. It is likely that the two lineages are not closely related and should be classified separately, as I have done here. The former lineage, *Glyptus et al.*, live in and around termite nests; their larvae are described and are predators on termites. One genus is recorded from Sri Lanka: *Parastriga*~1.

Tribe Lebiini (including Agrini, Eucheilini, Mormolycini, Miscelini) -The numerous members of this tribe's many genera are broadly adapted to all zones in all regions. They live in a variety of habitats including the soil humus layer and tree tops, and just about everywhere in between. Larvae of several groups are described, some are ectoparisitoid, some are specialized host-specific predators. Twenty-five genera are recorded from Sri Lanka, but undoubtedly new ones will be encountered when the taxonomy is completed; in these 25 genera, 69 species have been recorded. The only genus with more than 5 species is *Lebia*, which has at least 16 species compared to the 3 listed by Andrewes (1929).

Tribe Licinini - The many members of this tribe's several genera are tropically warm and cool-temperate adapted in all regions except South America. There, one obscure species placed in this tribe is said to occur in Chile. These beetles live both on the ground near water and in forests. Larvae are described. One genus is recorded from Sri Lanka: *Diplocheila*~2.

Tribe Masoreini (including Cyclosomini) - The members of this tribe's few genera are tropically warm and cool-temperate adapted in the pantropical regions and Palaearctic. They are diverse in habitat preference: many are found in dry soil under stones, in forest litter, and in caves. Others live on the ground in sandy, wet and sometimes salty areas. Larvae are described. Two genera are recorded from Sri Lanka: *Cyclosomus*~1, *Aephnidius* ~3.

Tribe Melaenini - The members of this tribe's single genus, *Melaenus*, are tropically adapted in the Oriental and Ethiopian Regions. They often are found in disturbed areas in decaying vegetation. Larvae are unknown. A single species, *Melaenus piger* Fabricius, occurs in Sri Lanka.

Tribe Morionini - The members of this tribe's three genera are pantropically distributed. They live in dead wood of fallen trees or with ants in the latter's refuse heaps. Larvae are known of at least two species. There is presently some controversy whether these beetles are related to the Scaratini or Pterostichini. One genus has been recorded from Sri Lanka: *Morion*~2.

Tribe Odacanthini - The members of this tribe's several genera are tropical and warm and cool-adapted in all regions. They live on and around grass and grass-like plants in several biotic zones. Larvae are described. Taxonomy is particularly confused in this tribe, even at the generic level. Six genera are recorded from Sri Lanka: *Lasiocera*~2, *Odocantha*~3, *Arame*~1, *Casnoidea*~4, *colliuris*~1, *Selina*~1.

Tribe Omophronini - The members of this tribe's two genera are tropically and warm and cool adapted in all continental regions except Australia and Antarctica. They live on the ground near water and make burrows in sandy soils. Larvae are described. A single species, *Omophron lunatus*, an Indian species recently discovered, occurs in Sri Lanka at low elevations.

Tribe Oodini (including Dercylini) - The members of this tribe's several genera are cool and warm-temperate adapted in all regions. They mostly are subaquatic or highly hygrophilous beetles, but some occur in the forest as geophiles and have become apterous (Erwin 1974). Larvae are described. Two genera are recorded from Sri Lanka: *Oodes*~3, *Anatrichus*~1.

Tribe Orthogonini - The members of this tribe's four genera are tropically adapted in the Oriental and Ethiopian Regions. The taxonomy and assignment of genera is at present arbitrary. *Orthogonius* and related taxa are depressed beetles with truncate elytra, while *Glyptus*, which normally is catalogued with *Orthogonius*, and its relatives are convex with rounded elytral apices (see above). It is likely that the two lineages are not closely related and should be classified separately as I have done here. One genus is recorded from Sri Lanka: *Orthogonius*~17.

Tribe Panagaeini - The members of this tribe's several genera are tropically and warm-temperate adapted in all regions. They live on the ground in open situations, but some occur also in forests. Larvae are described. Three genera are recorded from Sri Lanka: *Microcogmus*~1, *Craspedophorus*~6, *Peronomerus*~2.

Tribe Paussini - The members of this tribe's many genera are pantropical in distribution. These beetles live with ants and are highly adapted for such life. They are related to the more plesiotypic Ozaenini (see above) and Metriini (New World only). Larvae are known. Three genera are recorded from my specimens from Sri Lanka: *Cerapterus*~2, *Pleuropterus*~1, *Platyropalus*~1. However, Fowler (1912) also records 5 species of *Paussus*.

Tribe Pentagonicini - The members of this tribe's few genera are tropically and warm and cool-temperate adapted in all regions except the Palaearctic. They live on and around decaying vegetation, e.g. logs, leaf litter, and on certain lichens. Larvae are described. One genus is recorded from Sri Lanka: *Pentagonica*~5.

Tribe Pogonini (including Pogonopsini) - The members of this tribe's several genera are tropically, warm-temperate and sea coast adapted in the Holarctic, Ethiopian and northern Oriental Regions, and three species were recorded from the sea coast of South and Middle America. These beetles live on the ground in alkali or salty situations; at least one species is intertidal. Larvae are described. Two genera are recorded from Sri Lanka: *Pogonus*~1, *Sydenus*~1.

Tribe Pterostichini (including Microcheilini, Chaetodactylini, Anchonoderini, Chaetogenyini, Omphreini, Agonini, Lachnophorini) -The numerous members of this tribe's many genera are broadly adapted in all zones of all regions. Some species are found on oceanic islands. They mostly live on the ground or in decaying wood, and some are truly arboreal. Larvae of several lineages are described. Many groups are apterous or brachypterous. Sixty-one species in more than a dozen genera are recorded from Sri Lanka, but undoubtedly new ones will be encountered when the taxonomy is completed. A few apterous Agonina have been collected at upper elevations in the central mountains.

Tribe Rhysodini - The members of this tribe's numerous genera are tropically and warm-temperate adapted in all regions. They live in wood and eat fungus mycelia. Larvae are described. Many groups are apterous. Two species of *Arrowina* occur in Sri Lanka (Bell & Bell 1979).

Tribe Scaritini - The members of this tribe's numerous genera are adapted to all climates and occur in all areas. It is a dominant member of the family. Most species are burrowing forms. The larvae of several genera are described. Many groups are apterous or prone to brachyptery. Eight genera with a total of 47 species have been recorded in Sri Lanka, and it is likely that more will be found: *Coptolobus*~4, *Scarites*~10, *Oxylobus*~6, *Coryza*~3, *Clivina* ~18, *Psilus*~1, *Trilophus*~2, *Dyschirius*~3.

Tribe Siagonini - The members of this tribe's three genera are tropical and warm-temperate adapted in the Old world tropics and Palearctic Region. *Cymbionotum* and *Siagona* live on the ground in wet areas with decaying vegetation, or under bark. *Luperca* members have been found in a termitarium. Larvae are described. One species of *Cymbionotum* and 2 species of *Siagona* are known to occur in Sri Lanka at low to middle elevations.

Tribe Trechini - The members of this tribe's few genera are tropical and warm-temperate adapted in all regions. They live on the ground or in it and many species are alticolous, often brachypterous. These beetles and their larvae are predaceous. Larvae are known for many genera. The species of Sri Lanka are lowland forms belonging to a single genus: *Perileptus*~2.

Tribe Zuphiini - The members of this tribe's few genera are tropically and warm-temperate adapted in all regions. They live on the ground or in it. Larvae are unknown. Three genera with a total of 8 species are known to occur in Sri Lanka: *Zuphium*~4, *Patrizia*~1, *Planetes*~3.

Faunal connections

Tables 1 to 3 demonstrate the faunal connections among Sri Lanka and other geographic areas and these are graphically displayed in Figs. 2 to 4. Of the areas with well known faunas, New Guinea has only 2.4% of its fauna shared with Sri Lanka. Most of these are small widespread Tachyina. The other species, taxonomically a hodge-podge, are small to medium-sized in respect to body length. All of these species are highly vagile dispersants (Erwin 1979a) and belong to genera which are widespread or cosmopolitan. Although the Japanese fauna is not yet completely documented, Habu's (1967, 1973, 1978) treatment of Harpalini, 'truncatipennes', and Agonini indicates that 4.6% of the Japanese fauna is shared with Sri Lanka. These three large groups can be regarded as representative of the Japanese carabid fauna and it is likely that the per cent shared would not change much if all groups were analyzed. Even though Japan and New Guinea lie just outside the Oriental Region proper, they can be used for analysis of carabid range extent and a test of classical zoogeographic regions. These examples indicate that the carabid fauna of the Oriental Region is not generally composed of widespread species, that there are subregions, and that the faunal histories of these subregions are different. It is of interest that very few species are found among and between all three areas (Regions). Recent studies by George Ball (pers. comm.) indicate that *Celaenephes*, a lebiine genus, contains two or more species; *Celaenephes parallela* in the classic sense may be a composite of species, thus only *Sericoda ceylanicus* and *Dicranoncus queenslandicus*, both agonine species, *Paratachys faciatus, Polyderis truncatus* and *brachys, Elaphropus ceylanicus, acaroides* and *fumicatus, Tachyta umbrosus,* and *Tachys quadrillum,* all tachyines, *Perigonia plagiata,* a perigonine, *Oodes piceus,* an oodine, *Dolichoctis striata, Pentagonica pallipes* and *erichsoni,* both pentagonicines, *Cregaris labrosa,* a helluonine, are shared between Sri Lanka and New Guinea. These represent 3.8% of the catalogue fauna of Sri Lanka. On the other hand, 42% of the genera are shared. Because the Japanese fauna is not completely known or published in a convenient way it is more difficult to make a similar comparison, however in the groups covered by Habu (1967, 1973, 1978), I note 15 species are shared with Sri Lanka. Undoubtedly, more species will be discovered to be in common. Only two species, *Dolichoctis striata* and *Paratachys faciatus,* are shared among all three areas.

The single case of faunal imbalance, if that is what it is, concerns the paucity of Scaratini in New Guinea (Table 4). Sri Lanka seems to have a more normal Oriental fauna of that important group of fossorial beetles.

In a geographic analysis of all the species recorded by Andrewes (1928) and Fowler (1912) (Tables 1 and 2) it can be seen that a large percentage of the species are endemic (34.5%) although only 4.3% of the genera are endemic (Tables 1 and 2). Most species have distributions on the subcontinent (India and Burma) (31%) or throughout the Oriental Region (28%). Surprisingly, a mere 0.5% have ranges which extend to Africa, the same percentage as the cosmopolitan population. Of more interest and importance (Table 3), 68.5% of the endemic species of Sri Lanka belong to cosmopolitan or widespread genera; only 7.9% belong to endemic genera, 10.2% to genera confined to the subcontinent, 11% to Oriental Region genera.

Size

Darlington (1971) suggested that carabids of certain size classes might be selected against by ants, thereby making bimodal size histograms. His data consisted of carabid measures from Dobodura, New Guinea. Previously, I provided histograms of numbers of species per size class for Carabidae at several localities (Erwin 1979a, 1981a). Bimodal histograms are the rule for tropical and temperate lowland carabids according to may findings and bimodality decreases towards the north boreal zone as ant species decrease. Fig. 1 shows that Sri Lanka's fauna is a fairly smooth histogram with troughs at 5 mm and 9 mm respectively. In each fauna analyzed to date, ant size classes are unimodal and the peak falls near the trough of the carabid histogram. There is not always exact overlap, but the concordance is surprising and probably indicates some kind of interaction. Unfortunately, there are no data on ants of Sri Lanka except diverse original descriptions spread through the literature. When a size class histogram is available, I would predict that it has two peaks rather than one.

Discussion

The elements of the carabid fauna of Sri Lanka are in balance with those known from New Guinea, even though only 3.3% of the species are shared and the two areas widely separated. At the generic level, 2.5% of the genera are shared. Such a balance is confined to the Old World, as the New World has a much higher percentage of arboreal species and thus is relatively out of balance (Erwin 1979a). The fauna of Sri Lanka and New Guinea are much more like a temperate fauna (Erwin 1981a) in terms of taxon balance because of a lower number of arboricolous species and a higher number of hydrophilous/mesophilous species.

Most of the land mass of Sri Lanka has been dominated by mankind for several thousand years; the land has been cleared of forest, terraced, many forests cut over repeatedly, and water systems channeled and impounded. Nature on Sri Lanka has been 'unnatural' for a long time. This activity has imposed on the carabid fauna certain features, namely that the fauna is predominantly one of waterside dwelling, highly vagile, generalist groud-beetle species. However, overall it is in balance with New Guinea, a rather completely 'natural' island. Darlington (1971) suggests that the fauna of the latter is a recent one, that is a historically late pioneering group of taxa in Stage Two (post-invasion) of development and the fauna has moved in from the south (Australian Region) and the north (Oriental Region). It is possible that the very similar faunal balance of Sri Lanka has developed in response to man's activities rather than, as in New Guinea, the Pleistocene's.

There are no old carabid elements in the Sri Lanka fauna and none with direct connections to the South temperate; the number of endemic genera is very small (10), although at least 117 genera have endemic species (Table 3). The vast majority

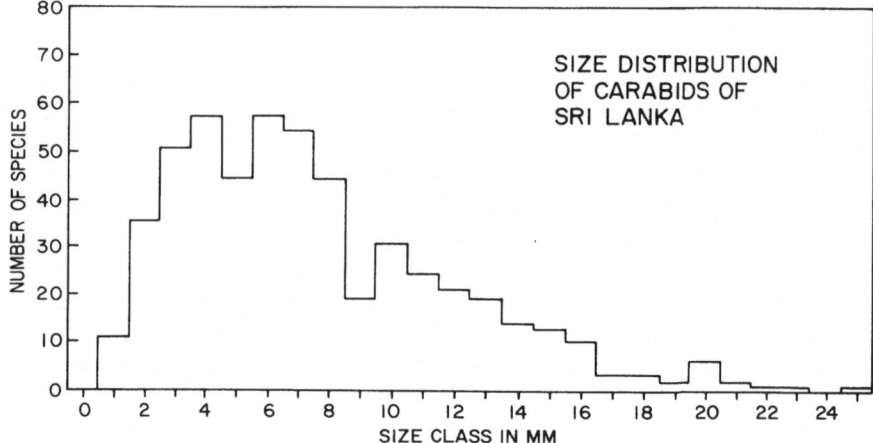

Fig. 1. Number of species of Carabidae per size class (1.0 mm increments) in Sri Lanka based on collection material.

of these genera with endemic species are Oriental or Cosmopolitan based (Fig. 4). It appears that the old *Gondawana fauna* that rode India across the equator, if any of its elements survived (and there is evidence for this on the Indian subcontinent), that they eventually were swamped in Sri Lanka by southward moving lineages from the north. There are no connections either with Madagascar or southern Africa except one, *Tyronia*, a bombardier beetle group occurring also in eastern and northeastern Africa.

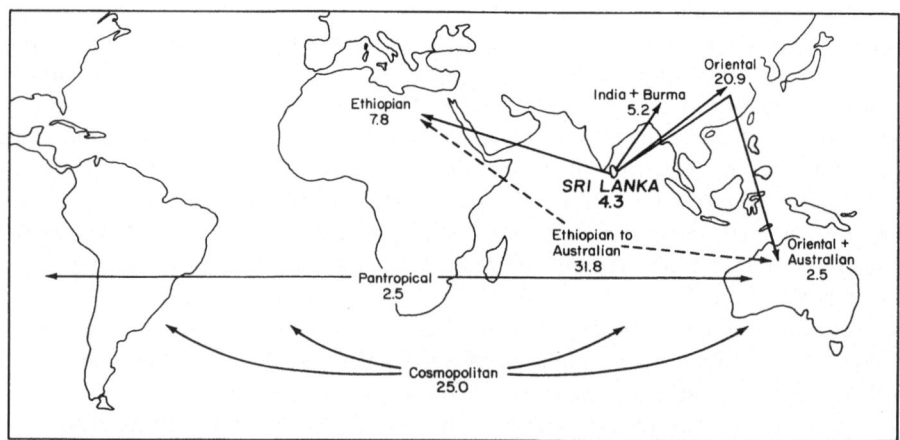

Fig. 2. Percent of genera either endemic to Sri Lanka or shared with other regions (from Table 1).

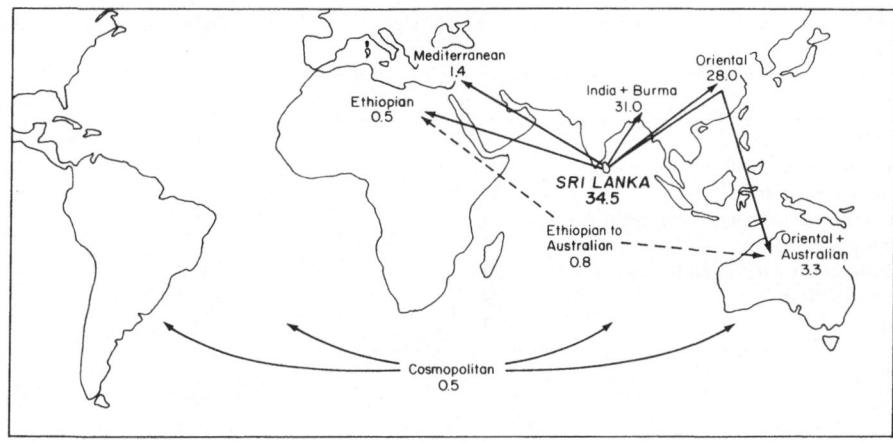

Fig. 3. Percent of genera either endemic to Sri Lanka or shared with other regions (from Table 2).

In summary, the ground beetle fauna of Sri Lanka has its origins in the north, mainly the Oriental Region. It is fairly well known now, at least collected, and it is time to begin the taxonomy and in-country biogeographical and ecological studies.

Acknowledgements

It is a pleasure to acknowledge the assistance of the following for aid in preparing the paper: Linda L. Sims, Gloria N. House, Lynne H. Kitagawa for preparation of

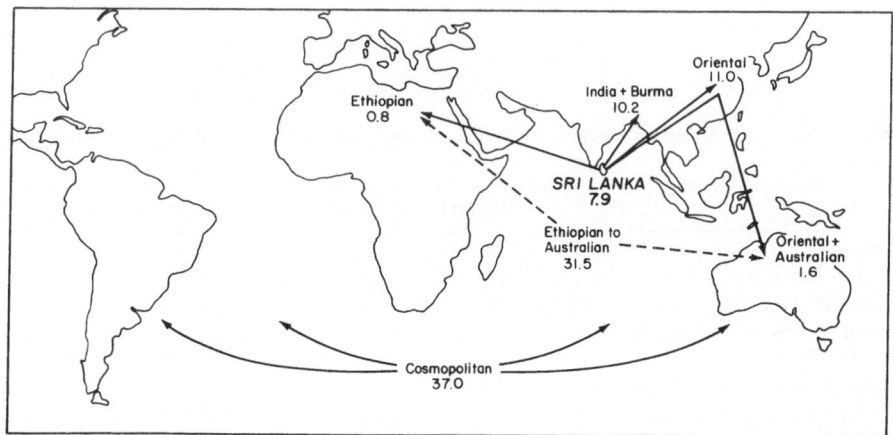

Fig. 4. Distribution and number of genera which have endemic Sri Lankan species (from Table 3).

Table 1. General distribution of genera known to occur in Sri Lanka (Figures based on catalogue counts).

Region	Number of genera	% Sri Lanka fauna
Endemic	5	4.3
Indian subcontinent + Burma	6	5.2
Oriental Region	24	20.9
Oriental + Australian Regions	3	2.5
Ethiopian Region	9	7.8
Ethiopian to Australian Region	37	31.8
Cosmopolitan	29	25.0
Pantropical	3	2.5
Total	116	100

Table 2. General distribution of species known to occur in Sri Lanka (Figures based on catalogue counts).

Region	Number of species	% Sri Lanka fauna
Endemic	127	34.5
Indian subcontinent + Burma	114	31.0
Oriental Region	103	28.0
Oriental + Australian Regions	12	3.3
Ethiopian Region	2	0.5
Ethiopian to Australian Region	3	0.8
Cosmopolitan	2	0.5
Mediterranean	5	1.4
Total	368	100

Table 3. General distribution of genera with endemic species in Sri Lanka (Figures based on catalogue counts).

Region	Number of genera	% of total endemics
Endemic	10	7.9
Indian subcontinent + Burma	13	10.2
Oriental Region	14	11.0
Oriental + Australian Regions	2	1.6
Ethiopian Region	1	0.8
Ethiopian to Australian Region	40	31.5
Cosmopolitan	47	37.0
Total	127	100

Table 4. Tribes of Carabidae in Sri Lanka and/or New Guinea with numbers of genera and species.

Tribe	Sri Lanka (collections)		New Guinea (literature)	
	# of genera	species	# of genera	species
Omophronini	1	1	0	0
Crepidogastrini	1	2	0	0
Brachinini	4	13	2	7
Ozaenini	3	3	1	1
Paussini	4	4	1	1
Siagonini	2	3	0	0
Carabini	1	1	0	0
Collyrini	4	12	1	1
Cicindelini	2	32	5	47
Scaritini	8	47	3	30
Morionini	1	2	1	2
Apotomini	1	1	0	0
Melaenini	1	1	0	0
Rhysodini	1	1	2	11
Trechini	1	2	2	3
Pogonini	2	2	0	0
Bembidiini	9	47	11	72
Psydrini	0	0	1	1
Pseudomorphini	0	0	3	8
Pterostichini	13	61	35	206
Panagaeini	3	9	5	6
Disphaericini	1	2	0	0
Idiomorphini	1	1	0	0
Callistini	3	31	1	12
Oodini	2	4	3	14
Licinini	1	2	5	8
Harpalini	23	92	15	53
Ctenodactylini	0	0	1	2
Perigonini	1	9	1	14
Tetragonoderini	2	4	0	0
Masoreini	2	4	4	4
Pentagonicini	1	5	3	15
Odacanthini	6	12	8	18
Lebiini	25	69	32	160
Dryptini	2	3	2	5
Zuphiini	3	8	3	8
Galeritini	1	1	0	0
Helluodini	2	6	1	10
Orthogonini	1	17	0	0
Totals	143	525	158	729
Land area (km^2)	65,610		141,691	

tables, graphs, and editorial assistance; Nigel Stork, Jean Menier, and Helene Perrin for assistance with type material; Lund University; Museum, Basel, Switzerland; Museum of Compartive Zoology, California Academy of Sciences, and the collecting efforts of Karl V. Krombein and his team, who provided the bulk of the material on which this paper is based.

References

Andrewes, H.E. 1928. A catalog of the Carabidae of Ceylon. Spol. Zeylan. 14:135-195

Andrewes, H.E. 1929. Coleoptera Carabidae Vol. I. Carabinae. pp. 391-397 In: Fauna British India, including Ceylon and Burma. Taylor & Francis Publ., London.

Andrewes, H.E. 1935. Keys to some Indian genera of Carabidae (Col.). V. The genera *Cyminoidea, Platytarsus,* and *Taridius.* Stylops 4:201-205.

Aneshansley, D., T. Eisner, J.M. Widom & B. Widom. 1969. Biochemistry at 100 °C: explosive secretory discharge of bombardier beetles (*Brachinus*). Science 165:61-63.

Ball, G.E. 1979. Conspectus of Carabid classification: history, holomorphology, and higher taxa. pp. 63-111 In: T.L. Erwin, G.E. Ball, D.R. Whitehead & A.L. Halpern (Eds.). Carabid beetles, their evolution, natural history, and classification. Proc. 1st Int. Symp. Carabidology. Dr. W. Junk, The Hague.

Bell, R. & J. Bell. 1979. Rhysodini of the World Part II. Revisions of the smaller genera (Coleoptera:-Carabidae or Rhysodidae). Quaest. Entomol. 15:337-446.

Bengston, S.-A., A. Fjellberg & T. Solhoy. 1975. Amara qhenseliiSchn. (Coleoptera:Carabidae) new to Svalbard. Norw. J. Entomol. 22:81-82.

Darlington, P.J. Jr. 1943. Carabidae of mountains and islands: data on the evolution of isolated faunas, and on atrophy of wings. Ecol. Monogr. 13:37-61.

Darlington, P.J. Jr. 1952-1971. The Carabid beetles of New Guinea. Parts 1-4. Bull. Mus. Comp. Zool. 1952. Part 2. 107:90-252; 1962, Part 1, 126:321-564; 1968. Part 3, 127:1-250; 1971. Part 4, 142:129-337.

Darlington, P.J. Jr. 1965. Biogeography of the Southern end of the world: distribution and history of the southern life and land, with an assessment of continental drift. Harvard Univ. Press, Cambridge. 236 pp.

Eisner, T., T.H. Jones, D.J. Aneshansley, W.R. Tschinkel, R.E. Silberglied & J. Meinsald. 1977. Chemistry of defensive secretations of Bombardier Beetles (Brachinini, Metriini, Ozeanini, Paussini). J. Insect Physiol. 23:1383-1386.

Erwint, T.L. 1974. The genus *Coptocarpus* Chaudoir of the Australian Region with notes on related African species (Coleoptera: Carabidae:Oodini). Smithsonian Contr. Zool. 184:1-125.

Erwin, T.L. 1979a. Thoughts on the evolutionary history of ground beetles: hypothesis generated from comparative faunal analysis of lowland forest sites in Temperate and Tropical regions. pp. 539-592 In: T.L. Erwin, G.E. Ball, D.R. Whitehead & A.L. Halpern (Eds.). Carabid beetles, their evolution, natural history, and classification. Proc. 1st Int. Symp. Carabidology. Dr. W. Junk, The Hague.

Erwin, T.L. 1979b. A review of the natural history and evolution of ectoparasitoid relationships in Carabid beetles. pp. 479-484 In: T.L. Erwin, G.E. Ball, D.R. Whitehead & A.L. Halpern (Eds.). Carabid beetles, their evolution, natural history, and classification. Proc. 1st Int. Symp. Carabidology. Dr. W. Junk, The Hague.

Erwin, T.L. 1980. Systematic and zoogeographic notes on Apotomini, an Old World faunal element new to the Western Hemisphere List, with a description of a new species from Brazil (Coleoptera:-Carabidae). Coleop. Bull. 34:99-104.

Erwin, T.L. 1981a. Natural history of Plummers Island, Maryland. 26. The ground beetles of a temperate forest site (Coleoptera:Carabidae): An analysis of fauna in relation to size, habitat selection, vagility, seasonality and extinction. Bull. Biol. Soc. Wash. 5:105-224.

Erwin, T.L. 1981b. Taxon pulses, vicariance, and dispersal: An evolutionary synthesis illustrated by carabid beetles. pp. 159-183 In: Vicariance Biogeography: a critique. G. Nelson & D. Rosen (Eds.). Columbia Univ. Press. N.Y.

Erwin, T.L. 1982 (in press). Seasonal, size and trophic patterns, and richness of Arthropods in the tropical arboreal ecosystem: methods of canopy fogging and the fauna of early dry season (with inundation) of four types of forest at Manaus, Brazil. Spec. Symp. Vol. Brit. Ecol. Soc.

Erwin, T.L., G.E. Ball, D.R. Whitehead & A.L. Halpern (Eds.). 1979. Carabid beetles, their evolution, natural history, and classification. Proc. 1st Int. Symp. Carabidology. Dr. W. Junk, The Hague. 635 pp.

Fowler, W.W. 1912. The fauna of British India, including Ceylon and Burma. C. Coleoptera. General introduction and Cicindelidae and Paussidae. Taylor & Francis Publ. London. pp. 1-529.

Jeannel, D.R. 1960. Sur les Anillini de L'inde et de Ceylan. Rev. Francaise d'Entomol. 27:16-23.

Habu, A.H. 1967. Fauna Japonica. Carabidae: Truncatipennes group (Insecta:Coleoptera). Tokyo Elec. Eng. Coll. Press, Tokyo. 337 pp.

Habu, A.H., 1973. Fauna Japonica. Carabidae: Harpalini (Insecta:Coleoptera). Tokyo Elec. Eng. Coll. Press, Tokyo. 430 pp.

Habu, A.H. 1978. Fauna Japonica. Carabidae: Platynini (Insecta:Coleoptera). Tokyo Elec. Eng. Coll. Press, Tokyo. 447 pp.

Lindroth, C.H. 1961-1969. The ground beetles (Carabidae, excl. Cicindelinae) of Canada and Alaska. Parts 1-6. Opscula Entomologica, 1192 pp. 1961, Part 2, Suppl. 20:1-200; 1963, Part 3, Suppl. 24:201-408; 1966, Part 4, Suppl. 29:409-648; 1968, Part 5, Suppl. 33:649-994; 1969, Part 6, Suppl. 34:945-1191; 1969, Part 1, Suppl. 35: i+xlviii.

Moore, B.P. 1979. Chemical defense in Carabids and its bearing on phylogeny. pp. 193-203 In: T.L. Erwin, G.E. Ball, D.R. Whitehead & A.L. Halpern (Eds.). Carabid beetles, their evolution, natural history, and classification. Proc. 1st Int. Symp. Carabidology. Dr. W. Junk, The Hague. Moore, B.P. & B.E. Wallbank. 1968. Chemical composition of the defensive secretion in carabid beetles and its importance as a taxonomic character. Proc. R. Ent. Soc. Lond. (B) 37:62-203.

19. Land snails: distribution and notes on ecology

R. Ratnapala

Introduction

The study of land snails of Sri Lanka has not had any long sustained interest since the monumental contributions of Blanford & Godwin-Austin (1908) and Gude (1914,1921). As a result, there is little information on their systematics, ecology and distribution in the different climatic zones of the Island.

Land snails are an important group of invertebrates. The climatic conditions to which the different species are adapted are wide and varied, yet they are extremely sensitive to even minor variations in climate outside their normal range. This climate sensitivity, together with the possession of a calcareous shell which bears distinctive conchological features and preserves the animals, renders them a most suitable group for the assessment of regional climatic changes over a period of time. Such an assessment is invaluable in determining the effects of anthropogenic and natural causes in the alteration of climatic conditions over short or long term periods.

Some land snails, particularly the introduced varieties, are important agricultural pests and others are a valuable source of food for birds, reptiles and mammals. The study of ecology and distribution of these animals will thus be important in the investigation of the effects of pesticides and the influence of the activities of man in altering the environment. Although investigations have not yet been carried out, some land snails are vectors of helminthic infections of vertebrates in Sri Lanka.

One of the most important prerequisites in the use of land snails in the above fields is the availability of precise information with regard to their distribution and ecology.

In this chapter, locality data available for land snails so far recorded from Sri Lanka, together with locality data of snails collected by the author from a variety of habitats on the Island, have been analysed to determine the distribution of land snails in the different climatic zones.

Climatically, Sri Lanka can be divided into three major zones, the dry zone, the intermediate zone and the wet zone. Each of these is further divided into subzones based on rainfall or altitude. The classification used in this study has been adapted from Gaussen *et al.* (1968) and Müller-Dombois (1968).

Dry Zone

Climate, Vegetation and Soils

This is the largest climatic zone and it covers an area of about two thirds of the island. It is characterized by periods of prolonged drought. The annual rainfall of this region is between 1,250-1,875 mm, which generally is received between November and February. For the rest of the year this region is subjected to hot, dry winds and inter-monsoonal droughts and monthly temperatures can be as high as 25-30 °C throughout the year. The vegetation consists of tropical dry mixed evergreen forests (Koelmeyer 1957). Swampy grasslands occur along certain floodplains. Recently, much of the forest area of this zone has been cleared for cultivation.

The dry zone can be divided into three subzones based on the rainfall. These are the arid subzone (F), the semi-arid subzone (A) and the lowland dry subzone (B) (Figure 1). The arid subzone covers the northernmost part and the islands off the main coast to the North.

The mean annual rainfall is less than 1,250 mm, mainly during the northeast monsoon. The period of drought extends over six months or more. The vegetation is sparse and consists of low open thorny scrub with stunted trees and shrubs. Human interference has resulted in extensive degeneration of forests. The soil is of Calcic Latosol type which is fairly dry and brittle with hardly any moisture, due to rapid penetration of water (Panabokke 1967). The semi-arid subzone has a shorter period of drought than the arid zone, and this period extends over the months of June to September; rainfall is moderate from April to June. The vegetation is a mixture of tropical thorn forest, dry evergreen forest and a small portion consists of moist deciduous forests. The soil is rich in calcium in the northern region. The soils in the remaining areas of this subzone consist of Red Yellow Latosols. The lowland dry subzone has a mean annual rainfall of 1,250-1,875 mm, mainly between March and June. June to September is a dry period, with a rainfall of less than 500 mm. The vegetation is a mixture of dry evergreen and moist deciduous forest.

Land Snails and Ecological Notes

The dry zone provides a variety of habitats such as scrub jungle, open habitats and cultivated land for its land snail fauna, and the locality data available indicate that this zone supports a fair number of species.

According to Evans (1972) snails adapted to living under dry conditions can be divided into two groups; (A) those for which dry habitats are only one of a number in which they can live (i.e. they are not restricted to dry places, but possess the necessary behavioural and physiological adaptations which allow them to do so)

Table 1. Distribution and land snails in the different climatic zones. The species have been presented according to the classification schemes for Prosbranchia and Pulmonata of Wenz (1928-1944) and Zilch (1959-1960) respectively. Some species listed have only non-specific localities (Sri Lanka, Hills, Northern, etc.).

| | Climatic zones | | | | | | | |
| | dry | | | intermediate | | wet | | |
	F	A	B	C	E	D_1	D_2	D_3
Streptaxidae								
Streptaxis								
*S. layardianus			+		+		+	+
*S. peroteti					+	+		
*S. cingalensis					+	+		
*S. gracilis					+			
*S. ravanea								
Ptychotrema								
*P. bicolor						+		
*P. planguncula								
Ariophantidae								
*Cryptozona								
*C. (Nilgiria) semirugata		+	+					
C. novella								
C. (N.) ceraria								+
C. (Xestina) bistrialis	+	+	+		+	+	+	
C. (X.) ceylanica			+		+	+	+	
C. (X.) taprobanensis								
C. (X.) cyix			+			+		
Hemiplecta								
H. chenui				+	+	+		
H. juliana					+	+	+	
H. ganoma						+		
Ravana								
R. politissima					+	+	+	
Euplecta								
E. subopaca					+	+	+	+
E. layardi		+	+			+		+
E. semidecussata							+	+
E. rosamonda						+		
E. travencorica					+	+	+	
E. praeeminens							+	
E. indica								+
E. albizonata								+
E. emiliana			+		+	+	+	+
E. laevis								
E. binoyaensis							+	
E. verrucula					+		+	
E. prestoni							+	
E. scobinoides		+					+	
E. gardneri					+	+		+

Table 1 (continued)

| | Climatic zones | | | | | | | |
| | dry | | | intermediate | | wet | | |
	F	A	B	C	E	D_1	D_2	D_3
E. partita						+	+	+
E. trimeni								+
E. acuducta								+
E. isabellina								
E. colletti							+	
E. hyphasma						+	+	
E. turritella				+			+	+
E. phidias					+			
E. concavospira						+		
Ratnadvipia								
R. irridians						+		+
R. edgariana						+		
Mariella								
M. idussumieri								
Satiella								
S. membranacea						+		
Sitala								
S. phyllophila					+	+		
S. pyramidula							+	
S. operiens							+	
Macrochlamys								
M. indica								
M. umbrina					+			
M. nepas								
M. woodiana								
M. vilipensa		+						
M. perfucta						+		
M. tratanensis							+	
M. kandiensis							+	
Microcystina								
M. bintenensis			+	+	+	+		
M. lita							+	
Kaliella								
K. barrakporensis								+
K salicensis							+	
K. colletti							+	
K. delectabilis							+	
K. leithiana								
Eurychlamys								
E. regulata				+		+		
Trochomorphidae								
Trochomorpha								
T. galerus						+		
T. hyptiocyclus				+				

Table 1 (continued)

	Climatic zones							
	dry			intermediate		wet		
	F	A	B	C	E	D_1	D_2	D_3
Endodontidae								
Thysanota								
T. eumita				+			+	
T. hispida				+				
T. elegans							+	
Philalanka								
P. secessa							+	
P. liratula								+
P. lamcabensis								+
P. trifilosa								
P. mononema						+		
P. thwaitesi							+	+
P. thwaitesi var. suavis							+	+
P. depressa							+	
P. circumsculpta							+	+
P. sinhila							+	
Ruthvenia								
R. clathratula					+	+	+	
R. clathratula var. compressa							+	
R. caliginosa							+	
R. biciliata					+			
Corillidae								
Corilla								
C. adamsi						+		+
C. adamsi var. hinidunensis								
C. beddomeae						+	+	+
C. gudei				+		+	+	
C. humberti							+	+
C. fryae							+	+
C. carabinata			+			+		
C. colletti						+		
C. erronea					+		+	+
C. lesleyae						+		
C. odontophora					+		+	
Vallonidae								
Pupisoma								
P. miccyla						+	+	
P. longstaffe						+		
Pyramidulidae								
Pyramidula								
P. halyi								+
Enidae								
Ena								
E. proletaria					+	+		

Table 1 (continued)

| | Climatic zones | | | | | | | |
| | dry | | | intermediate | | wet | | |
	F	A	B	C	E	D₁	D₂	D₃
E. stalix					+			+
E. panos				+				
Rachis								
R. adumbratus								
R. pulcher		+	+			+		
R. punctatus		+	+		+			
Pupillidae								
Pupoides								
P. coenopictus								
Pupilla								
P. cingalensis								
P. muscerda	+		+					
Chondrindae								
Gastrocopta								
G. mimula	+							
Acavidae								
Acavus								
A. haemastoma						+		
A. haemastoma var. melanotragus						+		
A. haemastoma var. conus								
A. haemastoma var. concolor								
A. fastosus						+		
A. prosperus			+	+		+	+	
A. phoenix		+		+		+		
A. superbus						+	+	
A. superbus var. grevillei						+	+	
A. roseolabiatus						+	+	
Oligospira								
O. waltoni						+	+	
O. waltoni var. polei						+		
O. skinneri					+			
Pleurodontidae								
Planispira								
P. fallaciosa	+	+						
P. vittata	+	+	+					
Beddomea								
B. ceylanicus						+	+	+
B. albizonatus						+	+	
B. albizonatus var. simoni						+		
B. intermedius						+	+	
B. trifasciatus		+	+			+	+	+
B. trifasciatus var. rufoptica						+	+	
Bradybaenidae								
Bradybaena								

Table 1 (continued)

| | Climatic zones | | | | | | | |
| | dry | | | intermediate | | wet | | |
	F	A	B	C	E	D$_1$	D$_2$	D$_3$
B. similaris					+	+	+	+
Aegista								
A. huttoni var. radleyi					+		+	+
Clausiliidae								
Phaedusa								
P. ceylanica					+			
Achatinidae								
Achatina								
A. fulica	+	+	+	+	+	+	+	+
Subulinidae								
Subulina								
S. octona								
Opeas								
O. gracile						+	+	
O. gracile var. panayaensis						+		
O. prestoni			+		+	+	+	
O. mariae						+		
O. layardi			+		+	+		
O. sykesi								
O. pussilus						+		
Zootecus								
Z. insularis								
Ferussacidae								
Glessula								
G. ceylanica						+	+	
G. inornata						+		
G. inornata var. minor								
G. lankana						+		
G. reynelli								+
G. reynelli var. immitis								
G. pallens						+		
G. prestoni								
G. parabilis						+	+	
G. punctogallana						+		
G. nitens						+		
G. fulgens						+		
G. panaetha				+				
G. serena				+		+		
G. sinhila								
G. layardi								
G. deshayesi	+	+				+	+	
G. pachycheila						+		
G. pachycheila var. taprobanica							+	
G. veruina				+				

Table 1 (continued)

	Climatic zones							
	dry			intermediate		wet		
	F	A	B	C	E	D_1	D_2	D_3
G. sattaraensis								+
G. capillacea						+		
G. collettae							+	+
G. pussila								
G. simoni								+
Digoniaxis								
D. cingalensis						+		
Succineidae								
Succinea								
S. ceylanica						+		
Vaginulidae								
Vaginulus								
V. templetoni						+		+
V. maculatus						+		+
V. reticulatus						+		+
Cyclophoridae								
Leptopoma								
L. semiclausum								
L. taprobanensis								
L. apicatum								
L. elatum								
Japonia								
J. binoyae							+	
J. occulta							+	
J. vesca							+	
Leptopomoides								
L. halophilus			+			+	+	
L. conulus							+	
L. flammeus								
L. orophilus				+				
L. poecilus								
Micraulax								
M. coeloconus			+					
Scabrina								
S. brounae								+
Theobaldius								
T. annulatus		+				+		
T. bairdi					+	+	+	+
T. cadiscus						+		
T. cratera								
T. cytopoma							+	
T. layardi								
T. liliputianus								
T. loxostoma								

Table 1 (continued)

| | Climatic zones | | | | | | | |
| | dry | | | intermediate | | wet | | |
	F	A	B	C	E	D_1	D_2	D_3
T. parpsis			+					
T. parma						+		
T. subplicatus						+	+	
T. thwaitesi						+		
Cyclophorus								
C. involvulus		+		+		+		
C. ceylanicus		+	+	+		+	+	
C. liratula								
C. menkeanus						+		
C. alabastrianus								
C. punctatus				+				
Aulopoma								
A. hofmeisteri						+	+	
A. grande		+	+			+	+	
A. helicinum		+	+			+	+	
A. itieri						+		
A. sphaeroideum								
Pterocyclus								
P. bifrons								
P. bilabiatus			+			+		
P. bilabiatus var. conica								
P. cingalensis			+	+			+	
P. cumingi								
P. troscheli			+					
Cyathopoma								
C. artatum							+	
C. album							+	
C. leptopomita							+	
C. mariae			+				+	
C. prestoni							+	
C. ceylanicum								+
C. colletti								+
C. conoideum							+	
C. innocens								+
C. ogdenianum							+	
C. perconoideum					+			
C. serendibense							+	
C. turbinatum							+	
C. uvanse					+			
Tortulosa								
T. aurea								
T. austeniana							+	
T. blanfordi						+		
T. colletti						+		

Table 1 (continued)

	Climatic zones							
	dry			intermediate		wet		
	F ·	A	B	C	E	D_1	D_2	D_3
T. congeneri								
T. connectens							+	
T. cumingi								
T. decora						+		
T. duplicata							+	
T. eurytrema								
T. greeni							+	
T. greeni var. robusta							+	
T. recurvata								
T. haemastoma								
T. layardi								+
T. marginata								+
T. marginata var. crenulata								
T. marginata var. notatal								
T. neville							+	
T. neville var. flaveola								
T. nietneri								
T. nietneri var. caperata						+	+	
T. nietneri var. unicolor							+	
T. prestoni								
T. pyramidata								
T. rugosa								
T. smithi								
T. sykesi								
T. templemani					+		+	
T. thwaitesi								
Nicida								
N. catathymia							+	
N. ceylonica								+
N. belectabilis						+		
N. lankaensis						+		
N. pedronis								+
N. prestoni							+	
Truncatellidae								
Truncatella								
T. ceylanica								

and (B) those which are virtually restricted to dry places. Some of these as well as being confined to dry places are inhibited by shade.

Those land snails that are distributed in the dry zone as well as the other climatic zones can be included in the first group (Table 1). It is possible that the majority of these species are more abundant in the moister areas but are also capable of inhabiting the dry zone, particularly subzones A and B, by confining themselves to the more secluded habitats.

Of the land snails recorded from Sri Lanka there are ten species which seem to be confined to the dry zone. These are: *Cryptozona (Nilgiria) semirugata, Macrochlamys vilipensa, Pupilla muscerda, Gastrocopta mimula, Planispira fallaciosa, P. vittata, Micraulax coelocomus, Theobaldius parapsis, Pterocyclus bilabiatus* and *P. troscheli.* Of these, *C. (N.) semirugata, Planispira fallaciosa* and *P. vittata* have been collected in large numbers and their restricted distribution in the dry zone appears definite. With regard to the rest of the species, only a few specimens have been collected in the dry zone and more information may be necessary before confirming them as well defined dry zone species.

Ecological notes on the species that are definitely restricted to the dry zone and have been collected by the author are given below.

C. (Nilgiria) semirugata

This is a large snail with a thick white shell. It is heliophilic and xerophilic in nature and is fairly common in open country, cultivated soils and also in areas of high lime content in the northern parts of the Island. They are also found close to human habitations in places such as crevices of walls and pillars and structures made of lime and coral.

Planispira vittata and P. fallaciosa

Snails of this genus are distributed in the arid and semi-arid regions. They are found in large numbers in many localities in the north, and also in the numerous islands off the coast of the Jaffna peninsula. Both species are found in similar habitats such as rocky areas, corals close to coastal areas and on the bark of trees such as *Thespia populnea.* They have thick solid shells with expanded peristomes (margin of shell aperture) and the inside of the aperture is deep smoky brown or violet in colour. Some snails have plain white shells while others of the same species have shells with one to many reddish brown spiral bands.

Generalized ecosystems are characterized by a wide variety of plant and animal species each of which (or the majority of which) is represented by a relatively small number of individuals, i.e. there is a fairly uniform distribution of numbers of individuals among the various species present. Specialized ecosystems on the other hand usually have a low diversity and are characterized by a small number of species, some of which are represented by a large number of individuals (Evans 1972). The dry zone is poor in the diversity of snail species represented compared with the other climatic zones. But the species that have been successful in colonizing areas of this zone are always found in very large numbers in any given place.

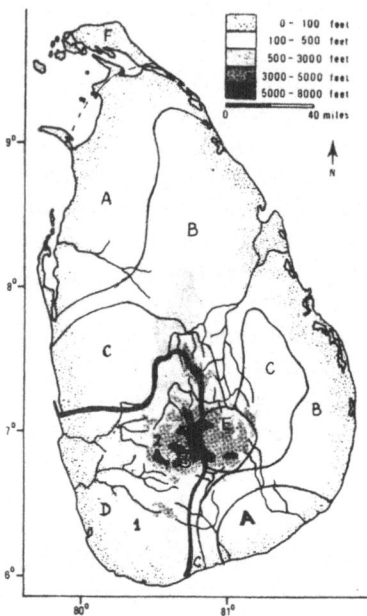

Fig. 1. Climate map of Sri Lanka (adapted from Gaussen *et al.* 1968 and Mueller-Dombois 1968).

Thus in this respect, the dry zone probably can be regarded as a specialized ecosystem.

It is not known what physiological adaptations enable dry zone snails to occupy unfavourable habitats. However, there are a few conchological features that are shared by most dry zone snails. These include a relatively thick shell which impedes the loss of water. Also, they usually possess shells with thick, expanded peristomes as illustrated by *Planispira* and *Nilgiria*. An expanded peristome when adpressed to the substratum seals off the interior of the shell resulting in the minimum loss of water (Machin 1967). Some dry zone snails, e.g. *Pupilla*, have shells with prominent surface sculpture, which impedes the circulation of air in contact with the shell, thus reducing the rate of loss of water. There are hairs on juvenile shells in *Micraulax* which may serve a similar purpose (Chatfield 1968). Another feature common to dry zone snails is the presence of shells which are either entirely white in colour as in *Nilgiria* or which have varying numbers of coloured spiral bands, as in *Planispira*. Coloured spiral bands are known to be helpful in the absorption of harmful ultraviolet radiation while white shells reflect light and help to keep the animal cool (Cain *et al.* 1969).

The intermediate zone

Climate, Vegetation and Soils

This zone is a climatically transitional area which links the dry and wet zone. It is less than 1,500 m in altitude and lies to the north and northeast of the wet zone. The intermediate zone has a mean annual rainfall of 1,875-2,500 mm and is divided into two subzones (C and E) based on altitude. The intermediate lowlands (Subzone C) are less than 1,000 m in altitude, and the vegetation is a mixture of moist semi-evergreen forests, wet semi-evergreen forests and villu and damana type grasslands. The mean annual temperature is about 26.6 ° C. The soil is Reddish Brown Earths. Large areas of this zone have been cleared for coconut cultivation. The intermediate uplands (subzone E) range from 1,000-1,500 m. Vegetation consists of areas of moist semi-evergreen forests, tropical savanna forests, dry patana grassland and montane temperate forests. The soil is modal Red-Yellow Podzol type which is loose and moist with a fair amount of organic matter. The average temperature in this subzone is about 28.3 ° C.

Land snails and ecological notes

There are only a few species, which according to available data are restricted to the intermediate zone. These are: *Streptaxis gracilis, Euplecta phidias, Macrochlamys umbrina, Trochomorpha hyptiocyclus, Thysanota hispida, Ena panos, Phaedusa ceylanica, Glessula panaetha, G. verruina, Cyclophorus punctatus, Cyathopoma perconoides, C. uvanse* and *Leptopomoidesor ophilus*. However, the data on all of these snails are inadequate to classify them as well defined intermediate zone species.

In general a majority of snails found in the intermediate zone are well represented in the other two climatic zones. There are also a number of species which are found predominantly in the dry zone or the wet zone, that are capable of tolerating a wider variation of climatic conditions and thus occupy niches of the intermediate zone close to the borders.

The following species are predominantly wet zone dwellers, but they have also been found to extend into the intermediate zone: *Streptaxis layardianus, S. peroteti, S. cingalensis, Hemiplecta chenui, H. juliana, Ravana politissima, Euplecta subopaca, E. travencorica, E. verrucula, E. gardneri, E. turritella, Sitata phyllophila, Eurychlamys regulata, Thysanota eumita, Ruthvenia biciliata, Corilla gudei, C. erronea, C. odontophora, Ena proletaria, E. stalix, Oligospira skinneri, Bradybaena similaris, Aegista huttoni var. radleyi, Glessula serena, Theobaldius bairdi* and *Tortulosa templemani*.

Short ecological notes on those species that have been collected by the author are given below.

The Genus *Streptaxis: S. layardianus, S. peroteti* and *S. cingalensis.*
Snails of this genus are found largely in higher altitudes of the wet zone and the intermediate zone. They are ground dwelling snails with small white shells. The shell is characteristic in having a body whorl which is eccentrically placed in relation to the rest of the spire. They move among decaying leaf litter in wet humus soil on forest floors and secluded areas.

The Genus *Hemiplecta: H. chemui* and *H. juliana.*
These are large snails with thin fragile shells. The shells are straw coloured with dark brown bands along the periphery and have a sharp lip to the aperture, as in most ground dwelling species. The snails are found on forest floors of the wet zone and intermediate uplands buried in a few inches in wet humus soil among decaying vegetation and leaf litter.

The Genus *Ravana: R. politissima.*
This snail, with a straw coloured globose shell is ground dwelling among leaf litter.

The Genus *Euplecta: E. subopaca, E. travencorica, E. verrucula* and *E. gardneri.*
Of these, *E. gardneri* has a globose, conoid shell. The shell is dark brown with a cream coloured band along the periphery of the body whorl. It is found on the bark and foliage of trees in lower forest strata. The other species are essentially forest floor or grassland dwelling. They are found in moist humus soil close to the base of grasses and in leaf litter close to the base of plants and trees.

The Genus *Corilla: C. gudei, C. erronea* and *C. odontophora.*
Snails of this genus have flat, discoidal shells. The aperture of the shell is extended and curved backwards. The characteristic feature of the shells of this genus is the presence of one or more lamellae entering the aperture. The snails are ground dwellers among rotting leaves and vegetation. They are active on surface layers of leaf litter in cold weather and just after rain.

The Genus *Oligospira: O. skinneri.*
This is a ground dwelling species, but they are also found on tree trunks up to a height of one or two feet. The shell is large and is characterized by the presence of a thick callus which is formed along the outer edges of the peristome. The peristome varies in colour from pale brown to deep violet. It is a common species in areas of subzone E, amongst fallen leaves on forest floors.

The Genus *Bradybaena: B. similaris.*
These snails have a thin, globose, conoid shells. They are found in large numbers in association with vegetation of cultivated areas, close to human habitations.

The Genus *Aegista: A. huttni* var. *radleyi.*
The snails are small with globose, conoid shells; the surface of the shell has well marked sculpture. These are ground dwelling in forests.

The Genus *Theobaldius: T. bairdi.*
This is a fairly common snail found in shaded areas of the Wet zone and intermediate uplands. These snails are usually found in large numbers in humus soil among decaying leaves, under fallen logs and on the bark of trees.

Of these wet zone species found in the intermediate zone, it is interesting to note that of a total of 27 species, 24 are found in the intermediate uplands, while only 3 are found as far as the intermediate lowlands.

Although the mean annual rainfall is ths same in both subzones of the intermediate zone, the uplands are at a lower temperature than the lowlands, as the former is at a higher elevation than the latter. The humidity of an environment is dependent upon the temperature and rainfall and to a lesser extent upon the vegetation, physical properties of the soil and the degree of soil disturbance (Evans 1972). Thus, subzone E is more humid and cooler than subzone C because it is at a higher elevation and also probably has been subjected to a lesser degree of athropogenic disturbances than subzone C, vast areas of which have been cleared for coconut cultivation. Thus, although the annual rainfall is the same, the other climatic and environmental conditions make subzone E a more favoured area for land snails than subzone C.

Dry zone snails in the intermediate zone

Rachis punctatus is a species which is well established and widely distributed in the dry zone, but it also has been recorded from subzone E of the intermediate zone. Such species probably could represent an intermediate stage in the transition of moisture and shade loving species into dry zone snails (Baker 1958). It is likely that they once were widely distributed in the intermediate zone, but with decreasing climatic stability due to repeated shifting cultivations and irregular droughts, there was a compensating shift in the physiological requirements of these snails. Such situations could force them to move into areas of the dry zone and change from indifferent xerophiles to obligatory xerophiles. The physiological basis for such behaviour is not fully understood, but the relatively high lime content of dry habitats is considered to be one of the factors responsible for the transition of moisture loving species into ones that are adapted to living in dry climatic conditions.

Wet Zone

Climate, vegetation and soil

The wet zone is confined to the southwestern quarter of the island and is smaller than the area covered by the dry and intermediate zones. Heavy and prolonged rains fall during the south-west monsoon periods (May to September) resulting in a well distributed annual rainfall of 2,500 mm or more. Since there are intermonsoonal thunderstorms in addition to the rains of the southwest and northeast monsoons, there is no marked period of drought. This region was at one time covered with tall luxurient tropical rain forests, but today most low lying districts and some of the hilly areas have been cleared for cultivation and human habitation. The temperature of this region ranges from 15.5-30 ° C. The wet zone is divided into three subzones based on altitude; the lowland subzone (D_1), the upland subzone (D_2) and the highland subzone (D_3). The lowland subzone is less than 1,000 m in elevation. The vegetation is of tropical wet evergreen forest type, but large areas are under cultivation and human habitation. The annual temperature ranges between 26.6-30 ° C. The upland subzone (D_2) has an altitude ranging from 1,000-1,500 m. The annual temperature is between 18.3-24 ° C and the vegetation is of submontane evergreen forest type and savannahs. The highland subzone (D_3) is between 1,500-2,600 m. The temperature ranges from 15.5-18.3 ° C. Formerly, this subzone was covered in dense montane temperate forests and wet patana grasslands. The trees in this subzone rarely exceed 10 m in height. The soil consists of Red Yellow Podzols, with peat deposits on the highest plateaus. Large areas of this subzone have been cleared for cultivation, but the remaining steep hillside jungles provide refuge for many of the endemic animals and plants.

Land snails and ecological notes

The wet zone by far exceeds the other two climatic zones in its diversity and density of land snails. Of about 265 species recorded from Sri Lanka, the locality data of 132 species are limited to the wet zone (Table 1). It is possible that with the availability of further locality data some of these snails might prove to be species that are distributed in the other climatic zones. There are certain genera and species which emerge as well defined wet zone snails. Ecological notes on some of these which the author has been able to collect are given below.

The Genus *Ratnadvipia: R. irridians* and *R. edgariana.*
This genus is confined to the wet zone. The snails have thin transparent shells which are dark brown in colour. The spire is almost flat and the body whorl is greatly enlarged with a large aperture. These snails are found at all altitudes. They require highly humid conditions as are found on forest floors, under grass cover

and in leaf litter. Occasionally they are found inhabiting cultivated areas, gardens and parks.

The Genus *Corilla: C. adamsi, C. beddomea, C. humberti, C. fryae, C. colletti* and *C. lesleyae.*
These snails are hygrophilic and are common at cooler temperatures between 1,690-2,500 m. elevation. They are ground dwelling and are found among rotting leaves and other vegetation in shaded environments close to hedges and the roots of plants. They are very active just after rains and come to the surface layers of leaf litter.

The Genus *Acavus: A. haemastoma, A. fastosus, A. superbus* and *A. roseolabiatus.*
Snails of this genus are found always at elevations less than 1,600 m. They are large snails which are arboreal and have shells with cryptic colour patterns. They show little mobility and their shells usually are covered with a green deposit. They remain stationary, attached to tree trunks. Barnacles (1962) attributes this to the watery secretions that they produce which do not allow them to attach firmly to the substratum when moving about. *A. haemastoma* has a globose conoid shell with broad dark brown bands and the lip of the aperture is a deep blood red. *A. fastosus* has a white shell with many narrow reddish brown bands. The lip of the aperture is rose pink. These two species have a similar geographical distribution. Both are found in fair numbers on trunks of trees in low lying, maritime districts. They also are found close to the crown of areca and coconut palms (Randles 1900). *A. superbus* has a shell which is characteristic in having oblique malleations on the surface of its body whorl. It is found attached to the moss covered trunks of large trees such as *Aleurites moluccana, Hygrophila spinosa* and *Digitaria adscendens,* growing in damp areas with thick vegetation. *A. roseolabiatus* is similar to *A. superbus* in shape and shell size, but the shell does not have malleations on the body whorl and the lip of the aperture is a brilliant pink. This snail has a very restricted distribution and is found high up on the trunks of large trees and areca palms.

The Genus *Oligospira: O. waltoni.*
These snails have large shells with a flat spire. The surface of the shell is dark brown with irregular yellow markings. They are found on forest floors buried under leaf litter which can be as deep as 6-12 inches in these areas.

The Genus *Beddomea: B. ceylanicus, B. albizonatus* and *B. intermedius.*
These have white, conical shells. The animal is bright green in colour. They live among foliage of trees close to the ground in forests. *B. albizonatus* is associated with *Adenanthera pavonica* (cardamon cultivations).

The Genus *Vaginulus: V. templetoni, V. maculatus* and *V. reticulatus.*
These are slugs. They are confined to the Wet zone and are found at all altitudes.
They are ground dwelling or live among branches and leaves of small trees.

The Genus *Tortulosa*

Except for one locality record of *T. templemani* from the intermediate uplands, all
species of this genus are recorded from the wet zone. They have white conical shells
with a characteristic projection from the margin of the aperture. The animal is
blackish-brown. Live snails are found on the upper surface of the foliage of trees
such as *Mesua* and also on the trunks and branches of *Macaranga digyna*. These
snails come out in large numbers just after a spell of rainy weather.

The occurrence of a large number of species in the wet zone undoubtedly is due
to their physiological dependence on climates of high humidity. In the distribution
analysis of wet zone snails, it was observed that the pattern of altitudinal distribution is one of decreased species diversity with increased altitude. Of the species
restricted to the wet zone, the majority inhabit altitudes below 1,600 m. This
perhaps could be due to the presence of optimum conditions of humidity at lower
altitudes.

Also, lower altitudes have more constant environmental conditions than higher
altitudes (Peake 1968). Higher altitudes are subjected to a wider temperature
variation due to the periodic changes in the temperature difference between the
lower strata and the canopy. High altitude and insulation due to heavy cloud cover
result in a decrease in temperature which creates a gradient between the canopy
and the forest floor. In the absence of cloud cover, the temperature in montane
regions follows the same pattern as observed for the lowland wet zone, but with a
lower mean maximum. The net result is a wider variation in microclimates in the
upper montane zone (Grubb & Whitmore 1966). Such unstable climates are not
favoured by most species of land snails. This probably could account for the
decrease in numbers and diversity of land snails in subzone D_3, as compared with
subzones D_1 and D_2. Another factor that could be contributory to this is the cooler
temperature at altitudes above 1,600 m. Such low temperatures could have an
effect on the physiological processes of some species. Also, soil acidity tends to be
higher at higher altitudes due to the lowered activity of micro-organisms. This
results in soils with poor lime content, which makes such environments unsuitable
for land snails.

Another contributory factor to lower densities of land snails at high altitudes is
the change in habitat structure in upper montane regions. At high elevations, there
is a decrease in the availability of suitable surfaces and firm substrates upon which
snails can crawl. This is because of the gradual increase in epiphytic mosses on
trees. Sometimes tree trunks are covered entirely with mosses saturated with water.
Leaf size also tends to be smaller at higher altitudes and this could also be a limiting
factor (Peake 1968). A further observation was the occurrence of large snails in the
upper arboreal strata as opposed to the lower aboreal and ground strata. This

gradation in size may be correlated with the variation in the structure of the habitat and associated changes in the microclimates. Generally, larger species are known to equilibrate with the external temperature at a lower rate than do smaller species (Peake 1968). This probably accounts for the occurrence of large snails such as *Acavus* in the upper strata of forests while smaller species predominate the ground and lower strata where microclimates are relatively stable.

Snails distributed in all three zones

Certain species were found in all climatic zones. *C. (Xestina) bistrialis* and *Achatina fulica* are found in large numbers in climatic conditions from the driest parts of subzone F to the cool temperatures of subzone D_3. Both species have been introduced and their habitats are invariably closely associated with human dwellings and cultivated areas. These snails have not been observed in primary forest areas such as the Sinharaja Reserve.

The following species, though not completely islandwide in distribution, are capable of inhabiting all zones, except perhaps subzone F of the dry zone. They are well established in the wet zone and the intermediate zone and also are capable of colonizing, to a certain extent, the cooler and more humid areas of the dry zone. These are: *Acavus prosperus, A. phoenix, Opeas prestoni, O. layardi, Microcystina bintenensis, Cyclophorus ceylanicus, C. cingalensis* and *Pterocyclus cingalensis*. There are certain other species which have been recorded from both the dry zone and the wet zone, but have no locality records from the intermediate zone (Table 1). Climatically, the intermediate zone is a transitional area between the wet and dry zones, therefore, it is assumed that if a species is tolerant of the climatic conditions of both zones, it should be capable of inhabiting the intermediate zone. However, large areas of the intermediate zone have been cleared for cultivation, which generally has a disturbing effect on the land snail fauna. Lack of disturbance is considered to be as important as optimum temperature and rainfall for the establishment and abundance of land snails. Thus, it is possible that species that were once nearly Islandwide in distribution have been eliminated from areas of the intermediate zone. Kerney (1966) points out, however that the effects of agriculture may, by creating a diversity of habitats, be more beneficial than harmful. A once rather uniform environment may give way to one with a variety of habitats, each harbouring a distinct population of snails. However, at this stage, it is not possible to attribute the apparent absence of certain species from the intermediate zone as compared with the dry and wet zones, to the disturbing effects of agriculture or inadequate collections. More extensive sampling needs to be carried out in this zone.

Endemic species

Endemicity is very high among land snails in Sri Lanka. A majority of endemic genera and species are concentrated in the central mountains of subzones D_2 and D_3 of the wet zone and subzone E of the intermediate zone uplands. Most of the other endemic species are restricted to the southwestern lowlands of the Island.

Conclusions

Although it is difficult to generalize distributional patterns of land snails in an area of such diverse climatic and vegetational conditions with a high degree of human disturbance, this preliminary analysis of locality data does demonstrate geographical and altitudinal patterns.

From an ecological standpoint, the largest density and species diversity of land snails is found in the wet zone. This is to be expected as land snails are best suited to environments of high soil moisture, shade and humidity. At the other end of the spectrum, the number of species characteristically found in the dry zone is very limited. It is interesting to note that certain definite patterns of distribution exist between the wet zone and the intermediate uplands. There seems to be a greater migration of species to the intermediate zone than to any other climatic zone. As the rainfall pattern over the entire intermediate zone is comparable, the difference in migration patterns have been discussed on the basis of temperature and vegetational differences and the degree of anthropogenic disturbances between subzones C and E. It must be borne in mind that the intermediate lowland is an area in which human interference with the natural environment has been greatest.

The land snails of Sri Lanka provide a rich source for a range of malacological studies. The present study is an attempt to provide distributional data for the land snails recorded. However, future studies on their distribution should be based on more intensive sampling, especially from the dry zone and intermediate zones. Such studies could form the base on which extensive studies on the biology, ecology and physiology of important taxa can be carried out.

References

Baker, H.B. 1958. Land snail dispersal. Nautilus 71:141-148.

Barnacles, G.A.S. 1962. Notes on the genus *Acavus* Montfort. J. Conch. 25:128-131.

Blanford, W.T. & H.H. Godwin-Austin. 1908. Mollusca-Testacellidae and Zonitidae. Fauna of British India, Mollusca 1. Taylor & Frances, London. 311 pp.

Cain, A.J., R.A.D. Cameron & D.T. Parkin. 1969. Ecology and variation of some helicid snails in northern Scotland. P.M.S. 38:269-299.

Chatfield, J.E. 1968. The life history of the helicid snail *Monacha cantiana* (Montagu), with reference also to *M. cartusiana* (Miller). P.M.S. 38:233-245.

411

Evans, J.G. 1972. Land snails in archaeology. Seminar Press, London. 436 pp.

Gaussen, H., M. Viart, P. Legris and L. Labrue. 1968. Explanatory notes on the vegetation map of Ceylon. French Institute, Pondicherry, India Government Press, Sri Lanka. 72 pp.

Grubb, P.J. & T.C. Whitmore. 1966. A comparison of montane and lowland rainforests in Equador II. The climate and its effects on the distribution and physiogomy of the forest. J. Ecol. 54:303-334.

Gude, G.K. 1914. Mollusca, Vol. II, Trochomorphidae, Janellidae. Fauna of British India. Taylor & Frances, London. 520 pp.

Gude, G.K. 1921. Mollusca, Vol. III, Land operculates. Fauna of British India. Taylor & Frances, London. 386 pp.

Kerney, M.P. 1968. Snails and man in Britain. J. Conch. 26:3-14.

Koelmeyer, K.O. 1957. Climatic classification and the distribution of vegetation in Ceylon. Ceylon Forester 3:144-163.

Machin, J. 1967. Structural adaptations for reducing water loss in three species of terrestrial snails. J. Zool. 152:55-65.

Mueller-Dombois, D. 1968. Ecogeographic analysis of a climate map of Ceylon with particular reference to vegetation. Ceylon Forester 3:39-58.

Panabokke, C.R. 1967. Soils of Ceylon and fertilizer use. Ceylon Assoc. Adv. Sci. Part 1. p. 74.

Peake, J.F. 1968. Habitat distribution of Solomon Island land Mollusca. Symp. Zool. Soc. Lond. and Malacol. Soc. Lond. 22:319-357.

Wenz, W. (1938-1944). Handbuch der Palaozoologie, Band 6, Gastropoda Teil I: Allgemeiner Teil und Prosobranchia. Borntraeger, Berlin. 834 pp.

Zilch, A. (1959-1960) Handbuch der Palaozoologie, Band 6, Gastropoda Teil 2: Euthyneura. Borntraeger, Berlin. 834 pp.

20. Ecology and biogeography of mammals

G.M. McKay

Introduction

Although a number of mammals from Sri Lanka had been described prior to the mid nineteenth century, the first systematic account of the fauna of the island was that of Kelaart (1851). Tennent (1859, 1861) gave a brief description of the mammals, but with a detailed account of one particular species, the elephant. Both these authors had resided in Sri Lanka and described only the fauna of that island. Blanford (1888, 1891), however, provided a systematic account of the mammals of the entire Indian region, correlating the island fauna with that of the mainland. Revisions in this century have been either local (Phillips 1935, Hill 1939, Eisenberg & McKay 1970) or wider in scope (Ellerman & Morrison-Scott 1966), or have dealt with only specific taxonomic groups (e.g. Ellerman 1947, Moore 1960, Musser 1981, Marshall 1977).

The purpose of this paper is twofold: to review recent literature on the status and ecology of the Sri Lankan mammals (pointing particularly to those groups which still lack information), and to discuss the relationships of the island mammal fauna to those of the mainland.

The mammals of Sri Lanka

This systematic treatment is not intended as a repetition, with amendments, of Eisenberg & McKay (1970), rather it is an addition to information presented there; subspecific names are not used and synonymies will be given (in parentheses) only when the name currently in use differs from the name employed in that paper. Nomenclature generally follows Honacki et al. (1982); departures from their system are discussed. The reader's attention is drawn to the fact that, for many species, the most recent source of ecological information other than anecdotal is still Philips (1935).

Order Insectivora
Family Soricidae
   Crocidura horsfieldi
   Crocidura miya-species endemic
   Feroculus feroculus-genus and species endemic
   Solisorex pearsoni-genus and species endemic
   Suncus murinus
   Suncus zeylanicus (S. murinus zeylanicus)-species endemic

This family has the highest percentage of genera and species endemic to Sri Lanka. *Feroculus* appears to be closely allied to *Suncus* and *Solisorex* to *Crocidura*.

Suncus zeylanica is here retained as a separate species contrary to Eisenberg & McKay (1970). *S. murinus* is highly variable and shows distinct clinal variation in Sri Lanka; specimens from the two localities where *S. zeylanicus* has been found show marked variation and have been referred to two subspecies. The five specimens of *S. zeylanicus* examined by the author are uniform in having the tail much longer than in any *S. murina* and, as noted by Phillips (1935), in having more strongly hooked upper teeth. Pending further study of Sri Lankan *Suncus* it seems wisest to retain specific status for this shrew. Consistent with opinionsexpressed in Honacki *et al.* (1982), *Podihik kura* is here considered to be the juvenile of *Suncus etruscus*.

Phillips (1935) considered *Feroculus* to be more carnivorous than other Sri Lankan shrews. This habit also is probably true of *Solisorex*; a female which the author observed in captivity for several days at Horton Plains accepted and ate frogs as well as worms.

Order Chiroptera
Family Pteropodidae
   Cyanopterus brachyotis-(included in next species)
   Cyanopterus sphinx
   Pteropus giganteus
   Rousettus leschenaulti (R. seminudus)

Family Emballonuridae
   Saccolaimus saccolaimus (Taphozous s.)
   Taphozous longimanus
   Taphozous melanopogon

Family Megadermatidae
   Megaderma lyra
   Megaderma spasma

Family Rhinolophidae
  Hipposideros ater (H. bicolor)
  Hipposideros fulvus (not included)
  Hipposideros galeritus
  Hipposideros lankadiva
  Hipposideros speoris
  Rhinolophus luctus
  Rhinolophus rouxi

Family Vespertilionidae
  Hesperoptenus tickelii
  Kerivoula hardwickei
  Kerivoula picta
  Miniopterus schreibersi
  Murina cyclotis
  Myotis hasselti
  Myotis peshwa
  Pipistrellis ceylonicus
  Pipistrellus coromandra
  Pipistrellus mimus
  Pipistrellus mordax
  Scotophilus heathi
  Scotophilus kuhli (S. temminki)

Family Molossidae
  Chaerephon plicata (Tadarida p.)
  Tadarida aegyptiaca (T. tragata)

References to most of the changes in bat nomenclature can be found in Honacki *et al.* (1982). Hill (1963) in his revision of the genus *Hipposideros* separated *ater* and *fulvus* from *bicolor*, and listed both as occurring in Sri Lanka. All three species are closely related and differ little in external morphology. Hill & Kitti (1972) considered the Sri Lankan *Myotis* to be *M. hasseltii* but Honacki *et al.* (1982) also list *M. peshwa* as possibly occurring in Sri Lanka. Mein & Tupinier (1977) have argued for placing *Miniopterus* in a separate family based on the possession of a vestigial anterior premolar, but in the absence of further evidence this does not seem warranted.

As with the shrews, the bats have received little attention in Sri Lanka since Phillips' surveys (see e.g. Phillips 1936); several species are represented in collections by only a few specimens. Among the insectivorous bats, the Emballonurids and Molossids are swift fliers which forage along with *Miniopterus* high above the canopy, Rhinolophids and most of the Vespertilionids forage in or below the canopy, and *Myotis* frequently feeds over water. The Megadermatids are slow-flying hunters, *M. lyra* being particularly carnivorous, preying upon birds, geckos and possibly other bats.

Order Primates
Family Lorisidae
   Loris tardigradus

Family Cercopithecidae
   Macaca sinica-species endemic
   Presbytis entellus
   Presbytis senex-species endemic

Groves (in Hanecki *et al.* 1982) considered *M. sinica* to be conspecific with *M. radiata* of peninsular India, and similarly *P. senex* to be probably conspecific with *P. johnii.* If these two endemic species are included with their peninsular relatives then all four of Sri Lanka's primates will have distributions limited to the Indo-Lankan area.

Petter & Hladik (1970) and Hladik (1975) report on the behaviour and ecology of *Loris.* This small nocturnal primate is dependent upon dense understory vegetation as it forages mainly at the 2-3 m level, rarely ascending larger trees. It feeds on repugnant insects, rarely chosen by other mammals, which it locates by smell. In suitable habitat *Loris* live at densities approximately 1 ha^{-1} with a basically solitary dispersion, one male home range probably overlapping with those of several females.

The diurnal Cercopithecids have received greater attention. Amerasinghe *et al.* (1971) examined the anatomy of the digestive tract, correlating the large stomachs and increased densities of argentaffin and goblet cells in the two species of *Presbytis* with their folivorous diet. Hladik (1975, 1978) and Hladik & Hladik (1972) compared diet with group size and structure, finding that *P. senex* had the smallest group size, smallest home range and most specialized diet. *P. entellus,* which feeds on a wider variety of plants, including herbs, ranges more widely in larger groups. The frugivorous *Macaca* forages even more widely.

The relationships between social behaviour and ecology of the monkeys have been reported extensively: Dittus (1975, 1977, 1979) on *Macaca,* Ripley (1967, 1970) on *P. entellus,* and Rudran (1970, 1973) on *P. senex.* Eisenberg *et al.* (1972) reviewed the evolution of social organization of primates in relation to habitat and adaptations. According to their classification of primate social systems, the unimale (*P. senex*) or age-graded male (*P. entellus*) social system, with limited male-male tolerance, is most common in arboreal, forest-dwelling species; the multi-male system as seen in *Macaca* is more common in terrestrial foragers. This analysis corroborates the findings of Hladik (1975, 1978) on relative sizes of home ranges and the particularly high densities achieved by *P. senex.*

Order Carnivora
Family Canidae
  Canis aures

Family Ursidae
  Melursus ursinus

Family Mustelidae
  Lutra lutra

Family Viverridae
  Paradoxurus hermaphroditus
  Paradoxurus zeylonensis-species endemic
  Viverricula indica

Family Herpestidae
  Herpestes edwardsi
  Herpestes fuscus
  Herpestes smithii
  Herpestes vitticollis

Family Felidae
  Felis chaus
  Felis rubiginosa
  Felis viverrina
  Panthera pardus

Honacki et al. (1982) retain the name *Panthera* for the large cats despite cytogenetic and other evidence (Wurster 1969, Van Gelder 1977).

At Wilpattu, jackals (*Canis*) live as bonded pairs together with their current year's offspring (Eisenberg & Lockhart 1972). Home ranges tend to be centred on permanent water. Jackals are opportunistic feeders, preying on birds and mammals up to the size of small fawns. Carrion also is a major source of food.

The sloth bear (*Melursus*) is restricted to the drier areas of the north and east and feeds primarily on termites and fruits, although Phillips (1935) reports carrion as being eaten 'not infrequently'. Eisenberg & Lockhart (1972) record females carrying young cubs on the back, a habit apparently peculiar to this species among bears.

The otter (*Lutra*) probably has become reduced in range and abundance since Phillips (1935) reported is as being common throughout the wet zone.

Eisenberg & Lockhart (1972) noted that the two species of *Paradoxurus* forage in the tree layer and are less strictly carnivorous than the terrestrial *Viverricula*. The four species of *Herepestes* appear to be diurnal equivalents of *Viverricula*, but although there is as yet no satisfactory evidence they must have different food preferences and foraging patterns to be able to live sympatrically over wide areas. Over much of the dry zone at least two species can be found in any one area. Eisenberg & Lockhart (1972) report *smithii* as mainly solitary and *fuscus* more gregarious. My own observations in the Gal Oya and Ruhunu areas tend to confirm and extend this view: there *H. smithii* and *edwardsi* usually are encountered as solitary individuals and *viticollis* is most frequently sighted in pairs. Observation of the foraging and feeding behaviour of these small carnivores would no doubt make a rewarding study in resource partitioning and competition.

The smaller cats are seldom seen, and their ecology little known. Phillips (1935) records that *F. viverrina* is active diurnally as well as nocturnally. During 1967 at Ruhunu National Park a single *F. rubiginosa* (presumed to be the same individual) was seen frequently hunting during daytime. This latter species is widespread but the two larger species effectively are allopatric, *F. viverrina* occurring in the south; *F. chaus* only in the northern arid region.

The behaviour and ecology of the leopard (*Panthera pardus*) have been studied by Muckenhirn & Eisenberg (1973) and Eisenberg & Lockhart (1972). Leopards are solitary with home ranges not overlapping those of other members of the same sex. Prey include large rodents, monkeys, hares and at least the smaller artiodactyls. At Wilpattu, *Cervus axis* and *Sus scrofa* appear to be preferred prey. Prey are not normally removed far from the site of the kill. Leopards will feed upon carcasses other than their own kills.

Order Proboscida
Family Elephantidae
  Elephas maximus

A detailed summary of the status and the biology of the Asiatic elephant including a bibliography is given by Shoshani & Eisenberg (1982).

The ecology and behaviour of the elephant in Sri Lanka are reported in detail by Eisenberg & Lockhart (1972), McKay (1973), Nettasinghe (1973), McKay & Eisenberg (1974), Kurt (1974) and Van Cuylenberg (1974, 1977). The largest, and in terms of biomass often most dominant, herbivore is now restricted to areas of the dry zone which have not yet been cleared for agriculture. Food consists of a wide variety of trees, shrubs, woody herbs and grasses, although some preferences are shown. The social organization is matriarchal with herds consisting of old adult females, their mature female offspring and juveniles of both sexes. At puberty, males leave the herd environment and either associate in labile and temporary unisexual groups or become solitary. At this stage, young males may move, or be forced to move, long distances. Home ranges of mature males overlap extensively and a dominance heirarchy may be formed. The home range of the herd will overlap those of a number of males and several herds may aggregate at particular areas (Lahugala Tank and the banks of the Mahaveli Ganga for example) mainly during the dry season. Male-male aggression seldom is observed except in the presence of oestrous females.

Reproduction in domesticated elephants has been studied by Jainudeen *et al.* (1971a, b) and Eisenberg *et al.* (1971) and the associated phenomenon of 'Musth' by Scheurmann & Jainudeen (1972) and Jainudeen *et al.* (1972). 'Musth' is a manifestation of the sexual and aggressive condition of the individual male, homologous to the 'Rut' of Artiodactyls but not restricted to any particular season.

A census of domestic elephants (Jayasinghe & Jainudeen 1970) located 532 individuals in captivity or approximately one quarter of the estimated maximum wild population of the island at that time (McKay 1973).

Order Artiodactyla
Family Suidae
  Sus scrofa

Family Tragulidae
  Tragulus meminna

Family Bovidae
  Bubalus bubalis

Family Cervidae
  Cervus axis (Axis axis)
  Cervus porcinus (Axis porcinus)
  Cervus unicolor
  Muntiacus muntjak

Wild pigs (*Sus*) occur throughout the dry zone and in undisturbed areas of the wet zone. Their social organization is one of a herd consisting of adult females and juveniles with peripheral, normally solitary, males (Eisenberg & Lockhart 1972). Pigs are omnivorous and in the southeast frequently were observed to feed on carrion. They also frequently cause damage to crops, some of which inexplicably is attributed to elephants.

Tragulids, the smallest of the ungulates, are restricted to areas where dense thickets provide adequate cover. They occupy a niche which is parallel to that occupied by caviomorph rodents in South America (Dubost 1968, Eisenberg & McKay 1974).

Among the cervids, social organization and feeding adaptations are in part related to size. The small *Muntiacus* is a solitary forest-dweller (Dubost 1971, Barrette 1977) which occurs throughout much of Sri Lanka; *C. porcinus,* also relatively small and solitary, is distributed only in the southwestern rainforest regions. Of the two larger species, *C. unicolor* is a forest living browser which either is solitary or may form small groups, but *C. axis,* which tends more toward grazing, lives in groups of 3-6 which aggregate in large herds in open grassland areas (Eisenberg & Lockhart 1972). In areas such as Wilpattu and Ruhunu National Parks where large areas of natural grassland allow *C. axis* to live at high densities, such large herds are common, but in other areas of forest or savannah, such as the Gal Oya Valley, it is at low density and only the smaller family units are seen.

There has been much debate about whether buffalo (*Bubalus*) are truly wild in Sri Lanka (see for example Phillips 1935). 'Wild type' water buffalo, distinguishable by colouration and horn morphology are to be found in remoter areas of Sri Lanka (Gal Oya, Wilpattu, parts of Eastern and North Central Provinces). Males mostly are solitary and female herds small. They do overalp with and probably mingle with domestic buffalo in Willpattu, Gal Oya and Ruhunu National Parks-wild males have been observed on several occasions with domestic herds in Gal Oya. In India, 'wild type' buffalo also are restricted to the most remote areas such as Manas Sanctuary; the so-called wild buffalo of Kaziranga display the widespread, exaggerated horns typical of several domesticated strains. This species undoubtedly was widespread formerly throughout the Indo-Lankan region, and, being tractable, was domesticated readily, probably without any major genetic change. As no apparent genetic barriers exist to interbreeding between wild and domestic stocks, and as domestic herds are quite commonly turned loose when not being milked, variation in the 'wild' stocks is inevitable. Thus the argument as to the wildness of free-living buffalo is at best academic.

No other wild bovids occur in Sri Lanka, although Tennent (1861) argued that the gaur (*Bos frontalis* = B. taurus) may have persisted until the seventeenth century (it cis recorded from the Pleistocene by Deraniyagala 1958, 1963).

Order Pholidota
Family Manidae
   Manis crassicaudata

Phillips (1935) said that the pangolin is found throughout the island at lower elevations, but specimen records would suggest that it is more common in the drier areas where the mound-building termites, its preferred diet, are more abundant.

Order Rodentia
Family Sciuridae
   Funambulus layardi
   Funambulus palmarum
   Funambulus sublineatus
   Petaurista petaurista
   Petinomys fuscocapillus
   Ratufa macroura

Family Cricetidae
Tatera indica

Family Muridae
Bandicota bengalensis
Bandicota indica
Coelomys mayori-(in Mus), species and genus endemic
Cremnomys blanfordi-(in *Rattus*)
Golunda ellioti
Mus cervicolor
Mus musculus
Mus fernandoi-species endemic
Rattus montanus-species endemic
Rattus rattus
Rattus norvegicus
Srilankamys ohiensis-(in *Rattus*) endemic genus
Millardia meltada
Vandeleuria nolthenii-(in *V. oleracea*) endemic species
Vandeleuria oleracea

Family Hystricidae
Hystrix indica

Agrawal & Chakraborty (1979) reviewed the taxonomy of the Sciurids.

Moore (1960) reviewed the geographical distribution of the tree squirrels, *Ratufa* and *Funambulus,* in the Indian region. He concluded that there was no evidence for prior continuous distribution of rainforest across northern India connecting the peninsula to the Himalayan or Indochinese regions and that the affinities of *Funambulus* were to the Ethiopian rather than to Oriental faunas. He presented a hypothesis for the evolution of the five Indo-Lankan species of *Funambulus* on the basis of repeated isolation and concurrence of the two sub-faunas during the Pleistocene. *F. layardi* and *F. sublineatus* are considered to have evolved in Sri Lanka and reinvaded the peninsula during later glaciations.

Nomenclature in the Muridae has undergone considerable change in the past decade, and this process is by no means completed. The reader is referred to Honacki *et al.* (1982) for detailed references to the changes listed above. I agree with Musser (as quoted in Honacki *loc. cit.*) that *Coelomys mayori* should be separated from *Mus* into a monotypic genus as it first was described. Marshall (1977) placed the Sri Lankan form of *Mus cervicolor (fulvidiventrsi)* in *M. booduga,* but given the lack of demonstrated karyological difference between *booduga, fulvidiventris* and *cervicolor,* I think it wisest to be conservative and to retain *cervicolor* for all the Indo-Lankan field mice until more concrete evidence for their separation is presented. On similar grounds, *M. musculus* should be retained for the commensal mice.

Rattus rattus is retained for the commensal and wild populations of the common rat in Sri Lanka. Baverstock *et al.* (unpubl.) describe two chromosomal forms - a 2n=40 form they attribute to *R. r. kandianus* (without precise locality) and a 38 form they attribute to *R. r. rufescens* from Trincomalee. Electrophoretically these two forms show 8% fixed differences, much less than the 24% between all *R. rattus* combined and *R. loseae* from Thailand.

Gatamiya weragami is considered a juvenile *Mus*.

Cremnomys and *Millardia* were reviewed by Mishra & Dhanda (1975) and *(Rattus) blanfordi* included in the former. Musser (1979) considers the two forms of *Vandeleuria* in Sri Lanka to represent different species, the larger highland form being endemic. Musser (1981) placed the unique *(Rattus) ohiensis* in a monotypic endemic genus, *Srilankamys*. He considered it to be a relict, possessing a number of primitive characters, and possibly related to the stock from which other Indo-Malayan rat-like genera evolved. *Rattus montanus* is considered by the author to be possibly a sample of extremely large *R. rattus kelaarti* males. Several large male rats trapped by the author and G.H. Manley at Horton Plains which would otherwise be attributable to *kelaarti* can be identified as *montanus* using Phillips' keys and descriptions.

Rodent ecology is yet another area which has been sadly neglected in Sri Lanka. A few species are abundant but many appear to be uncommon. Ecological and genetic information about all the species of Sri Lankan rodents should be given high priority.

Order Lagomorpha
Family Leporidae
   Lepus nigricollis

Hares are common through all parts of the island wherever the forest is interspersed with clearings. They are particularly abundant where grazing by large ungulates maintains short grassland.

Geographic relationships

The geographic affinities of Sri Lanka's mammal fauna can be analyzed in terms of shared taxa between the island and successively more distant regions of the mainland. The potentially at least, highly mobile bats show no discernable local effects at the generic level. Of 17 genera, 3 are pantropical, 2 occur widely in the Old World, 7 are Old World Tropical, and a further 5 occur widely throughout tropical Asia. None are endemic or even restricted to the Indian region.

Terrestrial mammals do, however, show patterns in their relationships as summarized in Fig. 1 and Table 1. The greatest affinity is, as expected, with the

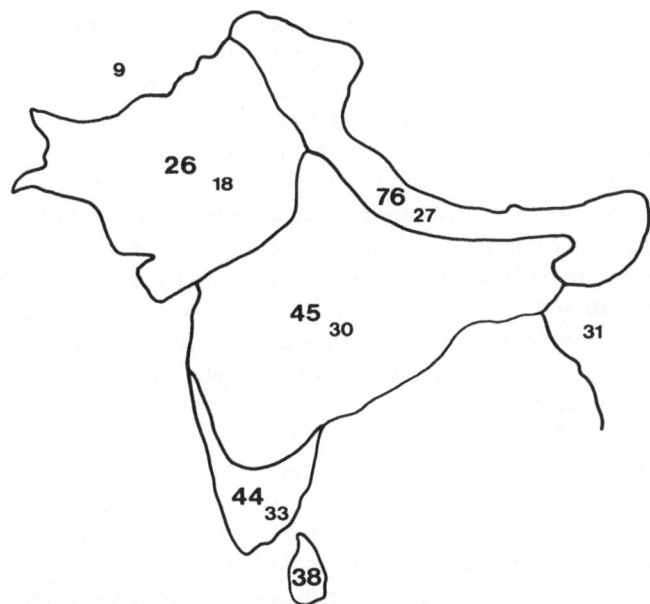

Fig. 1. The total numbers of terrestrial mammal genera in each sub-region of the Indian region and (small numerals) the number shared with Sri Lanka.

southern peninsula and the Deccan-Bengal areas. If a simple filter process were in operation it would be expected that the coefficients of association would continue to decrease proportional to increasing distance. Such a pattern is demonstrated by successive comparison between either Sri Lanka or the southern peninsula and the Deccan, Sind and Arabian faunas, but not when compared with the Himalayan region and Indochina.

This difference can be resolved by examining the distribution patterns of the individual genera involved. Twelve genera are widespread, occurring throughout all the Indian region and beyond, with four (*Suncus, Macaca, Herpestes,* and *Hystrix*) being primarily tropical or sub-tropical and eight (*Canis, Felis, Panthera, Lutra, Sus, Cervus, Mus, Lepus*) being even more widespread. A further eleven genera, although widespread in Asia, do not (except as commensals) extend to the Arabian region (*Presbytis, Viverricula, Paradoxurus, Muntiacus, Elephas, Petaurista, Ratufa, Bandicota, Millardia, Rattus* and *Vandeleuria*). Five genera have disjunct distributions: occurring in Sri Lanka and southern peninsular India, absent from northern India and appearing again in the Indochinese region (*Manis, Tragulus, Petinomys*); or occurring in Sri Lanka and northern India and beyond, but absent from the Peninsula (*Crocidura, Bubalus*). Only one genus (*Tatera*) has affinities through peninsular India and to the Arabian region alone.

Five genera (*Loris, Melursus, Funambulus, Cremnomys* and *Golunda*) are restricted to the peninsula and Sri Lanka, while a further four (*Solisorex, Feroculus, Coelomys* and *Srilankamys*) are endemic to the island. A similar analysis for the southern peninsula provides additional disjunct distributions (e.g. *Aonyx, Hemitragus* and *Martes*) and endemics (*Anathana, Diomys*).

The overall pattern is one of an island fauna derived mainly from that of the Indian-Indochinese regions, but with and overlay of more widespread forms. Most genera present in Sri Lanka are either mesic adapted (e.g. *Petinomys, Petaurista*) or adaptable to both mesic and arid environments (e.g. *Presbytis, Herpestes, Mus*). Although no one single factor can be used to explain the restriction of a relatively high proportion of genera to southern India and Sri Lanka, or the disjunctions in distribution, a formerly more widespread distribution followed by climatically-induced habitat changes would appear most likely (see, for example, the arguments presented by Moore, 1960).

Similar patterns of disjunction in distribution can be seen at the species level. As listed above there are twelve endemic species, in four of the families. Disregarding for the moment the four which are endemic at the generic level, the two Soricids and all three of the Murids have congeners present in Sri Lanka; the two primates and the *Paradoxurus* have closely related congeners in southern peninsular India. There is thus no particular geographic pattern to endemism at the species level, rather it appears to be related to isolation and the potential for rapid evolution in small species.

Distributions within Sri Lanka

Several efforts have been made to describe and classify the vegetation types of Sri Lanka in relation to soils (Fernando 1968) or climate (Mueller-Dombois 1968). Eisenberg & McKay (1970) followed Mueller-Dombois (1968) and the map of Gaussen *et al.* (1964) in using four classes (A to D) based on increasing humidity, and subdividing class D into three altitudinal zones. Analysis of mammal distribu-

Table 1. Geographic affinities of mammal genera of the Indian region. Coefficients of association between faunas[1].

	Sri Lanka	Peninsular India
Peninsular India	.63	—
Deccan-Bengal	.58	.78
Kutch-Indus	.33	.39
Arabian	.17	.22
Himalayan	.33	.28
Indochinese	.43	.35

[1] Jaccard index = a/(a + b + c) where: a = number of genera shared in the two faunas; b = number of genera restricted to the first fauna; c = number of genera restricted to the second fauna.

Table 2. Numbers of mammal species occurring in and restricted to major vegetation zones in Sri Lanka. Numbers of parentheses are species restricted to one zone.

Climatic/Vegetation Zone[1]	DRY			LOW WET		HIGH WET	
	A	B	C	D$_1$		D$_2$ D$_3$	
Order Insectivora	1			3	(1)	6	(4)
Chiroptera	18	(2)		23	(6)	8	(4)
Primates	4	(1)		3		3	
Carnivora	14	(3)		11		5	
'Ungulates'	7	(2)		6	(1)	4	
Pholidota	1			1		0	
Rodentia	15	(3)		17	(2)	13	(3)
Lagomorpha	1			1		1	
Total	61	(11)		65	(10)	40	(11)
Proportion restricted species		.18			.15		.28

[1] With zone classifications according to Eisenberg and McKay (1970).

tions using all six categories shows a large majority of species occurring over three or more such zones and only a few with more restricted distributions. By combining vegetation zones as in Table 2 some patterns can be seen. All orders other than the Insectivora have a reduced diversity in the high altitude wet zone. Bats and rodents are somewhat more diverse in the wet zones but carnivores, ungulates and primates have their highest diversity in the dry zone. The two most diverse groups (bats and rodents) have high numbers of species restricted to one area, the zone with the highest percentage of restricted species being the high altitude wet. Six of these high-wet zone restricted species are endemics. The generally low level of restriction to any particular zone or habitat apparent in all groups except the Insectivora is a further indication of the prevalence of species which can tolerate a large range of habitat conditions.

Mammal species showing variation within Sri Lanka

Ten species show sufficient variation within Sri Lanka to have been divided into two or more subspecies (Table 3). Of these, six show purely clinal variation with the two main factors being climatic (arid humid) and altitude. *Suncus murinus*, in addition to clinal variation in wild forms, is represented by one or possibly two introduced commensal forms.

Rattus rattus, or the group of semi-species to which this name is currently applied, shows a more complex pattern of variation (McKay, unpubl. data). Commensal forms can be found in towns and villages throughout the island. These have been referred variously to the named forms *rattus, ceylonicus* and *rufescens* and it is not unlikely that introduced rats came from more than one source. Wild

426

Table 3. Mammal species showing variation within Sri Lanka.

	No named forms[1]	Apparent type of variation		
		Clinal		
		Climate	Altitude	Other
Suncus murinus	4	—	+	commensal forms
Loris tardigradus	4	+	+	—
Macaca sinica[2]	3	+	+	—
Presbytis senex[2]	5	+	+	—
Herpestes fuscus	4	+	?	—
Funambulus palmarum	5	+	+	widespread polymorphisms
Ratufa macroura	4	+	+	—
Coelomys mayori[2]	2	—	+	—
Golunda ellioti	2	+	—	—
Rattus rattus	6	?	+	widespread polymorphisms

[1] Based on Eisenberg and McKay (1970); [2] species endemic.

forms show a partly clinal variation (altitudinal) from *kandianus* to *kelaarti* (the latter possibly including *R. montanus*; see discussion under Rodentia, above). Superimposed on this there are coat-colour polymorphisms affecting both dorsal and ventral pelage and variation in the degree of spinescence in the guard hairs. Until a detailed karyological and genetic analysis of a number of populations can be made, particularly to examine the nature of variation in the wild forms and the degree of interbreeding between wild and commensal forms, form conclusions clearly are impossible. P. R. Baverstock *et al.* (unpubl. data) have certainly indicated that there are sufficient differences to warrant further investigation.

Funambulus palmarum also shows a pattern of polymorphisms superimposed upon clinal variation (McKay unpubl. data). None of the characters used by Phillips (1935) (general hue, dorsal and lateral stripe colour, cap colour) nor external measurements show any consistent pattern that could be used to delimit significantly different forms in a sample of 118 adults examined in the Colombo and British Museums. Indeed a series of 37 individuals from presumably one population at Anasigalla, Matugama, have been identified as belonging to no less than four named forms. As this species is not complicated by the intrusion of commensal forms, it would present an interesting model for the study of genetic variation within and between population.

References

Agrawal, V.C. & S. Chakraborty. 1979. Taxonomic notes on some Oriental squirrels. Mammalia 43:161-172.
Amerasinghe, F.P., B.W.B. Van Cuylenberg & C.M. Hladik. 1971. Comparative histology of the alimentary tract of Ceylon primates in correlation with the diet. Ceylon J. Sci. (Bio. Sci) 9:75-87.

Barrette, C. 1977. Some aspects of the behaviour of muntjaks in Wilpattu National Park. Mammalia 41:1-34.

Blanford, W.T. 1888-1891. The fauna of British India, including Ceylon and Burma. Mammalia. (2 Parts). Vol. 1. Taylor & Francis, London. 617 pp.

Deraniyagala, P.E.P. 1958. The Pleistocene of Ceylon. Nat. Mus. Ceylon, Colombo. 164 pp.

Deraniyagala, P.E.P. 1963. Some mammals of the extinct Ratnapura fauna of Ceylon part V, with reconstructions of the hippopotamus and the gaur. Spol. Zeylan. 30:5-25.

Dittus, W.P.J. 1975. Population dynamics of the toque monkey, *Macaca sinica*. pp. 125-151 In. R.H. Tuttle (Ed.),Socioecology and psychology of Primates. Mouton, The Hague.

Dittus, W.P.J. 1977. The social regulation of population density and age-sex distribution in the toque monkey. Behav. 63:281-320.

Dittus, W.P.J. 1979. The evolution of behaviors regulating density and age-specific sex ratios in a primate population. Behav. 69:265-302.

Dubost, G. 1968. Les niches ecologiques des forêts tropicales sud-americaines et africaines, sources de convergences remarquables entre rongeurs et artiodactyles. Terre et Vie 22:3-28.

Dubost, G. 1971.Observations éthologiques sur le muntjak (*Muntiacus muntjak* Zimmerman, 1780 et *reevesi* Ogilby, 1839) en captivité et semi-liberté. Z. Tierpsychol. 28:387-427.

Eisenberg, J.F. & M. Lockhart. 1972. An ecological reconnaisance of Wilpattu National Park, Ceylon. Smithson. Contr. Zool. 101:1-118.

Eisenberg, J.F. & G.M. McKay. 1970. An annotated checklist of the recent mammals of Ceylon with keys to the species. Ceylon J. Sci. (Bio. Sci.) 8:69-99.

Eisenberg, J.F. & G.M. McKay. 1974. Comparison of ungulate adaptations in the new world and old world tropical forests with special reference to Ceylon and the rainforests of Central America. pp.585-602 In: V. Geist & F. Walther (Eds.), The behaviour of ungulates and its relation to management. I.U.C.N. Publ. New Ser. 24. Morges.

Eisenberg, J.F., G.M. McKay & M.R. Jainudeen. 1971. Reproductive behaviour of the Asiatic elephant (*Elephas maximus maximus* L.). Behav. 38:193-225.

Eisenberg, J.F., N.A. Muckenhirn & R. Rudran. 1972. The relation between ecology and social structure in primates. Science 176:863-874.

Ellerman, J.R. 1947. A key to the Rodentia inhabiting India, Ceylon and Burma, based on collections in the British Museum. J. Mammal. 28:249-278, 357-387.

Ellerman, J.R. & T.C.S. Morrison-Scott. 1966. Checklist of Palaearctic and Indian mammals. 2nd edit. Brit. Mus. Nat. Hist. Lond. 810 pp.

Fernando, S.N.U. 1968. The natural vegetation of Ceylon. Lake House Press, Colombo. 85 pp.

Gaussen, H., M. Viart, P. Legris & L. Labroue. 1964. International map of the vegetation of Ceylon. Cey. Surv. Dept. Colombo.

Hill, J.E. 1963. A revision of the genus *Hipposideros*. Bull. Br. Mus. (Nat. Hist.) Zool. 11:1-129.

Hill, J.E. & T. Kitti. 1972. Bats from Thailand and Cambodia. Bull. Br. Mus. (Nat. Hist.) Zool. 22:171-196.

Hill, W.C.O. 1939. A revised checklist of the mammals of Ceylon. Ceylon J. Sci. (B) 21:139-184.

Hladik, C.M. 1975. Ecology, diet and social patterning in old and new world primates. pp. 3-35 In: R.H. Tuttle (Ed.),Socioecology and psychology of primates. Mouton, The Hague.

Hladik, C.M. 1978. Adaptive strategies of primates in relation to leaf eating. pp. 373-395 In: G.G. Montgomery (Ed.),The ecology of arboreal folivores. Smithson. Inst. Press, Washington.

Hladik, C.M. & A. Hladik. 1972. Disponibilités alimentaires et domaines vitaux des primates à Ceylan. Terre et Vie 26:149-215.

Honacki, J.H., K.E. Kinman & J.W. Koeppl (Eds.). 1982. Mammal species of the world. Allen Press, Lawrence.

Jainudeen, M.R., J.F. Eisenberg & J.B. Jayasinghe. 1971. Semen of the Ceylon elephant, *Elephas maximus*. J. Reprod. Fert. 24:213-217.

Jainudeen, M.R., J.F. Eisenberg & N. Tilakeratne. 1971. Oestrus cycle of the asiatic elephant, *Elephas maximus*, in captivity. J. Reprod. Fert. 27:321-328.

428

Jainudeen, M.R., G.M. McKay & J.F. Eisenberg. 1972. Observations on musth in the domesticated Asiatic elephant (*Elephas maximus*). Mammalia 36:247-261.

Jayasinghe, J.B. & M.R. Jainudeen. 1970. A census of the tame elephant population in Ceylon with reference to location and distribution. Ceylon J. Si. (Bio.Sci.) 8:63-68.

Kelaart, E.F. 1851. Catalogue of the Mammalia of Ceylon. Ann. Mag. nat. Hist. ser. 2, 9:339-340.

Kurt, F. 1974. Remarks on the social structure and ecology of the Ceylon elephant in the Yala National Park. pp. 618-634 In: V. Geist & F. Walther (Eds.), Behavior of ungulates and its relation to management. I.U.C.N. Publ. New Ser. 24. Morges.

Marshall, J.T. Jr. 1977. A synopsis of Asian species of *Mus* (Rodentia,Muridae). Bull. Am. Mus. Nat. Hist. 158:173-220.

McKay, G.M. 1973. Behavior and ecology of the Asiatic elephant in southeastern Ceylon. Smithson. Contrib. Zool. 125:1-113.

McKay, G.M. & J.F. Eisenberg. 1974. Movement patterns and habitat utilization of ungulates in Ceylon. pp. 708-721 In: V. Geist & F. Walther (Eds.), Behavior of ungulates and its relation to management. I.U.C.N. Publ. New Ser. 24. Morges. Mein, P. & Y. Tupinier. 1977. Formule dentaire et position systématique du minioptère (Mammalia,Chiroptera). Mammalia 41:207-211.

Mishra, A.C. & V. Dhanda. 1975. Review of the genus Millardia (Rodentia:Muridae) with description of a new species. J. Mammal. 56:76-80.

Moore, J.C. 1960. Squirrel geography of the Indian subregion. Syst.Zool. 9:1-17.

Muckenhirn, N.A. & J.F. Eisenberg. 1973. Home ranges and predation of the Ceylon leopard. pp. 142-175 In: R.L. Eaton (Ed.), The worlds's cats. Vol. 1. World Wildlife Safari, Winston.

Mueller-Dombois, D. 1968. Ecogeographic analysis of a climate map of Ceylon with particular reference to vegetation. Ceylon Forester 8:39-58.

Musser, G.G. 1979. Results of the Archbold expeditions number 102. The species of *Chiropodomys*, arboreal mice of Indochina and the Malay Archipelago. Bull. Am. Mus. Nat. Hist. 162:377-445.

Musser, G.G. 1981. Results of the Archbold expeditions number 105. Notes on systematics of Indo-Malayan murid rodents, and descriptions of new Genera and species from Ceylon, Sulawesi and the Philippines. Bull. Am. Mus. Nat. Hist. 168:225-334.

Nettasinghe, A.P.W. 1973. The inter-relationship of livestock and elephants at Thamankaduwa Farm with special reference to feeding and environment. M.Sc. thesis, Univ. Ceylon, Peradeniya. pp.1-110.

Petter, J-J. & C.M. Hladik. 1970. Observations sur la domaine vitale et la densité de population de *Loris tardigradus* dans les forêts de Ceylan. Mammalia 34:394-409.

Phillips, W.W.A. 1935. Manual of the mammals of Ceylon. Ceylon J. Sci. and Delau & Co., London. pp.1-373.

Phillips, W.W.A. 1936. Survey of the distribution of mammals in Ceylon. Rept. 12. Ceylon. J. Sci. (B) 19:315-329.

Ripley, S. 1967. Intertroop encounters among Ceylon gray langurs (*Presbytis entellus*). pp. 237-253 In Altman, S.A. (Ed.), Social communication among primates. Univ. Chicago Press, Chicago.

Ripley, S. 1970. Leaves and leaf monkeys: the social organization of foraging in gray langurs (*Presbytis entellus thersites*). pp. 481-509 In J.R. Napier & P.H. Napier (Eds.), Old World monkeys. Acad. Press, N.Y.

Rudran, R. 1970. Aspects of ecology of two subspecies of purple-faced langurs (*Presbytis senex*). M.Sc. thesis, Univ. Ceylon, Colombo.

Rudran, R. 1973. Adult male replacement in one-male troops of purple-faced langurs (*Presbytis senex senex*) and its effect on population structure. Folia Primatol. 19:166-192.

Scheurmann, E. & M.R. Jainudeen. 1972. 'Musth' beim Asiatischen elefanten (*Elephas maximus*). Zool. Garten 42:131-142.

Shoshani, J. & J.F. Eisenberg. 1982. *Elephas maximus*. Mammalian Species 182:1-8.

Tennent, J.E. 1859. Ceylon, an account of the island. Vol. 1 Longman, Green, Longman & Roberts. pp.1-643.

Tennent, J.E. 1861. Sketches of the natural history of Ceylon. Longman, Green, Longman & Roberts, London. pp. 1-500.

Van Cuylenberg, B.W.B. 1974. The feeding behaviour of the Asiatic elephant in southeastern Ceylon. M.Sc. thesis, Univ. Ceylon, Peradeniya. pp. 1-407.

Van Cuylenberg, B.W.B. 1977. Feeding behaviour of the Asiatic elephant in south-eastern Sri Lanka in relation to conservation. Biol. Cons. 12:33-54.

Van Gelder, R.G. 1977. Mammalian hybrids and generic limits. Am. Mus. Novit. 2635:1-25.

Wurster, D.H. 1969. Cytogenetic and phylogenetic studies in Carnivora. pp. 310-329 In K. Benirschke (Ed.), Comparative mammalian cytogenetic

21. Man-made lakes: ancient heritage and modern biological resource

C.H. Fernando & S.S. de Silva

Introduction

The most substantial remains of the ancient civilization of Sri Lanka which date back to 600 BC, are the numerous man-made lakes. These comprised a well integrated irrigation system unmatched for its engineering sophistication in ancient times. This system of irrigation largely was abandoned around AD 1200 and was 'discovered' by observers only about 200 years ago. Large surface areas >1,000 ha[-1], sophisticated sluices for drawing water out, and well engineered channels and river dams reached construction standards which have excited the wonder of modern engineers. Rural settlements were centred around small reservoirs (called tanks after the Portuguese Tanqué) and the village came to be synonymous (in Sinhalese) with tank. The high density of reservoirs and adjacent ricefields indicate human densities which have been estimated at as high as 70 million. A more reasonable estimate of peak human density is around 4-6 million, still a very high density in ancient times for a country with only 65,000 km[2], of which about half was unoccupied hill country. Although large urban centres were few, they were well organized. The irrigation system was the basis for rice cultivation, and life was affluent.

Brohier (1934, 1935, 1936, 1975) has given the most comprehensive and critical account of the irrigation system in Sri Lanka. Brohier (1975) and A.D.N. Fernando (1979) have given detailed data on the agricultural history of Sri Lanka, and the role of irrigation, including the origin, development and use of man-made lakes.

Although written records are meagre regarding the use of man-made lakes in Sri Lanka, their role in supporting an advanced agricultural society from circa 200 BC till about 1200 AD is evident. These reservoirs served to store water for the irrigation of ricefields, raise the water table and formed a focus for settlements. The role of man-made lakes as a source of fish is not mentioned explicitly and till about thirty years ago only small quantities of fish were harvested. High fish yields is a very recent phenomenon, and coincides with the introduction of African lacustrine fish in 1952 (Fernando 1965a, 1971, 1973, 1977a, 1980b, Fernando & Ellepola 1969, Fernando & Indrasena 1969, de Silva & Fernando 1980).

432

In the present study we will examine critically the ancient system of man-made lakes in Sri Lanka with regard to its role in giving rise to the modern irrigation system. The place of the village reservoir and its social, cultural and biological impacts also will be examined. The modern irrigation system with its sophisticated engineering and multipurpose nature will be assessed critically with regard to its potential for fisheries and fish culture. As a backdrop to the efficient use of this biological resource, it is necessary to have a background of scientific data. We will therefore review and expand on what is known of the limnology and fisheries of man-made lakes in Sri Lanka. Lastly, the future of fisheries and fish culture in man-made lakes will be placed in perspective, taking into account advances made in the fields of limnology and fisheries on a global basis. The role of irrigation channels as a fishery resource also will be briefly discussed.

The ancient irrigation system

The irrigation reservoirs in Sri Lanka form part of an intricate, integrated, and sophisticated water storage and conveyance system. According to Brohier (1975), the origins of this system can be placed somewhere around 600 BC. Historical

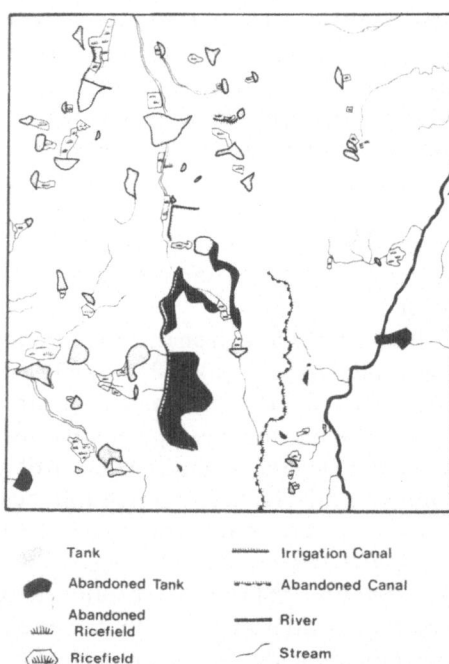

Tank		Irrigation Canal
Abandoned Tank		Abandoned Canal
Abandoned Ricefield		River
Ricefield		Stream

Fig. 1. An area in the Polonnaruwa district showing the high concentration of tanks. Each tank had a settlement in ancient times. The rice-fields were irrigated from water stored in the tanks.

evidence indicates that artificial irrigation systems originated in the valley of the Euphrates around 4,000 BC (Needham 1971). Rice cultivation arose in the southern China, Indo-Thai-Burma and Vietnam area at the foothills of the Himalayas about 3,000 BC (Chang 1976). Rice, an annual, needed wet conditions for its successful cultivation. Water storage became a necessary insurance for rice cultivation in arid areas where perennial rivers could not be harnessed for irrigation as was done in China, the Indo-Chinese area, and Indonesia. In Sri Lanka, the early civilizations adopted water storage in reservoirs on a wide scale. Each village had its small impoundment and an extent of ricefields that could be irrigated (Fig. 1). The major advances made by Sri Lankan engineers was the damming of rivers and large streams to provide water for storage in large reservoirs after conveyance by channels (Sinhalese-ela). These large reservoirs with surface areas of 1,500-3,000 ha^{-1} were first built around the first century AD, according to the Great Chronicle of Sri Lanka, the Mahavansa (Geiger 1960). The invention of a sophisticated sluice for control of outflowing water was indispensable. This was accomplished by Sri Lankan engineers (Brohier 1975). This irrigation system, which consisted of small village tanks, large reservoirs with their sophisticated intake and outflow channels and the associated engineering works like sluices, bunds, etc., remained in use in different parts of the country till the thirteenth century AD. The associated system of agriculture then was abandoned and cultivation shifted to the

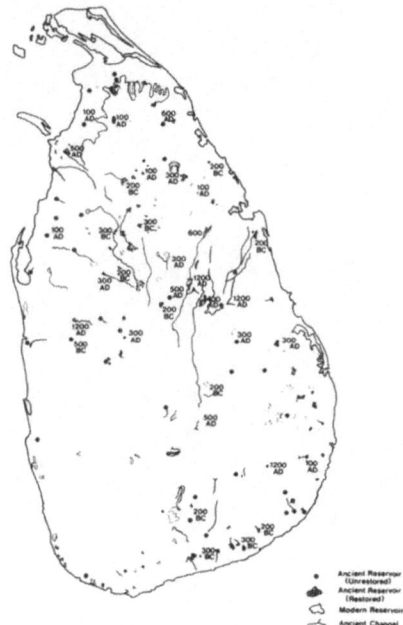

Fig. 2. Ancient and modern reservoirs and irrigation channels in Sri Lanka (after A.D.N. Fernando 1979).

foothills of the central mountains. The ancient irrigation system of Sri Lanka was well planned and consistently improved and extended during the period 600 BC to 1200 AD. Such a long ancient engineering history is truly an amazing achievement. In Fig. 2 we have shown the geographical distribution of ancient and modern reservoirs taken from a recent compilation by A.D.N. Fernando (1979). It is evident that the whole dry zone was under irrigation by an extensive and sophisticated irrigation system.

The small reservoir formed the backbone of the rural settlement of Sri Lanka from circa 600 BC to recent times (see Perera, this volume). The earliest agricultural settlements in Sri Lanka appear to have already practiced reservoir construction (Fernando 1979). Epigraphical evidence indicates that settlements where soils were of high quality (Perera 1978) had more dense populations in the dry zone (Brohier 1975). The association of a small reservoir with its cultivable ricefield area remained unchanged till very recent times (Ellepola 1955). This type of reservoir can be very dense in areas of intensive rice cultivation (Fig. 1). Anon (1955) records a density of >1 mile^{-2} in one river basin covering over 1,000 mile^{-2}. Brohier (1975) recorded that there were 11,200 small reservoirs in the northern province and the district of Anuradhapura. Brinck et al. (1971) showed the very high density of these village reservoirs taken from maps of what is now a National Park (Wilpattu) and a southern district. Fig. 3 gives the area and number of reservoirs in each district. Sri

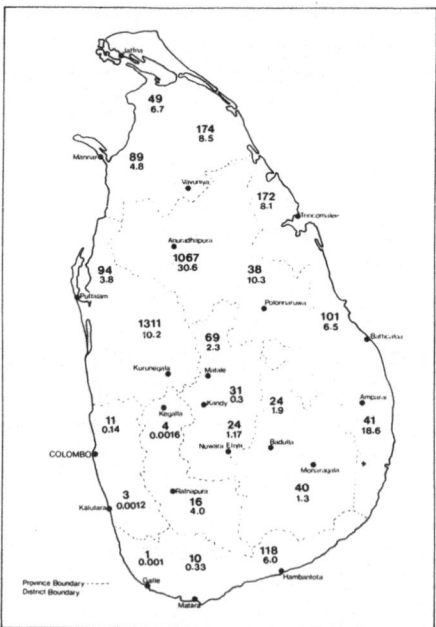

Fig. 3. Numbers of reservoirs (in larger type), and areas of reservoirs water surface (in thousands of ha), in the different districts of Sri Lanka.

Lanka has the highest density of dams in any country except in the ancient Nabatean civilization in the Negev desert. However, these dams were river dams to divert water for agriculture. Sri Lankan small reservoir distribution is shown in Fig. 4. These small reservoirs also were widespread in rice growing areas in southern India (Rao 1951). Small reservoirs (farm dams) still serve as a form of waterstorage for irrigation, fisheries and household use in many parts of the world. Hawley (1973) states that there are over 2 million small reservoirs in the U.S.A. Fernando (1980b) puts the figure for S.E. Asia at >500,000.

From the stage of small reservoirs to reservoirs with surface areas of >1,000 ha^{-1} was a step that Sri Lankan engineers seem to have taken ahead of all other ancient civilizations. Needham (1971) states that the dam building expertise of southern India came to its acme in large reservoirs with sophisticated outlets in Sri Lanka and not in any other ancient civilization. It is evident that dam building expertise came from Egypt, Mesopotamia, Greece and India (Hathaway 1958, Pigott 1952, Rao 1951, Snitter 1967, Smith 1971 and Van der Lippe 1951). De Camp (1974) states that the ancient irrigation system of Sri Lanka was the most elaborate of ancient times. The history of the Sri Lankan large reservoirs has been well document by Brohier (1933, 1934, 1935, 1975) and A.D.N. Fernando (1979). Poddubny (1976) quotes Soviet work on Sri Lankan reservoirs which mentions 10,000 ancient reservoirs of which 1,000 still are functional. The system of large ancient irrigation reservoirs was integrated and sophisticated. Its extent is shown in Fig. 4 in relation to modern reservoirs. Alagaratnam (1956) claimed, however, that the ancient irrigation system acts as a hindrance to modern multipurpose reservoir planning because it influences the restoration of inefficient reservoirs for sentimental reasons.

A key question regarding reservoirs as a biological resource, i.e. fish production, is their significance in this role in ancient times. Unfortunately the written evidence for fisheries in ancient reservoirs in Sri Lanka is very meagre. Brohier (1975), quoting an ancient source, only stated that ownership of fishing rights belongs to the owner of the body of water, be it reservoir or stream. Knox (1681) commented on the low level of fishing expertise of the people although he claimed that large fish were abundant in the Mahaweli river. Wiley (1910) described a primitive form of fish culture in the wet zone of Sri Lanka. This practice probably had ancient origins. However from Wiley's (1910) data and more recent observations, it is clear that the fishery resource of reservoirs was relatively meagre till about 1956. One of the authors (C.H.F.) has noted that reservoirs were fished only by rod and line and usually by juveniles and not for a living prior to 1956. Or if fishermen were involved, fishing usually was done by bailing out the water during the dry season from portions of the reservoir or stream. He noted also one case of fishing in a large reservoir with a seine net by migrant marine fishermen in 1951. The fish caught in all these cases consisted mainly of large carnivorous species. This catch composition is very similar to that in lakes and reservoirs in the temperate zone. This has changed to a predominantly herbivorous catch with the introduction of African cichlid fish (Fernando 1971).

The modern system of irrigation

The first modern reservoir system by damming a river was the gal oya scheme. This is shown in Fig. 5. The main reservoir, Senanayake Samudra, is the largest storage reservoir in Sri Lanka. It integrates the ancient and modern approaches to agriculture and land settlement.

The total surface water area of the ancient irrigation system probably was about 80,000 ha[≥1]. The present surface area is about 175,000 ha[≥1] and by the end of this century the area will be around 250,000 ha[≥1]. The present irrigation and power capacity of reservoirs is based on the blending of modern schemes such as the gal oya scheme with what can be salvaged from the ancient reservoir system. The distribution of ancient and modern reservoirs is shown in Fig. 4.

Biological studies

Limnology

Two of the earliest studies on tropical lakes or reservoirs are Apstein (1907, 1910) on two reservoirs in Sri Lanka. These studies were sophisticated for their time and

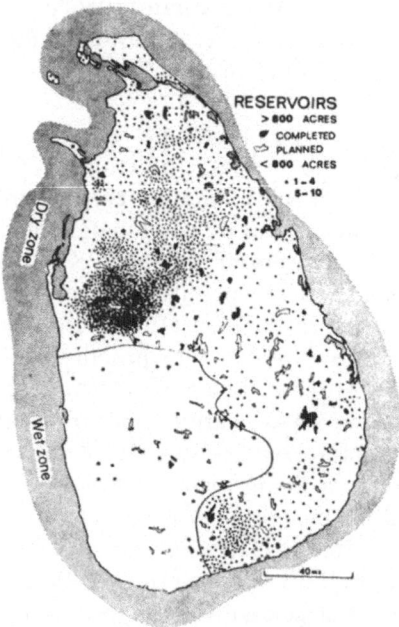

Fig. 4.   Distribution of reservoirs in Sri Lanka (after Fernando 1971).

dealt with the plankton and its seasonality in relation to temperature and rainfall. There were marked changes in species composition during different times of the year. The plankton was least abundant during the dry season in Colombo lake. The cooler Gregory's lake (2,300 m) had a similar periodicity. Holsinger (1955a, 1955b) studied the phytoplankton species composition and periodicity in three reservoirs. Two of them, in the lowlands, he considered tropical. They had a predominance of blue green algae while the upland reservoir (2,300 m) was considered temperate and had a predominance of diatoms. The upland lake also showed a greater fluctuation in numbers of phytoplankton seasonally. His findings agree with those of Apstein (1907, 1910) in that the greatest increase of plankton followed heavy rains and the dry periods showed a marked drop in plankton numbers. Mendis (1964) made a detailed study of the plankton and benthos of Colombo (Beira) lake. He found a high standing crop of plankton. The benthos at depths >3 m was poor but was rich elsewhere, consisting of Chironomidae, *Cyclestheria* (Conchostraca) amphipods (*Grandidierella megnae*), Gastropoda and oligochaetes in this order of numerical abundance. The fish yields of over 2,000 kg ha^{-1} annum^{-1} consisted almost entirely of *Saratherodon mossambicus*. Mendis (1965) investigated the plankton and benthos of 21 Sri Lanka lakes. He found standing crops of plankton ranging from about 4 times that of the highly eutrophic Colombo lake (Mendis 1964) to very low values. The benthos standing crops consisted of the same groups in Colombo lake except that *Corethra (Chaoborus)* replaced amphipods. The standing crops of benthos varied greatly from about 5 times that in Colombo lake to less than one tenth that value. Costa & de Silva (1969, 1978a, b, c, d, e) investigated various aspects of Colombo lake limnology from water chemistry and the seasonal fluctuations of the plankton to marginal fauna and the fauna associated with macrophytes. Fernando (1965b) studied the marginal fauna of 21 reservoirs and Fernando & Ellepola (1969) reported on the biology, plankton and littoral fauna of two small (seasonal) reservoirs. Fernando & Indrasena (1969) in their monograph on the freshwater fisheries of Sri Lanka summarized what was known about the limnology of Sri Lankan reservoirs.

Some data on physico-chemical factors are available in the literature mentioned above. In addition Sirimanne (1953), Geisler (1967) and Weniger (1972) provided additional data. Oglesby (1981) pointed out that the ionic concentration of waters increases from the uplands to the lowlands as the rivers accumulate more ions. There is a higher ionic concentration in the northern than in the southern rivers. The waters in reservoirs of the lowlands are well buffered. He also gives data he collected on pH, conductivity and alkalinity.

Fernando (1980c) made a comprehensive study of the zooplankton of all types of Sri Lankan freshwaters. There are relatively few crustacean zooplankters in the limnetic region and the Rotifera in this region are dominated by *Brachionus* spp. Also, the common crustaceans in the limnetic region are eurytopic species. There appear to be a few Rotifera which predominate in the limnetic zone. Fernando (1980d, 1983) has extended this study to cover the whole tropical region and finds

438

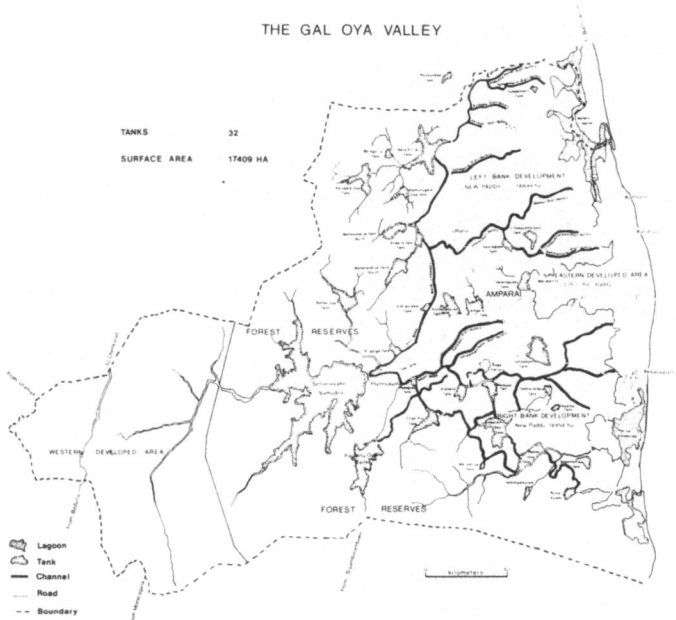

THE GAL OYA VALLEY

TANKS   32

SURFACE AREA   17409 HA

Lagoon
Tank
Channel
Road
Boundary

Fig. 5. The Gal-Oya irrigation system. A modern multipurpose water resource exploitation system.

the Sri Lankan limnetic zooplankton species composition is typically tropical. In addition, large zooplankters are absent or rare, and the genus *Daphnia,* which is a prominant member of the zooplankton in temperate regions, is rare and represented in the limnetic zone by a few small species. The most up to date and sophisticated study on limnology of Sri Lanka reservoirs is that of Schiemer (1980, 1983). Although these are preliminary reports on the work of the Austrian-Sri Lanka team of hydrobiologists, they deal with a single reservoir and provide reliable quantitative data and syntheses on water chemistry and nutrient cycling, benthos, plankton and primary production. Oglesby (1981) gave some data on plankton standing crops in reservoirs. Recently there has been a study of the seasonality and short- and long-term changes in the plankton of this reservoir (Duncan 1983, Duncan & Gulati 1981, Fernando & Rajapaksa 1983).

The Sri Lanka reservoirs, although of varied morphometry, do not include very large (>100 km^{-2}) or very deep (>40 m mean depth) reservoirs. The majority of low country reservoirs are shallow and <300 ha^{-1} surface area (Fig. 4). A classification was given by Fernando & Indrasena (1969). A slightly modified classification is shown in Fig. 6.

The studies on the limnology of Sri Lankan reservoirs can be seen in perspective when placed against the background of limnological studies of similar habits especially in tropical regions.

The Sri Lankan reservoirs are relatively small in surface area and shallow.

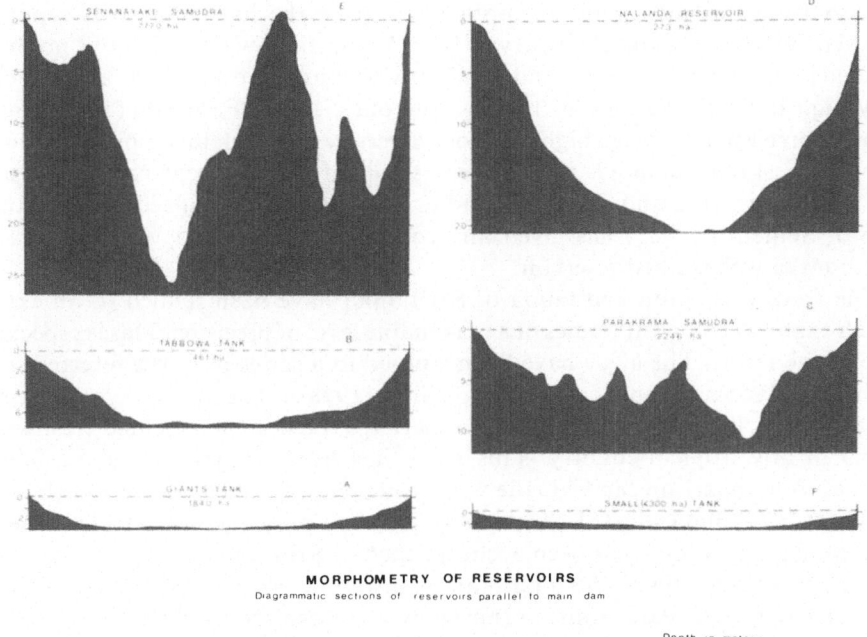

MORPHOMETRY OF RESERVOIRS
Diagrammatic sections of reservoirs parallel to main dam

Depth in meters
– – – – Waterline at full supply level

Fig. 6. Diagrams showing depth profiles of different types of Sri Lanka reservoirs (modified from Fernando and Indrasena 1969).

Although similar reservoirs have been constructed in other tropical and non-tropical regions, most of the extensive studies on reservoirs have been on large (>100 km^{-2}) reservoirs. Some general reviews of reservoir ecology are SCOPE (1972), Baxter (1977) and Petr (1978). Three symposia; Lowe-McConnell (1966), Obeng (1969) and Ackermann *et al.* (1973) have been devoted to the ecology of reservoirs.

Limnological studies on shallow and small reservoirs in tropical regions are meagre. Most of the recent studies for South East Asia have been summarized in Anon (1977) and Bhukaswan (1980). The more sophisticated limnological studies of Ruttner (1931, 1952), Lewis (1973, 1974, 1979) and Beadle (1974) have somewhat limited relevance to the shallow Sri Lankan reservoirs. The data available indicate that shallow reservoirs are highly productive in the tropics when suitable buffering and nutrient conditions exist. The rate of nutrient flux is very high and primary production values of >1,000 gC m^{-2} year^{-1} (gross) are commonplace. The zooplankton is poor in species diversity (Fernando 1980c, 1980d) and its production often is dominated by herbivorous cyclopoids and relatively low (Burgis 1974, Lewis 1979). Benthos productivity values are not available but standing crops appear to be uniform throughout the year (Darlington 1977). There is an indication that detritus may play an important role in the high fish yields reported (Bowen 1980). The water chemistry of S.E. Asian shallow reservoirs has not been studied in great detail. The pH and conductivity ranges are quite wide and indicate very high ionic concentrations to very low ones. Lai & Fernando (1981) found strong correlation between high calanoid diversity and high ionic concentrations. They suggest that calanoids can be used as indicators of eutrophication. Most of the S.E. Asian lake and reservoir studies have been summarized by Fernando (1983). Schiemer (1983) has given an account of the limnology of Parakrama Samudra, a well studied reservoir.

The freshwater flora and fauna of Sri Lanka have been studied to an extent which enables ecological studies of a reasonable level of precision. This is especially true of the fauna. The algae have been studied in a series of works referred to in Holsinger (1955a, 1955b) and Costa & de Silva (1978b). The macrophytes have not attracted specific studies. Sri Lanka has perhaps the best documented freshwater fauna of any tropical country. This work has been summarized in Fernando (1980e). A comparison can be made with the extent of similar studies in S.E. Asia (Dussart 1974) and India (Michael 1980). Although much remains to be done on taxonomy, a great deal has been accomplished in Sri Lanka.

Studies on other types of Sri Lankan freshwaters, especially streams, has been extensive in recent years. Some of this work is summarized in Brinck et al. (1971) and Costa & Starmühler (1972). The ecology of streams has a very close bearing on the ecology of reservoirs. A habitat which needs study urgently is the irrigation channel. The extent of irrigation channels in Sri Lanka is extensive and growing. These channels are the arteries and veins of reservoirs. They have a potential for fish production which has, in our opinion, been realized only to a small extent.

The present state of limnological knowledge on Sri Lanka reservoirs, although not impressive by comparison with that of the temperate regions, is nevertheless quite substantial. The flora and fauna is well known taxonomically; there are reliable data on water chemistry and data are accumulating on ionic composition and nutrient flux. The plankton composition and seasonality have been studied, though to a limited extent and we have figures for standing crops of plankton and benthos. The data available on productivity, though, are meagre. Phytoplankton productivity is very high in shallow eutrophic reservoirs of the low country. Data are needed on benthos productivity and turnover rates of plankton over a wide range of reservoir types.

Fish and fisheries

The main biological resource of reservoirs is the fishery. A detailed knowledge of the biology of the fishes is a necessary basis for maintaining a sustained high fish yield. At the present time this knowledge is not available. However some data are available for fishery managers and researchers. Some strategies for the management of tropical reservoir fisheries are given by Bhukeswan (1980) and Fernando (1980f).

Biology of indigenous fish

Sri Lanka has an indigenous fish fauna of 53 primary freshwater species (Fernando & Indrasena 1969, Munro 1955, de Silva *et al.* 1981). The number of secondary freshwater fish species is small. Munro (1955) lists 35 species as entering freshwater. Most of these are however found only in coastal areas. The low number of fish species appear to be a major factor in low fish yields. Sri Lanka has a depauperate fish fauna for a continental Asian tropical island. Singapore, which is about 100 times smaller than Sri Lanka had a freshwater fish fauna of over 70 species (Alfred 1961). Besides the small number of fish species, both Sri Lanka and southern India have few large herbivorous species. In Sri Lanka, small carps are the most numerous (19 species) and large carps are few (6). Eighteen species are facultative or obligate air breathing fishes and there are few large carnivores (Fernando & Indrasena 1969, Fernando 1971). In an attempt to enrich the fish fauna, 17 species have been introduced since 1982 (Fernando 1971). Of these, only the African Cichlidae have had a major impact on the fisheries. A similar attempt to enrich the freshwater fish fauna of New Guinea, which has an even poorer fish fauna than Sri Lanka has given the same result. 21 species were introduced and only two of them, the common carp and *Sarotherodon mossambicus,* an African cichlid, have become widespread (Glucksman *et al.* 1976). *S. mossambicus* is the mainstay of the capture fishery (comment by T. Petr in Fernando 1980a).

There are no true lacustrine fish in the indigenous fauna. The closest to a lacustrine fish is the estuarine *Etroplus suratensis* which became established only after introduction from the coastal lagoons (Fernando 1971). The other fishes can be divided into riverine (river and stream) and marsh dwelling species. The marshes, small streams and shallow ditches, have the largest spectrum of fishes. Fernando (1956b) found 35 fish species in the ricefields of the western lowlands alone. A large shallow reservoir, Parakrama Samudra, has 29 fish species (Fernando 1983), most of them riverine. The lacustrine fish fauna is seriously deficient and the riverine fauna is poor in species. The experience in Sri Lanka strongly suggests that high fish yields cannot be obtained necessarily by conservation and management of the indigenous fauna alone.

The food of fish has not been studied in any great detail. Fernando (1956a) studied the food of four common species. The feeding habits of all the species, obtained mainly from published work on the species outside Sri Lanka, are given in Fernando & Indrasena (1969). More recently, the food and feeding habits of some cyprinids has been dealt with by de Silva & Kortmulder (1977) and de Silva *et al.* (1977, 1980). Riverine fish in general are more catholic in their food habits than lacustrine fish and this applies to the indigenous fish in reservoirs and their channels. The marsh and riverine fish are adapted to feed only seasonally under recurring drought conditions. It would be interesting to study their food habits in the perennial reservoirs where they have become established. The low fish yields of indigenous fishes (Fernando 1971, 1973) indicates that the food available in reservoirs available via the high primary production was not being effectively utilized by commercially valuable fish. This situation has been remedied by the introduction of lacustrine African Cichlidae.

The breeding habits of the fish have an impact on the fish yields via the rate of recruitment of the fish stock available for fishing. Most riverine fish breed only seasonally during floods (Jhingran 1982, Lowe-McConnell 1975, Breder & Rosen 1966, Welcomme 1979). Marsh dwelling fish also generally breed during this season although they appear to be less affected by the floods. A little rain often initiates spawning and some of the species breed a number of times intermonsoonally (Fernando 1956b). Some aquarium fishes which are found in Sri Lanka in small streams and ponds can be bred easily in the laboratory throughout the year (Breder & Rosen 1966). This indicates very probably that they breed many times during an intermonsoonal season under natural conditions.

Biology of introduced fishes

Fernando (1965a, 1971) listed the introduced fish species. Only *Sarotherodon mossambicus* has had a marked impact on fish yields, hence it is reasonable to focus on the biology of this species and give only brief notes on the others.

Sarotherodon mossambicus is a mouth breeder. It builds nests in the form of circular depressions on the bottom. The density of these nests is a rough index of the intensity of breeding. Fernando & Indrasena (1969) mapped the nests of *S. mossambicus* in three reservoirs of varying morphometry. Nest density was inversely proportional to depth. The mean size of gill-netted fish showed a direct relation with increasing depth of the reservoir while fish yields were inversely related to depth. de Silva & Chandrasoma (1980) and Chandrasoma & de Silva (1981) investigated the breeding biology of *S. mossambicus* and *Tilapia rendalli*. They reported that *S. mossambicus* does not build nests at depths shallower than 1 m. They concluded that maintenance of more stable levels would reduce the breeding intensity of this species. These observations and conclusions are not compatible with previous work. It is likely that other factors, including the nature of the bottom (sand, soft mud, mud and sand) also influence nest building. Further studies are necessary to elucidate this aspect of the biology of *S. mossambicus*.

Sarotherodon mossambicus grows very rapidly under conditions prevailing in eutrophic waters. Bowen (1980, 1981) has shown the importance of detrital non-protein amino acids and periphytic detrital aggregate in inducing rapid growth in this species. Also, like *S. niloticus* (Moriarty 1973, in press), this species can assimilate blue greens (Abayasiri & Costa 1978). *S. mossambicus* grows more rapidly in saline than in freshwater (Canagaratnam 1966). The tilapias are noted for their ability to maintain dense (often too dense!) populations in ponds and shallow reservoirs. Their growth rate in eutrophic reservoirs also is extremely high.

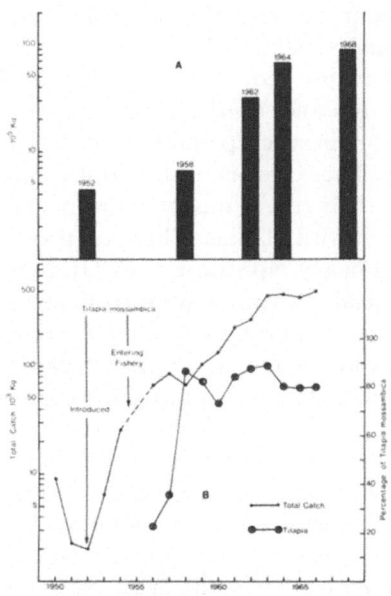

Fig. 7. Fish yields in Sri Lankan reservoirs for the period 1949-1968 (after Fernando 1971). A: Fish yields in Parakrama Samudra; B: Total fish yields for all the reservoirs.

Fisheries

The fish yields from Sri Lanka reservoirs has been well documented by Fernando (1965a, 1971, 1973, 1977a, b), de Silva & Fernando (1980), and Fernando & Indrasena (1969). The data available show clearly that prior to the introduction of *S. mossambicus* the fish yields from all reservoirs was low. The figure for this period of 10 kg ha^{-1} year^{-1} was given by Fernando (1971) for a shallow low-country lake, Parakrama Samudra. After the introduction of *S. mossambicus* in 1952 the fish yields from shallow low country reservoirs especially increased spectacularly (Fig. 7A). de Silva & Fernando (1980) gave a figure of about 400 kg ha^{-1}year^{-1} for Parakrama Samudra. The total fish yield from all reservoirs also has increased from about 400 tons to an estimated 20,000 tons a year during 1952-1982. The mainstay of the fishery is *S. mossambicus*. Other species of African Cichlidae; *S. niloticus, Tilapia rendalli* and *T. zilli* also have been successfully introduced but no reliable data are available on their contribution to fish yields. The rate of increase in fish yields from reservoirs in Sri Lanka is mirrored in the data from Parakrama Samudra, a shallow reservoir of 2,246 ha where detailed data on fish yields are available from 1948-1980. These data are shown in Figs. 7a, b and 8. The increase in fish yields in 1974 (Fig. 8) after the introduction of *Tilapia rendalli* may be coincidental, although Mendis (1977) claimed that this species had replaced *T.*

mossambica as a dominant species in the catch. This claim has not been substantiated (de Silva & Fernando 1980).

The other species of major economic value are *Labeo dussumieri, Puntius sarana, Etroplus suratensis* and *Wallago attu*. *Labeo dussumieri* and *Puntius sarana* are typical, large, riverine carp and have many similarities with the Indian carps of the same genera. These are of considerable economic importance in India. However, in reservoirs, their role is much less important because they have low standing stocks compared with tilapias. These carp ascend rivers to spawn in the floodwaters only during heavy monsoon rains (Jhingran 1982, Breder & Rosen 1966). Young from artificial spawnings are introduced into Indian reservoirs and the resulting fish yields are not particularly high. The cost of raising and introducing fry and juveniles may not be repaid by the fish caught. This same practice has been mooted for Sri Lanka. The possibilities for economic returns are even less likely than in India.

Etroplus suratensis, an indigenous fish, is of considerable importance in shallow and deep reservoirs in Sri Lanka (Fernando & Indrasena 1969). It does not ascend rivers for spawning. In a deep reservoir, Senanayake Samudra, it is of equal importance as *S. mossambicus* and *Labeo dussumieri*. In shallow reservoirs it is as important as indigenous carps (Fernando & Indrasena 1969). Of the carnivorous species *Wallago attu* is the most important fish commercially. Other carnivores include the snakeheads, *Ophiocephalus* spp., *Glossogobius giuris* and *Ompok bimaculatus*. Some omnivores like *Heteropneustes fossilis, Mystus* spp., *Mastacembelus armatus* and the eels *Anguilla nebulosa* and *A. bicolour* probably are poorly exploited at present. Their contribution to the fish catch can be enhanced by using special gear in reservoirs where they are common.

The fishing methods used in Sri Lanka reservoirs are detailed in Fernando & Indrasena (1969). Gill nets are the most commonly used gear. A detailed study of beach seine fisheries (Fernando 1967) showed that while the percentage of the total catch obtained by this method was small (3-15%), the fish composition was diversified. Long lines also have been used but not on a wide scale.

The fisheries in reservoirs contributes substantially to the fish caught in Sri Lanka. The present (1981) yield probably is about 20,000 tons. This accounts for about 18% of the total fish catch in Sri Lanka. It appears that the present fish yields from reservoirs can be increased substantially by more intensive exploitation (Oglesby 1981). Estimates of potential fish yields vary from a high of 134,000 tons to a low of 44,000 tons a year (Anon. 1968). Even the lower figure is a substantial quantity of fish. Taking into consideration that the reservoir area (mainly large reservoirs) will be increased from the present 175,000 ha to about 250,000 ha by the end of this century with the likelihood that present yields will be maintained (as they have been in Parakrama Samudra for 25 years), a figure of >50,000 tons a year seems to be a conservative estimate (de Silva & Fernando 1980).

Small reservoirs (<300 ha) comprise over 40% of the surface area of all reservoirs at present. They are not exploited for fisheries to any great extent. Many of

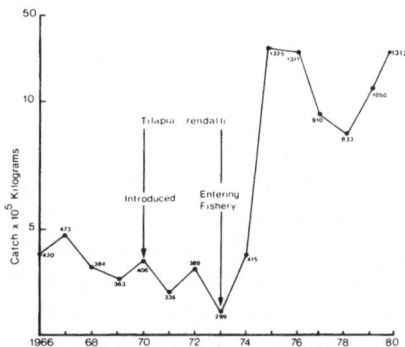

Fig. 8. Fish yields from Parakrama Samudra for the period 1966-1980.

them are very seasonal, lasting for 4-8 months, and a few are perennial (Fernando & Ellopola 1969). However they will be a very valuable fishery resource if a suitable system of stocking and harvesting is developed. Preliminary experiments indicate the feasibility of such a system (Fernando & Ellepola 1969, Oglesby 1981). Fast growing fish species (tilapias) are available in abundance as stocking material. An inexpensive means of harvesting and methods to reduce predation by birds, fish and reptiles should be developed. The use of common, Indian and Chinese carps as stocking material will increase costs and give poor returns. This approach should be avoided unless reasonable economic returns are likely.

Of the measures tried in Sri Lanka to increase fish yields in reservoirs, only the introduction of African Cichlidae has proved successful (Fernando 1965, 1971, 1973, 1977a, de Silva & Fernando 1980). A similar result has been reported from other tropical regions with poor lacustrine fish faunas; New Guinea (Glucksman *et al.* 1976) and tropical Brazil (Anon. 1979, Lovshin & Piexoto 1975). In other tropical regions annual or seasonal introduction of carps (India) or the harvesting of indigenous fish (Thailand) have both given lower fish yields in reservoirs than tilapias (Fernando 1980f).

In Sri Lanka, common carps have been raised in nurseries and attempts have been made to use these for fish culture, Also, common and Chinese carps have been introduced into reservoirs. Both in culture and capture fisheries these carps have proved an unmitigated failure in Sri Lanka. In the recent past Indian carps also have been suggested as candidates for pond culture and introduction into reservoirs.

The failure of carps in culture and reservoirs in Sri Lanka has been attributesd to several causes; use of small fry for stocking, low stocking densities, inadequate investment in culture fisheries coupled with poor technical know-how. These may all have been operative but perhaps the main reason for failure of the carps is their temperate and sub-tropical origin. Carp culture has not been successful in any tropical region. Low yields and high cost of fish is the result of raising carp in the tropics (Fernando 1980b).

The use of carps (Indian, common and Chinese) in tropical reservoirs to enhance fish yields has had little success. In India, where this technique is used extensively, fish yields are very low (Jhingran 1982). In Sri Lanka common carp have been introduced into reservoirs for 35 years with no returns. Even in New Guinea, where the common carp has become widespread (Glucksman *et al.* 1976), *Sarotherodon mossambicus* is the only important culture fish (Fernando 1980a). On the other hand African Cichlidae have made major contributions to enhanced fish yields in reservoirs in Asia, Africa and South America (Fernando 1977b, 1980a, 1980b).

Fernando & Holcíik (1982) concluded from a careful examination of data, that tilapias introduced outside Africa have exploited only the littoral zone, albeit more successfully than indigenous species in the tropics. The pelagic and deeper waters of most tropical lakes are almost fishless. They suggested the use of other African Cichlidae and pelagic Clupeidae to fill these niches (see also Eccles 1975 and Coulter 1981). If their contention is correct the management of reservoirs in the tropics can be placed on a more scientific basis than at present. The fish yields from reservoirs will be enhanced if all three major niches, i.e. littoral, deep water and the pelagic zone are utilized by fish rather than only the littoral zone as at present in many reservoirs and lakes.

Bird predation on fish in tropical lakes and reservoirs is known to be quite heavy. In Sri Lanka, Fernando & Indrasema (1969) noted the high density of fish-eating birds in a highly productive shallow reservoir, Parakrama Samudra. Schiemer (1980) gave the startling figure for fish consumption by birds in Parakrama Samudra as 5 tonnes a day. The fish yield from this reservoir is 2.5 tonnes a day (400 kg ha^{-1} year^{-1}). The role of fish-eating birds in influencing fish yields needs study. On the one hand reduced bird predation could increase fish yields. On the other, birds may take only small fish and thin out the population of the highly fecund Cichlidae, benefitting the fishery. Also, the contribution of bird faeces to eutrophication cannot be ignored.

Fisheries in the future

The present yield of about 20,000 tonnes of freshwater fish comes mainly from the shallow and very shallow large perennial reservoirs. The fishery is dominated by *Sarotherodon mossambicus*. Also contributing to the fishery are a few indigenous species: 2 large carps, and indigenous cichlid and one large carnivore. The introduction of tilapias (*S. mossambicus* and other cichlids) has not adversely affected indigenous species. The carps (common, Chinese and Indian) have made no contribution to fish catches. The small reservoirs which comprise about 40% of the total area of reservoirs have not been used for fisheries or fish culture to any great extent. The 4,000 km of rivers and >2,600 km of irrigation channels yield only very meagre quantities of fish at present.

The future development of freshwater fisheries should have short and long-term objectives. In the immediate future intensified fishing should be done in large reservoirs. Conventional gear including gill nets and beach seines should be supplemented by longlines, traps, etc. An effective gear technology urgently is required to obtain more fish more economically. Mechanization, though, should be avoided because it will increase costs, reduce employment and perhaps even damage the fishery by overfishing. Costs should be kept low by using the fish in a fresh state locally and by preservation cheaply using salt or smoking. Filleting and other measures to capture 'export' markets should not be undertaken lightly.

In the long term an effective monitoring should be done of fish stocks and the role of all the important fish species in the fishery. Research already initiated in limnology, fish biology and fisheries should be more effectively supported with funds and personnel, especially in the universities and at the technical level in the Department of Fisheries. Research should be directed in the following areas: (1) biology of economically important species, (2) the role of predatory birds on fish stocks and (3) general limnology).

Fish culture, which has been an unmitigated failure in Sri Lanka, should be reappraised in the light of the failure of the carps. The possibilities of using Cichlidae from Africa and other species for culture in small reservoirs should be explored. The regular droughts which occur in Sri Lanka do not augur well for fish culture unless fish culture practices are adapted to this water regime. Unless more effective water management can provide regular supplies of water, fish culture will have little chance of success in ponds. Ling (1962), who has done more to encourage fish culture in Sri Lanka than any other scientist, states that droughts were a major cause of failure. His words need heeding by prospective fish culturists. The high predation by birds also is a serious problem.

While focussing on the fish species giving high yields, the other species should not be neglected. Considerable quantities of protein are being harvested by subsistence fisheries of indigenous fish in reservoirs and irrigation channels. With better management this resource can be enhanced considerably. An intensive study of the fish in irrigation channels should give an indication whether the fish fauna can advantageously be increased by introductions. The ubiquitous *Caridina* and *Macrobrachium* can also perhaps be harvested in remote areas and used for food.

References

Abayasiri, R.R. & H.H. Costa. 1978. The hydrobiology of Colombo (Beira) lake. VII. The food and feeding ecology of *Tilapia mossambica*. Spol. Zeylan. 32:88-105.

Ackermannm W.E., G.F. White & E.B. Worthington (Eds.). 1973. Man-made lakes: their problems and environmental effects. Geophys. Monogr. 17:847 pp.

Alagaratnam, W.T.I. 1956. Some observations on restoration of ancient irrigation works. Proc. Ceylon Assoc. Adv. Sci. Part I, 85-95.

Alfred, E.R. 1961. Singapore freshwater fishes. Malay Nat. J. 15:1-19.

Anon. 1955. Six year programme of investment 1954/55 to 1959/60. Planning Secretariat Govt of Ceylon. 510 pp.

Anon. 1968. Fisheries in Asia and their development prospects. FAO, Rome 79507E/E. 25 pp.

Anon. 1977. Symposium on the development and utilization of inland fishery resources. Proc. Indo Pacif. Fish. Counc. 17 Sess. Sect. 3:1-500.

Anon. 1979. Quadros informicios sobre a administra cão de pesca em 102 acudes publicos controldos pelo DNOCS, No Ano de 1979. Fortaleza, Ceara, Brasil. 17 pp. (Mimeo.).

Apstein, C. 1907. Das plancton im Colombo see auf Ceylon. Zool. Jb. (Abt. Syst.) 25:201-214.

Apstein, C. 1910. Das plancton des Gregory see auf Ceylon. Zool.Jb. (Abt. Syst.) 29:611-680.

Baxter, R.M. 1977. Environmental effects of dams and impoundments. Ann. Rev. Ecol. Syst. 8:255-283.

Beadle, LC. 1981. The inland waters of tropical Africa. Longman, London. 475 pp.

Bhukaswan, T. 1980. Management of Asian reservoir fisheries. FAO Fish. Tech. Pap. 207:1-69.

Bowen, S.H. 1980. Detrital nonprotein amino acids are a key to rapid growth in *Tilapia* in lake Valencia, Venezuela. Science 207:1216-1218.

Bowen, S.H. 1981. Digestion and assimilation of periphytic detrital aggregate by *Tilapia mossambica*. Trans. Am. Fish. Soc. 110:239-245.

Breder, C.M. & D.E. Rosen. 1966. Modes of reproduction in fishes. Nat. Hist. Press, N.Y. 941 pp.

Brinck, P., H. Andersson & L. Cederholm. 1971. Report No. 1 from the Lund University Ceylon expedition in 1962. Ent. Scand. Suppl. 1:3-36.

Brohier, R.L. 1934-36. Ancient irrigation works in Ceylon: part 1, 37 pp., Part 2, 43 pp., Part 3, 46 pp. Ceylon Govt Press, Colombo.

Brohier, R.L. 1975. Food and the people. Lake House, Colombo. 200 pp.

Burgis, M. 1974. Revised estimates for the biomass and production of zooplankton in lake George, Uganda. Freshwat. Biol. 4:535-541.

Canagaratnam, P. 1966. Growth of *Tilapia mossambica* Peters in different salinities. Bull. Fish. Res. Stn Ceylon 19:1-4.

Chandrosoma, J. & S.S. de Silva. 1981. Reproductive biology of *Puntius sarana*, an indigenous species and *Tilapia rendalli*, an exotic in an ancient man-made lake in Sri Lanka. Fish Mgmt 12:7-28.

Costa, H.H. & S.S. de Silva. 1969. Hydrobiology of Colombo (Beira) lake. 1. Diurnal variations in temperature, hydrochemical factors and zooplankton. Bull. Fish. Res. Stn Ceylon 20:141-149.

Costa, H.H. & S.S. de Silva. 1978a. The hydrobiology of Colombo (Beira) lake. II. Seasonal variation in physico-chemical characteristics. Spol. Zeylan. 32:19-34.

Costa, H.H. & S.S. de Silva. 1978b. The hydrobiology of Colombo (Beira) lake. III. Seasonal fluctuations of plankton. Spol. Zeylan. 32:25-53.

Costa, H.H. & S.S. de Silva. 1978c. The hydrobiology of Colombo (Beira) lake. IV. Seasonal fluctuations in aquatic fauna living on water plants. Spol. Zeylan. 32:54-70.

Costa, H.H. & S.S. de Silva. 1978d. The hydrobiology of Colombo (Beira) lake. V. Season study of marginal fauna. Spol. Zeylan. 32:71-81.

Costa, H.H. & S.S. de Silva. 1978e. The hydrobiology of Colombo (Beira) lake. VI. Seasonal variations in primaryproductivity. Spol. Zeylan. 32:83-92.

Costa, H.H. & F. Starmühlner. 1972. The Austrian-Ceylonese Hydrobiological Mission 1970. I. Preliminary report: introduction and descriptions pf the stations. Bull. Fish. Res. Stn Sri Lanka (Ceylon) 23:43-76.

Coulter, G.W. 1981. Biomass, production and potential yield of the Lake Tanganyika pelagic fish community. Trans. Am. Fish. SOc. 110:325-335.

Darlington, J.P.E.C. 1977. Temporal and spatial variation in the benthic fauna of lake George, Uganda. J. Zool. Lond. 181:95-111.

De Camp. L.S. 1974. The ancient engineers. Ballantyne Books, N.Y. 450 pp.

de Silva, S.S. & J. Chandrasoma. 1980. Reproductive biology of *Sarotherodon mossambicus* and introduced species in an ancient man-made lake in Sri Lanka. Env. Biol. Fish. 5:253-259.

de Silva, S.S., P.A.T. Cumaratunge & C.D. de Silva. 1980. Food, feeding ecology and some morphological characters related to food intake of four co-occurring cyprinids. Neth. J. Zool. 30:54-73.

de Silva, S.S. & C.H. Fernando. 1980. Recent trends in the fishery of Parakrama Samudra, and ancient man-made lake in Sri Lanka. pp. 927-937 In J.I. Furtado (Ed.) Tropical ecology and development. Proc. 5th. Int. Symp. Trop. Ecol.

de Silva, S.S. & K. Kortmulder. 1977. Some aspects of the biology of three species of *Puntius* (=Barbus) (Pisces:Cyprinidae) endemic to Sri Lanka. Neth. J. Zool. 27:182-194.

de Silva, S.S., K. Kortmulder & P. Maitipe. 1981. The identity of *Puntius malamphyx sinhala* (Duncker, 1911)(Pisces, Cyprinidae). Neth. J. Zool. 31:

de Silva, S.S., K. Kortmulder & J. Wijeratne. 1977. A comparative study of the food and feeding habits of *Puntius bimaculatus* and *P. titteya*. Neth. J. Zool. 27:253-263.

Duncan, A. in press. The composition, density and distribution of the zooplankton of Parakrama Samudra during 1980. In F. Schiemer (Ed.), Limnology of Parakrama Samudra, Sri Lanka: a case study of an ancient man-made lake in the tropics. Developments in Hydrobiology. Hydrobiologia.

Duncan, A. & Gulati, R.D. 1981. Parakrama Samudra (Sri Lanka) Project, a study of a tropical lake ecosystm. 3. Composition, density and distribution of the zooplankton in 1979. Verh. Internat. Verein. Limnol. 21:1007-1014.

Dussart, B.H. 1974. Biology of inland waters in humid tropical Asia. Nat. Resources Res. 12:331-353. UNESCO, Paris.

Eccles, D.H. 1975. Fishes of the African Great Lakes as candidates for introduction into large tropical impoundments. J. Fish. Biol. 7:401-405.

Ellepola, D.P. 1955. Changing trends in village Ceylon. Proc. Ceylon Assoc. Adv. Sci. Sect. 2:121-146.

Fernando, A.D.N. 1979. Major ancient irrigation works of Sri Lanka. J. Roy. Asiatic Soc. (Sri Lanka Branch) 22:1-24.

Fernando, C.H. 1956a. On the food of four common freshwater fish of Ceylon. Ceylon J. Sci. (C) 7:201-217.

Fernando, C.H. 1956b. The fish fauna of paddy fieldsand small irrigation ditches in the western lowlands of Ceylon: and a bibliography of references to fish in paddy fields. Ceylon J. Sci. (C) 7:223-227.

Fernando, C.H. 1965a. The development of Ceylon's fisheries 11; the role of inland waters in relation to the development of Ceylon's fisheries and a note on the pearl oyster fishery. Bull. Fish. Res. Stn Ceylon 17:291-297.

Fernando, C.H. 1965b. A preliminary survey of 21 Ceylon lakes. 3. Parasites and predators, food of fish and marginal fauna. Bull. Fish. Res. Stn Ceylon 18:17-28.

Fernando, C.H. 1967. The beach seine fishery in Ceylon's freshwaters. Indo-Pacif. Fish. Counc. Curr. Aff. Bull. 50:1-21.

Fernando, C.H. 1971. The role of introduced fish in fish production in Ceylon's freshwaters. pp. 295-310 In E. Duffey & A.S. Watt (Eds.) The scientific management of animal and plant communities for conservation.

Fernando, C.H. 1973. Man-made lakes of Ceylon: a biological resource. pp. 664-671 In W.C. Ackermann, G.F. White & E.B. Worthington (Eds.) Man-made lakes:their problems and environmental effect. Amer. Geophys. Monogr. 17.

Fernando, C.H. 1977a. Reservoir fisheries in South East Asia: past, present and future. Proc. Indo. Pacif. Fish. Counci. 17th Sess. Sect. 3:475-498.

Fernando, C.H. 1977b. Fisheries of natural lakes and man-made reservoirs. Proc. Indo Pacif. Fish. Counci. 17th Sess. Sect. 3:9-12.

Fernando, C.H. 1980a. Tropical reservoir fisheries: a preliminary synthesis. pp. 883-892 In J.I. Furtado (Ed.) Tropical ecology and development. Proc. 5th Internat. Symp. Trop. Ecol.

Fernando 1980b. Tropical man-made lakes, African fish and cheap protein. ICLARM newsletter, Manila, 3:15-17.

Fernando, C.H. 1980c. The freshwater zooplankton of Sri Lanka, with a discussion of tropical freshwater zooplankton composition. Int. Rev. ges. Hydrobiol. 65:85-125.

Fernando, C.H. 1980d. The species and size composition of tropical freshwater zooplankton, with special reference to the Oriental Region (South East Asia). Int. Rev. ges. Hydrobiol. 65:411-426.

Fernando, C.H. 1980e. The freshwater invertebrate fauna of Sri Lanka. Spol. Zeylan. 35:15-42.

Fernando, C.H. 1980f. The fishery potential of man-made lakes in Southeast Asia and some strategies for its optimization. BIOTROP Anniv. Publ. Bogor, Indonesia. pp. 25-38.

Fernando, C.H. 1983. Reservoir and lake ecosystems in South East Asia (Oriental Region). In F.B. Taub (Ed.) Lake and reservoir ecosystems. Elsevier, Amsterdam.

Fernando, C.H. & W.B. Ellopola. 1969. A preliminary study of two village tanks (reservoirs) in the Polonarruwa area, with biological notes on these reservoirs in Ceylon. Bull. Fish. Res. Stn Ceylon 20:3-13.

Fernando, C.H. & J. Holcíik. 1982. The nature of fish communities: a factor influencing the fishery tial of tropical lakes and reservoirs. Hydrobiologia 97:127-140.

Fernando, C.H. & H.H.A. Indrasena. 1969. The freshwater fisheries of Ceylon. Bull. Fish. Res. Stn Ceylon 20:101-134.

Fernando, C.H. & R. Rajapaksa. 1983. Some remarks on long-term and seasonal changes in the zooplankton of Parakrama Samudra. In F. Schiemer (Ed.) Limnology of Parakrama Samudra, Sri Lanka: A case study of an ancient man-made lake in the tropics. Developments in Hydrobiology, Hydrobiologia.

Geiger, W. (Transl.) 1960. The Mahavamsa or the great chronicle of Ceylon. Ceylon Govt. Inform. Dept. 323 pp.

Geisler, R. 1967. Limnologische - Ichthyologische beobachtungen im Sudwest Ceylon. Int. Rev. ges. Hydrobiol. 52:559-572.

Glucksman, J., G. West & T.M. Berra. 1976. The introduced fishes of Papua New Guinea with special reference to *Tilapia mossambica*. Biol. Conserv. 9:37-44.

Hathaway, G.A. 1958. Dams, their effects on some ancient civilizations. Civil. Eng. (Jan. 1958):58-63.

Hawley, A.J. 1973. Farm ponds in the United States: a new resource for farmers. pp. 746-749 In W.C. Ackermann, G.F. White & E.B. Worthington (Eds.), Man-made lakes: their problems and environmental effect. Amer. Geophys. Monogr. 17.

Holsinger, E.C.T. 1955a. The plankton algae of three Ceylon lakes. Hydrobiologia 7:8-24.

Holsinger, E.C.T. 1955b. The distribution and periodicity of the phytoplankton of three Ceylon lakes. Hydrobiologia 7:25-35.

Jhingran, V.G. 1982. Fish and fisheries of India. 2nd edit. Hindustan Publ. Corp., New Delhi. 666 pp.

Knox, R.I. 1681. A historical relation of Ceylon. Tisara Prakasakayo, Dehiwela, Sri Lanka. 356 pp.

Lai, H.C. & C.H. Fernando. 1981. The freshwater Calanoida (Crustacea:Copepoda) of Thailand. Hydrobiologia 76:161-178.

Lewis, W.M. 1973. The thermal regime of Lake Lanao (Philippines) and its theoretical implications for tropical lakes. Limnol. Oceanogr. 18:200-217.

Lewis, W.M. 1974. Primary production in the plankton community of a tropical lake. Ecol. Monogr. 44:377-409.

Lewis, W.M. 1979. Zooplankton community analysis. Springer Verlag, N.Y. 163 pp.

Ling, S.W. 1962. Project of inland fishery development. Report to the Govt of Ceylon. FAO Rep. 1527:43 pp.

Lovshin, L.L & J.T. Peixoto. 1975. *Tilapia* sp. in the northeast of Brazil. Economic and ecological considerations. Symp. Limnol. Fisheries Fish. Cult. Belo Horizonte, Brazil. Sept. 1975. 13 pp. (Mimeo.).

Lowe-McConnell, R.H. (Ed.) 1966. Man-made lakes. Academic Press, Lond. 218 pp.

Lowe-McConnell, R.H. 1975. Fish communities in tropical freshwaters. Longmans, Lond. 337 pp.

Mendis, A.S. 1964. A contribution to the limnology of Colombo lake. Bull. Fish. Res. Stn Ceylon 17:213-220.

Mendis, A.S. 1965. A preliminary study of 21 Ceylon lakes. 2. Limnology and fish production potential. Bull. Fish. Res. Stn Ceylon 18:7-16.

Mendis, A.S. 1977. The role of man-made lakes in the development of freshwater fisheries in Sri Lanka. pp. 247-254 In Proc. Indo-Pacif. Fish. Counc. 17th Sess. Sect 3.

Michael, R.G. 1981. A historical resume if Indian limnology. Hydrobiologia 72:15-20.

Moriarty, D.J.W. 1973. The physiology of digestion of blue-green algae in the cichlid *Tilapia nilotica*. J. Zool. Lond. 171:25-39.

Moriarty, D.J.W. in press. Digestion of microorganisms in animals. In M. Recheigl (Ed.) CRC Handbook, Series Nutrition and Food.

Munro, I.S.R. 1955. The marine and freshwater fishes of Ceylon. CSIRO, Canberra. 351 pp.

Needham, J. 1971. Science and civilization in China. Vol. 4 part 3. Physics and physical technology. Cambridge Univ. Press. 931 pp.

Obeng, L.E. (Ed.) 1969. Man-made lakes: the Accra Symposium. Ghana Univ. Press, Accra. 398 pp.

Oglesby, R.T. 1981. A synthesis of the reservoir fisheries of Sri Lanka. FAO Field Document 2; FL/TCP/8804, Rome. 30 pp.

Perera, N.P. 1978. Early agricultural settlements in Sri Lanka in relation to natural resources. Ceylon Hist. J. 25:58-73.

Petr, T. 1978. Tropical man-made lakes - their ecological impact. Arch. Hydrobiol. 81:368-385.

Pigott, L. 1952. Prehistoric India. Penguin Books. 293 pp.

Poddubny, A.G. 1976. Ecological topography of fish populations in reservoirs. Transl. from Russian. New Delhi 414 pp.

Rao, K.L. 1951. Earth dams, ancient and modern in Madras State. ICOLD, New Delhi 1:285-301.

Ruttner, F. 1931. Hydrographische und hydrochemische beobachtungen auf Java, Sumatra und Bali. Arch. Hydrobiol. Suppl. 8:197-454.

Ruttner, F. 1952. Planktonstudien der Deutschen Limnologischen Sunda Expedition. Arch. Hydrobiol. Suppl. 21:1-274.

Schiemer, F. 1980. Parakrama Samudra, Sri Lanka Limnology Project:Interim Report. Inst. Internat. Coop. Vienna, Austria. 112 pp.

Schiemer, F. (Ed.) 1983. Limnology of Parakrama Samudra - Sri Lanka: A case study of an ancient man-made lake in the tropics. Developments in Hydrobiology, Hydrobiologia.

SCOPE. 1972. Man-made lakes as modified ecosystems. Rome. 77 pp.

Sirimanne, C.H.L. 1953. A survey of potable and industrial waters of Ceylon. Proc. Ceylon Assoc. Adv. Sci. Part 2 (1951):5-36.

Smith, N.A. 1971. A history of dams. Peter Davies, London. 279 pp.

Snitter, N.J. 1967. A short history of dam engineering. Water Power 19:142-148.

Van de Lippe, P. 1951. Ceylon restores its ancient irrigation works. Civil Engineering (Sept. 1951):41-44.

Welcomme, R.L. 1979. Fisheries ecology of floodplain rivers. Longman, Lond. 317 pp.

Weniger, G. 1972. Hydrochemical studies on mountain rivers in Ceylon. Bull. Fish. Res. Stn Sri Lanka (Ceylon) 23:77-100.

Wiley, A. 1910. Note of the freshwater fisheries of Ceylon. Spol. Zeylan. 7:88-106.

22. Natural resources, settlements and land use

N.P. Perera

Introduction

A resource is a means of supplying a want, or a stock that can be drawn on. As such, a resource is determined by human concepts of what is useful and required. Human needs alter with diversification and change of socio-economic conditions and the advancement of technology. Natural resources mainly are concerned with the attributes of the land, soil, climate, hydrology and the biota. These features are physically a part of a habitat. In other words, habitat is one aspect of the complex of physical, chemical and biological processes together with their resultant products which is referred to as nature. Among these, land is central in importance. Land is not only physical space, but also has the characteristics that govern its uses. These qualities include the nature of the underlying rock, shape of the terrain, the nature and fertility of the soil, the availability of water and the suitability of the climate.

Land use is a permanent or cyclic human intervention to satisfy human needs, either material, spiritual or both, from a complex of natural or artificial resources which together are called land (Vink 1975). Building a shelter or an abode constitutes a settlement. Human settlements are not evolved in isolation, but are shaped by complex processes of cultural transfer, acculturation, absorption and replacement. These characteristics of a settlement are reflected in its site. Sites are thus situated and related to their mutually adjusting ecological system or systems, with which they are reciprocally linked.

The aim of this study is to focus attention on the available evidence, and discern how, over the years, the availability of natural resources or the attributes of the land have evolved into a distinctive pattern or patterns of human occupance in Sri Lanka. It is also the objective of this study to examine the way by which improvement and advancement of technology and socio-economic conditions have diversified the resource, to the recognition and use of these resources.

In and attempt to examine the relationship between natural resources, settlement and land use, it is proposed rather arbitrarily to divide the long history of man's tenure in Sri Lanka into periods such as pre-historic, proto-historic, Dry

Zone Civilization (5 BC-12 AD), late medieval (1200-1500), western colonial (1500-1947) and finally the post-independence from 1948. This division is discretionary in the sense that there was no really clear cut and complete break and change of man's activities in any two consecutive periods. In reality, it was a gradual process which later overwhelmed and submerged the preceding one, thus giving each period a dominant and predominant character.

Pre-historic period

There is still unresolved controversy and debate about the origin and time of pre-historic man in Sri Lanka. The foundations for these studies were laid in the early decades of the 20th century, by Sarasin & Sarasin (1907), Seligman & Seligman (1908), Hartley (1911), Wayland (1917) and Noone & Noone (1940, 1945). They were followed closely by the Director of the Colombo Museum, P.E.P. Deraniyagala, whose published work covers a period of over 30 years.

The pre-historic period usually is divided into the Paleolithic, Mesolithic and the Neolithic. Their relative duration and time scale, however, varies in different parts of the world. The Congress of Asian Archaeology held in New Delhi in 1961 proposed a division of the Stone Age into only the Middle and Later Stone Age, and suggested that this would better fit the situation in the Indian subcontinent (Boisselier 1979). Most pre-historians are of the opinion that the advent of pre-historic man in Sri Lanka begins with the Middle Stone Age, and this fits well into the chronology proposed at the Congress. This Middle Stone Age period is called the Ratnapura culture phase, as these artefacts first were identified in the environs of the Ratnapura valley. Characteristic types of such stone implements comprised bifaces, flakes and worked cores of approximately 4.5 to 8 cm in length. With these were a fairly diversified range of bone implements (Deraniyagala 1953). These folk were mostly cave dwellers, the caves being situated in such places within easy access to hunting, fishing and a gathering way of life. Little is known about the animals hunted or the types of fish caught for food, nor of the varied kinds of edible leaves, fruits and roots gathered from the tropical rain forest ecosystem that prevails in this area.

As Sri Lanka is a continental island, it is reasonable to surmise that these folk entered from the Indian subcontinent. According to Boisselier (1979), the upper levels of this culture show contact with the Neolithic culture of South India.

Artefacts pertaining to Neolithic man have been found in a number of places, and especially on the slopes of the hills and knolls in the region of the Uva highlands. It appears that such situations have been preferred as working sites for tool making, as the raw materials needed, in the form of quartz veins, pebbles and blocks of chert frequently are found in this area, According to Noone (1945), these implements are more refined that those of the Ratnapura culture phase and formed a highly diversified microlithic industry. Common implements found in the Bandawawela area consist of knives, a variety of scrapers and borers together with small arrowhead forms and points. These tools, which mostly are of quartz, were fashioned using fire. The presence of charcoal and fired pottery provides evidence of the knowledge and the use of fire. Implements such as these undoubtedly would have helped in hunting and other domestic chores as well as in fighting and defence.

The hill top sites other than affording raw materials needed for making stone implements, also were preferred for other reasons. They were very much drier and leech free, and consequently were healthier. These sites also provided security and afforded look-outs for watching the movements of game and possibly of enemies. Noone (1945) called this the Bandarawella culture phase. Implements pertaining to Neolithic man have been found in the lowlands too. Deraniyagala (1960) found a number of skeletons together with implements at Bellanbendipelessa on the southern flanks of the Uva highlands. He termed this the Balangoda culture phase. Others such as Wayland (1917) and de Alwis & Pluth (1976) reported on a number of sites in the Red Earth formation of Quaternary age, in the Puttalam-Manaar areas. These sites in the Red Earth formation generally are on the hollows and lower slopes of the undulating surface of sand material, and hence close to water. It thus appears that the Neolithic culture phase in the island was widespread.

Skeletons of 25-35 year olds found at Bellanbendipellesa indicate predominantly Australoid characteristics. Features such as these suggest affinities with the aboriginal **Veddha**. These people, as is characteristic of the Neolithic, may have engaged in a form of rudimentary shifting (slash and burn) cultivation. There is no evidence of what plant species they domesticated. A typical Neolithic settlement probably was a group of rudimentary hutments on a sheltered slope, surrounded by a shallow entrenchment. They probably would move to another area as game became rather scarce.

Frequent firing, clearing and cultivation of land would induce soil erosion on the steeper slopes, a condition that would favour invasion by cosmopolitan species of weeds, to the detriment of the regeneration of forest species. The resultant reversal of soil forming processes and the invasion of new plant species tolerant to fire, inevitably would induce and bring about change of the forest vegetation to a grassland type, dominated by pantropic grasses such as *Cymbopogon confertiflorus* (**mana**). The patana grasslands (sub-montane grasslands) were formed by such a process (Perera 1969, 1976). The evidence thus suggests that the biological degeneration of at least the original sub-montane rain forest on the Uva highlands was begun by these Neolithic folk. A few dates proposed for Balangoda Man is from $5,000 \pm 700$ BC to 144 ± 200 AD (Boisselier 1979). This phase of Neolithic culture appears tp be related to the 'teri industry', found in the red sand dunes of Tinnevelly district in South India (Allchin 1958). Thus the clearance of the forest and shrub for cultivation or other purposes, represents a major intervention by man in nature. From this stage it appears man not merely accepted the natural environment as he found it, but adapted it to his own purpose.

Proto-historic period

The proto-historical period also may be termed the Iron Age. It is often knwon as the Megalithic period, and is rather widely represented in Sri Lanka (Deraniyagala 1958). The word 'proto-historical' to name this period seems appropriate, for it was a period which as yet is not dated with any degree of historicity. Glimpses of life in this period nevertheless probably could be surmised by references to these folk in early proto-historical or historical writings like the Indian epics, Ramanayana and Mahabharata and the Mahavamsa (a chronicle compiled about 6 AD), giving historical information based on earlier records in ancient Sri Lanka (see Geiger 1960 for translation).

The Iron Age of Sri Lanka has close affinities with a similar phase in South India, typefied by sites such as Adichanllur (Kirk 1975). Nevertheless, more advanced aspects of this culture than in South India did not develop in the Island, probably because of the rapid progress of Buddhism during this period. The best known site in Sri Lanka is at Pomparripu, near the N.W. coast. At this site, human and animal bones, bronze objects, iron blades and well shaped terra cotta urns, some with markings suggesting a very primitive Brahmi script have been found (Deraniyagala 1958). Other sites of this culture have been located in widely separated areas such as Kadiraveli (north of Batticoloa), Gurugalhinna (Anuradapura district), Galsohonkanatta (Kurunegala district) and at Padiyagampola (Kegalle district). The finding of such dispersed sites indicates that much of the northcentral plains, parts of the eastern seaboard and the northern parts of the foothills have been settled by folk of this culture phase. Whether these folk were the Yakkas and Nagas mentioned in the Ramanayana and Mahavamsa is a matter of conjecture. It is certain, however, that the Indo-Aryans led by the legendary Vijaya did not come to an empty land about 500 BC. If any credence is given to the Mahavamsa reference that when Vijaya landed on the N.W. coast, Kuveni (supposed to be the queen of the area) was seen spinning cotton, suggests that more permanent and diversified forms of agricultre and settlement already had been developed. According to the Mahavamsa, the Indo-Aryan settlements were founded where water was easily available. Therefore, like successful farmers, they probably understood the supreme importanceof land and water, and set out to discern and use the easily worked and productive soils. As the Indo-Aryans penetrated inland, these native Iron Age folk were probably driven more and more to remote and inaccessible areas before they gradually were absorbed and/or exterminated. In this episode one fact stands out clearly, that for about 1,000 years up to the middle of the first millenium BC, there was a period when a relatively advanced civilization was built up in Sri Lanka with improved and diversified agricultural activities.

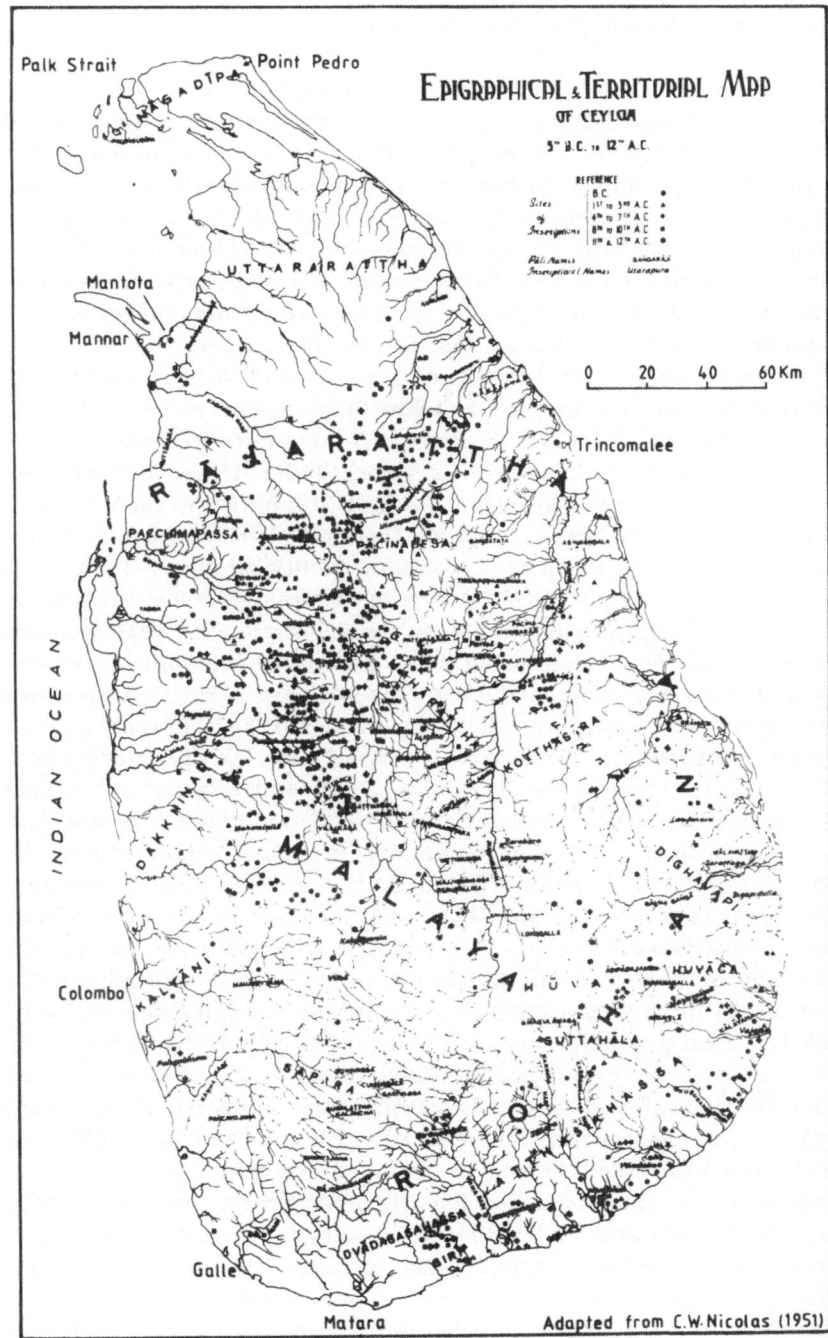

Fig. 1. Epigraphical and territorial map of Sri Lanka.

The dry zone civilization (500 BC-1200 AD)

The availability of written historical records especially lithic (rock inscriptions) places the beginning of this period with the establishment of Buddhism in Sri Lanka (Fig. 1). However it is proposed that this period commenced with the colonization of the island by the Indo-Aryans, associated with Vijaya, the first king of the Sinhalese. This event is regarded as a watershed in the island's history, although much of the information till the much later advent of Buddhism still is mainly mythical and legendary. Nevertheless, it marks the beginning of a long saga, first transmitted by oral tradition, over the years, from generation to generation and later transferred into written history. The earliest Sinhala inscriptions prove the historicity of the Mahavamsa, as the context of six of the earliest inscriptions refer to royal persons who reasonably can be assigned to the second half of the 3rd century BC (Paranavitana 1970). Data from legend, corroborated with linguistic and anthropometrical evidence, indicate that the Sinhalese were people of Aryan origin (Indo-European/Indo-Iranian) who came to the island from North India about 500 BC. The exact location of their original home in India has not be determined with any degree of certainty. Siddhartha (1935), Geiger (1937) and later Paranavitana (1952) were of the opinion that it was in northwest India. This view is supported by a study of the anthropometry of the Sinhalese and Sri Lanka Tamils by Chanmugam (1949). These early colonists appear to have landed and firstsettled in parts of the northwest coast, and also in the southeast.

These later were followed by settlements on the west coast (Kelaniya). Except in the west, it was mainly in the dry zone that these early Sinhalese settlements were established. It appears that they first were riverine in character, and rice perhaps was the staple crop. Cultivation was during the period of the northeast monsoon and cyclonic rains from October to January. It should be pointed out however, that these two areas of the country (NW and SE) where these early settlements were founded are the driest parts of the island and the two areas where the variability of rainfall is most marked (Cooray 1948, Farmer 1956). Further, as the soil map (Fig. 2) indicates, the soils along the northwest coast are unsuitable for rice cultivation, as these soils are either too sandy or too alkaline. The only soils suitable for cereal growing are the alluvial soils. The extent of such soils necessarily is limited, due to the rivers in this region being mostly ephemeral, as they rise in the dry zone itself. (A full analysis of these environmental limitations and hazards will follow). It was only when the Sinhalese discovered and mastered the use of the Reddish Brown Earth soils and particularly the characteristics of this type when-considered in its catenary sequence, that major settlements sustained by irrigation projects were established. Till the shift to the wet zone in 12 AD, the dry zone was the cradle, and witnessed the golden age, of Sinhalese civilization. Today, the Sinhala people form about 73% of the total population.

The largest minority in Sri Lanka today are the Sri Lanka Tamils. Like the Sinhalese, there is no firm evidence concerning the establishment of early Tamil settlements in the island. According to Maloney (1970), Tamil and other literary sources indicate the establishment of substantial urban and trading settlements in South India by 300 BC. By 237 BC it is possible that they had captured much of Rajarata (comprising a major part of the northern dry zone) and ruled from Anuradapura for twenty two years till the Sinhalese recaptured their throne (Ray 1959). Nevertheless, by the early centuries of the Christian era, the Tamils not only lived in Sri Lanka but asserted themselves from time to time by becoming rulers. However, their numbers and settlements were few when compared with the Sinhalese (Kanapathipillai 1948). It would thus appear that the Indo-Aryan settlements and colonization preceded the arrival of Dravadian settlers by several centuries. In this sense, Sri Lanka has been from early times in recorded histories, a multiracial society, in which there was a distinct Dravadian element which could not, however, alter the basic Aryan or North Indian character of its population. Although it is true that the modern Sinhalese are no more pure Aryan than the modern Tamils are pure Dravadian, a complete amalgamation of the two races never took place (Tambiah 1968). The Sinhalese, at least, preserved their distinctive language. At the same time, it should be noted that as Sri Lanka occupies a commanding and strategic position in the northern Indian Ocean, it was accessible from both the east and west, which could have helped the Sinhalese in obtaining North Indian support when necessary (de Silva 1977). Indeed, as Nichols (1963) cryptically states, 'It will probably never be resolved satisfactorily why the South Indians did not resist the North Indian colonization of Ceylon, or if they did, how the resistance was overcome, by the 4th century BC if not earlier by the Indo-Aryan speaking people'. Today, the Sri Lanka Tamils are concentrated mainly in the Northern Province and the coastal belt of the Eastern Province. They comprise about 11 percent of the population.

During this long period of over 1,500 years, the Dry Zone Civilization was able by an immense and increasing struggle to clear the forests, aswedumize (layout) paddy fields, construct reservoirs (tanks) of varied proportions and complexity, terrace hill slopes, drain marshes and build stupendous and highly artistic religious monuments. By such processes the limits of agriculture and of settlement were extended into different milieux of the country.

At this stage it is considered appropriate and apposite to consider the characteristics of the dry zone environment in some detail, and examine how the etablishment and spread of settlements were influenced by these and other attributes.

The dry zone environment

In the dry zone the potential for agricultural development and the type of settlement pattern, depends to a considerable extent on the edaphic and climatic characteristics. In addition, relief and hydrology also are significant as contributory factors. Within these different variables, a number of possibilities arise for the

460

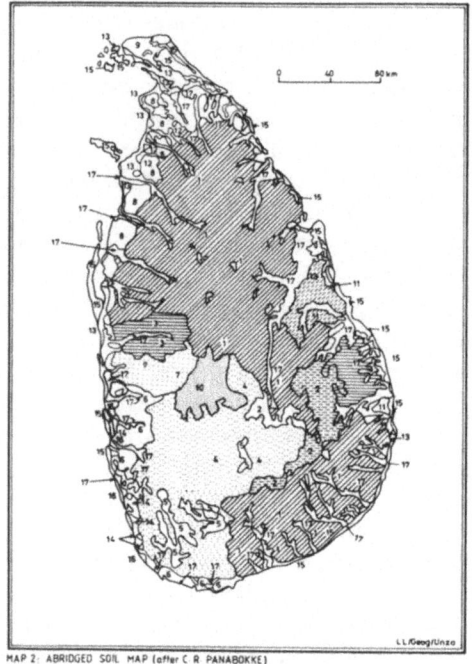

MAP 2: ABRIDGED SOIL MAP (after C R PANABOKKE)

Fig. 2. Abridged soil map (after C.R. Panabokke) numeral index – 1-17.
1. Reddish brown earths and low humic gley soils, undulating terrain; 2. Reddish brown earths, non calcic brown soils and low humic gley soils, undulating terrain; 3. Reddish brown earchs with moderate amount of gravel in sub soil and low humic gley soils, undulating terrain; 4. Red yellow podzolic soils, steeply dissected hilly and rolling terrain; 5. Red yellow podzolic soils with semi-permanent Al horizon, hilly and rolling terrain; 6. Red yellow podzolic soil with soft and hard laterite, rolling and undulating terrain; 7. Red yellow podzolic soils with slightly mottled sub soil and low humic gley, rolling and undulating terrain; 8. Red yellow latosols, flat and slightly undulating terrain; 9. Calcic red yellow latosols, flat terrain; 10. Red yellow latosol with slightly mottled sub soil and low humic gley soils, rolling and undulating terrain; 11. Soladized solonetz and solonchaks, flat terrain; 12. Grumusols, flat terrain; 13. Soils on recent marien calcereous sediments, flat terrain; 14. Bog and half bog soils, flat terrain; 15. Regosols on recent beach and sand dunes, flat terrain; 16. Regosols on recent beach sands, flat terrain; 17. Alluvial soils of variable drainage and texture, flat terrain.

utilization of these resources. In the case of the Dry Zone civilization, the ability of the early Sinhalese people to recognize and utilize the different facets of the environment, particularly of the soil catena in relation to the soil-water requirements of crops, may be regarded to be the most important. Thus, with the coming of the Sinhalese, one arrives quite suddenly at a period in the study of the landscape, when the impact of civilized life in the fullest sense of the term, is felt for the first time.

The Dry Zone Civilization developed basically in that part of the dry zone which is underlain by crystalline metamorphic basement (hard) rocks of Precambrian age. This region is surrounded by Miocene limestone terrain to the north and by Pleistocene sands in the west and east (Cooray 1967). The general aspect of the hard rock surface is undulating, made up mostly of shallow but well defined valleys. Normally, the distance from the crest of the ridge to the floor of the adjacent valley, is frequently of an average length of about one kilometre. The recognition of this as a fundamental land unit, and planning a land use system based on its characteristics, can be considered an important contribution by those early settlers. The usual and more regular pattern of the topography is at times broken by a number of low ridges, which radiate in a northwest or northeast direction from the northern part of the central highlands. More conspicuous and characteristic of the general relief, are the number of numerous isolated hills or monadnocks of varying height and extent, that dot this undulating surface. In addition to these, the surface frequently is strewn with outcrops of low bare rock mounds shaped like turtle backs. These mounds have been sometimes used to build part of the bunds of the village tanks (Cooray 1967). The northwest coastal area where the early Sinhala settlers landed is characterized by sand-dunes, broad sandy beaches and brackish water lagoons. Adjacent to this in an inland direction is the area of Red Yellow Latosol soils. In the southeast part of the coastal sand dune belt is narrower and frequently merges with an extensive parkland. This parkland is littered with large and picturesque rock outcrops, buttes and hills in marked contrast to the northwest. The most important perennial river that flows through the dry zone is the Mahaweli ganga and its principal tributary the Amban ganga. The rest of the dry zone rivers, of which there are many, have only a seasonal flow of varying duration. It has been found that the dry weather flow in second order streams of most of these dry zone rivers, usually ceases after about 45 consecutive dry days. A striking feature of the dry zone rainfall is that most areas receive 65 to 75 percent of their annual rainfall between October and early January, although some heavy rains do occur during late March and early April. Consequently, the seasonality of rainfall is so marked that three or four months without any rain at all is common, even in years of normal rainfall (Sirinanda 1979). Thus the rainfall is seasonally unreliable, and the surface runoff during the rains is a scarce resource, providing only limited potential for varied users. This is an agro-climatic criterion of supreme importance which has to be noted, particularly in devising a farming system. According to Murphey (1957), available evidence suggests that climatic conditions basically were similar nearly 2,000 years ago. In such a precarious and risky rainfall situation, surface water storage for irrigation and domestic purposes is of paramount importance, if crops like rice with a growth period of longer duration than the 3-4 month west season rainfall, can be grown successfully.

Fig. 2 indicates that the soil type that covers much of the dry zone is classified under the great soil groups as the Reddish Brown Earths, or under the 7th approximation as Alfisols (de Alwis & Panabokke 1972). In a typical catenary sequence of this type, from the crest of the ridge to the valley bottom; its drainage associates the Rhodstalfs and the Haplustalfs occupy the well drained parts of the gently convex relief, while the Plinthustalfs occupy the imperfectly and poorly drained parts of the slightly concave and flat landscape. The Reddish Brown Earths are considered the most important soil type of the country for a number of reasons. Basically they are more fertile and cover the largest areal extent than any other soil type found in Sri Lanka. A significant plant nutrient shortage is that it has a low phosphate content. If properly managed, these soils provide the necessary edaphic conditions for a variety of crop plants. Further, the imperfectly drained Plinthustalfs, or low Humic Gley soils, when saturated with water could very conveniently be tilled by the traditional bullock drawn, wooden or iron plough. The most serious limitations of the physical properties of these soils for agriculture are its high bulk density values, low macro-aggregate stability and the very hard consistency in the dry state (Panabokke 1958). The Reddish Brown Earths lying in association with the Non-Calcic Brown soils, and immature Brown Loams limit agriculture to a variable extent. More important of the drawbacks of these soils are their shallowness, poor fertility and high erodability (Panabokke 1978). Elsewhere in the dry zone, the Regosols, Solodized Solonetz and Latosols generally are unsuitable for crops, especially rice. These soils have a number of deficiencies, such as a high percentage of salts; are unsatisfactorily drained, or lack suitable soil structure. The small areas of Grumusols also require special treatment for cultivation (de Alwis & Panabokke 1972). It can be surmised that although the above soil types were first encountered by the Sinhala immigrants along the northwest coast, hardly any large settlements, except port settlements, were found in these areas (Perera 1978). The natural vegetation developed on the Reddish Brown Earths are the tropical evergreen seasonal rain forests, producing higher biomass than any other type of dry zone vegetation (Perera 1975) (Fig. 3). This type of forest has a two tiered tree strata with a shrub undergrowth. Some tree species are deciduous, nevertheless the general physiognomy of the forest as a whole is evergreen. The tree strata contain valuable timber species such as *Manilkara hexandra* (**palu**)m *Vitex pinnata* (**milla**), *Berrya cordiflora* (**halmilla**), *Diospyros ebenum* (**ebony**), *Chloroxylon swietenia* (satin wood) and *Euphorbia longana* (**mora**). Much of the undergrowth is occupied by *Drypetes separiaria* (**wira**), but this species is of little economic value. Seeds of *Madhuca longifolia* (**mee**), a tree of the lower stratum, provides oil used for cooking and illumination. Most of the fine timber used in the construction of ancient palaces, temples, homes and bridges were obtained from these forests. The heterotrophic component in such an ecosystem would include the larger herbivorous mammals like *Elephas maximus maximus* (elephant), *Bubalus bubalus* (buffalo), *Cervus unicolor* (sambhur), *Sus scrofa cristatus* (wild boar), *Axis axis* (spotted deer), *Macaca sinica sinica* (Macaque monkey), *Presbytis entellus thersites* (grey langur), and *Melursus ursinus* (sloth bear). Important carnivorous animals are *Canis aureus lanka* (jackal), *Panthera pardus fusca* (leopard) and *Felis viverrina* (fishing cat) (Phillips 1942). In the region of the Reddish Brown Earths and Non-Calcic soils, particularly in the eastern interior and higher elevations of the dry zone, the vegetation type is upland savanna. This is a grass and tree mixed plant community often having the appearance of an abandoned orchard. The grass element is mostly the coarse tussocky types such as *Imperata cylindrica* (**illuk**), *Cymbopogon nardus* var. *confertiflorus* (**mana**), *Anthistivia argeas* and *Sorghum nitidium*. The tree species in the association include *Terminalia chebula* (**aralu**), *Terminalia belerica* (**bulu**), *Phyllanthus emblica* (**nelli**), *Pterocarpus marsupium* (**gammalu**) and *Diospyros melanoxylon* (**kadumberiya**). These trees provide a number of useful products commonly used in Ayurveda (indigenous) medicine. The use of such savanna products may have contributed to a general belief, still held by some of the local populace, that the upland savanna was planted during the time of the Dry Zone Civilization as medicinal plantations. Frequent firing often is employed to facilitate the collection of these products, a practice which has ill effects on the soil. This has contributed in a noteworthy degree to the perpetuation of the upland savanna vegetation in its present state (Perera 1969).

Two plant communities of smaller areal extent in the dry zone are the lowland savanna or **Damana**, and the lowland **Villu** grasslands. The Damana savanna are found particularly in the Polonnaruwa district. These exist side by side with strips of the dry mixed seasonal rain forest. It is a mixed community of grass and herb species with scattered groves of low stunted trees of average height 2-5 m. The

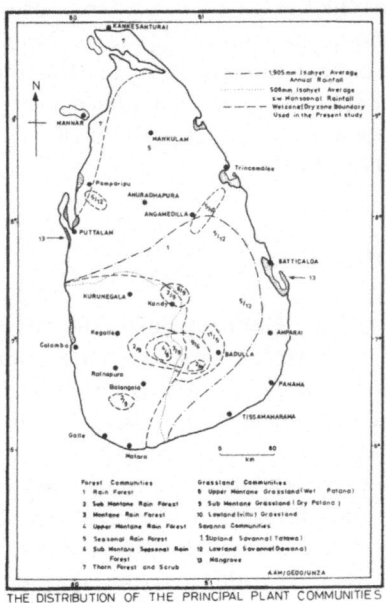

THE DISTRIBUTION OF THE PRINCIPAL PLANT COMMUNITIES

Fig. 3. Distribution principal plant communities.

groves of trees are found on elevated, rather circular areas usually formed of old and new termite mounds. The general picture is of small islands of trees scattered in a sea of poor grassland. These trees mostly are of species found in the canopy stratum of the seasonal rain forest, but are very much stunted and gnarled and of poor form. The typical grass species are *Aristida setacea* and *Imperata cylindrica*. The soil is shallow with a layer of rock or superficial decaying rock. The ground also is flooded during the rainy season and on the whole, both porosity and water percolation of the soil tend to be poor. Nevertheless the soil becomes extremely dry and hard during the long dry season. The areas covered by these soils are unsuitable for cultivation and would have been little used except for rough grazing by domestic and wild animals. Holmes (1951) was of the opinion that the Damana areas were flourishing paddyfields in the hey-day of the Dry Zone Civilization period, but were abandoned later due to increasing salinization caused by faulty irrigation techniques. This view, however, is untenable as these soils were not suitable for cultivation in the first place, because of their shallowness and low fertility. Furthermore, conditions conducive to salinization do not occur in a region of heavy seasonal rainfall.

The Villu grasslands also are found mainly in the Polonnaruwa district on either side of the Mahaweli ganga. It occupies the area of old Alluvial soils. The Villu grasslands have relatively few trees usually widely spaced on the river levees. The grass colour in the dry season is a rich verdant green, presenting a striking contrast to the dark green of the adjacent forest. These areas are flooded for nearly four months of the year, but as the flood water recedes, it provides excellent succulent fodder during the dry season. The dominant grass is *Brachiaria mutica*. Historically, these grasslands are reputed to have been used to maintain the Royal cattle herds of Parakramabahu I (1153-1215). These herds were largely for draught and milk purposes. At the present time it is the habitat of a giant race of elephant - the swamp elephant (Jayasekera 1969).

The evolution of the land use pattern

The response of the Dry Zone Civilization to the environmental conditions is evident today in the land use pattern it created. According to Needham (1969), the indigenous irrigation system on which it was based was unique in the whole of South Asia, both in its scale and the technological ability it displayed, details of which will be discussed later.

It has been shown that there is a close correlation between the major soil types, the distribution of the irrigation reservoirs (tanks) (Fig. 4) and the siting of lithic inscriptions. If the site of an inscription is assumed to indicate that it usually was on or near a settlement of that period, then the spatial extent and date of the inscription can be taken as an indication of the spread and extent of settlement. Such data become all the more revealing when they are plotted on a map, and considered in relation to each other. They also afford an opportunity for the study of archaeological sites as an integral part of the mutually adjusting environmental and ecological systems with which they once were adaptively integrated (Perera 1978). Such correlations indicate that the irrigation tanks mostly were in the Reddish Brown Earth soil area. Elsewhere, the density of tanks is low, as in mixtures of the Reddish Brown Earths and the immature Brown Loams, Reddish Brown Earths and Non-Calcic Brown soils on Old Alluvium and Alkali soils. These soils besides being inferior in fertility to the Reddish Brown Earths, are on terrain that either is subjected to flooding or is rugged and rough, and occupied by numerous monadnocks and low rock outcrops. Consequently the settlements in such soil areas were not impressive not dense.

It could be argued that adaptation to the distinct qualities and needs of the attributes of the Reddish Brown Earth soils, necessitated the development and growth of the threefold land use pattern. This ensured the maximum conservation and utilization of rain water within each mini catchment area. Essentially, it enabled the use of irrigable and unirrigable land in a single integrated land use system, by the artifial creation of soil moisture conditions favourable to the growth of a number of drought susceptible perennial crops.

The landscape of the Dry Zone Civilization shows a unique combination of natural beauty and man-made features. It is almost impossible to imagine such a landscape without the glistening expanse of water of the small village tank and its upraised bund, the chequer-board pattern of the paddy fields in the valley bottom and the homesteads clustered on slightly higher ground (**Gangoda**) mainly surrounded by tree crops. The picture is completed by the tapering spire of a bubble shaped **Dagoba** of a Buddhist temple, usually perched on a prominent spot, near the homesteads; and the patches of shifting cultivation (**Hena/Chena**) fields amid the dark green of the forest on the upper unirrigable slopes. In this sense, the countryside would have looked as a sort of Agrarian Archipelago, with innumerable islands of cultivation, set in a sea of forest and scrub, as is indeed reminiscent of many areas today.

Development of water conservation

The pattern of water resource development and use indicate several progressive trends in improvement and complexity. The earliest were primarily local or at village level. In the course of time it was expanded to regional and national. Irrigation development thus progressed in a gradation from the relatively simple single purpose project to the more complex multi-pupose one. If Anuradhapura is taken as a core and nucleus of the Sinhala settlement and government, the gradual spread of irrigation activity corroborated by inscriptional evidence is indicative of the spread of settlement from this core. Fernando (1979), relying mainly on legend, postulates the irrigation activity based on the construction of the tank was a pre-Sinhala innovation. Such a view is difficult to accept as Gunawardana (1971) stated that rice culture was introduced by the early Sinhala immigrants and irrigation was essential if a civilization based on rice culture was to succeed and expand in the dry zone environment. Taking into consideration the hazards of rice culture in the dry zone rain fed environment, it can be surmised that water conservation and utilization by means of a tank was mainly a Sinhala development. This view also is strengthened by the fact that historical, inscriptional and other records like the chronicles, indicate that the earliest hydraulic works recorded seem to have been rather unsophisticated. It is therefore considered that tracing of the gradual development and expansion of the tank system over this long period is pertinent to this study.

The first and simplest stage in irrigation development was, perhaps, in a small pond in a valley, in which water was collected during the rainy season. With a primitive apparatus, water from such a pond was lifted and used to irrigate small patches of cultivation beside it. This may be the type that existed in pre-Sinhala times.

The second stage was the development of low artificial embankments or weirs (**Avarana**) built across the valley of small ephemeral rivulets. Such a tank when full usually is about 2.5 m deep immediately behind the bund. Very roughly the full tank covers much the same area of ground as the land below it, which it is capable of irrigating. This tank is followed downstream by yet another tank and paddy field (**Yaya**). The process continues down the valley until the topography becomes too flat for gravity distibution of the impounded water into the fields below. This is the type mentioned in the inscriptions as **Gamaka Vapi** (village tank). Being small, their construction and upkeep did not demand more labour than the village community could provide. Some of these, however, also were owned by private individuals (**Vapi-hamika**), or tank owner (Paranavitana 1970). Such a system of water conservation and utilization brought about a society based on what may be referred to as a 'one tank-one village' ecological pattern. Most place names of such tank villages end in the word **Wewa** (Sinhalese) or **Kulam** (Tamil). One of the earliest of such tanks mentioned in the Mahavamsa is the Basawakkulama in Anuradhapura, built by King Pandukahabaya in about 3 BC.

The next stage was an improvement of the second type, the low primitive bunds of some of the second tank type were stengthened, and the limitation on capacity of water storage and extent of irrigable land improved. Though large, they were not part of a complex network of tanks. Examples of this type include Dighavapi built by Saddhatissa (150 BC), Vavunikulam by Dutu Gemunu (150 BC), Kuda Wilachchiya by Vasaba (65 AD), and the Kalawewa and Balaluwewa before their bunds were stengthened and the two joined to form one large tank (**Mahavapi**). In this way most of the important large tanks have, over the centuries, been rebuilt, bunds strengthened, enlarged and augmented with more secure supplies of water.

The fourth stage was damming of the bed of comparatively large, though not perennial rivers, such as the Kal-oya during the reign of Dhatusena about 459 AD.

In such instances, rock outcrops in the river frequently were utilized as foundation for the **Anicut** (dam across a narrow part of a river to raise the water level. The water is then led off laterally via a channel for irrigation or to augment the water supply of a tank). From such a large tank (**Danavapi**), water was transferred for a considerable distance. For example, Kelawewa fed the Anuradhapura city tanks and its environs through a 86.4 km long canal, the Jaya ganga. This canal meanders gently, and transports the water over astonishly easy gradients. Some idea of this can be garnered from the fact that for the first 29.2 km the fall is not much more than 1.5 cm km^{-1}. By such means the Kal-oya and Walwatu oya catchments were linked, and an area of about 450 km^{-2} in extent irrigated (Ievers 1887).

The fifth stage was the construction of reservoirs on large, though ephemeral rivers and tributaries, and linking these to Anicuts built on rivers having catchments in areas of perennial water supply in the wet zone. One such early construction was the Anicut at Elahera, across the Amban ganga. The Elahera canal fed the huge Minneriya tank which was built across thecomparatively small tributary. Here again, though it was first built by Vasaba (65-109 AD) it was later enlarged and diversified by Mahasena (276-303 AD). Brohier (1938) gives a vivid description of some of the many remarkable features of the Elahera canal, which among others illustrate the skill and indomitable perserverance of the ancient engineers. The first section of the canal, from Elahera to Minneriya (approximately 40 km) runs through ranges of hills and ridges on the west and east respectively. Anticipating erosion on the bund by hill streams which flow down the face of these ridges, stone foundations and stone pitchings were constructed at intervals on the inner face of the bund of the canal. Without such constructionm the canal would easily have been broken at places and destroyed. Surveys done before modern reconstruction of this canal was undertaken showed that many of the ancient bunds at these points had remained intact. Apart from precautions of this type, several **Galvanas** or rock spills were built into the bund at intervals. Some of these were as large as 7.5 m wide. These Galvanas permitted an outflow of the canal, and appear to have been distributed at points where there was a possibility that the water might rise too high for the safety of the bund. In this manner too, damage to the bund had been lessened. After filling Minneriya tank, the Elahera canal was carried 32 km further to the northeast to feed a second large reservoir, Katalai. Before this, one more large tank, the Kaudulu wewa, also was supplied with water. The total area of these three tanks is >12,000 ha.

Similar diversions were made on the main Mahaweli ganga itself. Before the river eventually merges onto the plains from the hill country, its waters negotiate the steepestdescent in a narrow gorge, not more than 9 m wide. At this point, two rocks united by masonry intercepted the flow, and raised the water to a great height. A channel leading from this point carried the water in a northerly direction a a much higher level than the river but parallel to it, to irrigate large tracts of land. This was the Manimekhala canal (Minipe) built by Aggabodi I (575-608). Further downstream on the Mahaweli ganga was the Pabbatana canal, to carry water to the area around Dimbulagala in the Egoda Pattu of Tamankaduwa.

In the construction of large tanks needed as feeder and storage, one problem was the necessity to control the outflow of water from the tank to the canals. This was achieved by an ingenious method that was developed to overcome pressure of water, and control the quantity of the outflow of water, when it was released from the reservoir to the canal. The invention and development of the value-pit or Bisokotuwa, made this possible. The earliest reference to a Bisokotuwa, is in the Samanthapasadika of Buddhagosha, written in the 5th Century A.D. (Gunawardana 1971).

The sixth phase may be considered as the transbasin transfer of water from a perennial catchment to an emphemeral catchment based reservoir. Such an attempt was the linking up of the Amban Ganga catchment with the Kala-Oya catchment. In this linkage, water from the Demada Oya, a tributary of the Amban Ganga was transferred to the Welimito Oya, a tributary of the Dambulu Oya, the principal tributary of the Kala Oya. A weir at Nalanda on the Demada Oya, carried the water through a low saddle in the dividing ridge of the two opposite water sheds, and passed it down to the Welimiti Oya. This was constructed by Mahinda II (777-797 A.D.) (Nicholas 1963). Another such transbasin diversion, was the Gomati Ela built by Mahesena in 275 A.D., at Kalinga Nuwara on the main Mahaweli Ganga, to transfer water to the Maduru Oya basin (Brohier 1937). The achievement of mastering the transfer of water from the Amban Ganga and the Maheweli Ganga to feed the reservoirs in Nuwara-Kalawiya and Tamankaduwa districts respectively, were all probably completed by the end of the 8th century A.D. This helped to establish firmly the two core areas of the Sinhala civilization in the Rajarata. Thus the two cities of Anuradhapura and Polonnaruwa located in them, were the two foci of economic and cultural activity during this period. The growth and spread of population and of land development as early as the 6th century in the Polonnaruwa region, made this area so important as to be placed under the authority of the heir apparent. There was also a third Sinhala core area, as old as Anuradhapura, in the dry zone of the southeast in Ruhuna. Here too an irrigation system though of lesser complexity and scale than in Rajarata was developed. This was centered at Mahagama (Tissamaharama). Elsewhere in the dry zone, but more particularly in the area of Reddish Brown Earth soils, distant from the two core areas, irrigation activity was extended in all directions. For example, the Padaviya tank was constructed on the Ma Oya, towards the north eastern extremity of these soils, by Moggallana II (535-555). Similarly, towards the north western limits was built the Vavunikulam tank by Vasaba in 65 A.D. and the Yodawewa (Giants tank) by Dhatusena in 459 A.D. The Kurunegala district to the south of Nuwara-Kalawiya district was also developed. After the expulsion of the Colas, from South India, by Vijaya Bahu I in 1070, and especially under Parakrama Bahu I (1153-1186), a remarkable series of irrigation works, including the massive Parakrama Samudra tank was either constructed or restored. Further, the techniques of tank construction, also showed marked improvement (Nicholas

1954/55). It should be realized that the development of irrigation and of agricultural land use on such an enormous scale in a vast area, needed an organization for control, and maintenance of such activity (Paranavitana 1958). A government department called the Dolos Naha Vatana was entrusted with the care of twelve major reservoirs, though these twelve great tanks have not been mentioned.

Paranavitane (1970) is of the opinion that terms in the early inscriptions such as Anonika and Adikaya connotes the meaning of an irrigation engineer or a functionary in charge of works relating to canals. Numerous agricultural officers referred to as Vel Vassan and Vel-Kamiyan etc, were in charge of the administrative work in connection with rice fields (Paranavitana 1958). By the end of the 8th century, settlement had spread to most parts of the Reddish Brown Earths in the dry zone, except perhaps to the more rocky and poorer soils of the Non-Calcic Brown Earths, and the sandy and limestone areas to the north and northwest. On the other hand, there seems to have been a depopulation and abandonment of some areas in the dry zone were mainly on the coastal belt regions of Solodized Solonetz and Latosol soils. The Jaffna peninsula, underlain by limestone rocks has underground water, which could be tapped by wells. The Mahavamsa mentions that the Bodhi Tree arrived at Jambukola in 246 B.C., which Nicholas (1963) identified as being Sambilturai near Kankesanturai. This was an important port settlement at that time. References in the Mahavamsa to the construction of Buddist temples in the Jaffna Peninsula as early as 2 A.D. testify to considerable settlement in north Sri Lanka too. These references are also collaborated by epigraphical evidence such as the gold plate inscription in Sinhala of Vasaba (67-111) found at Vallipuram, together with excavations of Buddhist temples to substantiate the view that the Jaffna area was settled from early times (Pieris 1917). The wet zone uplands consisting mostly of the ancient Malay region and the southwest lowlands (Mayarata) were unpopulated, except perhaps, by small settlements such as in Kelaniya and the foothills in the Matale and Kegalle districts. On the other hand, the intermediate zone which is more humid than the dry zone, especially in the Kurunegala and Matara districts were being increasingly penetrated and settled.

Population and settlement types

The number of tanks in the dry zone may be taken as a preliminary index of the intensive land use, especially in the region of the Reddish Brown Earth soils. Brohier (1934) estimated that for the North Central Province (Anuradhapura and Polonnaruwa districts) alone, the number of historic tanks identified by the topographical survey of Sri Lanka in 1873 at 2,877. An exact count of the numerous village tanks does not seem at all possible, as some are yet covered with dense jungle, while others have been obliterated and erased due to erosion or have been submerged under newer constructions. Cook (1931), attempted to map the

470

density of village tanks, and (Fig. 4), is based on her work. When compared with the soil map it indicates the close correlation that exist between the area of the high density of village tanks and the distribution of the Reddish Brown Earths and its associated soils. The mutual relationship of these two are also indicated when they are compared with the distribution of the sites of lithic and other inscriptions (Fig. 1).

Such empirical studies testify to the fact that, areas of the Reddish Brown Earths and their associated soils, were densely settled and populated during the period of the Ancient Sinhala Civilization. Today in the Anuradhapura district, there are over 3,000 tanks in an area of 7,752 km². Some of these tanks are so small, that they run completely dry if there is no rain for about two months (Tennakoon 1974). Taking into consideration the extention of large scale irrigation activity and cultivation, it can be estimated that the spatial expansion of the Dry Zone Civilization could have achieved its optimum development between 400 B.C. and 900 A.D. It appears that the development of irrigation activities also enabled the gathering of two or even three harvests per year, as mentioned in the Tonigala inscription of Sri Meghavarna (303-331). There is no evidence, however, to show that multiple cropping of rice on this scale was the norm, rather than the exception. It is clear however, that the labour resources for the construction of the large

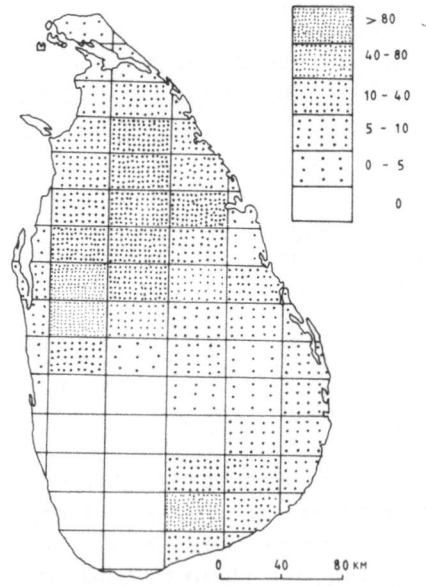

The Relative number of tanks (in use and abandoned)
The rectangles indicate the different sheets under which
Sri Lanka is mapped on the scale of one inch to a mile.

Fig. 4.   Irrigation reservoirs (tanks) in Sri Lanka.

irrigation projects and religious monuments that were built during this period, were ensured of a regular and adequate food supply, though droughts and famines were not unknown. In fact, large scale irrigation facilities would have represented the differences between a subsistence economy and the agricultural surplus, necessary to maintain the large section of the population, not engaged in food production, which is characteristic of a mature civilization (Boserup 1965). The total population during the period of Sinhala Civilization is not known. It is normal to assume that during the times of foreign and civil war, long drought periods and crop failures, and consequent famines, coupled with the scourge of epidemics, whole populations would have been decimated.

Nevertheless, irrigation development on the scale achieved, would not have been possible without a sizable growth in population over the intervening years. One of the earliest estimates of population during the Dry Zone Civilization period was by Tennent (1860) who reckoned it to be 17 million for the 12th century. Denham (1912) put it as 14 million, while Murphey (1957) estimates a number between 2-4 million. On the other hand, the estimate of 70.5 million inhabitants for this period contained in an old palm leaf (ola) manuscript found in a temple at Tissava, seems improbable (Domros 1974). Evidence from the economic and land use pattern, the ancient road system, descriptions left by contemporary travellers and also of some places mentioned in the inscriptions and ecological factors, provide material for the reconstruction and assessment of the major settlement types establishhed during this period. Probably the hierarchy of rural settlements commences with the isolated house whose occupants eked out a rather precarious existence by shifting cultivation. Next in the hierarcy and the most commonest, was the village or (Gama), consisting of a tank, paddy fields, garden land and areas of shifting cultivation (Hena). It usually had a number of families belonging to the same kingship and occupational group (Codrington 1938). A number of such villages were linked by roads and footpaths to a larger village or Nigama. This was usually a marketing centre. A settlement of this type was the village of Hopitigamu, probably close to Sorabora wewa north of Mahiyangana; as mentioned in the Budulla Pillar Inscription of Udaya III (935-938). A still larger and more urbanised settlement was described as a Patungama. Usually it was also a port, where merchants both local and foreign mixed and traded. The two ports of Mahatitta (Mantota) and Gokonna (Trincomalee) are examples. These were linked to major towns and cities by roads, the existence of which are confirmed by the discovery of paved roads and stone bridges. The capital city usually had the ending Pura or Nagara, for example Anuradhapura. It was often the seat of the king and government. Most of its activities were centered on the needs of the royal palace, the court, the Buddhist religious establishments and the army, together with the considerable bureaucracy with which the king surrounded himself. The development of crafts, industries and commerce was geared in the main to this internal market. The capital was connected by roads to the port and other important towns. Fa-Hsien the Chinese monk who visited Anuradhapura in the 5th century A.D.

has left evidence of its flourishing state. He mentions, 'main streets and side streets level and well kept'.

No traces however, exist of an outer ring of walls enclosing both citadel and city whose area in the 10th century, extended to nearly 32 km² (Giles 1923). Archaeological evidence corroborates such descriptions, to provide considerable insight into the perceptions and behavioral environments of town and country planning in this period. Fortress settlements were also common, especially close to fording points on rivers, and possibly Polonnaruwa began as such a settlement. Sigiriya is perhaps the best example of a fortified capital, built in a magnificently impressive situation.

Land classification and agriculture

Agriculture in the dry zone is by and large, determined by the physical characteristics of the land. As mentioned earlier, the slightly undulating surface of the typical valley and the soil catena associated with it from the top of the interflue to the valley bottom, is divided into: a) soils on upper slopes (Rhodustalfs) which are well drained during both the wet and dry seasons. These are dark reddish brown in colour and with a texture akin to a sandy clay loam. It is used for slash and burn (Hen/Chena) rainfed cultivation. b) The relatively poorly drained soils during the wet season (Haplustalfs) on middle slopes. They are yellowish brown in colour, and sandy loam to sandy clay in texture. They are used as areas of homesteads (Ge-watta) and garden lands (Arub). Such soils are planted with perennial crops and sometimes used for semi irrigated farming too. c) The poorly drained soils during both seasons (Tropoqualfs/ Gley) in the valley bottom. They are usually dark grey in colour with a texture of sandy clay loam to clay. It is generally used as part of the tank bed and for rice cultivation (Yaya). Each micro- habitat in the landscape was thus recognised as a natural unit in land classification. This ensured its optimum use and management. Inscriptional and other sources name the different varieties of crops cultivated. Small variations in the micro-relief and soil moisture conditions even within a Chena, would be made use of to plant crops that would benefit by such micro variations. Chena crops were mainly herbaceous varieties. Cereals such as kurakkan (*Eleusine coracana*) hill paddy (*Oryza sativa*), millets such as Meneri and Tanahal, types of lentils (*Ulundu*), green gram (*Phaseolus aureus*), oil seeds (*Gingerly*) and various types of vegetables were grown. The practice of shifting cultivation (Hen) means the necessity to ensure for natural restoration of soil fertility. This restoration recognizes the nature of plant succession and soil development in an intergrated vegetation - soil complex. Terms used to identify such vegetation - soil complexes, indicate that change of plant species during stages of plant succession were appreciated and utilized as indicators for land classification. Such terms as Kanatte for an abandoned Hen plot, Lande for a 3-4 year old abandonded plot, Hirilanda for a 5-6 year old abandoned plot, Attandawa for a 10-20 year old jungle, suitable for Chena; and Mukalana for high jungle illustrates this. In a typical home garden, the different plants were grown not in an orderly crop row manner, but rather in a random fashion. Such an arrangement resembles the natural forest it replaced with a mixture of species growing in its own niche and partial habitat. By such means, the adaptation and requirements of the different species especially for light, and of the roots for soil moisture were recognized and utilized. Exposure of the soil to sun and rain is also largely eliminated, and weeding brought within reasonable dimensions. The homestead gardens were cultivated with perennial tree and shrub crops like coconut (*Cocos nucifera*), palmyrah (*Borassus flabellifer*), mango (*Mangifera indica*), citrus, arecanut (*Areca catechu*) betel (*Piper betel*), jak (*Artocarpus heterophylum*), tamarind (*Tamarindus indica*), bananas and a variety of medicinal plants. Being mostly used for subsistence consumption, the number of each species planted in a garden would not have been many. It could be expected that home gardens were fertilized by the household garbage and the droppings of domesticated animals, notably cattle and buffalo. The Yaya was the area of flooded rice culture.

Such a method of cultivation helps to maintain soil fertility more than any other permanent system of land use. Little soil is lost by erosion from the levelled or gently sloping bunded fields. Soil and nutrients flow in with irrigation water, or in the water washed down from adjacent highland. Nutrient loss by leaching is reduced, while crop residues and weeds supply considerable amounts of humus. A significant amount of nitrogen is also fixed by the bacteria in such waterlogged soils (Kalpage 1976). On the other hand, problems caused by siltation of tanks and salinization of soils due to faulty cultivation, were not completely unknown (Gunawardana 1971).

Land holdings according to the Kondavattavam Pillar Inscription of Dappula IV (924-935) were divided in the 10th century into four main types such as heritable land (Pamunu) subject to nominal rent, ones own (Ninda) or the entire property of the owner, land given for maintenance in lieu of salary (Jivel), and leased land (Patta). A large portion of the land was owned by the great monasteries such as the Mahavihara etc, and the tenants of these lands performed various services to these establishments. In the matter of taxation too, land was divided into three catergories such as Utte (best), *Manda* (middling) and *Passe* (worst). These were most probably based on the yield, which depended on the fertility of the soil and the availability of water (Codrington 1938).

The decline of the dry zone civilization

From about the middle of the 13th century began the depopulation of the dry zone with the breakup of the Sinhala kingdom at Polonnaruwa. A number of views have been expressed as to the cause of this more or less complete demise, and the population migration to the west zone. Nicholls (1921) put it to the spread of malaria, Codrington (1960) to wars and the destruction of the large irrigation system. Murphey, (1957) to malaria and foreign invasions, Paranavitana (1960) to results of foreign invasions and internal discord, leading to the disruption of the elaborate administrative setup, that was necessary for the construction and maintenance of the complicated irrigation system. Liyanagamage (1968) to the cumulative result of the operation of several factors such as foreign invasions, internal warfare, dissensions and the spread of diseases, Indrapala (1965) to the growth and importance of the Tamil Kingdom in the northern part of the island, not far from Polonnaruwa, while Roberts (1971), gives an important place to the soils becoming more and more infertile due to salinity and erosion, necessitating the colonization of new lands. The natural choice for migration and colonization was the wet zone. Another reason and a very pertinent one, may be the change of state income, due to the export of certain new agricultural commodities like cinnamon, growing only in the wet zone. In this instance, it appears that the change was not so much the reasons mentioned above or the inherent modification of physical characteristics of the environment; but to the impact of external ideas and culture contacts in a manner not hitherto experienced.

Late medieval period (1200-1500)

The collapse of the dry zone civilization in the core areas during the three hundred years from 1200-1500, constitute a period of what de Silva (1977) describes as 'uniterrupted decline'. It was during this period that there was a movement and shift to the wetter southwestern part of the country. In this migration, the first area to be settled on a large scale was naturally, the part just adjoining the dry zone, in the so called intermediate zone. It has an average annual rainfall of between 1, 250mm-500mm per year. It is also the part that most resembles the environment of the dry zone, and where especially in dry years, irrigation was essential to ward off famine. The dominant soils type being mostly Non-Calcic Brown soils, occur in association with the Reddish Brown Earths, on the crests and the upper and middle slopes. The Humic Gley soils occur on the lower slopes and valley bottoms. Hence the migration would not have made a big difference on the type of agriculture practised. Further, this movement was not to some unknwon and unfamiliar region of the country; rather it was to a familiar location, which offered not merely security, but also the potential for more than modest economic growth. As such the capital was at first in this intermediate zone in places as Dambadeniya, Yapahuwa and Kurunegala. From here the capital was moved more to the wet zone. One branch was set up at Gampola in the hill country, and the other at Kotte near Colombo. This polarization of people and the capital, was further completed with the establishment of the Jaffna Kingdom in the 14th century. The shift of the capital to Kotte, may also be linked with the increasing dependence on foreign trade that was centered on the export of cinnamon. At first, there was no need to cultivate it for its fragrant bark, as it grew in abundance in the southwest coastal areas. This trade in cinnamon was much in the hands of Arabs and other foreigners and it was during this period that another race to the already plural society was added. This category being the Muslims, who constitute 7 per cent of the present population. Today they are concentrated close to the port settlements, while the rural population of the Batticoloa and Amparai districts have a large percentage of Muslims. The southwest lowlands of the wet zone are nearly everywhere under 200 meters in elevation. A characteristic of the relief, is that it consists of a series of alternating low ridges and valleys. This gives the area a dissected undulating appearance. The rainfall is heavy being over 2000 mm and is even throughout the year. On this upland surface, the characteristic soil type is the Red Yellow Podzolic type, with Gley and Alluvial soils in the velleys. Unlike in the dry zone, the problem of cultivation of rice in the wet zone velleys, would have been more of the drainage of excess water and protection from the hazards of floods. Consequently, the dominant factor which governs the utilization of these Gley soils is in controlling the high water table which develops on the impermeable stratum, during the excessively wet rainy season from May to August. Hence the main rice cultivation season in the wet zone (Maha) is from October-March, while the (Yala) is from April to September. During the period under discussion, valleys were cultivated with rice, while the uplands supported home gardens growing coconut, arecanut, and a variety of fruit and vegetables. The uplands were used for Chena cultivation too. The frequency of place names ending with Hena, in the wet zone lowlands is an indication that during this period Chena cultivation was also important. In the mid hill country often termed the Kandy plateau, the environmental conditions were further different and diverse, from the wet zone lowlands. This necessitated a different adaptation to its special features. Usually the hill country rises somewhat abruptly from the surrounding coastal lowland areas, forming more or less an immense natural fortress. It thus served as a bulwark to safeguard the Sinhala Kingdom in the hills, from foreign invasions. The establishment of Kandy as the capital, was undoubtedly due to its natural fortifications and command of routes. Routes from the northern dry zone via the Matale valley, from the wet zone lowlands by the Kadugannawa pass and the eastern dry zone lowlands of the Mahaweli ganga valley through the Nugatenna Gap near Teldeniya, converged on it. As the valleys are narrow and insufficient for an extreme paddy cultivation, most of the hill sides were skillfully terraced, and water led to these fields from perennial streams. Highland areas which could not usually be terraced and culti-vated with rice were used to build houses and grow garden crops.

Land still above these were forested with could be used for Chena cultivation or for grazing and the collection of forest products. Such land was termed (Aduttu Deval) or appurtenances to the village fields (Pieris 1956).

The period of western colonial rule (1905-1947)

By the time that the first European colonial power came in contact with Sri Lanka, the wet zone lowlands had been settled. Besides settlement at Kelaniya, dating from pre-historical times, other important historical sites settled were Attanagalla, Wattala, Jayawardanapura (=Kotte) and Dalugama. According to Nicholas (1963), Colombo as a port can be traced back to the year 949 A.D. when Muslim traders settled there. Kalutara was the capital of aders settled ta Sinhala prince when the Colas (993-1070) occupied much of the dry zone. Diyagama, in the lower Kaluganga basin, is mentioned in a 5th century inscription. Keselhenawa (near Anguruwatota) further up the river, together with Velmilla and Pokunuwita in Panadura Totamune, were important settlements. The Sandesa style of Sinhala poetry, which was introduced during the later part of the previous period, makes special reference to numerous place names situated on the routes described in these poems. The spatial distribution of such place names, gives an indication about the close network and density of settlement that had taken place with the drift to the wet zone lowlands.

In the present day Galle district, most areas of which are in the lowland wet zone, the earliest inscription found is of the 10th century. Nevertheless, during the late Medieval period, settlement had taken place in around such places as Bentota, Hikkaduwa and Totagamuwa. As a matter of fact, Totagamuwa was a flourishing and famous centre of learning, under the great Sinhala poet, Totagamuwe Sri Rahula, during the time of Parakrama Bahu VI (1412-1467).

The two districts of the wet zone lowlands without a seaboard are Kegalla and Ratnapura. Certain parts of the Kegalle district were settled as early as pre-Christian times, but it was during the late Medieval period that settlement increased, as it lay through the important route linking Kandy with Kotte. On the other hand, most of the Ratnapura district, though undoubtedly the main source of gems from ancient times, was not much settled. The earliest inscription found in this district is of the 10th century. This suggests that there was no permanent settlement in that area before this period, as much of its terrain was covered by dense and impenetrable tropical rain forest, bogs, and marshes.

When the Portuguese arrived in Colombo in 1505, there were three kingdoms; Jaffna, Kotte and Kandy. By 1591 Jaffna came under Portuguese rule, and Kotte under the protection of the Portuguese in 1543; but nearly two thirds of the Kotte kingdom remained under the ruler of Sitawaka (Avissawella). Such conquest meant that the Portuguese held most of the western and southern lowlands of the country, leaving the Kandyan kingdom with only part of the eastern seaboard as its lowland territory. As the Portuguese were interested mainly in the control of commerce, especially the island's cinnamon trade, occupation of the wet zone lowlands ensured that no rival interfered in this activity. It also meant, that the traditional subsistence agriculture in the region was not much affected.

A crop that became prominent and could be readily cultivated with success in the lowlands was coconut, and mention is made of large scale coconut plantations even in the previous period. However, it was to the successors of the Portuguese, the Dutch (1658-1796) who are credited with the organization of planting cinnamon and coconut for export. Other crops introduced by either the Portuguese or the Dutch could be the new world tropical food plants, such as cassava, (*Manihot ultisima*), breadfruit (*Artocarpus nobilis*) and potato, as well as tobacco. The Dutch are also credited with the introduction of teak (*Tectona grandis*), and planting it as a source of timber. In the hill country, but more especially in most of the mid country or the region approximately below 1,000 meters, comprising much of the present day Matale, Kandy and Badulla districts, were settled before the British conquest of the lowlands in 1796. Colonization of higher altitudes in the above districts was probably hampered in pre-British times as rice cultivation, was not successful, at these higher elevations.

It was only after the fall of Kandy in 1815, and the British gaining control over the whole island, that steps were taken to radically alter the economic base of the country; through the expansion of the plantation system in the wet zone. The heavy rainfall, well distributed throughout the year, together with the decrease in temperature with altitude, provides the ideal environmental conditions needed for the cultivation of a variety of plantation crops. Besides the climate, the Red-Yellow Podzolic soils which are the dominant soil type throughout wet zone lowlands and uplands, are well suited for the growth of perennial tree crops, if anti-erosion measures are taken (Domros 1974).

The growth of plantation agriculture

By the beginning of the British occupation, the staple of the island's export economy, cinnamon, declined. This stimulated a period of experimentation with other plantation crops, and from the 1830s coffee was cultivated on a plantation scale. It thus replaced cinnamon as the chief agricultural product in trade. Coffee was not a new crop, as it was already grown in peasant gardens in the Kandyan hills, not so much for using the berries as a beverage, but rather for its while and sweet smelling flowers, as temple offerings, and the leaflets were used as a vegetable. It was also a crop whose cultivation was undertaken not only by foreign and local capitalists, but by some of the peasants as well. Nevertheless, most peasants grew coffee in a casual manner. They simply planted a few bushes in their gardens and allowed them to mature, not bothering either to weed or fertilize them. This was in contrast, to the attention paid by the large plantations to management techniques, of selection of seed for planting, fertilizer application and processing. Cultivation of coffee on the scale of large plantations demanded a reliable and disciplined labour force, and this was supplied by Tamil immigrants from South India. Large scale Indian immigration began from about the early 1840s. This added the latest large scale component of people, to the already existing plural society of Sri Lanka. Today, they form nearly 9 per cent of the population.

Divergent views have been expressed as to the reason why the local Kandyan Sinhala labour was excluded. Possibly, the Kandyan Sinhalese were not attracted by the wage incentives and terms of employment that were offered by the planters. They were probably also not entirely prepared to forgo the traditional village life, to the regimented life of work on the estates.

Thus the plantations largely utilized just one important indigenous resource, extensive and fertile land. This affected the traditional agricultural sector in two very important ways. Firstly, it took away land, to which the peasants felt they had traditional rights; land which was communally owned, such as grazing land and Chena land. Secondly, with the growing dependence on plantation crops, scant attention was made to encourage traditional agriculture, with the result that its returns enabled its practitioners to only eke out a bare livelihood. Furthermore, the most damaging blow to the development of this (traditional) sector, was that the surplus profits produced by the plantation sector, were either reinvested in the plantation sector itself or paid as dividends to foreign investors (Sondgrass 1966).

It has been argued that it was the divergent economic policies and attitudes between plantation agriculture and the traditional, that created two broad sectors which came to be known later as the dual economy. Each sector differed radically from the other, in patterns of resource use and technology, and had relatively few economic interrelations. Thus arose a dichotomy in the economic and political structure of the country, creating in a sense, two nations, which successive governments especially in the post independence (1948 onwards) period, have been battling hard to eliminate.

Coffee plantations were mainly established in the mid hill country, though from early British times, adventurous officers had penetrated into the higher elevations of the hill country, and discovered areas having a remarkably cooler climate. Not long afterwards, in the 1830s, the hill station of Nuwara Eliya was founded.

Coffee was infected with a leaf disease caused by a fungus (*Hemileia vastatrix*) and was more or less completely wiped out by 1880. With the demise of the coffee industry, planters were left with land lying idle, a trained labour force and management expertise accumulated over half a country, which could easily be used in growing other crops. In the low country, coconut had emerged as an important plantation crop, though this was mainly cultivated in the relatively drier parts of the Colombo, Kurunegala and Puttalam districts, on the sandy loams of the Non-Calcic Brown earth and Red-Yellow Podzolic soils. It had little competition for land from coffee. In the hill country proper, two crops, tea (*Camellia sinensis*) and rubber (*Hevea braziliensis*) became the dominant plantation crops from the 1890s. Tea cultivation expanded from the mid country Kandy plateau region, to much higher and also comparable elevations in the Kandy, Nuwera Eliya and Badulla districts. Probably the only limiting factor that restricted tea cultivation in the area above 2,135 m was frost, which has an adverse effect especially on the leaves (Domros 1974).

Rubber needing a higher rainfall and also higher temperatures, was cultivated especially in the Kalutara, Kegalle and Ratnapura districts. Other plantation crops such as cocoa (*Theobroma cacao*), cinchona, citronella and cardamon (*Flettaria cardamomum*) were also introduced, though not cultivated on a major scale. It should also be noted that the growth of the coffee plantation industry, together with the expansion of the network of roads and railways, broke the isolation of the Kandyan kingdom. This was indeed a positive step, that brought the Kandyan Sinhalese to the modern sector of the economy. Thus as de Silva (1977) remarks, the plantations helped in consolidating the process of unification set by the administrative and judicial reforms of 1832, when the separate existence of the Kandyan provinces was abolished, and they were amalgamated with the maritime provinces in a single political and administrative structure for the whole island. When the shift to the wet zone took place after the 13th century, population rapidly declined in the dry zone, with the exception of the Jaffna peninsula, the coastal areas near Batticaloa and the borders of the dry zone within the Southern and North Western provinces. Hence by the 19th century, the vast expanse of the dry zone was sparsely populated, and largely the domain of the jungle and its beasts, and a fertile seed bed for malaria and yaws (Roberts 1972). Few of the large irrigation works built during the period of the Dry Zone Civilization remained intact. Several village tanks remained however, to support small hamlets. The economics of these dispersed hamlets were stagnant with their inhabitants leading a precarious existence, in which Chena cultivation played a vital role.

Till the 1850s little concern was paid to the welfare and encouragement of traditional agriculture by the British. A change began under Governor Ward (1855-60), with regard to the restoration of irrigation works and an attempted rehabilitation of the dry zone. Although at first such efforts brought about only a modest increase in the amount of land cultivated, it marked the tendency to mitigate the neglect of peasant agriculture. The achievement of this effort was that its programme of irrigation activity was to convert irregularly cultivated land to regular cultivation; rather than open new areas to peasant agriculture (de Silva 1977). This restoration of irrigation works did not touch the wet zone, where by the 1890s peasant agriculture and plantation activity were in unequal competition. Plantation crops brought more income than rice cultivation. One factor that continued rice cultivation in spite of uneconomical prices, was probably that, the wet Gley soils on which it was cultivated, could not be used to grow any plantation crop.

However, as the plantation economy prospered with increasing numbers of emigrant labourers settling in the plantations and the creation of an elite class among the native population, rice had to be imported. This imported rice was also cheaper, and as a consequence, rice production was not viable. Frequent references were made in the published official documents to famines, near famines, chronic rural poverty, destitution and above all, to starvation in many parts of the country, in the last quarter of the 19th century and the first decade of the 20th century. Clearly, the living standards of the rural population in many parts of the country had shown no improvements, after nearly a century of British rule (de Silva 1977). The attention paid to the upliftment of the local population may be easily seen by comparing the natural increase of the local population, with the new immigrants that came in the wake of the plantation industry.

The above census figures (Table 1) indicate the impact of South Indian migrants on the population of the country. During the 20 years from 1871 when the first census was recorded, to 1891 the migration increase of this category was more than the natural increase of population of the other groups. In real terms, until the beginning of the 20th century, migratory increase constituted over 50% of the increase in total population (Sondgrass 1966). Thereafter it declined rather substantially. Nevertheless, this influx of new migrants brought about a radical change in occupance and employment in a dimension which were to adversely affect the economic conditions of the Kandyan Sinhalese in particular. In the cultivated areas of the hill country, devoted to plantation agriculture, a new pattern of rural settlements was established. It was something not unlike that which existed under the feudal system in England. The planter (superintendent) had his large bungalow, normally commanding a beautiful view, perched on the slopes or top of a hill; while the labourers and their families were huddled together in one or two roomed apartments, built in a back to back row, and hence called 'lines'.

In the plantation areas, the development of roads and the railway promoted the growth of numerous route centers of greater or lesser importance, for the collection and distribution of goods, administration and the provision of other infrastructural facilities. These were the central places that were linked with scattered plantations in the periphery. In the plantation areas these services were provided mostly by low-country Sinhalese, and Sri Lanka Tamils. The Kandyan Sinhalese, were yet to avail themselves of the economic opportunities that the new plantation economy afforded. Before the advent of the money economy and the subsequent attention paid to health and hygiene, the rate of population increase did not very much hamper the requirement of the peasant for land, to lead a subsistence existence. With the growth of the plantations, local as well as foreign capitalists were able to acquire under freehold title, large extents of land. The wet zone lowlands, by the time of the British occupation, was closely settled, except in some remote areas from the coast. Cultivation at first of cinnamon and later of coconut and rubber, increased the desire to posses land in freehold in these areas, by local capitalists. The neglected and impoverished peasantry at the mercy of moneylenders was easy prey to them. This brought about the beginning of the problem of landlessness. Prior to the advent of western influence, lands in the dry zone were held and administered by the village society. Hence absolute ownership of land by individuals in traditional villages, is a relatively recent development, and the practice of issuing title deeds of land began only at the end of the 19th century (Leach 1971). At first, it was more a problem in the wet zone, but later one, after the exodus to the dry zone began with the elimination of the malaria hazard in the 1940s, this became increasingly accenuated in the dry zone as well.

481

Table 1. Population of Sri Lanka.

Census year	(1) Total population	(2) Intercensal increase	(%)	(3) Natural increase	(4) Migration increase	(5) 3 as % of 2	(6) 4 as % of 2
1871	2,400,380						
1881	2,759,738	359,358	(14.9)	119,792	239,566	33.3	66.7
1891	3,007,789	248,051	(8.9)	144,260	103,791	58.1	41.9
1901	3,565,954	558,165	(18.5)	225,406	332,759	40.4	59.6
1911	4,106,350	540,396	(15.1)	356,147	184,249	65.9	34.1
1921	4,498,605	392,255	(9.5)	310,410	72,845	81.4	18.6
1931	5,306,871	808,266	(17.9)	656,990	151,276	81.3	18.7
1946	6,657,339	1,350,468	(25.4)	1,280,916	69,552	94.8	5.2
1953	8,097,895	1,440,556	(21.6)	1,328,199	112,357	92.2	7.8
1963	10,583,064	2,484,169	(30.6)	2,513,248	-29,079	101.2	-1.2
1971	12,711,143	2,129,079	(20.1)	2,208,061	-78,982	103.7	-3.7

Sources: Department of Census and Statistics: Census data, Ceylon Year Book, 1970; The Population of Sri Lanka, Colombo, 1974.

Re-colonization and settlement in the dry zone

Under the 1931 Donoughmore constitution, Sri Lanka was granted a substantial measure of freedom in planning and managing her own destiny. It came at a time when the country was being affected by the great economic depression which slowed down development projects. It was also the time when especially the dry zone was riddled with the scourge of malaria. By the late 1920s the claims of the peasant sector began to receive support and strength. Two factors that prompted such thinking were firstly, the increase in population in the wet zone, and secondly, the increase in the volume of rice imports (Sondgrass 1966). It was universally believed that one of the best ways of establishing a prosperous type of peasant proprietors was to develop irrigation facilities in the dry zone. Restoration of some of the tanks, building canals and clearing up the jungle for settlers were just one step in achieving the desired objective of reclaiming the abandoned dry zone. The important component of this operation, such as selecting the right type of settler, based on his experience in farming and his ability to adjust to new sociological relationships, in a new environment affected by disease, was however not given much consideration. As a sequel, nearly all these early attempts to attract migrants to the dry zone failed to achieve its aims.

The real breakthrough came in 1932 when D.S. Senanayake became the first Minister of Agriculture and Lands. Under Senanayake's guidance, planned peasant colonization schemes in the dry zone became the order of the day. After the restoration of the Minneriya Tank in 1935, the government took upon itself the task of clearing the land alloted for the establishment of peasant settlements and also bore the cost of establishing infrastructural facilities. The land development ordinance of 1935, gave the allottes in these restored schemes, tenure in perpetuity, with careful restrictions to prevent the sub division of land. It is doubtful that without such inducements, it would have been possible to attract peasants from the wet zone to the malaria infested dry zone. Restoration of large irrigation schemes resulted in the gradual migration of especially the Sinhalese from the wet zone back to the dry zone. The new settlements established were more of the dispersed type, rather than the clustered or nucleated type of the traditional dry zone village settlements. At first this new development did not affect the old settlers (the decendants of those who remained back during the time that the dry zone was depopulated) as the help and inducements these peasants received was only minimal, compared with the facilities offered to the new settlers. Thus arose the distinction between the Purana or old settlers and the modern settlers.

Post independence period (after 1948)

By the time of independence in 1948, much of the wet zone had been opened up for cultivation. The alluvial valleys and flood plains in this region were cultivated with rice, while the upland surfaces were planted primarily by the three major perennial crops of tea, rubber and coconut. A comparison of the land cultivated with tea, rubber and coconut with the climatic potential lands for these crops, showed that the optimal climatic lands were already completely utilized. An area where further expansion of cultivation of these types was possible was the Sabaragamuwa hill country (Domros 1974).

The Research Institutes for these three crops (tea, rubber and coconut) at Talawakele, Agalawatta and Lunuwila respectively, have been greatly instrumental in helping the development and expansion of these three crops. Such assistance included production of high yielding clones, and hybrid varieties. Pests and diseases of these cultivars, like tea-tortix caterpillar, the tea leaf disease caused by *Exobasidium vexas,* blister blight in the case of rubber, and the coconut caterpillar have been contained or completely eliminated. Much of this had been achieved by biological methods. Likewise, the Rice Research Institution at Batalagoda have evolved numerous varieties which paved the way for the green revolution from the fifties. In the case of rice in the wet zone, the areas that still await drainage and cultivation are the marshy lands like Muturajawela.

By comparison with the wet zone the proportion of cultivated land is extremely low in the dry zone. Badulla District shows the highest level in the dry zone (31.5 percent), but this is conditioned by its partial linkage with the intensely cultivated wet zone 'island' of the Namunukula Massif; it is followed by the Jaffna District (26.5 percent) and Puttalam/Chilaw District (29.1 percent). The lowest percentage of cultivated land is found in the Monaragala District (5.9 percent). Over the greater part of the dry zone the average percentage of cultivated land amounts to less than 10 percent, as compared to a proportion of cultivated land in the wet zone ranging from about 55 to 75 percent (land utilization report 1968). Exotic timber trees were first introduced by the Dutch when in 1680 teak (*Tectona grandis*) was introduced by Van Rhede. Systematic forestry plantations as an integral part of planned land use were begun by the British, in the last decades of the 19th century. The common species introduced were those suitable for the upcountry areas. These exotics included the gums *Eucalyptus* species, *E. globulus, E. robusta, E. saligna,* Mahogony, *Pinus* species such as *Pinus caribaea* and *Pinus petula, Cupressus macrocarpa, Acacia melanoxylon, Acacia mollisima* and the bamboo *Dendrocalamus strictus.* These introduced varieties are useful sources of timber, pulp, tannin and oil (Perera 1962).

Extension of settlement and cultivation in the dry zone

In the 1940s and 1950s the prime objective in dry zone colonization was to get land developed and settled as fast as possible. Economic viability and maximisation of production were not important considerations, and little thought was given to the establishment of integrated communities. A study made by the ARTI in 1976, showed that the results of this situation were several.

First, following the traditional pattern of cultivation in both wet and dry zones, prime emphasis was given to paddy cultivation, to the exclusion of other crops, with the result that highland soils often more fertile than the lowland, were neglected and both land and labour was such, that efficient management of water resources, and measurement of water issues, to individual farms, proved very difficult. Thirdly, the practice of siting farmer's homesteads along the irrigation channels rather than in village clusters, led to ribbon type of settlements, where farmers are isolated from each other; community development and co-operation are inhibited, and the cost of providing services greatly increased. Fourthly, the tendency to adopt 'crash programmes', starting new schemes and bringing in new settlers, even when adequate plans had not been made for their reception and supervision with unsuitable holdings being allocated proved disastrous. Unsuitability due to fields being on poor soils and slopes which hinder gravity irrigation were also common.

By the time that the first multipurpose scheme, the Gal-Oya scheme begun in 1949, as opposed to the earlier schemes of restoration and/or enlargement of the ancient tanks, these shortcomings had surfaced and attempts were made to overcome them. This resulted for the first time with the new Gal-Oya project, which was placed under a separate board to take overall responsibility for planning and development. An evaluation of its activities in 1969 by a committee however, censured it heavily on the grounds of faulty planning and lack of cost consciousness. These findings hastened the need for a settlement planning and development board, with Project Managers in charge, and teams of specialists in agriculture, irrigation, soils, land use, and social development. The Mahaweli Development Board established in 1977 has made use of earlier experience in land settlement.

The green revolution

Together with the expansion of the opening up of new areas for cultivation, an impressive breakthrough in evolving rice varieties suitable to local conditions have been achieved.

Efforts to breed fertilizer responsive, high yielding varieties of rice appears to have been made in Sri Lanka as far back as the 1950s. By 1958 two new high-yielding varieties, H4 and H5 with a yield potential of 5,156 kg ha⁻¹ year⁻¹ were introduced. These were followed by H7 and H8 during 1964-68. IR8, the variety developed in the International Rice Institute in the Philippines was introduced in 1968. Experiments and trials conducted in different parts of the country indicate that there was 'little to choose' between IR8 and H4 - the local cross-bred variety at low (45 kg/ha) to moderate (90 kg/ha) level of nitrogen application and management. At higher levels of nitrogen application (135 to 185 kg/ha), however, IR8 has given better yields than the local H4 variety. But due to its susceptibility to certain diseases and unsuitability for some regions of the country, IR8 was not widely cultivated. Since 1970, two other locally developed high yielding varieties are said to represent marked improvements in several cases over IR8. In all, the high yielding varieties evolved in Sri Lanka such as H4, H7, H8, LD66, BH11-11 etc., are widely in use and exotic varieties such as IR varieties and Taichung have not made much headway (Abdul Hameed, 1977). Since the introduction of the H-series varieties in 1958 the total area devoted to high yielding varieties of food grains has recorded a steady increase, representing about 60 percent of the total area planted with rice.

The cumulative effect of the introduction of new varieties, bringing additional land under paddy production and the development and strengthening of the institutions and infrastructural facilities, especially by promoting the adoption of new biological and chemical inputs, has led to several optimistic remarks about the possibilities of self sufficiency in rice. One of the main obstacles to this being achieved in the vagaries of the weather pattern in successive years. Another is that provision of irrigation is still basically confined to the utilization of the Reddish Brown Earth soils and their associates. The need for exploring and using the ground water resources especially in the Red Yellow Latosol soils of the north western part of the dry zone for cultivation has assumed urgency. It also means that rice cultivation will expand not only in the traditional rice cultivation areas of the Reddish Brown Earths of the dry zone, but also into the Non-Calcic Brown Earths in the Maduru Oya and Deduru Oya basins. In this upsurge of the increase in production using the high yielding varieties and high inputs, usually referred to as the 'Green Revolution', the position of Sri Lanka is different from many parts of the developing countries. In this context, the small farmer in Sri Lanka has not been left behind, either for want of resources or for reasons such as monopoly by the elite farmers over the use of new inputs and techniques of production. The reasons for this are firstly, that the rice economy of the country is characterized by the predominance of a peasant form of production without much disparities in the size of holding; and secondly, that the existing institutionalized arrangements, provide equal access to all sections of the farming community (Abdul Hameed 1977).

The Mahaweli project

The diversion of the waters of the Mahaweli Ganga (river) to irrigate the dry zone, is by far the most ambitious project undertaken to develop the dry zone. The Mahaweli region covers the area bounded by the watershed of the Mahaweli Ganga in Central Highlands to the south, the Maduru Oya in the southeast, the Kala Oya basin in the southwest, the area of Latosol soils in the northwest, and the east coast in the northeast. The original plan envisaged full development of the project to be completed in about 30 years. The inauguration of Stage 1 of this project was ushered in 1976, by the diversion of the river at Polgolla near Kandy, under the Sirimavo Bandaranaike government. After the election victory of the United National Party, later in the same year, under the leadership of J. R. Jayawardana; the original plan was modified to be completed within seven to eight years. This modified scheme was to be known as the Accelerated Mahaweli Project.

The gross area of the project will comprise about 55 percent of the dry zone, or 39 percent of the whole island. Within the project the land classified as being suitable for irrigated agriculture covers some 375,000 ha of which about 300,000 ha are underdeveloped forest lands owned by the state, and 75,000 ha currently under cultivation. The water resources available after full development would however, only be enough for year around irrigation of about 200,000 ha. A major limiting factor therefore is the availability of water and not of land.

The programme envisaged under the Accelerated Mahaweli Project provides for the development of 128,000 ha in four systems in the following manner.
a) System C, 29,200 ha in the areas on the right bank of the Mahaweli Ganga between Mahiyangana and the Polonnaruwa district.
b) System B, 49,200 ha in the Polonnaruwa, Batticaloa and Badulla districts.
c) System A, 36,000 ha in the Trincomalee and Batticaloa districts.
d) System D, 16,000 ha in Polonnaruwa and Trincomalee districts.

The major dams to be constructed will be the following: Victoria near Teldeniya, Maduru Oya near Welikande, Kotmale, Randenigala and Moragahakanda between Victoria and Minipe and Ulhitiya. These reservoirs will enable the development of about 500 megawatts of hydro-electric power (Anon 1979).

Stage 1 of the original project consisted of the diversion of the river at Polgolla by means of a dam, which links with the Sudu Ganga, a tributary of the Amban Ganga, through a 63 m diameter tunnel, 8 km long. At the end of the tunnel will be a power house at Bowatenna producing 50 megawatts of power. A high dam across the Amban Ganga at Bowatenna will divert a portion of the augmented flow of the Amban Ganga to the Dambulu Oya, a tributary of the Kala Oya; to replenish the Kalawewa-Rajangana system. A small portion of the water at Bowatenna, will be led through a feeder channel, to Kandalama tank near Dambulla. The Elahera anicut will be raised so that the flow of water in the Elahera-Minneriya Yoda Ela will be increased. Thus completion of Stage 1 of the old programme, which has been incorporated into the Accelerated Mahaweli Project, will help in the rehabilitation of the existing irrigation schemes in the Kandalama - Kalawewa - Ranjangana systems, the Elahera - Minneriya - Kaudulla - Kantalai systems, and the Anuradhapura city tanks. Today (1983), this stage of the Mahaweli Project has been almost completed. In addition, the Kotmale, Victoria, Ulhitiya and Madura Oya reservoirs are under construction. Simultaneously with these constructions, work in proceeding space with the clearing of the land, building of new settlements and other infrastructural facilities, for the new colonists and settlers.

The pattern of settlement is of the clustered type. Each settled farmer is given a plot of 1 ha of irrigated land. A cluster consisting for 400-500 settlers, will as far as possible correspond to an irrigation block and will comprise not more than 5 sub units or villages equally clustered. As an alternative, a single cluster without separate sub units will be organised, when compelled to do so by topography, and the requirements of the irrigation layouts. Four or five of these clusters will form a sub area. At sub area level, a semi urban centre will function as a service centre, with services and facilities of a higher order, concentrated in a township (NEDCO) 1979).

One of the basic aims of the project is the maximisation of agricultural production. This however will depend on soil characteristics, agronomic considerations, climatological factors and market limitations. The ecological impact of the Mahaweli Project on the preservation and conservation of wild and forestry resources is being examined. An assessment of the consequences of diminished river flow on the vegetation, particularly of the villu grassland will be made. At the same time, the effect of environmental pollution and connected hazards posed by the increased use of fertilizers, herbicides and insecticides, and eutrophication particularly on the water and the effects on the fishery resources is under investigation.

Unlike in many other countries, the fauna also thrives outside the boundaries of National Parks and Nature Reserves in Sri Lanka. The country has to a considerable extent managed to preserve is cultural heritage of coexistence between humans and wildlife. It is indeed amazing to find that the same village tank which is the artery of life for the local population, is shared with the crocodile, turtle, elephant, heron, etc. It clearly demonstrates that at least in the traditional dry zone agricultural system, human and animal interests were hardly conflicting, but with time, when they are competing for the same resources, there are bound to be conflicts. The Sri Lanka elephant which is listed as an endangered species by the International Union for the Conservation of Nature and Natural Resources is estimated at 2,500 to 6,000 (Hoffmann 1975). One problem in the preservation of the elephant is that they have an ecological range which necessitates movement along traditional migrating routes. With agricultural development the habitat of these animals is bound to diminish substantially. It has been proposed that jungle corridors be established. How effective such corridors will be to preserve the elephants, who will now be confined to small reserves, scattered within the newly developed areas, has yet to be seen.

Population and settlements

Population growth averaged about 1.4 percent annually from 1888 through 1939. At the same time the annual rate of growth for real growth of Gross Domestic Product per capita was about 1.8 percent during this period (Sondgrass 1966). There is thus evidence that over the last century the trend of the country's domestic product, has been distinctively upward and at a rate faster than population growth. In 1939 the birth rate was very high, 34.9 per thousand and the death rate was also substantial, 22.0 of every 1,000 live births. The population was about 85 percent rural and most of them relied heavily on traditional cultivation of rice; yet yield per acre in this crop, was amongst the lowest in the world. The average paddy holding was just over 0.4 ha and with supplemental holdings of other kinds, the typical peasant family had no more than perhaps 1.5 ha to work with.

The population distribution till the early 1940s was such that the wet zone, lowlands and highlands were generally thickly populated. This region comprises about 40 percent of the country's total area and also has the highest percentage of cultivated land.

Many urban settlements had developed especially in the wet zone coastal belt and also inland in the administrative and route centers. The rural - urban fringe area expanded significantly, especially in the Colombo district. In the dry zone, on the other hand, dense settlement with high densities of population was only confined to the Jaffna peninsula and the coastal belt from Batticaloa to Kalmunai. Elsewhere in the dry zone, the urban centres were mainly administrative, and consequently grew up as route centres. Of these Anuradhapura, Hambantota, Trincomalee, Manaar and Puttalam, were the most important.

In the dry zone the proportion of cultivated land is much less than in the wet zone. The exploited land and the settlements associated with it, extends in dispersed form with only a few agriculturally exploited areas standing out. For example, the percentage of cultivated land in relation to the total district area in the dry zone districts in 1962, varied from 26.5 percent in the Jaffna district to 11.2 percent in the Anuradhapura and Polonnaruwa districts to only 5.9 in the Moneragala district (Anon 1962). The comparison of the proportion of cultivated land between the districts on the one hand, and the whole of Sri Lanka on the other, leads to the striking result that those districts with a proportion of cultivated land higher than the national average (28.8 percent) are in the wet zone. While all districts with smaller percentage of cultivated land lie in the dry zone.

The eradication of malaria with DDT in the 40s in the dry zone, hastened and fascilitated the resettlement, of the traditional Dry Zone Civilization part of the tank country. The Sri Lanka population almost doubled during the twenty five year period (1946-71), increasing from 6.6 million to 12.7 million. This is in marked contrast to the increase during the previous 75 years, since the first national census in 1871, which registered a population of 2.4 million. The present trends characterised by a declining birth rate, and a death rate that has already reached low levels with little likelihood of any major changes, indicate a new phase in Sri Lanka's demographic transition (Balakrishnan and Gunasekera 1977).

A significant change in the spatial pattern of distribution also came about after 1946, when the population of all districts increased considerably. More striking of this trend was that the rate of increase in the sparsely populated dry zone districts was much higher than in the wet zone districts. The major colonization and land settlement schemes established in the dry zone had been the chief cause of this increase of population. It has contributed to the movement of population away from many of the densely populated wet zone areas, where the pressure on land had become more acute, with the swift growth of population. For the island as a whole, density of population in terms of agricultural land increased from 1.56 persons per 0.5 ha in 1946, to 2.23 in 1962 (Jones and Selvaratnam 1972).

Overview and future outlook

Bereft of any significant mineral resources other than perhaps gems, the two most important natural resources of Sri Lanka are her soils and water supplies. When the Mahaweli project is completed in the near future, the only large area that remains to be developed, will be the area of Latosol soils in the northwest part of the country. As a consequence of this extensive land development and settlement, careful considerations to the planning, execution and rational use of nature reserves will have to be given serious thought.

The conservation of nature may take the form of land utilization where particular pieces of terrain are set aside for the protection of plants, animals, physiographic features or whole ecosystems. The idea that every reserve ought to be as large as possible under the prevailing circumstances has undergone some modifications recently (Simmons 1978). In the case of islands, it has been argued that the biogeographical theory which leads to the designation of the largest single area as a refuge, is incorrect under a variety of biologically feasible conditions (Simberloff and Abele 1976). This may be found practicable in Sri Lanka, where obviously, the percentage of land available for nature conservation will necessarily be small.

Some argue that protected landscapes and ecosystems are natural resources that contribute little to the wealth producing economy of the country. Traditionally they have by and large been considered as luxury items by operational planners. On the other hand, the protection of species, communities and habitats is usually undertaken for one of several anthropocentric reasons. One of these is recreational values. Another, involves the possibility of the discovery of useful plants and animals either now or in the future. These areas may also contain genetic material likely to be of use to human societies as new domesticates, or as reserve pools of genetic diversity of current domesticates. The use of natural systems as environmental baselines, against which to measure the effects of human activity in intensifying agriculture or disposing of toxic wastes, etc. is yet another. It should be realised that nature often performs, free of charge, services which otherwise would have to be paid for in cash through energy and material inputs. In many cases therefore, the monetary value of 'undeveloped land' is much higher than when developed. Sri Lanka has to make careful judgements on the proportion of land to be nature conserved and the amount that can be developed for land use in all its varied aspects, considering the needs of her population now and in the future; in the context of her available natural resources.

490

References

Abdul Hameed, N.D. 1977. Edit. Rice Evolution in Sri Lanka. United Nations Research Institute for Social Development, Geneva.

Allchin, B. 1958. The late stone age of Ceylon. J. Royal Anthropological Institute of Great Britain and Ireland. 99:179-201.

Anon. 1962. Census of agriculture, department of census and statistics. Ceylon Government Press, Colombo.

Anon. 1968. Report of the land utilization committee 1967, sessional paper 11. Ceylon Government Press, Colombo.

Anon. 1970. Report of the Gal-Oya project evaluation committee, 1967-68 sessional paper 1. Ceylon Government Press, Colombo.

Anon. 1974. The population of Sri Lanka, department of census and statistics, Colombo.

Anon. 1976. Land settlement in Sri Lanka, a review of major writing on the subject. ARTI (Agricultural Training and Research Institute), Colombo.

Anon. 1979. Facts about Sri Lanka. Department of Information. Sri Lanka Government Press, Colombo.

Balakrishnan, N. and Gunasekera H.M. 1977. A review of demographic trends, In K.M. de Silva (ed.), Sri Lanka, a survey: 109-130, C. Hurst & Company, London.

Boserup, Ester. 1965. The conditions of agricultural growth. Aldine Publishing Company, Chicago.

Boisselier, J. 1979. Archaeologia Mindi Ceylon, Nagel, Geneva, 206 pp.

Brohier, R.L. 1934. Ancient irrigation works of Ceylon. Parts 1, 2, 3. Ceylon Government Press. 37; 43 and 46 pp.

Brohier, R.L. 1937. The inter-relation of groups of ancient reservoirs and channels in Ceylon. J. Roy Asiatic Soc. (Ceylon Branch), 34:64-77.

Chanmugam, P.K. 1949. Anthropometry of Sinhalese and Ceylon Tamils. Ceylon J. Science, 4:1-18.

Codrington, H.W. 1938. Ancient land tenure and revenue in Ceylon. Ceylon Government Press, 77 pp.

Codrington, H.W. 1960. The decline of the medieval Sinhalese kingdom. J. Roy. Asiatic Soc. (Ceylon Branch), 3:39-42.

Cook, E.K. 1931. A geography of Ceylon. Colombo Apothecaries Ltd., Colombo.

Cooray, P.G. 1948. Effective rainfall and moisture zones in Ceylon. Bull. Ceylon Geogr. Soc. 3:39-42.

Cooray, P.G. 1967. An introduction to the geology of Ceylon. National Museum of Ceylon Publications, Colombo. 324 pp.

de Alwis, K.A. and Panabokke, C.R. 1972. Handbook of the soils of Sri Lanka, J. Soil Science Soc. Sri Lanka, 11:3-83.

de Alwis, K.A. and Pluth, D.J. 1976. The Red Latosols of Sri Lanka. 1. Macromorphological, physical and chemical properties; genesis and classification. J. Soil Sci. Soc. Amer. 40:912-920.

Denham, E.B. 1912. Ceylon at the census of 1911. Ceylon Government Press, Colombo.

Deraniyagala, P.E.P. 1953. The Stone Age of Ceylon. J. Roy. Asiatic Soc. (Ceylon Branch) 3:113-124.

Deraniyagala, P.E.P. 1958. Report of the Archaeological survey of Ceylon for 1957. Ceylon Government Press, Colombo.

Deraniyagala, P.E.P. 1958. The pleistocene of Ceylon. Ceylon Government Press, Colombo, 164 pp.

Deraniyagala, P.E.P. 1960. Open air habitation sites of *Homo sapiens*. Balangodadensis. Spolia Zeylan. 29:95-109 and 1963, 30:86-110.

de Silva, K.M. 1977. Historical survey: 31-85. In Sri Lanka. A survey. Edit. K.M. de Silva. The Institute of Asian Affairs, Hamburg, C. Hurst and Co., London.

Domros, M. 1974. The Agroclimate of Ceylon. Franz Steiner Verlag, GMBH, Wiesbaden, 359 pp.

Farmer, B.H. 1956. Rainfall and water supply in the dry zone of Ceylon: 223-268. In G.A. Fisher and R.W. Steel (ed.), Geographical essays on British tropical lands, London.

Fernando, A.D.N. 1979. Ancient irrigation works of Sri Lanka. J. Roy. Asiatic Soc. (Sri Lanka Branch), 22:1-24.

491

Geiger, W. 1937. The linguistic character of Sinhalese. J. Roy. Asiatic Soc. (Ceylon Branch), 34:16-41.

Geiger, W. 1960. The Mahavamsa, Colombo, 323 pp.

Giles, H.A. 1923. The travels of Fa-Hsien. Cambridge University Press, 70 pp.

Gunawardana, R.A.L.H. 1971. Irrigation and hydraulic society in early medieval Ceylon. Past and Present. 53:3-57.

Hoffman, T.W. 1975. Elephants in Sri Lanka, their number and distribution. Loris, 35:278-280.

Holmes, C.H. 1951. The grass, fern and savannah lands of Ceylon, their nature and ecological significance. Imperial Forestry Institute, paper No. 28, Oxford University Press. 95 pp.

Ievers, R.W. 1887. Notes on tanks lying below the Kalawewa Yoda-Ela. Sessional paper 29 of 1886. Ceylon Government Press. Colombo.

Indrapala, K. 1965. The invasions from South India and the abandonment of Polonnaruwa, In Dravidian settlements in Ceylon and the beginnings of the kingdom of Jaffna. Unpublished Ph.D. thesis, Univ. London.

Jayasekera, R.D.B. 1969. The elephants of the Mahaweli plains. Loris, 11:354-356.

Jones, G.W. and Selvaratnam, S. 1972. Population growth and economic development in Ceylon, Colombo, 199 pp.

Kalpage, F.S.C.P. 1976. Tropical soils, Macmillan, London.

Kanapathipillai, K. 1948. Ceylon and its contributions to Tamil. In S. Sanmuganathan (ed.), The pagent of Ceylon: 39-51. Daily News Press, Colombo.

Kirk, William. 1975. The role of India in the diffusion of early cultures. Geogr. J. 141:19-34.

Leach, E.R. 1959. Hydraulic Society in Ceylon. Past and Present. 15:2-6.

Liyanagamage, A. 1968. The Disintegration of Polonnaruwa: 67-75. In Amaradase Liyanagamage, The Decline of Polonnaruwa and the Rise of Dambadeniya, Colombo.

Maloney, C. 1970. The beginning of civilization in South India. J. Asian Studies. 29:603-616.

Murphey, R. 1957. The ruin of Ancient Ceylon. J. Asian Studies. 16:181-200.

Needham, J. 1969. The grand tradition, science and society in East and West, London, 209 pp.

Nicholas, C.W. 1954-55. The irrigation works of King Parakrama Bahu 1, Ceylon Historical J., 4:52-68.

Nicholas, C.W. 1963. Historical topography of ancient and medieval Ceylon. J. Roy. Asiatic Soc. (Ceylon Branch), New Series, 6, Special number 232 pp.

NEDCO. 1979. Mahaweli ganga development programme, implementation and strategy study. Kingdom of Netherlands, Ministry of foreign affairs, Vol. 5.

Nicholls, Lucius. 1921. Malaria and the lost cities of Ceylon. Indian medical Gazette. 54:9-13.

Noone, N.A. and Noone, H.V.V. 1940. Stone implements of Bandarawella. Ceylon J. Science, 3:9-24.

Noone, H.V.V. 1945. Stone age relicts at Bandarawella. Loris 4:263-267.

Panabokke, C.R. 1958. A pedologic study of dry zone soils. Tropical Agriculturist, 64:151-170.

Panabokke, C.R. 1978. A case study of tropical alfisols in Sri Lanka, In Soil resource data for agricultural development, pp. 155-162. D. Swindale (ed.), Hawaii Agricultural Experiment Station, College of Tropical Agriculture, Univ. Hawaii.

Paranavitana, S. 1952. The evolution of the Sinhalese language, 1181-1199. In W. Sorata Maha Thera, Sri Sumangala Sabdakosaya. Mahabodi Press, Colombo.

Paranavitana, S. 1958. Glimpses of the political and social conditions of medieval Ceylon. Sir Paul Pieris Felicitation Volume: 69-74.

Paranavitana, S. 1960. Civilization of the period: economic and social conditions, In H.C. Ray (ed.), University of Ceylon, History of Ceylon. 1:713-720.

Paranavitana, S. 1970. Inscriptions of Ceylon. Archaeological survey of Ceylon, Ceylon Government Press.

Perera, N.P. 1969. The ecological status of the montane grasslands of Ceylon. Ceylon Forester, 19:27-51.

Perera, N.P. 1969. The ecological status of the savannah of Ceylon. Tropical Ecology 10:207-221.

Perera, N.P. 1975. A physiognomic vegetation map of Sri Lanka, J. Biogeography, 2:185-203.

Perera, N.P. 1976. The human influences on the vegetation of the Sri Lanka highlands. Vidyodaya Univ. J. Arts, Science and Letters, 5:17-40.

Perera, N.P. 1978. Early agricultural settlements in Sri Lanka in relation to natural resources. Ceylon Historical J. 25:58-73.

Perera, W.R.H. 1962. The development of forest plantations in Ceylon since the 17th century. Ceylon Forester, 12:142-151.

Phillips, W.W.A. 1942. Distribution of the mammals of Ceylon, with special reference to the need for wildlife sanctuaries. Loris, 3:4-9.

Pieris, P.E. 1917. Nagadipa and Buddhist remains in Jaffna. J. Roy. Asiatic Soc. (Ceylon Branch), 26,1:11-44.

Pieris, P.E. 1919. Nagadipa and Buddhist remains in Jaffna. J. Roy. Asiatic Soc. (Ceylon Branch), 28,4:40-67.

Pieris, Ralph. 1956. Title to land in Kandyan law. Sir Paul Pieris Felicitation volume: 92-113.

Ray, H.C. 1959. University of Ceylon, History of Ceylon. University of Ceylon Press.

Roberts, Michael W. 1971. The ruin of ancient Ceylon and the drift to the south-west. pp. 99-109. In K. Indrapala (ed.), The Collapse of the Rajarata Civilization.

Roberts, Michael W. 1972. Irrigation policy in British Ceylon during the nineteenth century. South Asia - J. of South Asian Studies. 2:47-63.

Siddharatha, R. 1935. The Indian languages and their relation with the Sinhalese language. J. Roy. Asiatic Soc., (Ceylon Branch), 33:124-145.

Seligman, G.C. and Seligman, B.Z. 1908. An itinerary of the veddha country. Spolia Zeylanica, 5:155-170.

Simberloff, D.S. and Abele, L.G. 1976. Island biogeography, theory and practice. Science, 191:285-286.

Simmons, I.G. 1978. Resource management and conservation. Progress in human geography. 2:330-335.

Sirinanda, K.U. 1979. Water balance types and water resource development in the dry zone of Sri Lanka. J. Tropical Geogr. 44:55-71.

Sondgrass, Donald R. 1966. Ceylon: an export economy in transition. Richard D. Irwin, Inc., Homewood, Illinois, 416 pp.

Tambiah, H.W. 1968. Sinhala Laws and customs. Lake House Investments, Colombo.

Tennakoon, M.U.A. 1974. Rural settlement and land use in north central Sri Lanka. Unpub. M.A. thesis, Department of Geography, Syracuse Univ.

Tennent, J.E. 1859. Ceylon. Sixth Edition. Vol 1 and 2. Tisasara Prakasakayo Dehiwela, 543 pp.

Vink, A.P.A. 1975. Land use in advancing agriculture. Springer-Verlag, Berlin.

Wayland, E.J. 1917. Outlines of the stone age of Ceylon. Spolia Zeylan. 11:37-46.

Generic index

504